W9-AZU-063

INTRODUCTION TO ORGANIC AND BIOCHEMISTRY

Second Edition

INTRODUCTION TO ORGANIC

WILLARD GRANT PRESS
BOSTON, MASSACHUSETTS

AND BIOCHEMISTRY

William H. Brown
Beloit College

Printed in the United States of America.

Second printing: June 1977

Library of Congress Cataloging in Publication Data

Brown, William Henry.
Introduction to organic and biochemistry.

Bibliography: p.
Includes index.
1. Chemistry, Organic. 2. Biological chemistry.
I. Title. [DNLM: 1. Biochemistry. 2. Chemistry, Organic. QU4 B881i]
QD253.B73 1976 547 76-41885
ISBN 0-87150-719-6

Designed by Kathleen K. Riley and the staff of Willard Grant Press. Composed by European Printing Corporation Limited, Dublin, on the Linotron 505C in Times Roman and Helvetica. Illustrations by Magnuson & Larson, Inc. and Vantage Art, Inc. Printed and bound by Halliday Lithograph Corporation. Cover photograph courtesy of George S. Sheng.

PREFACE

Like the first edition, this text provides an introduction to organic and biochemistry. It assumes as background one quarter or one semester of general chemistry. Although students take this course for different reasons, most have one thing in common. They are aiming toward careers in the health and life sciences, but few if any intend to become professional organic or biochemists.

Although the basic approach and content of the first edition is retained, suggestions and comments from users have been incorporated in this second edition. Throughout, many problems have been revised and the number of problems has been increased substantially.

In overall organization, the book can be divided roughly into three sections. Chapters 1-9 lay the foundation in organic chemistry by discussing the structures and typical reactions of the important functional groups that will be encountered throughout the rest of the text. Major additions to Chapter 1 include a table of these important functional groups, a discussion of structural and functional isomerism, and presentation of the valence-shell electron-pair repulsion model for predicting bond angles and the shapes of molecules. The treatment of hybridization and covalent bond formation by the overlap of atomic orbitals is retained. However, instructors may omit this material without affecting the balance of the book.

Revisions in the other chapters of this section are aimed at integrating the organic and biological chemistry more fully. For example, in Chapter 3 a section on terpenes and terpene hydrocarbons has been included. A new section in Chapter 4 discusses the significance of asymmetry in the biological world. In Chapter 6, reduction of NADH has been added, and is illustrated by the reduction of acetaldehyde to ethyl alcohol, the final step in alcoholic fermentation. The chapter on carboxylic acids contains new sections on the essential fatty acids and acid–base balance in blood plasma.

Chapters 10–14, the second major section, provide a more comprehensive coverage of the structure and function of four key classes of biomolecules: the carbohydrates, amino acids and proteins, lipids, and nucleic acids. In Chapter 10, a section on ascorbic acid and its role in the cross-linking of collagen fibers has been included. The treatment of phospholipids in Chapter 11 has been expanded, and a section on cell membranes, including the fluid-mosaic model, has been added. This chapter also contains a discussion of the general characteristics of hormones and hormone action. In subsequent sections, specific types of hormones are presented. Chapter 12 now includes titration of amino acids, isoelectric point and isoelectric precipitation, paper electrophoresis, and denaturation of proteins. Enzymes are treated separately in a new chapter.

Chapter 15 is a transition between the previous discussions of organic and bioorganic molecules and the chapters on metabolism that follow. It presents an overview of the oxidation of foodstuffs and the central role of ATP in the transfer of energy in biological systems.

Chapters 16–18, the third major section, have been revised and enlarged to show how the metabolism of carbohydrates, fatty acids, and amino acids is interrelated and precisely regulated. Several important new figures present overviews of glucose and amino acid metabolism, and show glycolysis, the tricarboxylic acid cycle, the urea cycle, and the fates of carbon skeletons from amino acid degradation. Ketone bodies are treated more fully, and the importance of the insulin–glucagon ratio in the regulation of carbohydrate and fat metabolism is discussed.

The Mini-Essays, which were enthusiastically received in the first edition, have been retained, and a new one on prostaglandins has been added. These short, optional articles are included to demonstrate that organic and biochemistry are highly exciting and creative activities, and to offer a glimpse of the human involvement in research and discovery.

A new edition of the *Study Guide* also has been prepared. Its purpose is to emphasize the important points in the text, to guide students in their approach to problem solving, and to provide complete and detailed solutions to all the problems in the text. It also contains a self-evaluation test for each chapter.

For their many comments and suggestions that have helped shape the second edition, special thanks are due users of the first edition at over one hundred schools, particularly George Washington University, Iowa State University, University of Manitoba, University of Minnesota, Montgomery Community College, University of North Dakota, Pacific Union College, San Francisco State University, University of Vermont, and Washington State University.

I am indebted to Dr. Leodis Davis, University of Iowa, Dr. William Epstein, University of Utah, and Dr. Scott Mohr, Boston University, for their careful reading of the entire manuscript and their many valuable contributions.

I am grateful to all those who provided encouragement and support during this project, especially to the students of Beloit College. Finally, I wish to express my personal thanks to Kay Riley of Willard Grant Press for the creative insight and diligent hard work she brought so consistently to all stages in the preparation of this text.

William H. Brown
Beloit, Wisconsin

CONTENTS

3 UNSATURATED HYDROCARBONS

4 STEREOISOMERISM AND OPTICAL ACTIVITY

5 ALCOHOLS, PHENOLS, AND ETHERS

1

THE COVALENT BOND AND THE GEOMETRY OF MOLECULES

1.1 INTRODUCTION

An introduction to organic chemistry and biochemistry must begin with a review of atomic structure and bonding. Much of organic chemistry is the chemistry of carbon and a few other common elements: hydrogen, nitrogen, and oxygen. Therefore, our initial discussion of chemical bonding will focus on these four elements.

There are some simple rules for chemical bonding:

1. Atoms bond with other atoms to attain electronic configurations similar to the inert gases, which have filled valence shells.
2. Like charges repel (for example, electron pairs).
3. Unlike charges attract (for example, a positively charged nucleus and a negatively charged electron).

We shall proceed with these and only a few other simple ground rules to describe how atoms combine to form molecules by sharing electron pairs, and then to describe the three-dimensional shape of molecules. After discussing these aspects of structure and bonding in organic molecules, we will turn in subsequent chapters to the reactions they undergo, the conditions under which certain bonds can be broken and new ones formed, and the ways to convert one molecule into another.

Although discussions of this type can become quite sophisticated, we must not lose sight of the fact that they are based on the application of a few fundamental and logical principles. By the time you finish this course you should have a good working knowledge of these principles.

1.2 ELECTRONIC STRUCTURE OF ATOMS

You should already know certain fundamental principles about the electronic structure of atoms and ionic and covalent bonding from a previous chemistry course. Some of these concepts will be reviewed here.

An atom of an element consists of a dense nucleus surrounded by electrons. The nucleus bears a positive charge that is numerically equal to the number of electrons that surround it. The mass of an atom is concentrated in the nucleus and is equal to the sum of the masses of the protons and neutrons in the nucleus.

The electrons of an atom are concentrated in certain regions called orbitals. The electron orbitals are further grouped in shells or principal energy levels identified by the principal quantum numbers 1, 2, 3, and so on. These are also sometimes referred to by the letters K, L, M, etc. The principal quantum number 1 shell consists of a single orbital, the $1s$ orbital; the 2 shell consists of four orbitals, the $2s$, $2p_x$, $2p_y$, and $2p_z$; the 3 shell consists of nine orbitals, one $3s$ orbital, three $3p$ orbitals, and five $3d$ orbitals.

Besides knowing the number and kind of orbitals in each shell, we need to remember that an orbital can accommodate a maximum of two electrons. Thus the maximum number of electrons that can occupy the first shell is two; the second shell, eight; and the third shell, eighteen. If we "build" an atom by surrounding the nucleus with just enough electrons to neutralize its positive charge, the first orbital to fill will be the $1s$, that is, the orbital of lowest energy (the one closest to the nucleus). Next to fill will be the $2s$ orbital, then the $2p$, etc. Table 1.1 shows the electronic configuration of the first eighteen elements of the periodic table.

Table 1.1 *The electronic configuration of the first 18 elements.*

Element	Atomic Number	Orbital								
		$1s$	$2s$	$2p_x$	$2p_y$	$2p_z$	$3s$	$3p_x$	$3p_y$	$3p_z$
H	1	1								
He	2	2								
Li	3	2	1							
Be	4	2	2							
B	5	2	2	1						
C	6	2	2	1	1					
N	7	2	2	1	1	1				
O	8	2	2	2	1	1				
F	9	2	2	2	2	1				
Ne	10	2	2	2	2	2				
Na	11	2	2	2	2	2	1			
Mg	12	2	2	2	2	2	2			
Al	13	2	2	2	2	2	2	1		
Si	14	2	2	2	2	2	2	1	1	
P	15	2	2	2	2	2	2	1	1	1
S	16	2	2	2	2	2	2	2	1	1
Cl	17	2	2	2	2	2	2	2	2	1
Ar	18	2	2	2	2	2	2	2	2	2

We generally focus our attention on the electrons in the outermost or valence shell for it is these electrons that participate in chemical bonding and reaction. The valence electrons are represented by one or more dots surrounding the usual symbol for the atom. Each dot represents one valence electron. In the second-period elements, lithium through neon, we consider the nucleus and the two $1s$ electrons as a unit called the kernel. In the third-period elements, sodium through argon, we consider the two $1s$ electrons, the two $2s$, and the six $2p$ electrons as the kernel. Thus we would write the valence electrons for the first eighteen elements as shown in Table 1.2.

You should compare these valence electron representations with

Table 1.2 *Valence electrons for the first 18 elements.*

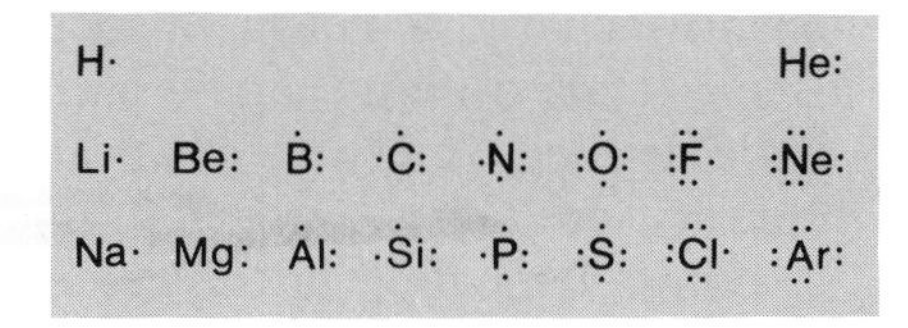

H·							He:
Li·	Be:	Ḃ:	·Ċ:	·Ṅ:	:Ö:	:F̈·	:N̈e:
Na·	Mg:	Ȧl:	·Ṡi:	·Ṗ:	:S̈:	:C̈l·	:Är:

the electron configurations given in Table 1.1. For example, notice that in Table 1.2 beryllium is shown with two paired valence electrons; these are the two paired $2s$ electrons listed in Table 1.1. Carbon is shown with four valence electrons, two of which are paired and two of which are unpaired; these represent the two paired $2s$ electrons and the single $2p_x$ and $2p_y$ electrons listed in Table 1.1. Notice also that carbon and silicon each have four valence electrons, nitrogen and phosphorus each have five valence electrons, oxygen and sulfur each have six valence electrons, and fluorine and chlorine each have seven valence electrons.

While the number of valence electrons for each of these pairs of atoms is the same, the shells in which these valence electrons are found are different. For C, N, O, and F, the valence electrons belong to the principal quantum number 2 shell. With eight electrons this shell is completely filled. For Si, P, S, and Cl, the valence electrons belong to the principal quantum number 3 shell. This shell is only partially filled with eight electrons; the $3s$ and $3p$ orbitals are fully occupied but the five $3d$ orbitals can accommodate an additional ten valence electrons. Because of this difference between the number and kind of orbitals in shells 2 and 3, we should expect differences in the covalent bonding of oxygen and sulfur, and of nitrogen and phosphorus. Such differences do exist, as we shall soon see.

1.3 THE LEWIS MODEL OF THE COVALENT BOND

In 1916 Gilbert N. Lewis, professor of chemistry at the University of California, came up with a beautifully simple hypothesis that unified many of the apparently disparate facts about reactions of the chemical elements. Lewis pointed out that the chemical inertness of the noble gases indicates a high degree of stability of the electronic complements of these elements: helium with a shell of two electrons, neon with shells of two and eight electrons, and argon with shells of two, eight, and eight electrons.

According to the Lewis model of bonding, elements other than the noble gases can achieve a greater degree of stability by gaining, losing, or sharing electrons, and all atoms will tend to undergo reactions to acquire a noble gas electronic configuration in their outer (valence) shell. For example, sodium (which has one too many valence electrons for the electronic configuration of neon) and chlorine (which has one too few valence electrons for the electronic configuration of argon) can achieve a mutually advantageous state by the transfer of one electron from the sodium atom to the chlorine atom.

$$\mathrm{Na}\cdot + \cdot\ddot{\underset{\cdot\cdot}{\mathrm{Cl}}}\!: \longrightarrow \mathrm{Na}^+ + :\!\ddot{\underset{\cdot\cdot}{\mathrm{Cl}}}\!:^-$$

As a result of this transfer, sodium acquires a unit positive charge and eight electrons in its outermost filled shell; chlorine acquires a unit negative charge and eight valence electrons in its outermost filled shell. Because oppositely charged particles attract, in the crystal of sodium chloride there is a force of attraction between the sodium ion and the chloride ion. This electrostatic attractive force is called an ionic bond.

For elements near the center of the periodic table, the complete transfer of electrons is not energetically favorable; such a transfer would result in too great a concentration of either positive or negative charge on the atom. For example, if carbon were to form purely ionic bonds, it would have to either gain four electrons to become C^{4-} (and achieve an electronic configuration resembling that of neon) or lose four electrons to become C^{4+} (and achieve an electronic configuration resembling that of helium). Instead, atoms near the center of the periodic table tend to acquire filled outer shells by a process of sharing electrons.

The chemical bond arising from the sharing of electrons is called a covalent bond. The simplest example of the formation of a covalent bond is found in the case of the hydrogen molecule. When two hydrogen atoms combine, the single electrons from each combine to form an electron pair.

$$\mathrm{H\cdot} + \mathrm{\cdot H} \rightarrow \mathrm{H{:}H} + 104\ \mathrm{kcal/mole}$$

In terms of our model of the covalent bond, we consider this pair of electrons to function in two ways simultaneously; it is shared equally by the two hydrogens and at the same time fills the outer shell of each hydrogen. In other words, for the purposes of acquiring a noble gas electron configuration, we consider each atom to "own" completely all electrons it shares in covalent bonds with other atoms. The gain in stability of the combination is evidenced by the large quantity of energy liberated (as heat) when the hydrogen molecule is formed from two separate atoms. The covalent bond is a strong bond, for it would require 104 kilocalories to dissociate one mole of H_2 into hydrogen atoms.

What is the reason for this great stability of the covalent bond? We can still imagine each electron as belonging to one nucleus. In the Lewis picture, the electron pair occupies the region between the two nuclei and interacts with both of them. The electron pair forming the covalent bond serves to shield one positively charged nucleus from the repulsive force of the other nucleus, and at the same time the electron pair attracts both nuclei. In other words, putting an electron pair in the space between two nuclei bonds them together and fixes the distance between atoms to within very narrow limits. We call this distance the bond length.

By way of additional examples of chemical bond formation through electron pair sharing, the molecules HCl, H_2O, NH_3, CH_4 and C_2H_6 can be formulated as shown in Figure 1.1. In these compounds, each nucleus has an electron atmosphere resembling that of a noble gas. For example, each hydrogen atom of the water molecule shares one electron pair with the oxygen atom and therefore has a filled outer shell resembling that of helium. The oxygen atom of water is surrounded by eight electrons, and therefore has a filled outer shell resembling that of neon. Note that of the

$H\cdot + \cdot\ddot{\underset{\cdot\cdot}{Cl}}: \longrightarrow H:\ddot{\underset{\cdot\cdot}{Cl}}:$		hydrogen chloride
$2H\cdot + \cdot\ddot{\underset{\cdot\cdot}{O}}\cdot \longrightarrow H:\ddot{\underset{\cdot\cdot}{O}}:H$		water
$3H\cdot + \cdot\ddot{\underset{\cdot}{N}}\cdot \longrightarrow H:\ddot{\underset{\cdot\cdot}{N}}:H$ (with H below N)		ammonia
$4H\cdot + \cdot\dot{\underset{\cdot}{C}}\cdot \longrightarrow H:\underset{\cdot\cdot}{\ddot{C}}:H$ (with H above and below C)		methane
$6H\cdot + 2\cdot\dot{\underset{\cdot}{C}}\cdot \longrightarrow H:\underset{\cdot\cdot}{\ddot{C}}:\underset{\cdot\cdot}{\ddot{C}}:H$ (with H above and below each C)		ethane

Figure 1.1 *Electronic formulas for some simple molecules.*

eight electrons surrounding oxygen, two pairs are involved in covalent bonding and are called bonding electrons. The other two pairs of electrons are not involved in bonding and are called nonbonding electrons, or unshared electron pairs.

Indicating each individual electron pair in a covalently bonded compound, as we have done in Figure 1.1, is sometimes tedious. Therefore it is common practice to represent a shared electron pair (a covalent bond) by a dash. In addition it is common to show only shared electron pairs. Thus we customarily represent the molecules H_2O and NH_3 as

H—O—H (water)

H—N—H with H bonded below N (ammonia)

Two atoms may share more than a single pair of electrons. Since one shared pair of electrons represents a single covalent bond, the sharing of more than one pair represents a multiple covalent bond between the atoms involved. If two pairs of electrons are shared, we speak of a double bond. If three pairs of electrons are shared between two atoms, we speak of a triple bond. Even in the case of compounds containing multiple bonds we still find carbon with its characteristic four bonds, oxygen with two bonds, and nitrogen with three bonds.

ethylene	acetylene	formaldehyde	nitrous acid
$H_2C{=}CH_2$	H—C≡C—H	$H_2C{=}\ddot{O}:$	$H{-}\ddot{\underset{\cdot\cdot}{O}}{-}\ddot{N}{=}\ddot{O}:$

HNO_2

1.4 ATOMIC VALENCES

As a guide in writing structural formulas for organic compounds, it is useful to define valence for the various elements. Simply stated, the valence of an element is the number of bonds that it can form. That the concept of valence applies to organic as well as inorganic compounds was first recognized and stated in 1859 by the brilliant German chemist, August Kekulé. Subsequently G. N. Lewis pointed out in 1916 that this characteristic valence can be correlated with the electronic structure of the element, if it is postulated that in the process of bonding each atom acquires a complete shell of electrons. Hydrogen has one valence electron and in the process of bonding acquires one more electron; hydrogen has a valence of 1. Oxygen has six valence electrons and in the process of bonding acquires two electrons; oxygen has a valence of 2. By similar reasoning, nitrogen has a valence of 3, carbon a valence of 4, and the halogens a valence of 1. As we have already seen, each of the valences must be used in bonding, and they are used by some appropriate combination of single, double, and triple bonds. In the case of carbon, there are four different ways the valence of 4 may be satisfied (Table 1.3).

Table 1.3 *The tetravalence of carbon, satisfied by appropriate combinations of single, double, and triple bonds.*

Type	Example	Name
—C— (four single bonds)	CH_4, $CH_3—CH_3$	methane, ethane
>C=	$CH_2{=}CH_2$, $H_2C{=}O$	ethylene, formaldehyde
=C=	O=C=O	carbon dioxide
—C≡	H—C≡C—H	acetylene

Knowing just these simple rules will help you in writing proper structural formulas. For example, you would expect methane, CH_4, to be a stable molecule because the normal tetravalence of carbon is satisfied. However, you would not expect compounds of formula CH_2 or CH_3 to be stable molecules because they do not satisfy the normal tetravalence of carbon. Your expectations are fully justified in fact. Compounds of these formulas have been shown to exist, but they are so highly unstable that except under extraordinary circumstances they exist for only fractions of seconds.

1.5 FORMAL CHARGES

Sometimes a covalent bond is formed in which one atom provides both electrons. In this case the atom providing the pair of electrons acquires a formal positive charge. Examples are the reaction of a water molecule with a proton to form the hydronium ion, H_3O^+, and the reaction of an ammonia molecule with a proton to form the ammonium ion, NH_4^+.

$$H{-}\ddot{O}{:}(H) + H^+ \longrightarrow H{-}\ddot{O}^+{-}H\,(H)$$

water → hydronium ion

$$H{-}\ddot{N}(H)_2 + H^+ \longrightarrow H{-}N^+(H)_2{-}H$$

ammonia → ammonium ion

While it is obvious that the hydronium and ammonium ions are positively charged, we must carry our thinking a bit farther and ask, "Which *atom* in each of these molecules bears the positive charge?" Formal positive or formal negative charges are derived by assigning all unshared (nonbonding) electrons and half of the shared (bonding) electrons to a particular atom. Comparison of this number with the normal complement of valence electrons in the neutral, unbonded atom gives the formal charge.

In the case of the hydronium ion, the formal charge on oxygen is derived by assigning five electrons (two nonbonding and half of three bonding pairs) to oxygen. This is one fewer than oxygen's normal complement of six valence electrons and accordingly the oxygen atom is said to have a formal positive charge. By applying the same electron bookkeeping you should be able to show that the nitrogen atom in the NH_4^+ ion bears the formal positive charge.

Problems

1.1 Give a simple electronic structure for the following compounds and ions. Show all valence electrons.

a	H_2O_2	**b**	N_2H_4	**c**	OH^-
d	H_3O^+	**e**	CH_3OH	**f**	CH_3NH_2
g	CH_3NHCH_3	**h**	CH_3Cl	**i**	CH_3OCH_3
j	C_2H_6	**k**	C_2H_4	**l**	C_2H_2
m	CO_2	**n**	H_2CO_3	**o**	CO_3^{2-}
p	CH_3CO_2H	**q**	CH_3COCH_3	**r**	CH_3CHCH_2
s	CH_3NNCH_3	**t**	HCN	**u**	HNO_2

1.2 Following the rule that each atom of carbon, oxygen, and nitrogen reacts to achieve a complete outer shell of eight valence electrons, add unshared pairs of electrons as necessary to complete the valence shells in the following molecules or ions. Assign formal positive or negative charges to each as appropriate.

a H—O—C(=O)—O

b CH_3CH_2—O

c CH_3—N(CH_3)(CH_3)—CH_3

d H—C(H)(H)—C(H)(H)

e H—N(H)(H)—C(H)(H)—C(=O)—O

f H—C(H)—C(=O)—H

1.3 As early as the beginning of the 19th century it was recognized that chemical substances isolated from living materials contained carbon and hydrogen, and in fact, organic chemistry was defined as a branch of chemistry dealing with compounds obtained from plant and animal sources. Since organic substances were derived from living organisms, they were thought to be different from inorganic substances in that they contained some kind of essential "vital force."

In 1828 the German chemist, Friedrich Wöhler, discovered that urea could be made by heating ammonium cyanate. Urea had previously been obtained only from urine, whereas ammonium cyanate was a typical inorganic or "nonorganic" salt. This experiment was significant in the history of organic chemistry for it demonstrated the interconversion of inorganic and organic substances.

Draw Lewis formulas for urea and ammonium cyanate and describe the differences in structure between the two.

$$H_2N-\overset{\overset{\displaystyle O}{\|}}{C}-NH_2 \qquad NH_4^+OCN^-$$

urea — ammonium cyanate

1.6 COMMON ORGANIC FUNCTIONAL GROUPS

In the preceding sections we have examined the different types of covalent bonds formed by carbon with hydrogen, nitrogen, and oxygen. These bonds combine in various ways to form certain unique structural features known as functional groups. The concept of the functional group is important to organic chemistry for several reasons. First, functional groups serve as a basis for nomenclature (naming) of organic compounds. Second, they serve to classify organic compounds into families based on the presence of one or more functional groups. Finally and perhaps most important, they are the sites of chemical reaction or function. A particular functional group will have very similar chemical and physical properties whenever it is found in an organic molecule.

Collected in Table 1.4 are the functional groups that will be of most importance in this introductory course. Keep in mind that this list does not include all of the functional groups in organic chemistry. However, by mastering the chemistry of these groups, you will acquire a good grounding in organic chemistry and will be in a position to apply this understanding to the biochemistry that follows. Table 1.4 is repeated on the front endleaf for easy reference.

1.7 STRUCTURAL ISOMERISM

A molecular formula is simply a listing of the number and kind of atoms in a substance. By referring to the examples given in Table 1.4 and counting the number of atoms, we can see that the molecular formula of ethyl alcohol is C_2H_6O. The molecular formula of dimethyl ether is also C_2H_6O. Dimethyl ether and ethyl alcohol have the same molecular formula but different structural formulas; that is, the order of attachment of the atoms in each is different. Notice also that propanal and acetone have the same molecular formula, C_3H_6O, but different structural formulas. Likewise propanoic acid and methyl acetate each have the molecular formula $C_3H_6O_2$ but different structural formulas. Substances which have the same molecular formula but different structural formulas are called structural isomers.

Table 1.4 *Important functional groups.*

Functional Group Name	Functional Group (attached to carbon)	Example	
acid anhydride	$-C(=O)-O-C(=O)-$	$CH_3-C(=O)-O-C(=O)-CH_3$	acetic anhydride
alcohol and phenol	$-OH$	CH_3CH_2-OH	ethyl alcohol
aldehyde	$-C(=O)-H$	$CH_3CH_2-C(=O)-H$	propanal
alkene or olefin	$>C=C<$	$H_2C=CH_2$	ethylene
alkyne	$-C\equiv C-$	$H-C\equiv C-H$	acetylene
amide	$-C(=O)-N<$	$CH_3-C(=O)-NH_2$	acetamide
amine, primary	$-NH_2$	CH_3-NH_2	methylamine
amine, secondary	$-NH-$	$CH_3-NH-CH_3$	dimethylamine
amine, tertiary	$-N<$	$CH_3-N(CH_3)-CH_3$	trimethylamine
carboxylic acid	$-C(=O)-O-H$	$CH_3CH_2-C(=O)-O-H$	propanoic acid
ester	$-C(=O)-O-$	$CH_3-C(=O)-O-CH_3$	methyl acetate
ether	$-O-$	CH_3-O-CH_3	dimethyl ether
halide	$-X$	CH_3CH_2-F	ethyl fluoride
ketone	$-C(=O)-$	$CH_3-C(=O)-CH_3$	acetone
thiol or mercaptan	$-SH$	CH_3CH_2-SH	ethanethiol
sulfide	$-S-$	$CH_3CH_2-S-CH_2CH_3$	diethyl sulfide
disulfide	$-S-S-$	$CH_3-S-S-CH_3$	dimethyl disulfide
sulfonic acid	$-S(=O)_2-OH$	$CH_3-S(=O)_2-OH$	methanesulfonic acid

Problems

1.4 Shown below are five structural formulas. By counting the number of atoms and noting the order of attachment, verify for yourself that each has the same molecular formula, $C_3H_6O_2$, but a different structural formula. Name the functional groups present in each of these structural isomers.

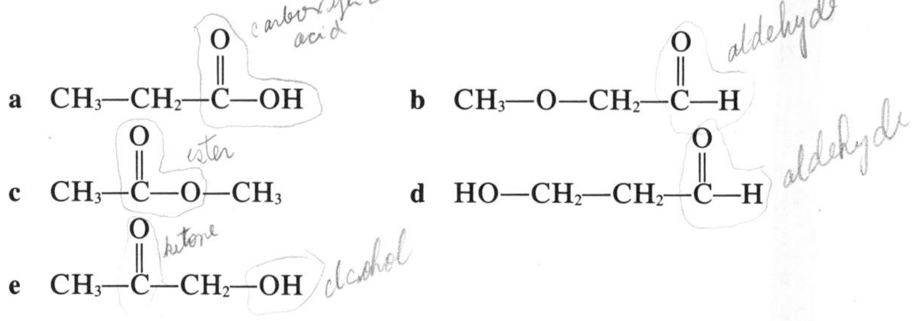

a $CH_3{-}CH_2{-}\overset{\overset{\displaystyle O}{\|}}{C}{-}OH$ **b** $CH_3{-}O{-}CH_2{-}\overset{\overset{\displaystyle O}{\|}}{C}{-}H$

c $CH_3{-}\overset{\overset{\displaystyle O}{\|}}{C}{-}O{-}CH_3$ **d** $HO{-}CH_2{-}CH_2{-}\overset{\overset{\displaystyle O}{\|}}{C}{-}H$

e $CH_3{-}\overset{\overset{\displaystyle O}{\|}}{C}{-}CH_2{-}OH$

1.5 Write a structural formula for a compound of molecular formula C_3H_7NO that contains the following functional groups:

a an amide
b a ketone and a primary amine
c an aldehyde and a primary amine
d an aldehyde and a secondary amine

1.6 Write a structural formula for a compound of molecular formula C_4H_8O that contains the following functional groups:

a a carbon-carbon double bond and an alcohol
b a ketone
c an aldehyde
d a carbon-carbon double bond and an ether

1.7 Write structural formulas for all:

a amines of molecular formula $C_4H_{11}N$
b alcohols of molecular formula $C_5H_{12}O$
c ethers of molecular formula $C_5H_{12}O$
d aldehydes of molecular formula $C_6H_{12}O$
e ketones of molecular formula $C_6H_{12}O$
f carboxylic acids of molecular formula $C_6H_{12}O_2$
g esters of molecular formula $C_6H_{12}O_2$

1.8 PREDICTING BOND ANGLES

Thus far, we have developed the concept of the electron pair as the fundamental unit of the covalent bond and, using this concept and the Lewis model of bonding, we have been able to write simple electronic formulas for a variety of small molecules. Now let us take up the next question—how can we predict bond angles and molecular geometry? For example, what is the H—C—H bond angle in CH_4, the H—N—H bond angle in NH_3, or the H—O—H bond angle in H_2O?

To answer this type of question chemists have devised several theories or models. Of these we shall study two, the valence-shell electron-pair repulsion model and the molecular orbital model. Each of these models gives correct predictions of bond angles and molecular geometries. If our interest in this course were only in studying organic molecules to the extent of being able to predict bond angles and molecular geometry, we would probably use just the valence-shell electron-pair repulsion model, for it is easier to learn and apply. However, our interests go considerably beyond this. We also want to develop some understanding of the relationships between molecular structure and chemical reactivity, and for this we will need to discuss the molecular orbital model.

Let us begin with the valence-shell electron-pair repulsion model for, as we have already said, it is less complex to learn and easier to apply. This model proposes that the arrangement of atoms or groups around any central atom is determined primarily by the repulsive interactions among all the electron pairs in the valence shell of the central atom. These pairs of valence electrons will be spread as far away from each other as possible so as to minimize their repulsive interactions.

In order to apply this model, you must first determine the number of electron pairs in the valence shell of the atom in question. A direct way to do this is to look at a Lewis electronic formula and count the pairs of valence electrons. Second, you must calculate the geometric arrangement that will minimize electrostatic repulsion among these electron pairs.

As a first example consider the molecule methane, CH_4. The central atom, in this case carbon, has four separate pairs of valence electrons to be arranged around it. How can these four pairs be arranged so that the repulsive interactions between them are minimized? A little experimentation with molecular models or a little paper-and-pencil calculation involving some solid geometry will convince you that these pairs will be distributed around carbon in the form of a regular tetrahedron with the hydrogen nuclei at the corners of the tetrahedron and the carbon nucleus embedded in the center of the tetrahedron (Figure 1.2).

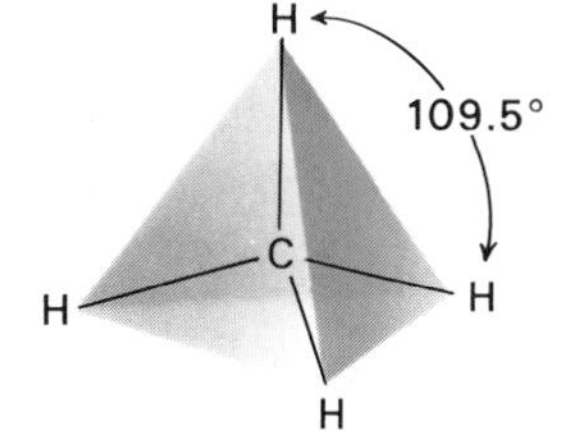

Figure 1.2 *Bond formation in methane, CH_4. The four bonds of carbon are directed toward the corners of a regular tetrahedron.*

The H—C—H bond angle in this tetrahedral arrangement is 109.5°. How does this prediction compare with the observed angle of 109.5°? The two are identical and we therefore conclude that our prediction is a good one. A more general prediction also emerges from what we have just said about CH_4. Any time there are four separate pairs of valence electrons around an atom, we will predict tetrahedral distribution.

In ammonia there are four pair of valence electrons to be distributed around nitrogen (Figure 1.3). Therefore we predict 109.5° for the H—N—H bond angle in ammonia. However, because there are only three atoms bonded to nitrogen, we say that this molecule is pyramidal in shape.

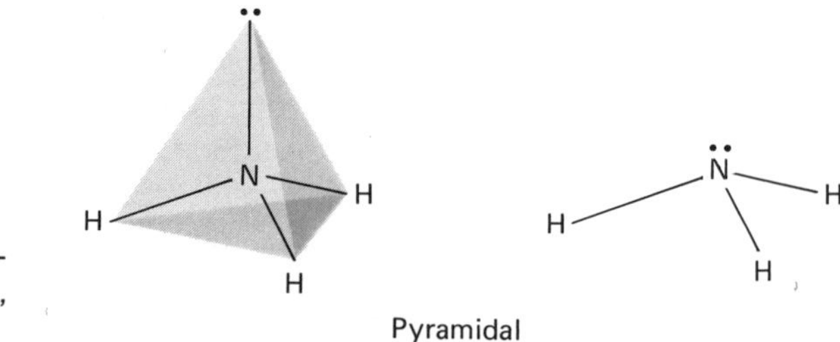

Figure 1.3 Bond formation in ammonia, NH_3.

In water, there are also four pairs of valence electrons to be distributed around oxygen (Figure 1.4). Therefore we predict 109.5° for the H—O—H bond angle in water. Because there are only two atoms bonded to oxygen, we say that this molecule is angular.

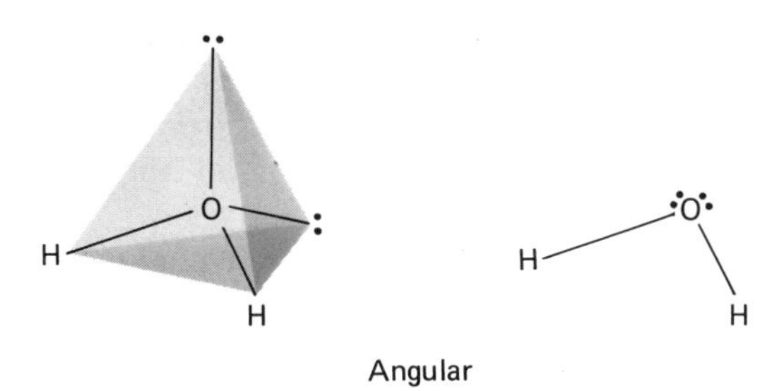

Figure 1.4 Bond formation in water, H_2O.

How do the predicted bond angles for ammonia and water compare with observed angles? The experimentally determined H—N—H angle in ammonia is 107.3°, and the H—O—H angle in water is 104.5°. These small differences between the predicted and observed bond angles can be accounted for by proposing that the unshared (nonbonding) electron pairs repel adjacent electron pairs more strongly than do bonding pairs. Note that the distortion from 109.5° is larger in H_2O, which contains two unshared pairs of electrons, and is smaller in NH_3, which contains only one unshared pair of electrons. There is no distortion in CH_4.

Molecules that contain a double bond can be treated by considering the double bond as just one region or "bundle" of electron density to be accounted for. In formaldehyde, CH_2O, there are three regions of electron density to be spread around the central carbon atom. Experimentation with molecular models should convince you that the two single pairs and the one double pair of bonding electrons will be arranged in a triangular-coplanar structure. The carbon will lie at the center of an equilateral triangle with the oxygen and hydrogen atoms at the corners

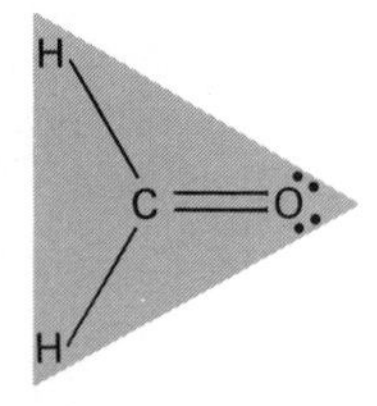

Figure 1.5 Bond formation in formaldehyde, CH_2O.

of the triangle (Figure 1.5). Given this triangular-coplanar geometry, you would predict 120° for the H—C—H and H—C—O bond angles.

In ethylene, C_2H_4, there are also three regions of electron density to be spread around each carbon atom. Again it is a problem of arranging two single pairs and one double pair of valence electrons about a carbon atom so as to minimize repulsive interactions, and again you would predict 120° for all bond angles in the molecule (Figure 1.6).

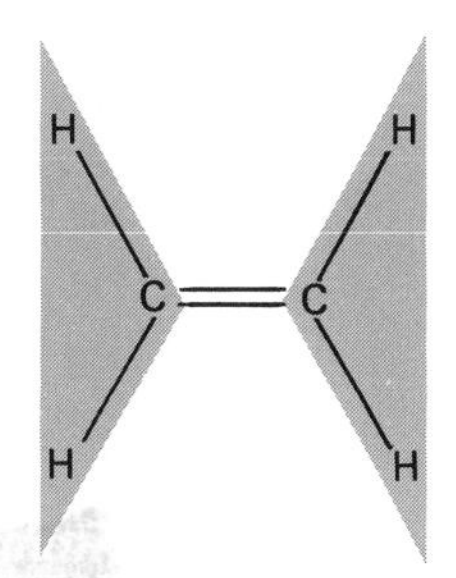

Figure 1.6 *Bond formation in ethylene, C_2H_4.*

Molecules that contain a triple bond can also be treated by considering the triple bond as just one region of electron density. In acetylene, C_2H_2, there are two regions of electron density to be spread around each carbon atom. Experimentation with molecular models will convince you that the one single pair and the one triple pair of bonding electrons will be arranged in linear fashion (Figure 1.7).

Figure 1.7 *Bond formation in acetylene, C_2H_2. The bond angle in this linear arrangement is 180°.*

By examining together the Lewis electronic structures you drew in answer to Problem 1.1 and our discussions in this section on the geometry of covalent bonding, we can draw certain generalizations. For stable, uncharged molecules containing atoms of carbon, hydrogen, nitrogen, and oxygen:

1. Carbon forms four covalent bonds. Bond angles on carbon are approximately 109.5° for four attached groups, 120° for three attached groups, and 180° for two attached groups.
2. Nitrogen forms three covalent bonds and has one unshared pair of electrons. Bond angles on nitrogen are approximately 109.5° for three attached groups, and 120° for two attached groups.
3. Oxygen forms two covalent bonds and has two unshared pairs of electrons. Bond angles on oxygen are approximately 109.5° for two attached groups. We cannot specify a bond angle for only one attached group (as for example, about the oxygen atom in CH_2O) since it requires three nuclei to determine a bond angle.

Thus, by using the valence-shell electron-pair repulsion model, we can make good predictions of bond angles in a simple and straightforward way. Yet as much as this model has helped us to understand something of molecular geometry, it does not really explain the relationships between molecular structure and chemical reactivity. For example, a carbon-carbon double bond is quite different in chemical reactivity from a carbon-carbon single bond. Most carbon-carbon single bonds are quite unreactive, but carbon-carbon double bonds react with a wide variety of reagents under a variety of experimental conditions. To discuss modern organic and biochemistry, at even the most minimal level, we must have a clear understanding of how the chemist accounts for these differences. Therefore let us now approach the question of covalent bonding on a different and more sophisticated level—in terms of atomic orbitals and covalent bond formation by the overlap of atomic orbitals.

1.9 THE EXTENSION OF ATOMIC ORBITALS IN SPACE

Atomic orbitals are essentially regions of space centered around the nucleus of an atom. In this course we will consider only the sizes and shapes of *s* and *p* orbitals, for these are the only ones used in covalent bonding in compounds of hydrogen, carbon, oxygen, and nitrogen. Yet in considering the covalent bonding of just these few atoms, we can cover a great deal of organic and biological chemistry.

Figure 1.8 shows the shape of an atomic 1*s* orbital. Note that this orbital is a sphere with its center at the nucleus. Also shown in Figure 1.8 is a cross-section of a 1*s* orbital.

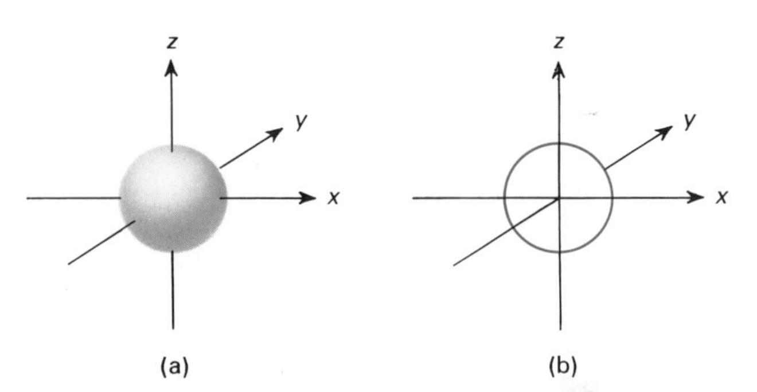

Figure 1.8 *Atomic orbitals: (a) 1s orbital; nucleus at the center. (b) Spherical cross section of 1s atomic orbital.*

The 2*s* orbital is also a sphere with its center at the nucleus. The 2*s* orbital is several times as large as the 1*s* orbital.

Next are three 2*p* orbitals. Each of these is dumbbell shaped with the center of the dumbbell at the nucleus. The axis of each 2*p* orbital is perpendicular to that of the other two. They are designated $2p_x$, $2p_y$, and $2p_z$, where *x*, *y*, and *z* refer to the three coordinate axes. The three 2*p* orbitals are shown schematically in Figure 1.9.

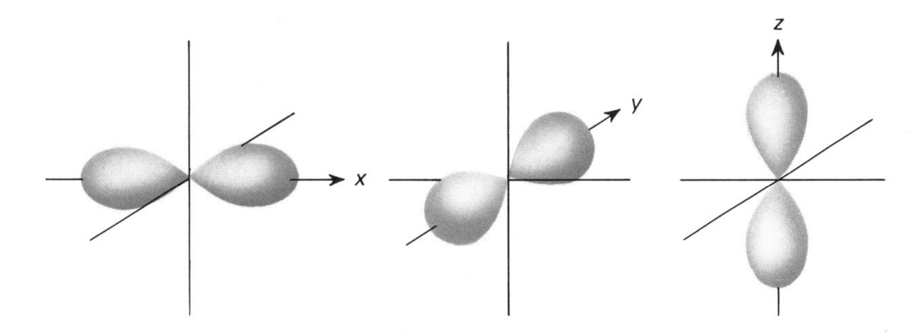

Figure 1.9 *Atomic orbitals: p orbitals. Axes mutually perpendicular.*

1.10 COVALENT BOND FORMATION BY THE OVERLAP OF ATOMIC ORBITALS

In Section 1.3 we described the covalent bond of the hydrogen molecule in terms of the Lewis formulation. Recall that according to the Lewis model each hydrogen atom donates one valence electron to form a two-electron covalent bond.

$$H\cdot + \cdot H \rightarrow H{:}H \qquad \text{or} \qquad H—H$$

Now let us re-examine this covalent bond and see how it can be interpreted in terms of our more sophisticated model.

The formation of a covalent bond between two atoms amounts to bringing the atoms up to each other in such a way that an atomic orbital of one atom overlaps with an atomic orbital of the other atom. For example, in forming the covalent bond in the molecule H_2, the two hydrogen atoms approach each other so that their $1s$ atomic orbitals overlap (Figure 1.10).

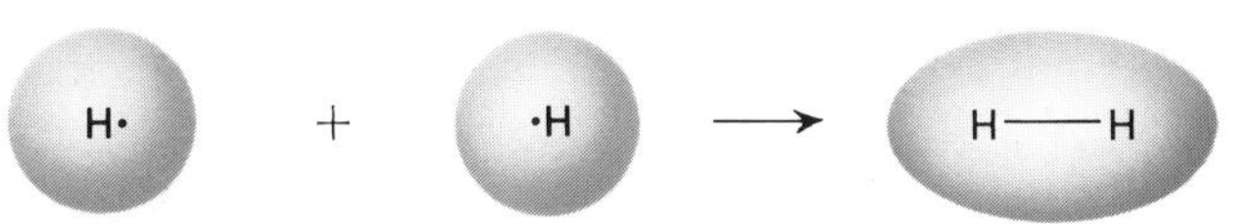

Figure 1.10 *Bond formation. The overlap of two 1s atomic orbitals to form a molecular orbital.*

This new orbital formed by the overlap of atomic orbitals encompasses both nuclei and is called a molecular orbital. Like an atomic orbital, a molecular orbital can accommodate at most two electrons. This is why the single covalent bond involves the sharing of only two electrons. The molecular orbital resulting from the overlap of two $1s$ atomic orbitals is cylindrically symmetrical about the axis joining the two nuclei. Molecular orbitals which have this electron distribution are called sigma (σ) orbitals, and the bond is called a sigma bond.

Note that the sigma bond in the hydrogen molecule is formed by the overlap of atomic $1s$ orbitals. Such bonds can also be formed by the overlap of other combinations of s and p orbitals as we shall see very soon. The essential feature of the sigma bond is the overlap or fusion of two orbitals that lie on the bond axis, that is, on the line joining the two nuclei which are bonded together.

1.11 THE TETRAHEDRAL CARBON ATOM—THE sp^3 HYBRID

The formation of four bonds to carbon is one of the central facts of organic chemistry. That the configuration of compounds of the type C_{abcd} is tetrahedral was postulated on quite empirical grounds by Jacobus van't Hoff and Joseph Le Bel in 1874. And indeed we arrived at this same conclusion using the valence-shell electron-pair repulsion model.

Clearly, bonding by one spherically symmetrical 2*s* orbital and three highly directional 2*p* orbitals at angles of 90° does not correspond to the actual geometry of the molecule. How can we reconcile the observed tetrahedral angle with our orbital theory of bonding? We do it by devising a new set of four equivalent, tetrahedrally oriented atomic orbitals. If we assume that atomic orbitals on the same atom can combine with each other, then the mathematical combination of one 2*s* orbital and three 2*p* orbitals results in the formation of four new and equivalent atomic orbitals (Figure 1.11).

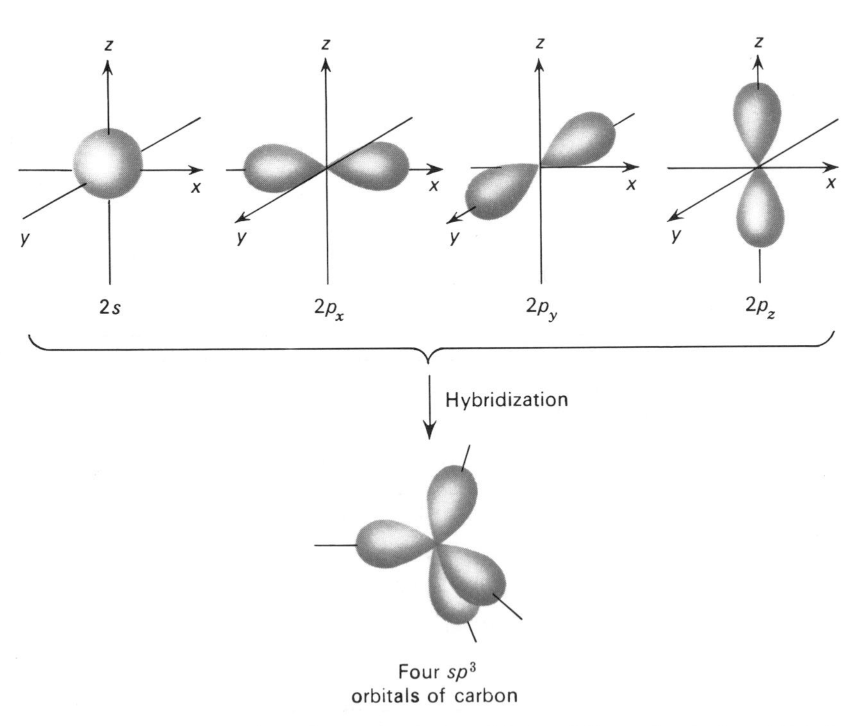

Figure 1.11 *Hybridization of one atomic 2s orbital and three atomic 2p orbitals to form four sp^3 hybrid orbitals.*

This process of combining atomic orbitals is called hybridization. The new orbitals in this case are called sp^3 orbitals. These sp^3 hybrid orbitals are directed from the carbon nucleus toward the corners of a regular tetrahedron, and the resulting molecular orbitals that bond each carbon and hydrogen are also tetrahedrally oriented. Accordingly we would predict all bond angles in methane to be 109.5°, a prediction in complete agreement with the experimentally measured angle.

1.12 COVALENT BONDING IN WATER AND AMMONIA

Now we shall again consider the bonding in H_2O, this time by using sp^3 hybrid orbitals. First let us consider oxygen with its six valence electrons. Filling two sp^3 orbitals accounts for four of the six valence electrons, and placing one electron in each of the other two sp^3 orbitals accounts for the remaining two. Each partially filled sp^3 orbital forms a molecular orbital with a $1s$ orbital of hydrogen and the hydrogen atoms will occupy two corners of a regular tetrahedron. The other two corners of the tetrahedron will be occupied by the unshared pairs of electrons (Figure 1.12). Thus, we predict the H—O—H bond angle of 109.5°. Although the experimentally measured angle of 104.5° is somewhat less than we had predicted, we can rationalize this difference by postulating that the sp^3 orbitals containing the unshared pairs are somewhat larger than those containing bonding pairs.

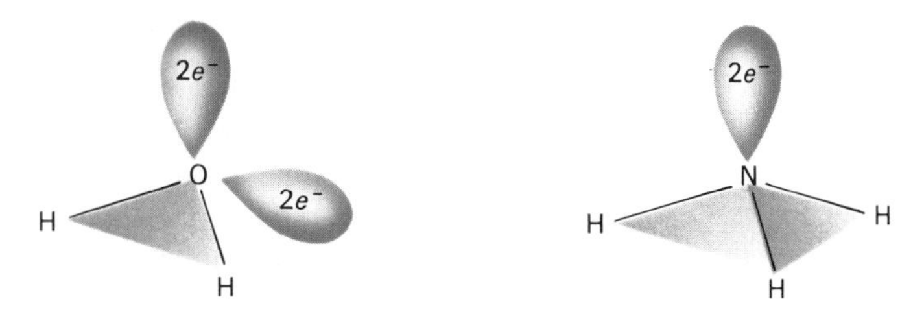

Figure 1.12 Bond formation in water and ammonia showing the unshared electron pairs in filled sp^3 hybrid orbitals.

Finally let us look at the covalent bonding in ammonia, NH_3, in terms of hybridized atomic orbitals. The nitrogen atom has five valence electrons. In the bonding state, one of the sp^3 orbitals is filled with a pair of electrons while each of the other three sp^3 hybrids has but one electron. Overlapping each of these three sp^3 hybrid orbitals with a $1s$ orbital of hydrogen gives the molecule NH_3. The fourth sp^3 hybrid orbital contains the unshared pair of electrons. Figure 1.12 shows a representation of the NH_3 molecule.

Because we have used sp^3 hybrid orbitals of nitrogen for bonding, we would predict the H—N—H angle to be 109.5°. From experimental measurements the H—N—H bond angle in ammonia is known to be 107°. This small difference between the predicted and the measured bond angle can be accounted for by assuming, as we did in our discussion of bonding in the water molecule, that the single pair of nonbonding electrons fills a larger region in space than the bonding pairs.

1.13 CARBON-CARBON AND CARBON-OXYGEN DOUBLE BONDS

To form bonds with three other atoms, carbon uses three equivalent sp^2 hybrid orbitals formed by mixing the one $2s$ and two $2p$ orbitals (arbitrarily designated $2p_x$ and $2p_y$ orbitals). After hybridization we have one $2p_z$ orbital and three equivalent sp^2 orbitals which lie in a plane and are directed toward the corners of an equilateral triangle; the angle between the sp^2 hybrid orbitals is 120°. This trigonal arrangement maximizes the separation of the hybrid orbitals and accordingly

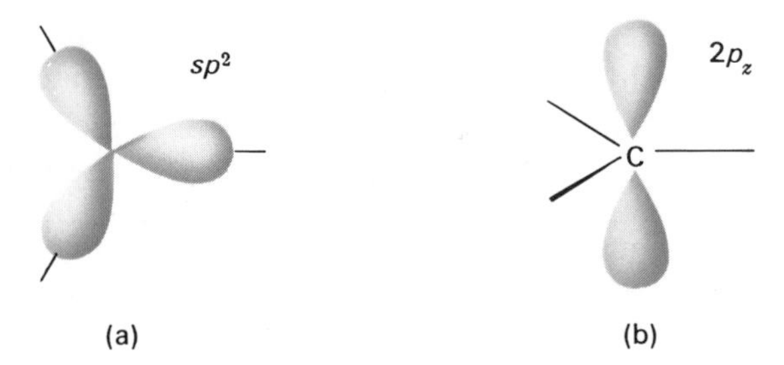

Figure 1.13 Atomic orbitals: hybrid sp^2 orbitals. (a) The three equivalent sp^2 hybrid orbitals lying at angles of 120°. (b) The unhybridized $2p_z$ orbital lying perpendicular to the sp^2 hybrid orbitals.

minimizes their electrostatic interaction. The remaining $2p_z$ orbital, which is not involved in the hybridization, consists of two lobes lying perpendicular to the plane of the sp^2 hybrid orbitals. Note that this arrangement puts the unhybridized $2p_z$ orbital the maximum distance from the three sp^2 hybrid orbitals. Figure 1.13 shows the three equivalent sp^2 hybrid orbitals and the $2p_z$ orbital of carbon.

To form a carbon-carbon double bond as in the ethylene molecule, C_2H_4, we arrange the two carbons and the four hydrogens so that the carbons are bonded to each other by overlapping sp^2 orbitals and so that each carbon is bonded to two hydrogens by the overlap of an sp^2 orbital of carbon and a 1*s* orbital of hydrogen. Each of these bonds is cylindrically symmetrical about the line joining the nuclei and is called a sigma bond.

The remaining $2p_z$ atomic orbitals overlap and the two electrons of this new molecular orbital form a second bond, called a pi (π) bond. The pi bond consists of two sausage-shaped regions of electron density, one on either side of the plane formed by the carbon and hydrogen atoms of ethylene. Because of the lesser degree of orbital overlap, the pi bond joining the two carbons is weaker than the sigma bond. As we can see from Figure 1.14, this overlap of $2p_z$ orbitals to give the pi bond can occur only if the $2p_z$ orbitals are parallel. Hence all six atoms of the ethylene molecule must lie in a plane.

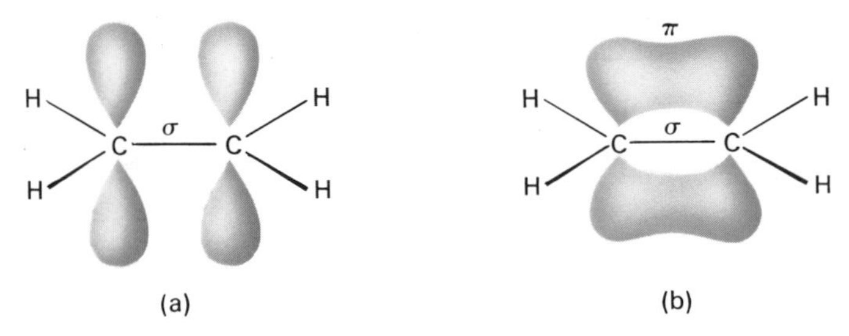

Figure 1.14 The ethylene molecule and the carbon-carbon double bond. (a) Sigma bonds shown; $2p_z$ orbitals not overlapping. (b) $2p_z$ orbitals overlap to form the pi bond above and below the plane of the atoms.

To form a carbon-oxygen double bond as in the formaldehyde molecule, CH_2O, the carbon atom is joined to three other atoms by overlapping the three equivalent sp^2 hybrid orbitals with two hydrogen

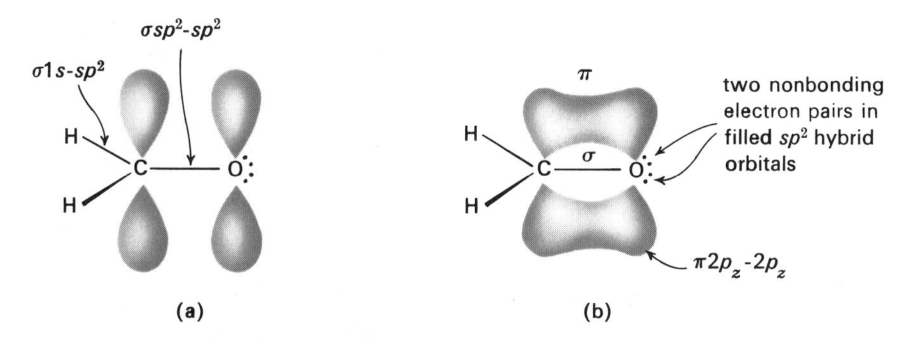

Figure 1.15 *The formaldehyde molecule. (a) Sigma bonds shown; $2p_z$ orbitals not overlapping. (b) $2p_z$ orbitals overlap to form a pi bond.*

$1s$ orbitals and one sp^2 orbital of oxygen (Figure 1.15a). The remaining $2p_z$ orbital of carbon overlaps with a $2p_z$ orbital of oxygen to form the pi bond (Figure 1.15b). Here, as in ethylene, the double bond consists of two types of molecular orbitals: sp^2 orbitals overlapping to form sigma bonds and $2p_z$ orbitals overlapping to form a pi bond.

Before going on, we should note that both the direct, almost intuitive valence-shell electron-pair model and the much more sophisticated hybridized atomic orbital model give us equally good semiquantitative predictions of bond angles in the small molecules we have examined so far. However, it is the hybridized atomic orbital model which provides us with a much clearer picture of the nature of the carbon-carbon and carbon-oxygen double bonds. The double bond is not just a combination of two identical bonds. Rather, according to this model, the double bond consists of one sigma bond and one pi bond. It is just this approach to bonding that we shall use in subsequent chapters to help us understand the chemistry of compounds containing double bonds.

1.14 ORGANIC CHEMISTRY—THE UNIQUENESS OF CARBON

According to the most simple definition, organic chemistry is the chemistry of compounds of carbon. While the term "organic" itself does remind us that a great many of the compounds of carbon are of either animal or plant origin, by no means is that the limit. With ever increasing skill and ease, man has synthesized thousands upon thousands of new "organic" compounds in the laboratory. It is estimated that there are over 3,000,000 known organic compounds, either isolated from nature or synthesized in the laboratory, and this number is growing with increasing speed.

Why consider organic chemistry as a separate branch of chemistry? Or, why single out the chemistry of one atom for study as a special branch of chemistry? Put another way, what is so unique about the chemistry of carbon? Of course the answer is in part due to the exceptionally large number of compounds of carbon. While the number of known organic compounds is over 3,000,000 at the present time, the number of inorganic compounds is approaching 100,000. The answer is also due in part to the tendency of carbon atoms to bond together in long chains and the possibilities for structural isomerism. But at a more fundamental level, the answer rests on the special stability of compounds containing carbon-carbon and carbon-hydrogen single bonds. To see this in more concrete terms, let us look at Table 1.5 and some representative bond dissociation energies (BDE).

Table 1.5 *Representative bond dissociation energies (BDE).*

Molecule	BDE (kcal/mole)	Bond	Average BDE (kcal/mole)
H_2	104	C—C	83
Cl_2	58	N—N	50
Br_2	46	O—O	34
I_2	36	C—H	94
		C—N	73
		C—O	85

For diatomic molecules, the bond dissociation energy is defined as the energy necessary to split one mole of gaseous molecules into separate atoms at 25°C and at 1 atmosphere pressure. For more complex polyatomic molecules only average bond dissociation energies are listed. While the BDE for the C—C bond is given as 83 kcal/mole, the BDE for any particular C—C bond may be larger or smaller depending on the location of the C—C bond in the molecule and on the other atoms attached to the carbons. But while there is a certain amount of variation in particular bonds, we can nonetheless make two generalizations: (1) With the exception of C—C and H—H, single bonds between identical atoms are relatively weak. Compare C—C and H—H with N—N, Cl—Cl, etc. (2) C—H bonds are on the average stronger than C—N and C—O bonds. Another way of stating generalization (1) is that single bonds between carbon (C—C) are notably stronger than single bonds between atoms with unshared pairs of electrons (N—N, O—O).

We might wonder this is so. One possible explanation lies in the fact that the unshared pairs of electrons of nitrogen and oxygen repel each other and thereby substantially weaken the respective single bonds. The fact that the O—O bond is considerably weaker than the N—N bond is consistent with this explanation.

Another fact of organic chemistry is that C—C and C—H single bonds show little tendency to participate in chemical reactions. The C—O and C—N bonds, on the other hand, readily undergo a variety of chemical reactions. Why? Again the explanation lies in the presence of unshared pairs of electrons on the nitrogen and oxygen atoms. It is the presence of these unshared pairs of electrons that makes oxygen and nitrogen atoms susceptible to attack by electron-deficient atoms. Carbon, on the other hand, when it is tetrahedrally bonded by four single bonds to four other carbons or hydrogens, has no unshared pairs of electrons and is not susceptible to attack by electron-deficient atoms. Furthermore, since tetravalent carbon has a complete outer shell of electrons, it is not susceptible to attack by electron-rich reagents.

Herein lie the reasons for the uniqueness of carbon compounds and of organic chemistry: the particular strength of the carbon-carbon single bond, and the resistance of carbon-carbon and carbon-hydrogen bonds to attack by electron-deficient or electron-rich reagents.

Problems

1.8 Explain how the valence-shell electron-pair repulsion model can be used to predict bond angles in compounds containing atoms of C, N, O, and H.

1.9 Define the terms:

a atomic orbital
b molecular orbital
c sigma bond
d pi bond
e hybrid atomic orbital

1.10 According to the atomic orbital theory of covalent bonding, sigma and pi bonds are similar in that each is formed by the overlap of atomic orbitals of adjacent atoms. In what way(s) do sigma and pi bonds differ?

1.11 What shape would you predict for the NH_4^+ ion? for the H_3O^+ ion? Explain the basis for your predictions.

1.12 Structural formulas for the following molecules are shown in Table 1.4. Using the valence-shell electron-pair repulsion model, predict all bond angles. Next describe the bonding in each molecule in terms of the orbitals involved, and based on this orbital description, predict all bond angles. Finally, compare your predictions of bond angles according to each model. They should be the same.

a acetic anhydride
b ethyl alcohol
c propanal
d ethylene
e acetamide
f methylamine
g dimethylamine
h trimethylamine
i propanoic acid
j methyl acetate
k dimethyl ether
l acetone

1.13 Define the term bond dissociation energy.

1.14 How might you account for the fact that C—C single bonds are notably stronger than N—N or O—O single bonds?

1.15 How might you account for the fact that compounds containing C—O and C—N single bonds react quite readily with H^+, whereas those containing only C—C and C—H single bonds do not?

2

ALKANES AND CYCLOALKANES

2.1 INTRODUCTION

The compounds consisting solely of carbon and hydrogen are called hydrocarbons. If the carbon atoms in a hydrocarbon are joined together only by single covalent bonds, the compounds are called saturated hydrocarbons or alkanes. If any of the carbon atoms of the hydrocarbon are bonded together by one or more double or triple bonds, the compounds are called unsaturated hydrocarbons. In this chapter we shall discuss the saturated hydrocarbons. They are the simplest organic substances from the structural point of view and therefore a good place to begin the study of organic chemistry. We shall discuss unsaturated hydrocarbons in Chapter 3.

The terms saturated and unsaturated classify hydrocarbons according to the presence or absence of carbon-carbon double and triple bonds. Another set of terms classifies hydrocarbons according to the presence or absence of carbon atoms joined together in such a way as to form rings:

1. Aliphatic hydrocarbons are composed of chains of carbon atoms and do not contain carbon rings. Aliphatic hydrocarbons are sometimes called open-chain or acyclic hydrocarbons.
2. Alicyclic hydrocarbons contain one or more carbon rings. They are also called cyclic hydrocarbons.
3. Aromatic hydrocarbons are derived from benzene.

We have already encountered the first two members of the alkane family, methane and ethane. Remember that the representation of ethane as in Figure 2.1 makes no attempt to depict bond angles. All bond angles in ethane are 109.5° rather than either 90° or 180° as the diagram might suggest. In even more abbreviated form, we can write a condensed structural formula for ethane as $CH_3—CH_3$ or CH_3CH_3.

By increasing the number of carbon atoms in the chain, we can

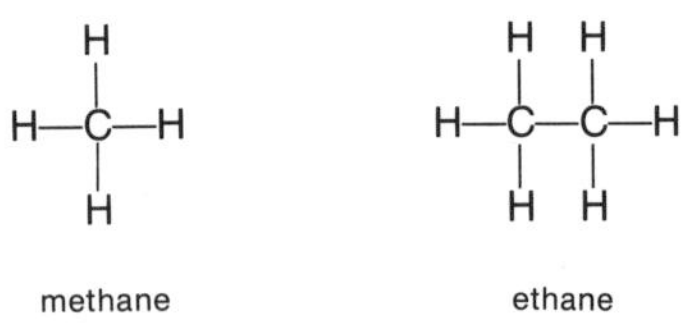

methane ethane

Figure 2.1 *Methane and ethane. All bond angles predicted to be 109.5°.*

form the next members of the series: propane, C_3H_8; butane, C_4H_{10}; and pentane, C_5H_{12}.

$$\underset{\text{propane}}{CH_3CH_2CH_3} \qquad \underset{\text{butane}}{CH_3CH_2CH_2CH_3} \qquad \underset{\text{pentane}}{CH_3CH_2CH_2CH_2CH_3}$$

If we examine the structures of these alkanes, we see that ethane contains one C and two H more than methane; propane contains one C and two H more than ethane; etc. A series of compounds wherein each member differs from the previous member by a constant increment is called a homologous series. In the case of the alkanes, the constant increment is the unit $—CH_2—$, the methylene group. We refer to the members of such a series as homologs.

Inspection of the molecular formulas of the alkanes reveals that each member of the series has the general formula C_nH_{2n+2}. As we shall see in subsequent chapters, each homologous series has its own characteristic general formula. The names of the first ten alkanes in Table 2.1 should be committed to memory for they are the basis of the systematic nomenclature of alkane derivatives.

Table 2.1 *Names, molecular formulas, and condensed structural formulas of the first ten alkanes.*

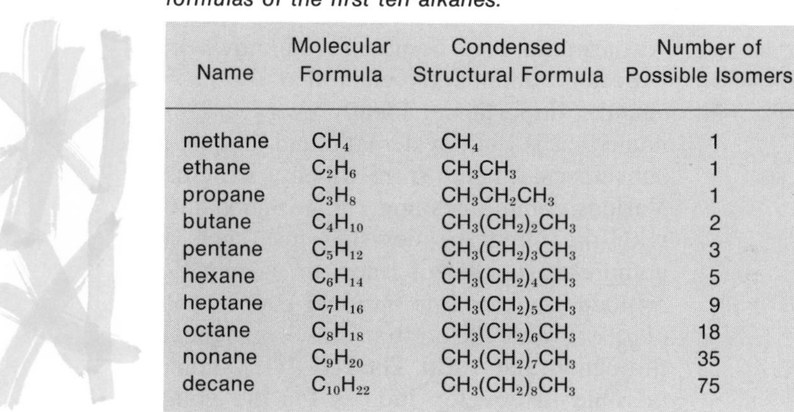

Name	Molecular Formula	Condensed Structural Formula	Number of Possible Isomers
methane	CH_4	CH_4	1
ethane	C_2H_6	CH_3CH_3	1
propane	C_3H_8	$CH_3CH_2CH_3$	1
butane	C_4H_{10}	$CH_3(CH_2)_2CH_3$	2
pentane	C_5H_{12}	$CH_3(CH_2)_3CH_3$	3
hexane	C_6H_{14}	$CH_3(CH_2)_4CH_3$	5
heptane	C_7H_{16}	$CH_3(CH_2)_5CH_3$	9
octane	C_8H_{18}	$CH_3(CH_2)_6CH_3$	18
nonane	C_9H_{20}	$CH_3(CH_2)_7CH_3$	35
decane	$C_{10}H_{22}$	$CH_3(CH_2)_8CH_3$	75

2.2 STRUCTURAL ISOMERISM IN ALKANES

We have already encountered examples of structural isomerism (Section 1.7). Recall that two or more compounds having the same molecular formula but a different order of attachment of atoms are called structural isomers; the phenomenon is known as structural isomerism.

For methane, ethane, and propane there is only one structural formula that can be drawn once the molecular formula has been established. However, as the number of carbon atoms increases, so too does the possibility for structural isomerism. Butane has the molecular formula C_4H_{10} and we can draw two different structural formulas corresponding to this molecular formula. In one of the structures of composition C_4H_{10}, the four carbons are attached in a chain; in the other they are attached three in a chain with the fourth carbon as a branch on the chain.

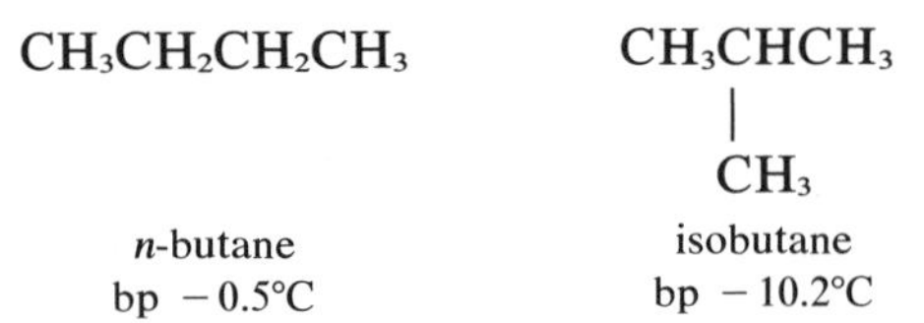

We distinguish between these two structural formulas by the names *n*-butane (normal butane) and isobutane. The prefix "normal" or "*n*-" is used to designate compounds containing only a single, continuous and unbranched chain of carbon atoms. Notice that the boiling points of the isomeric butanes differ by over 9 degrees.

There are 3 structural isomers of C_5H_{12}, 18 of C_8H_{18}, and a total of 75 for $C_{10}H_{22}$. It should be obvious that for even a rather modest number of carbon and hydrogen atoms, a very large number of structural isomers is possible. In fact, the potential for structural and functional individuality open to nature from just the basic building blocks of carbon, hydrogen, oxygen, and nitrogen is practically limitless. We shall continue to develop this theme of structural isomerism in this and later chapters.

2.3 NOMENCLATURE OF ORGANIC COMPOUNDS

As the number and complexity of known organic compounds increased, it became abundantly clear that the problem of nomenclature is of singular importance. Ideally every organic substance should have a name that is both systematic and unique, and, at least for the simpler substances, a name from which a structural formula can be deduced. Various committees and commissions have met over the last century with the intent of devising just such a system. A thorough and comprehensive set of rules for systematic nomenclature was recommended by the International Union of Pure and Applied Chemistry (IUPAC) and the system has been generally accepted by chemists throughout the world. The IUPAC system of nomenclature has made it possible to develop indices for the voluminous catalogs of factual information about organic compounds and thereby facilitate chemical communication.

But in spite of the precision of the IUPAC system, routine communication in organic chemistry still relies on a hodgepodge of trivial, semisystematic, and systematic names. The reasons for this situation are rooted in both convenience and historical development.

For example, there are three acceptable names for the relatively simple compound $CH_3CHOHCH_3$: isopropyl alcohol, 2-propanol, and dimethylcarbinol.

$$\begin{array}{c} CH_3{-}CH{-}CH_3 \\ | \\ OH \end{array}$$

isopropyl alcohol
2-propanol
dimethylcarbinol

The first of these, and the name most commonly used, is derived from a semisystematic system which is easily applicable to only low-molecular-weight hydrocarbons and their derivatives. The second is the IUPAC name. The third is a derived name and is rarely if ever used today.

Also, many of the organic compounds isolated from the biological world have been given names derived from the natural source of the material (e.g., penicillin from the mold *Penicillium notatum*).

We shall strive as far as possible to use systematic names in this text, but quite unavoidably there will be some crossing between the systems.

In the older, semisystematic nomenclature, the total number of carbon atoms in the alkane, regardless of their arrangement, determines the name. The first three alkanes are methane, ethane, and propane. All alkanes of formula C_5H_{12} are called pentanes, all alkanes of formula C_6H_{14} are hexanes, etc. For those alkanes beyond propane, "normal" or the prefix "*n*-" is used to indicate that all carbons are joined in a continuous chain. The prefix "iso-" indicates that one end of an otherwise continuous chain terminates in the $(CH_3)_2CH—$ group. The prefix "neo-" indicates that one end of an otherwise continuous chain of carbon atoms terminates in the $(CH_3)_3C—$ group. Examples using this system of nomenclature are shown below.

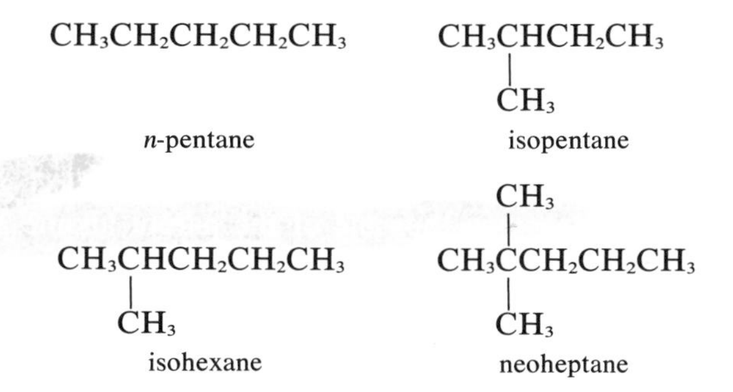

This semisystematic system has no good way of handling other branching patterns, and for more complex alkanes it is necessary to use the more flexible IUPAC system of nomenclature.

2.4 THE IUPAC SYSTEM OF NOMENCLATURE

In this section we will study the IUPAC system for naming alkanes. The rules themselves are quite simple and straightforward. It is important that you learn them well for they apply not only to alkanes but also to most derivatives of alkanes.

1. The general name of the saturated hydrocarbon is alkane.
2. For branched-chain hydrocarbons, the hydrocarbon derived from the longest continuous chain of carbon atoms is taken as the parent compound and the root name is that of the parent compound.

3. Each substituent group is given a name and a number. The number refers to the carbon of the parent compound to which the substituent is attached. The name of the hydrocarbon substituent is derived from the alkane of the same number of carbons by changing the -ane ending to -yl. Saturated hydrocarbon groups are commonly called alkyl groups and are commonly represented by the symbol R—.
4. If two or more substituents are present, the numbering of the parent alkane starts from the end that gives the lower number to the substituent that is encountered first.
5. If the same substituent occurs more than once, the number of the carbon atom to which each is attached is given. The number of identical groups is indicated by the Greek prefixes di-, tri-, tetra-, etc.
6. If there are several different alkyl substituents, they may be named either (a) in order of increasing complexity or (b) in alphabetical order. Whichever of the two systems is used, the names are placed in order before the name of the parent compound.
7. The last alkyl group named is prefixed to the name of the parent alkane, forming one word.

2.5 ALKYL GROUPS

As we have indicated in rule 3, alkyl groups are named simply by dropping the -ane of the parent alkane and adding -yl. The first two alkyl groups are methyl and ethyl (Figure 2.2).

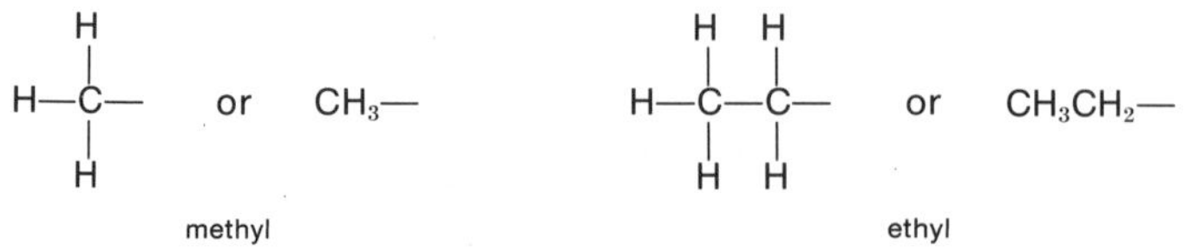

H—C(H)(H)— or CH_3— (methyl)

H—C(H)(H)—C(H)(H)— or CH_3CH_2— (ethyl)

Figure 2.2 *Alkyl groups: methyl and ethyl.*

There are two isomeric alkyl groups of molecular formula C_3H_7—, namely *n*-propyl and isopropyl (Figure 2.3). There are four isomeric

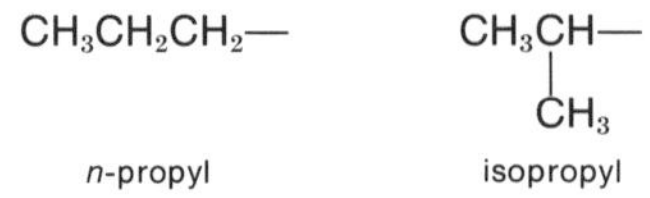

$CH_3CH_2CH_2$— (*n*-propyl)

CH_3CH(—CH_3)— (isopropyl)

Figure 2.3 *Alkyl groups: n-propyl and isopropyl.*

butyl groups, two derived from the straight-chain butane and two derived from branched chains (Figure 2.4).

```
                                                                       CH3
                                                                        |
CH3CH2CH2CH2—      CH3CH2CH—          CH3CHCH2—                    CH3C—
                         |                |                            |
                         CH3              CH3                          CH3
   n-butyl          sec-butyl          isobutyl                  tert-butyl
                 (secondary butyl)                             (tertiary butyl)
```

Figure 2.4 *Alkyl groups: the four butyl groups.*

Beyond butyl the number of isomeric alkyl groups derived from each alkane becomes so large that it is impractical to designate them by prefixes such as *n*-, iso-, *sec*-, *tert*- and neo-. Thus, for more complicated alkyl group substituents, IUPAC names are used.

Below are some examples of nomenclature of alkanes.

```
 1     2     3     4
CH3—CH—CH2—CH3                          2-methylbutane
     |                                  not 3-methylbutane
     CH3

 1     2     3     4     5     6
CH3—CH—CH2—CH—CH2—CH3                   2-methyl-4-ethylhexane
     |           |                      not 3-ethyl-5-methylhexane
     CH3         CH2CH3

     CH3
 1  2|   3     4     5                  2,2,4-trimethylpentane
CH3—C—CH2—CH—CH3                        not 2,4-trimethylpentane
     |         |                        not 2,2,4-methylpentane
     CH3       CH3

                                CH3
 1     2     3     4     5      |
CH3—CH—CH2—CH2—CH—CH—CH3                2-methyl-5-isopropyloctane
     |                 |                not 2,6-dimethyl-3-n-
     CH3              6CH2                propylheptane
                       |    8
                      7CH2—CH3

                   CH2—CH2—CH3
 1     2     3    4|   5     6     7    4-n-propyl-4-isopropyl-
CH3—CH2—CH2—C—CH2—CH2—CH3                 heptane
                   |                    not 2-methyl-3,3-di-n-
                   CH—CH3                 propylhexane
                   |
                   CH3
```

Problems **2.1** Write names for the following structural formulas:

```
a  CH3CHCH2CH2CH3              b  CH3CHCH2CH2CHCH3
      |                               |        |
      CH3                             CH3      CH3

c  CH3CH2CHCH2CHCH3            d  (CH3)3CH
           |     |
           CH3   CH2CH3
```

e $CH_3CH_2CHCH_2CH_2CH_2CH_3$
$\quad\quad\quad |$
$\quad CH_3CHCH_3$

f $CH_3CH_2CH_2CHCH_3$
$\quad\quad\quad\quad |$
$\quad\quad\quad CH_2CH_2CH_3$

2.2 Write structural formulas for the following compounds.

a 2,2,4-trimethylhexane
b 1,1,2-trichlorobutane
c 2,2-dimethylpropane
d 2,4,5-trimethyl-3-ethyloctane
e 2-bromo-2,4,6-trimethyloctane
f 2,4-dimethyl-5-*n*-butylnonane
g 4-isopropyloctane
h 3,3-dimethylpentane

2.3 Draw structural formulas for all isomeric alkanes of formula C_6H_{14}. Name each.

2.4 Explain why each of the following names is incorrect. Write a correct name.

a 1,3-dimethylbutane
b 4-methylpentane
c 2,2-diethylbutane
d 2-ethyl-3-methylpentane
e 4,4-dimethylhexane
f 2-*n*-propylpentane
g 2,2-diethylheptane
h 5-*n*-butyloctane
i 2-dimethylpropane
j 2-*sec*-butyloctane
k 4-isopentylheptane

2.5 There are 35 structural isomers of formula C_9H_{20}. Name and draw the 6 isomers that have five carbons in the longest chain.

2.6 CYCLOALKANES

So far we have considered only linear arrangements of carbons atoms, often having one or more branches. The ends of these chains can be folded and joined together to form a ring or cycle of carbon atoms. Molecules of this type containing only carbon atoms in the ring are called cyclic hydrocarbons. Further, when all the carbons of the ring are saturated, the molecules are called cycloalkanes. The use of carbon bonds to close the ring means that the cycloalkanes contain two hydrogen atoms fewer than the corresponding open-chain alkanes.

Cyclic hydrocarbons of ring size from three to over thirty are found in nature and in principle there is no limit to ring size. Five-membered rings (cyclopentanes) and six-membered rings (cyclohexanes) are especially abundant in nature and therefore have received special attention.

Once you have mastered the nomenclature of the alkanes, the naming of cycloalkanes will pose no new problems. Simply prefix the name of the corresponding open-chain hydrocarbon by cyclo-. Each substituent on the ring is given a name and a number to indicate its position.

As a matter of convenience the organic chemist does not usually write out the structural formulas of the cycloalkanes in the manner shown on the left side of Figure 2.5. Rather the rings are represented by the appropriate polygon, as shown at the right.

The chemical and physical properties of the cycloalkanes resemble very closely the properties of the corresponding linear hydrocarbons. However, some significant differences do arise because of the

1,1-dimethylcyclopropane

cyclobutane

isobutylcyclopentane

1-chloro-2-methylcyclohexane
or
1-methyl-2-chlorocyclohexane

Figure 2.5 *Examples of cycloalkanes.*

reduced flexibility when the carbons are constrained in a cyclic structure. In Sections 2.8–2.9 we shall discuss some of the consequences of this constraint.

2.7 PHYSICAL PROPERTIES OF ALKANES

Table 2.2 lists the boiling points, melting points, and densities for a number of alkanes. As can be seen, the melting points and boiling points increase as the number of carbons increases for the linear-chain alkanes. Up to a point, the densities increase with the size of the alkanes and then tend to level off at about 0.8 grams per milliliter. Hence all of the alkanes are less dense than water. Those that are liquids or solids between 0°C and 100°C will float on water.

Table 2.2 *Physical properties of alkanes.*

Name	Formula	mp (°C)	bp (°C)	Density at 20°C
methane	CH_4	−182	−164	
ethane	CH_3CH_3	−183	−88	
propane	$CH_3CH_2CH_3$	−190	−42	
butane	$CH_3(CH_2)_2CH_3$	−138	0	
pentane	$CH_3(CH_2)_3CH_3$	−130	36	0.626
hexane	$CH_3(CH_2)_4CH_3$	−95	69	0.659
heptane	$CH_3(CH_2)_5CH_3$	−90	98	0.684
octane	$CH_3(CH_2)_6CH_3$	−57	126	0.703
nonane	$CH_3(CH_2)_7CH_3$	−51	151	0.718
decane	$CH_3(CH_2)_8CH_3$	−30	174	0.730

Isomeric branched-chain alkanes do exhibit some differences in physical properties. Table 2.3 lists the physical properties of the five isomeric hexanes.

Notice that each of the branched-chain isomers has a boiling point lower than hexane itself and that the more branching there is, the lower

Table 2.3 *Physical properties of the isomeric hexanes.*

IUPAC Name	bp (°C)	mp (°C)	Density at 20°C
hexane	68.7	−95	0.659
2-methylpentane	60.3	−154	0.653
3-methylpentane	63.3	−118	0.664
2,3-dimethylbutane	58.0	−129	0.661
2,2-dimethylbutane	49.7	−98	0.649

the boiling point. Particularly striking is the difference of 19° between the boiling points of hexane and its isomer 2,2-dimethylbutane. Why should branching have the effect of lowering the boiling point? Perhaps the most reasonable explanation is that with branching the shape of the molecule tends to become more compact and spherical. As this happens, surface area decreases and the degree of intermolecular interaction between molecules decreases. Thus it will require less energy (lower temperature) to separate a molecule from its neighbors in the liquid state. This effect of branching on boiling point is observed in all families of organic compounds.

2.8 CONFORMATIONS OF ALKANES AND CYCLOALKANES

In recent years, the actual shapes of molecules defined by distances between atoms and the angles between bonds have assumed greater and greater importance in the thinking of chemists and biochemists. Structural formulas such as we have been using so far are at best inadequate and frequently downright misleading with respect to the three-dimensional molecular geometry. At this point, therefore, as we take our first look at conformational questions, we must simultaneously introduce new graphical representations.

In a simple molecule such as ethane, there is an infinite number of possible arrangements of the atoms depending on the angle of rotation of one carbon with respect to the other. Such arrangements are called conformations. Two possible conformations are shown in Figure 2.6. The conformations of ethane are interconvertible by rotation about the carbon-carbon bond. At room temperature, ethane molecules undergo collision with sufficient energy that rotation about the carbon-carbon single bond from one conformation to another is easily possible. In

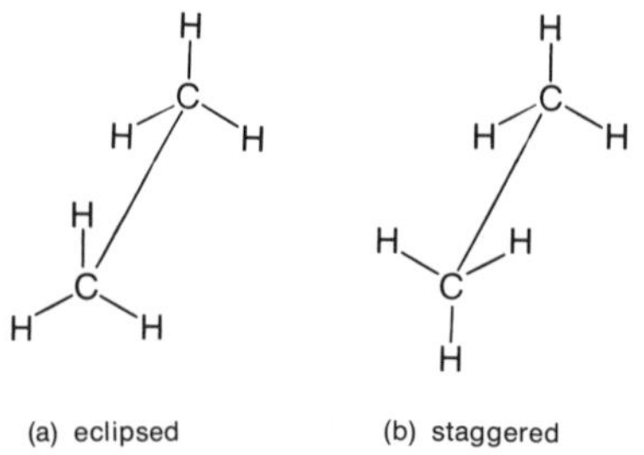

(a) eclipsed

(b) staggered

Figure 2.6 *Two possible conformations of ethane: (a) eclipsed and (b) staggered.*

fact, for a long time chemists believed there was completely free rotation about the carbon-carbon single bond. However, more recent studies of ethane and other molecules have revealed that the rotation is not completely free, but rather is hindered by the size and the electrical character of the atoms joined to one or both of the connected carbon atoms. In the case of ethane, there is a preference for the staggered conformation over the eclipsed because the interaction between the hydrogen atoms is minimized. Three possible conformations of butane are shown in Figure 2.7.

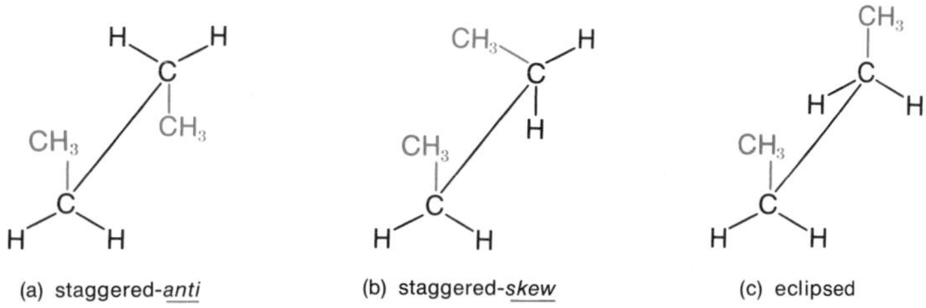

(a) staggered-*anti* (b) staggered-*skew* (c) eclipsed

Figure 2.7 *Three conformations of the butane molecule.*

The Newman projection, an alternative type of graphic representation, is especially convenient for showing molecular conformations. In this representation the molecule is viewed along the axis of one (arbitrarily chosen) carbon-carbon bond.

The directed valences of the carbon atom nearer the eye are designated by radii placed at angles of 120°, whereas the valences of the carbon farther from the eye are indicated by a circle with lines extending from the circumference. Remember, of course, that the bond angles about each of these carbon atoms are 109.5°.

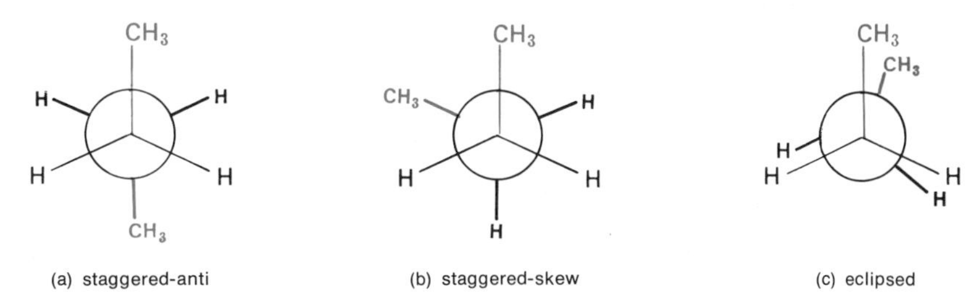

(a) staggered-anti (b) staggered-skew (c) eclipsed

Figure 2.8 *Newman projection representations of three conformations of butane.*

Of the three conformations shown for butane in Figure 2.8, the staggered-anti form (a) is preferred because it minimizes the repulsive forces between the larger methyl groups. The eclipsed form, conformation (c), is least preferred since it maximizes the interactions of the methyl groups and of the eclipsed hydrogens.

We shall now discuss the conformations of cyclohexane and its simple derivatives. With an understanding of the conformational aspects of this simple carbocyclic ring, we will be prepared for later discussions of naturally occurring molecules containing six-membered rings. In particular, we will apply these concepts to our discussion of the three-dimensional shapes of carbohydrates (Chapter 10).

Thus far we have represented cyclohexane as a regular hexagon (Figure 2.5), and if this molecule were a regular hexagon, it would have C—C—C bond angles of 120°. Distortion of the regular tetrahedral angle to 120° would result in considerable angle strain. Yet inspection of a molecular model of cyclohexane will reveal that the carbon atoms need not lie in a plane, and that there need be no distortion of the regular tetrahedral angles. Molecular models show that a six-membered ring may take up two distinct nonplanar conformations in which all bond angles are 109.5°. One of these is called the chair conformation, the other the boat conformation. Both conformations may be readily interconverted without breaking the bonds in the molecule (Figure 2.9).

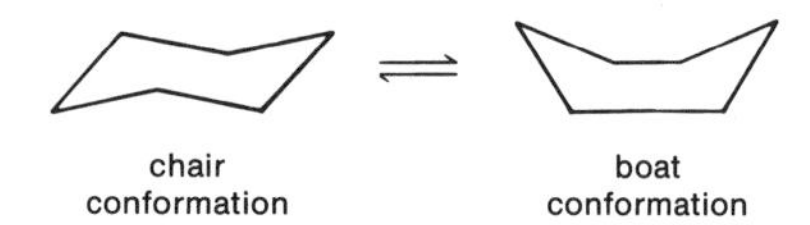

Figure 2.9 Two conformations of cyclohexane. In each nonplanar conformation, the C—C—C bond angles are 109.5°.

When the arrangement of hydrogens on the chair conformation is considered it can be seen that there are hydrogen atoms in two different geometrical situations—those projecting in the plane of the ring, called equatorial hydrogens, and those projecting perpendicular to the plane of the ring, called axial hydrogens (Figure 2.10).

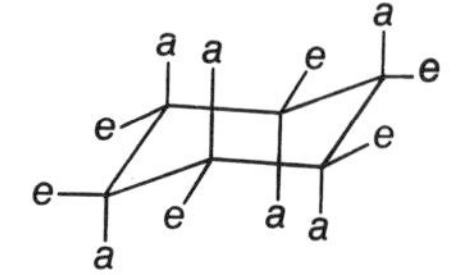

Figure 2.10 Chair conformation of cyclohexane. Equatorial hydrogens indicated by e and axial hydrogens indicated by a.

Note that in the chair conformation adjacent pairs of hydrogens are completely staggered with respect to each other just as they are in the ethane molecule (Figure 2.6b).

Drawing accurate chair conformations of cyclohexane rings sometimes presents a problem in that you must represent a three-dimensional structure on a two-dimensional or flat surface. How can you do this in such a way that another person will see what you are trying to indicate? Two simple guidelines should help.

1. Notice that bonds to carbon atoms on opposite sides of the ring are parallel. For example, the axial bonds on carbons 1 and 4 are parallel, the equatorial bonds on carbons 2 and 5 are parallel, etc.

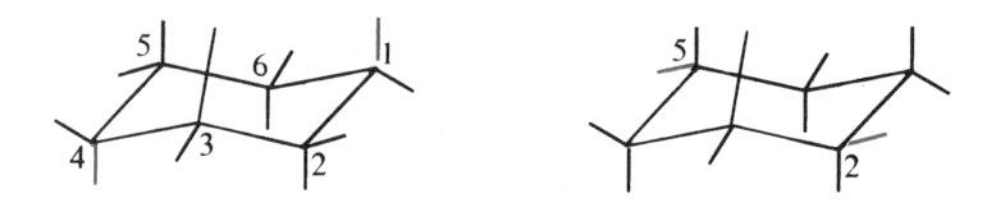

2. Notice also that each equatorial bond is parallel to two other C—C ring bonds once removed from the equatorial bond you are looking at. For example, the equatorial bond on carbon 1 is parallel to the

bonds between carbons 2–3 and 5–6; the equatorial bond on carbon 2 of the ring is parallel to the ring bonds between carbons 3–4 and 6–1.

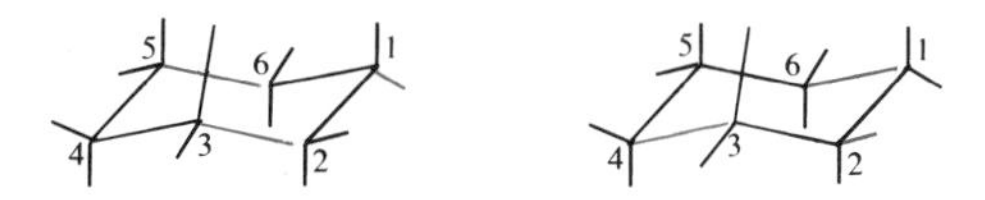

When one chair is converted to the other chair (Figure 2.11), the equatorial hydrogens become axial and vice versa. Thus in cyclohexane itself, where the two chair forms are readily interconverted, each hydrogen atom is axial half of the time and equatorial the other half of the time.

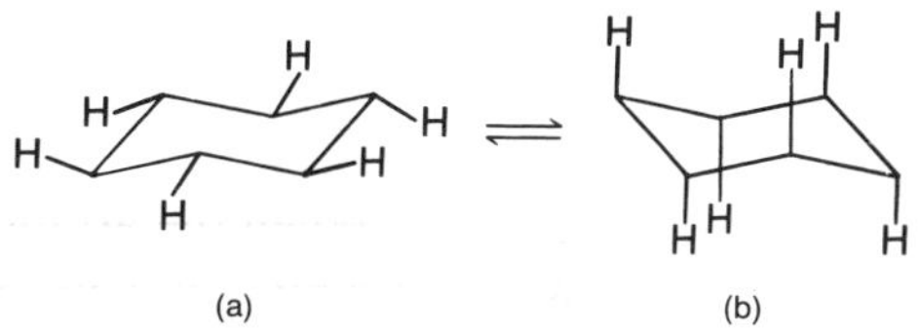

Figure 2.11 *Two chair conformations of cyclohexane (the intermediate boat conformation is not shown). Hydrogen atoms equatorial in conformation (a) are axial in (b).*

(a) (b)

In the boat conformation (Figure 2.12), the hydrogens of carbons 2–3 and 5–6 are eclipsed (as in ethane) and the hydrogens of carbons 1 and 4 jut forward toward each other. Because of these interactions the boat form of cyclohexane is of higher energy than the chair form.

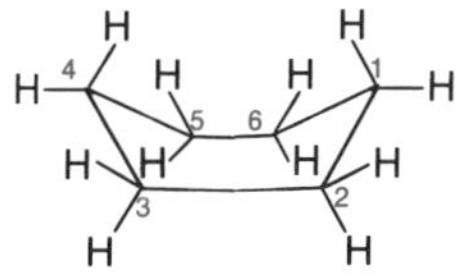

Figure 2.12 *Interactions in the boat form of cyclohexane.*

If one of the hydrogen atoms in cyclohexane is replaced by a methyl group or other substituent, the group will occupy an axial position in one chair and an equatorial position in the other chair. This means that the two chair conformations are no longer equivalent and are no longer of equal stability. In general, bulky substituents do not occupy axial positions because of the strain arising from the interaction with the two other axial groups (hydrogens in the case of methylcyclohexane itself) on the same side of the ring. On the other hand, a substituent group in an equatorial position is as far away as possible from the other atoms of the ring, causing the chair form with the substituent equatorial to be favored at equilibrium. When the substituent is methyl, most of the molecules exist in the chair conformation with the —CH_3 in the less crowded equatorial position, and only a small percentage (about 5%) of the molecules of methylcyclohexane have the methyl group in the axial position. The two chair conformations of methylcyclohexane are shown in Figure 2.13.

As the size of the substituent is increased, the preference for the conformation with the group equatorial is increased. When the group is

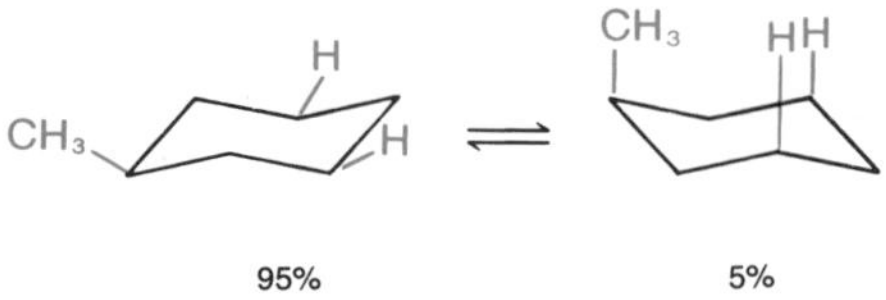

Figure 2.13 *The two chair conformations of methylcyclohexane.*

95% ⇌ 5%

as large as *tert*-butyl, the equatorial conformation is 10,000 times more abundant at room temperature than the axial and, in effect, the ring is "locked" into this chair conformation.

2.9 GEOMETRIC ISOMERISM IN CYCLOALKANES

Because of the restricted rotation about the carbon-carbon single bonds imposed by the ring structure, cycloalkanes show a type of isomerism called stereoisomerism. To put this type of isomerism in perspective, recall our discussions of structural isomerism. Structural isomers have the same molecular formula but different orders of attachment of the atoms. With stereoisomerism we deal with compounds of the same molecular formula, the same order of attachment of the atoms, but different arrangements of the atoms in space.

The term "geometric isomerism" is applied to the type of stereoisomerism that depends on the arrangement of substituent groups, either on a cyclic structure as we shall discuss here, or on a double bond as will be shown in the following chapter.

The principles of geometric isomerism in cyclic structures can be illustrated by looking at models of 1,2-dimethylcyclopentane. (For our purposes in describing geometric isomerism it is quite sufficient to consider the cyclopentane ring as a planar pentagon.)

H_3C CH_3
cis-1,2-dimethylcyclopentane

CH_3 CH_3
trans-1,2-dimethylcyclopentane

There are two geometric isomers of 1,2-dimethylcyclopentane. In one of the isomers, both of the —CH_3 groups are on the same side of the ring, whereas in the other isomer they are on opposite sides of the ring. The prefix *cis-* indicates that the substituents are on the same side of the ring; *trans-* indicates that they are on opposite sides.

The *cis* and *trans* forms cannot be superimposed on each other, and no amount of rotation about the carbon-carbon bonds will convert one into the other.

Problem **2.6** Draw structural formulas for the *cis* and *trans* isomers of 1,3-dimethylcyclopentane.

All cycloalkanes show this type of isomerism. In the case of cyclohexane and the larger rings, the situation is somewhat more complicated by the existence of various ring conformations. We will consider only the case of cyclohexane. There are two geometric isomers of 1,2-dimethylcyclohexane. In the *cis* isomer both —CH_3 groups are on the same side of the ring, and in the *trans* isomer they are on opposite sides. If the cyclohexane ring is represented as a planar hexagon, these isomers are as shown below.

H_3C CH_3 | CH_3 CH_3

cis-1,2-dimethylcyclohexane *trans*-1,2-dimethylcyclohexane

Yet as we saw in Section 2.8, a planar hexagon does not accurately represent the cyclohexane ring. Rather we should use the chair representation.

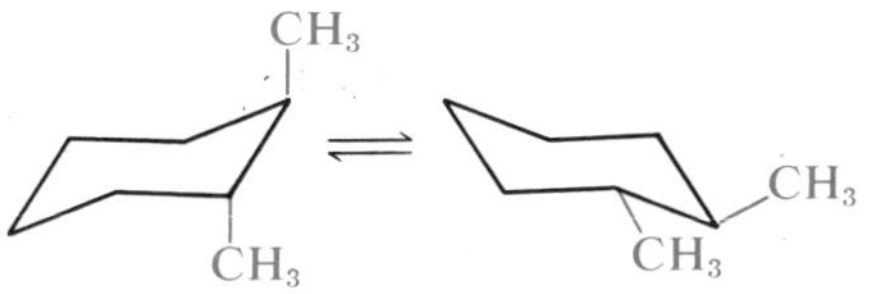

CH_3 CH_3 ⇌ CH_3 CH_3

trans-1,2-dimethylcyclohexane

Clearly when the two —CH_3 groups are in axial positions, they are *trans* to each other. Since one chair is readily interconvertible to the other chair by rotation about carbon-carbon bonds, the two —CH_3 groups are also *trans* to each other when they are attached to 1,2-diequatorial positions. Of these two conformations, the *trans* diequatorial is by far the more stable.

In the *cis* configuration, one of the —CH_3 groups will occupy an equatorial position on the ring, and the other will occupy an axial position.

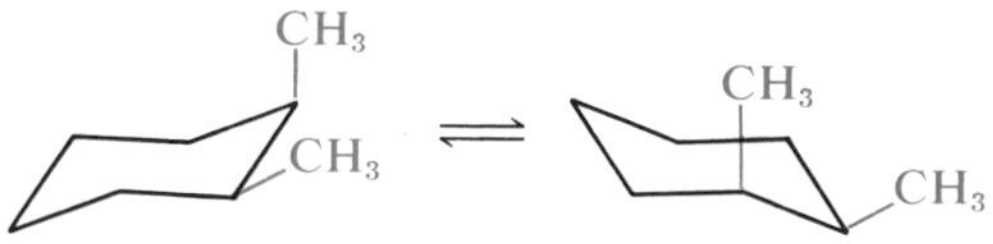

CH_3 CH_3 ⇌ CH_3 CH_3

cis-1,2-dimethylcyclohexane

These chair conformations are of equal stability.

Problems

2.7 Draw the alternative chair conformations for the *cis* and *trans* isomers of 1,2-dimethylcyclohexane; of 1,3-dimethylcyclohexane; and of 1,4-dimethylcyclohexane. Label axial and equatorial positions.

a For which isomers are the two chair conformations of equal stability?

b For which isomers is one chair conformation more stable than the other chair?

2.8 Draw and name the *cis* and *trans* isomers of dimethylcyclopropane.

2.9a How many geometric isomers are there of 2-isopropyl-5-methyl-1-cyclohexanol? One of these isomers is the fragrant oil menthol.

2-isopropyl-5-methyl-1-cyclohexanol

b Draw a chair conformation of the isomer you predict to be the most stable, i.e., the isomer with minimal interactions between the atoms of the molecule. (If you have drawn the correct answer, you have drawn the structural formula of menthol.)
c Draw the less stable chair conformation of the isomer shown in part **b**.

2.10 "Benzene hexachloride," more properly named 1,2,3,4,5,6-hexachlorocyclohexane, is a mixture of various geometric isomers. The crude mixture is sold as the insecticide benzene hexachloride (BHC). The insecticidal properties of the mixture arise from one isomer known as the γ-isomer (gamma isomer) which is marketed under the name Lindane or Gammexane. Below is drawn a representation of the gamma isomer with the cyclohexane ring shown as a planar hexagon.

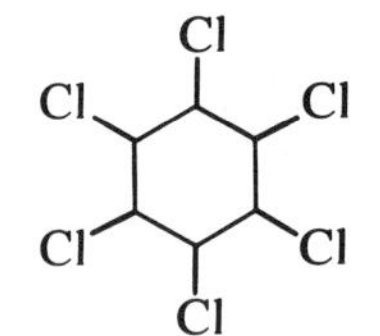

benzene hexachloride (BHC)

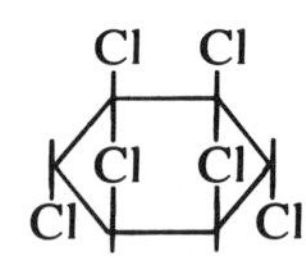

the γ-isomer of BHC

a Draw a chair conformation of the γ-isomer and label the chlorine substituents either axial or equatorial.
b Draw the other chair conformation of the γ-isomer and again label chlorine substituents axial or equatorial.
c Which of these two chair conformations of the γ-isomer would you predict to be the more stable? Why?

2.10 REACTIONS OF ALKANES

As we saw in Section 1.14, the saturated hydrocarbons are quite inert because (1) carbon-carbon and carbon-hydrogen bond dissociation energies are high and the bonds themselves are nonpolar, and (2) since carbon has no unshared pairs of electrons, it is not susceptible to attack by electron-deficient or electron-rich reagents. Saturated hydrocarbons are quite resistant to attack by most strong acids and bases, and to powerful oxidizing and reducing agents. However, the saturated hydrocarbons do react with halogens and, when ignited, with oxygen. We will consider both of these reactions, the first because it is useful in the preparation of substituted alkanes and the second because it is the basis for the use of saturated hydrocarbons (and unsaturated hydrocarbons as well) as fuel.

2.11 HALOGENATION OF ALKANES

If a mixture of an alkane and chlorine gas is kept in the dark at room temperature, no detectable change will occur. However, if the mixture is heated or exposed to light, a reaction will begin almost at once with the evolution of heat. The products are chloroalkanes and hydrogen chloride. What occurs is a substitution reaction—the replacement of one or more hydrogen atoms on the alkane by chlorine and the production of an equivalent amount of hydrogen chloride. Recall from Section 2.4 (rule 3) that the symbol R— represents an alkyl group.

$$R{-}H + Cl{-}Cl \xrightarrow{\text{heat or light}} R{-}Cl + HCl$$

Both methane and ethane yield only one monochlorination product.

$$CH_4 + Cl_2 \xrightarrow{\text{light}} \underset{\text{methyl chloride}}{CH_3Cl} + HCl$$

$$CH_3CH_3 + Cl_2 \xrightarrow{\text{light}} \underset{\text{ethyl chloride}}{CH_3CH_2Cl} + HCl$$

If methyl chloride is allowed to react with more chlorine, further chlorination produces a mixture of methylene chloride, chloroform, and carbon tetrachloride, each of which is widely used as a solvent.

$$\underset{\text{methyl chloride}}{CH_3Cl} \xrightarrow{Cl_2} \underset{\text{methylene chloride}}{CH_2Cl_2} + \underset{\text{chloroform}}{CHCl_3} + \underset{\text{carbon tetrachloride}}{CCl_4}$$

These various chlorination products of methane have different boiling points and may be readily separated from one another by distillation.

Both propane and butane can yield two isomeric monochlorination products.

$$CH_3CH_2CH_3 + Cl_2 \xrightarrow{\text{light}} \underset{\substack{\text{1-chloropropane} \\ n\text{-propyl chloride} \\ 48\%}}{CH_3CH_2CH_2Cl} + \underset{\substack{\text{2-chloropropane} \\ \text{isopropyl chloride} \\ 52\%}}{CH_3\underset{|\atop Cl}{C}HCH_3} + HCl$$

$$CH_3CH_2CH_2CH_3 + Cl_2 \xrightarrow{\text{light}} \underset{\substack{\text{1-chlorobutane} \\ n\text{-butyl chloride}}}{CH_3CH_2CH_2CH_2Cl} + \underset{\substack{\text{2-chlorobutane} \\ sec\text{-butyl chloride}}}{CH_3CH_2\underset{|\atop Cl}{C}HCH_3} + HCl$$

Problems

2.11 Name and draw structural formulas for:

a the two monochloro derivatives of 2-methylpropane
b the two dichloro derivatives of ethane
c the four dichloro derivatives of propane
d the one monochloro derivative of cyclopentane
e all monochloro derivatives of hexane
f all monochloro derivatives of 2-methylpentane
g all monochloro derivatives of 2,3-dimethylbutane
h all monochloro derivatives of 2,2,5-trimethylhexane

2.12 An alkane of formula C_5H_{12} gives only one monochlorination product when reacted with chlorine gas at 300°C.

a What is the structure of the starting alkane; of the monochlorinated product, $C_5H_{11}Cl$?
b If the product of the first chlorination is now subjected to further chlorination, how many dichloro substitution products would you expect?

As you might expect, higher alkanes and those with more branching yield more complex mixtures of products, and for this reason halogenation of higher alkanes is not of great synthetic use in the laboratory.

2.12 COMMERCIALLY IMPORTANT HALOGENATED HYDROCARBONS

Because of their physical and chemical properties, several of the halogenated hydrocarbons have found wide commercial use. These include applications as commercial solvents, refrigerants, dry cleaning agents, local and general anesthetics, and insecticides.

Carbon tetrachloride, CCl_4, is a dense, nonflammable liquid, bp 77°C, which is remarkably inert to most common reagents and laboratory conditions. It is immiscible with water but is a good solvent for oils and greases, and at one time found wide use in the dry cleaning industry. It is somewhat toxic, readily absorbed through the skin, and like all organic solvents should be used only with adequate ventilation. Prolonged exposure to carbon tetrachloride vapors results in liver and renal damage.

Chloroform, $CHCl_3$, is a colorless, dense, rather sweet-smelling liquid, bp 61°C. It too is a widely used solvent for organic substances. In the past chloroform was used extensively as a general anesthetic for surgery but it is rarely used for this purpose now because it is known to cause extensive liver damage.

Ethyl chloride, CH_3CH_2Cl, is used as a fast-acting, topically applied local anesthetic. It owes its anesthetic property more to its physical than chemical characteristics. It boils at 13°C, and unless under pressure, is a gas at room temperature. When sprayed on the skin, it evaporates rapidly and because evaporation is a cooling process, it cools the skin surface and nerve endings.

Halothane, $C_2HBrClF_3$, is a recently discovered and now widely used inhalation anesthetic. It has distinct advantages over other general inhalation anesthetics (as for example diethyl ether and cyclopropane) in that it is nonflammable, nonexplosive, and causes minimum discomfort to the patient. Although there have been a few cases of liver damage caused by its use, its record as a safe anesthetic is impressive.

The more complex chlorinated hydrocarbons are often poisons, and as you well know some are very widely used as insecticides. The most famous of the polychlorinated hydrocarbon insecticides is undoubtedly DDT. The history of the development of this powerful agent is discussed in the mini-essay "DDT." Other polychlorinated hydrocarbon insecticides including Aldrin, Dieldrin, and Chlordane are also presented in that essay. The structure of Lindane (BHC) was shown in Problem 2.10.

Of all the fluorinated hydrocarbons, those manufactured under the trade name Freon (du Pont) have had the most dramatic impact. The first of the Freons was developed in a search for new refrigerants. Of all the refrigerants in use prior to 1930, none was without serious disadvantage. Some, like ethylene, were flammable. Others like sulfur dioxide were corrosive and quite toxic. Ammonia combined all three of these hazards. Carbon dioxide would have been ideal except that it had to be used at relatively high pressure and the equipment required for its use was almost prohibitively bulky. It was against this background that Thomas Midgley, Albert Henne, and others at General Motors set out to find an ideal refrigerant—a compound that would be nontoxic, nonflammable, odorless, and noncorrosive. In 1930 they announced the discovery of just such a compound, dichlorodifluoromethane, which was marketed under the trade name Freon-12.

The Freons are manufactured by reacting a chlorinated hydrocarbon with hydrofluoric acid in the presence of an antimony pentafluoride or antimony chlorofluoride catalyst. Both Freon-11 and Freon-12 can be prepared from carbon tetrachloride. Freon-22, monochlorodifluoromethane, is made from chloroform.

$$CCl_4 \xrightarrow[SbF_5]{HF} \underset{\text{Freon-11}}{CCl_3F} \xrightarrow[SbF_5]{HF} \underset{\text{Freon-12}}{CCl_2F_2}$$

The new refrigerant, Freon-12, went into commercial production in 1931 and very shortly thereafter the refrigeration industry began to modify its equipment to use the new product. By 1935 the product line included five Freons, three derivatives of methane and two of ethane. A major new use of the Freons came during World War II with the development of aerosol insecticides in which they serve as propellants. By 1974, U.S. production of Freons had grown to more than 1.1 billion pounds annually, almost one-half the world production.

Concern about the environmental impact of fluorocarbons like Freon-11 and Freon-12 arose in 1974 when Drs. Sherwood Rowland and Mario Molina of the University of California, Irvine, announced their theory of ozone destruction by these substances. When used as aerosol propellants and refrigerants, these fluorocarbons escape to the lower atmosphere but because of their general inertness do not decompose there. Slowly, they find their way to the stratosphere where they absorb ultraviolet radiation from the sun and then decompose. As they decompose, they set up a chemical reaction that may also lead to the destruction of the stratospheric ozone layer. What makes this a serious problem is that the stratospheric ozone layer acts as a shield for the earth against excess ultraviolet radiation. The depletion of this shield and the increased levels of ultraviolet radiation striking the earth could result in damage to certain crops and agricultural species, climate modification, and even increased incidence of skin cancer in sensitive individuals. Controversy continues and for this reason it is critical that the scientific community, along with government and consumer agencies, take steps to determine the real potential of fluorocarbons for ozone depletion and its impact on the environment.

2.13 SATURATED HYDROCARBONS FOR HEAT AND POWER

The reaction of alkanes with oxygen to form carbon dioxide and water—and most important, heat—is the basis for the use of hydrocarbons as a source of heat (natural gas and fuel oil) and power (fuel for the internal combustion engine).

$$CH_4 + 2O_2 \longrightarrow CO_2 + 2H_2O \qquad \Delta H = -212 \text{ kcal/mole}$$

$$\underset{\text{"isooctane"}}{CH_3\overset{\overset{\displaystyle CH_3}{|}}{\underset{\underset{\displaystyle CH_3}{|}}{C}}CH_2\overset{\overset{\displaystyle CH_3}{|}}{C}HCH_3} + \frac{25}{2}O_2 \longrightarrow 8CO_2 + 9H_2O \qquad \Delta H = -1304 \text{ kcal/mole}$$

These hydrocarbons are obtained commercially from natural gas, coal, petroleum, and from shale rock.

Natural gas consists of methane mixed with varying amounts of ethane, propane, butane, and isobutane. The last three components can be liquefied under pressure at room temperature. Liquid propane (LPG) can be stored easily and shipped in metal tanks and is a convenient source of gaseous fuel.

2.14 PETROLEUM

Petroleum is a liquid mixture of literally thousands of substances, most of them hydrocarbons, formed from the decomposition of marine plants and animals. The impact of petroleum-derived products on the economy is enormous. They fuel automobiles, aircraft and trains; they provide heat for buildings and fuel for electric generating plants; they provide most of the greases and lubricants required by the machinery of our highly industrialized society; and they provide close to 90% of the organic raw materials for the synthesis and manufacture of synthetic fibers, plastics, detergents, drugs, and a multitude of other products.

Problems

2.13 Define and illustrate the following terms:

a hydrocarbon
b saturated hydrocarbon
c unsaturated hydrocarbon
d homologous series
e cycloalkane
f conformation
g alkyl group
h alkane
i structural isomerism
j stereoisomerism
k geometric isomerism
l chlorinated hydrocarbon

2.14 What is the major component of natural gas? of bottled or LP gas?

2.15 What generalization can you make about the densities of alkanes relative to that of water?

2.16 In a handbook of chemistry or other suitable reference, look up the densities of methylene chloride, chloroform, and carbon tetrachloride. Which of these substances are more dense than water; which are less dense?

2.17 What straight-chain alkane has about the same boiling point as water? (Refer to Table 2.2 for data on the physical properties of alkanes.) Calculate the molecular weight of this alkane and compare it with water.

2.18 Account for the fact that saturated hydrocarbons are quite inert, i.e., they are resistant to attack by most strong acids and bases as well as most oxidizing and reducing agents.

2.19 Complete and balance the following combustion reactions. Assume that each hydrocarbon is converted completely to carbon dioxide and water.

a propane $+ O_2 \rightarrow$
b octane $+ O_2 \rightarrow$
c 2,2,4-trimethylpentane $+ O_2 \rightarrow$
d benzene $(C_6H_6) + O_2 \rightarrow$

2.20 Draw structural formulas for Freon-11, Freon-12, and Freon-22. Explain why Freons such as these have become so widely used as refrigerants and aerosol propellants.

2.21 In 1974, Drs. Sherwood Rowland and Mario Molina proposed that the Freons used as aerosol propellants and refrigerants may have a very harmful effect on the environment. Explain the basis for this concern.

3

UNSATURATED HYDROCARBONS

3.1 INTRODUCTION

There are two homologous series of compounds generally included in the classification of unsaturated hydrocarbons—the alkenes and the alkynes. The compounds of each of these classes contain one or more multiple bonds and accordingly have fewer hydrogens than the corresponding alkanes. The alkenes contain one or more carbon-carbon double bonds, and alkynes contain one or more carbon-carbon triple bonds.

We study the alkenes in some detail for two reasons. First, they are widely distributed in nature, and more specifically for our purposes, compounds containing carbon-carbon double bonds are key intermediates in the biosynthesis and metabolism of fats and carbohydrates. Second, the alkenes are a convenient point to introduce the concept of reaction mechanisms: what they are, how they are deduced, and why they are of immense importance to the practicing organic chemist. Actually these concepts could be introduced at a number of points in the development of organic chemistry. We do so at this point because the chemistry of the carbon-carbon double bond is quite well understood and the important concepts are relatively straightforward.

Also in this chapter we look at the structures of some aromatic hydrocarbons. Although these compounds also contain carbon-carbon multiple bonds, they are more conveniently differentiated from other organic compounds on the basis of their unique reactivities.

3.2 ALKENES

Ethylene, $CH_2{=}CH_2$, is the simplest member of the class of compounds known alternatively as olefins, unsaturated hydrocarbons, or alkenes. The term olefin is derived from early observations that certain of the higher homologs of ethylene react with chlorine gas to produce oily liquids. These hydrocarbons were accordingly characterized as olefiant (quite literally, oil-forming) substances. The name was shortened to olefin and, as is typical of organic nomenclature, this purely descriptive, nonsystematic name has persisted as a general name for this class of compounds. The reaction of an olefin with chlorine is illustrated with ethylene.

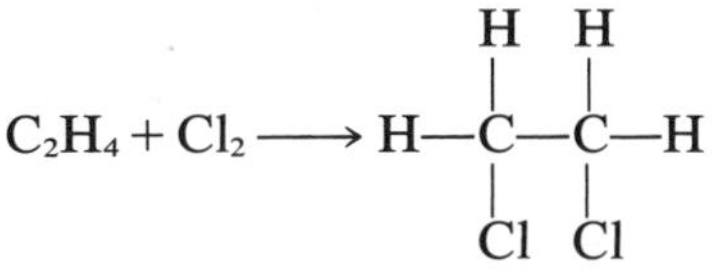

$$C_2H_4 + Cl_2 \longrightarrow H{-}\underset{\large Cl}{\overset{\large H}{\underset{|}{\overset{|}{C}}}}{-}\underset{\large Cl}{\overset{\large H}{\underset{|}{\overset{|}{C}}}}{-}H$$

ethylene (gas) — 1,2-dichloroethane, ethylene chloride (bp 83.5°C)

This reaction involves addition to the double bond and as we shall see in this chapter, addition reactions are characteristic of olefins.

Because ethylene and its homologs have fewer hydrogen atoms than the alkanes of the same number of carbons, they are said to be unsaturated. The "saturation" of ethylene and its homologs is ordinarily taken to mean addition of hydrogen to the double bond forming an alkane.

In the IUPAC system of nomenclature, ethylene and its homologs are known as alkenes.

3.3 STRUCTURE OF ALKENES

Ethylene or ethene, C_2H_4, is the first member of the alkene family. The next member is propylene or propene, C_3H_6.

$$CH_2{=}CH_2 \qquad CH_3{-}CH{=}CH_2$$

ethene, ethylene — propene, propylene

For the next member of the alkene family, there are four alkenes of formula C_4H_8. The structural formulas for three of these should be obvious; two isomers with four carbons in a continuous chain and differing only in the position of the double bond, and one isomer with a branched chain. According to the IUPAC rules, these are named 1-butene, 2-butene, and 2-methylpropene, respectively.

$$CH_3{-}CH_2{-}CH{=}CH_2 \qquad CH_3{-}CH{=}CH{-}CH_3 \qquad CH_3{-}\underset{\large CH_3}{\underset{|}{C}}{=}CH_2$$

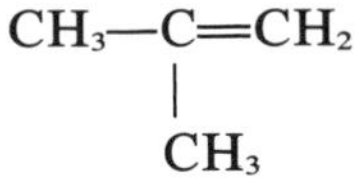

1-butene — 2-butene — 2-methylpropene, isobutylene

The existence of the fourth isomer, or more accurately the third and fourth isomers, depends on the spatial arrangement of the atoms about the carbon-carbon double bond. There are two isomers for the structure represented as 2-butene. These are designated as *cis*-2-butene and *trans*-2-butene.

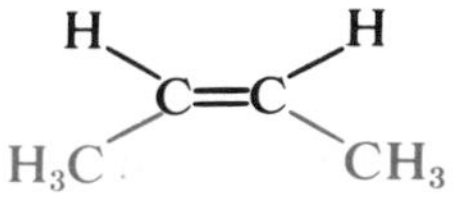

$$\begin{matrix} H & & & & H \\ & \diagdown & & \diagup & \\ & & C{=}C & & \\ & \diagup & & \diagdown & \\ H_3C & & & & CH_3 \end{matrix} \qquad\qquad \begin{matrix} H & & & & CH_3 \\ & \diagdown & & \diagup & \\ & & C{=}C & & \\ & \diagup & & \diagdown & \\ H_3C & & & & H \end{matrix}$$

cis-2-butene 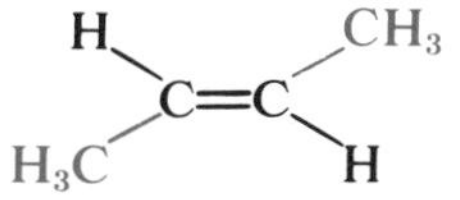 — *trans*-2-butene

Recall our discussion in Section 1.13, where we noted that, for the pi bond to have a maximum stability, the two $2p_z$ orbitals must have maximum overlap. This necessity for orbital overlap places a constraint on the flexibility of the molecule and the result is hindered rotation of the two carbons of the double bond. The compounds *cis*-2-butene and *trans*-2-butene are not readily interconvertible, and they are not superimposable. They are geometric isomers. Table 3.1 lists the boiling and melting points of the isomeric alkenes of formula C_4H_8.

Table 3.1 *Physical properties of the isomeric alkenes of formula C_4H_8.*

Name	bp (°C)	mp (°C)
2-methylpropene	−7	−141
1-butene	−6	−185
trans-2-butene	1	−106
cis-2-butene	4	−139

We have now seen geometric isomerism in two classes of organic compounds—the cycloalkanes and the alkenes. Remember that geometric isomers have the same empirical formula, the same molecular formula, and the same order of attachment of the atoms; geometric isomers differ in the orientation of the atoms in space. The key structural feature responsible for the existence of geometric or *cis-trans* isomerism is restricted rotation about a bond. This type of restraint may be introduced into a molecule in either of two ways: by the presence of a carbon-carbon double bond as we have just seen, or by the presence of a cyclic structure as we saw in the previous chapter.

It is possible to convert a *cis* isomer to a *trans* isomer and vice versa. By using sufficient energy (often by heating or by exposing the molecule to ultraviolet radiation) it is possible to rupture the pi bond and yet not disrupt the whole molecule. It is just this type of light-induced double-bond isomerization that is the primary event in vision (see the mini-essay "The Chemistry of Vision").

The alkenes form a homologous series in which the general formula is C_nH_{2n} and the constant increment is $—CH_2—$. The number of isomeric alkenes increases rapidly as the number of carbon atoms increases, for in addition to variations in chain length and substitution, there are variations in both the position of the double bond and the geometry about the double bond.

3.4 NOMENCLATURE OF ALKENES

The two commonly used methods of naming alkenes are illustrated in Table 3.2. Many alkenes, particularly the smaller ones, are known almost exclusively by their common names, e.g., ethylene, propylene, isobutylene. The use of common names is generally avoided for

Table 3.2 Naming of alkenes. The name preferred in common usage is starred.

Formula	Common Name	IUPAC Name
$CH_2{=}CH_2$	ethylene*	ethene
$CH_3CH{=}CH_2$	propylene*	propene
$CH_3CH_2CH{=}CH_2$	α-butylene	1-butene*
$CH_3CH{=}CHCH_3$	β-butylene	2-butene*
$(CH_3)_2C{=}CH_2$	isobutylene*	2-methylpropene

alkenes of more than four carbons because of the large number of isomers possible. The IUPAC system is the most versatile of the two nomenclature systems and is readily extended to substances having more than one double bond.

The IUPAC names of alkenes are formed by changing the -ane of the parent alkane to -ene. Hence C_2H_4 becomes ethene. There is no chance for ambiguity in naming the first two members of this homologous series because ethene and propene can contain a double bond in only one position. Since in butene and all higher alkenes there are isomers that differ in the location of the double bond, a coding system must be used to indicate the location of the double bond. According to the IUPAC system, the longest continuous carbon chain that contains the double bond is numbered in such a manner as to give the doubly bonded carbons the lowest possible numbers. The position of the double bond is then indicated by the number of the first carbon of the double bond. Branched or substituted alkenes are named in a manner similar to alkanes. The carbon atoms are numbered, substituent groups are located and named, the double bonds are located, and the main chain is named.

$$\overset{6}{C}H_3\overset{5}{C}H_2\overset{4}{C}H_2\overset{3}{C}H{=}\overset{2}{C}H\overset{1}{C}H_3$$

2-hexene
(*cis* and *trans*)

$$\overset{6}{C}H_3\overset{5}{C}H_2\overset{4}{C}H(CH_3)\overset{3}{C}H{=}\overset{2}{C}H\overset{1}{C}H_3$$

4-methyl-2-hexene
(*cis* and *trans*)

Note that alkenes are named by selecting the longest continuous carbon chain that contains the double bond. In the following compound there is a continuous chain of five carbon atoms, but the longest chain that contains the double bond is a four-carbon chain. The proper name for this compound then is 2-ethyl-3-methyl-1-butene.

$$\overset{4}{C}H_3{-}\overset{3}{C}H(CH_3){-}\overset{2}{C}(CH_2CH_3){=}\overset{1}{C}H_2$$

2-ethyl-3-methyl-1-butene

In naming cyclic alkenes, we number the carbons of the ring in such a direction that the carbons of the double bond are numbered 1 and 2, and the substituents are given the smallest numbers possible.

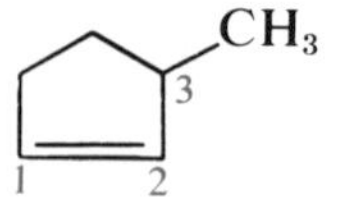

3-methylcyclopentene

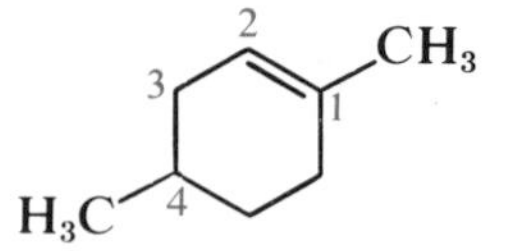

1,4-dimethylcyclohexene
(not 2,5-dimethylcyclohexene)

Alkenes that contain more than one double bond are called alkadienes, alkatrienes, or more simply dienes, trienes, etc.

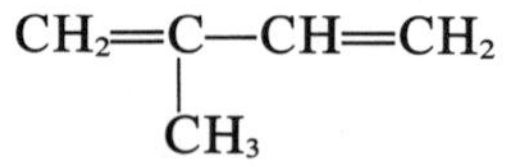

2-methyl-1,3-butadiene
(isoprene)

$CH_2{=}CH{-}CH_2{-}CH{=}CH_2$

1,4-pentadiene

In complex molecules that contain one or more double bonds, we generally pick the longest continuous chain containing the double bond and designate the geometry of the molecule based simply on whether the groups making up the main chain are on the same side (*cis*) or the opposite side (*trans*) of the double bond.

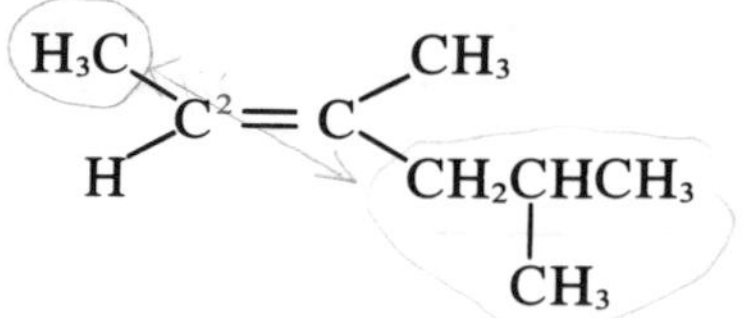

3,5-dimethyl-*trans*-2-hexene

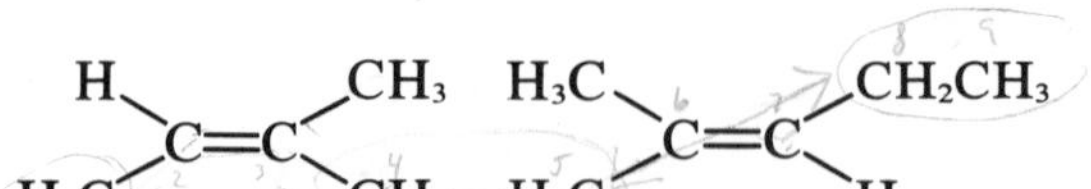

3,6-dimethyl-*cis*-2-*trans*-6-nonadiene

Problems **3.1** Draw structural formulas for the following alkenes. Indicate which compounds will show geometric isomerism, and draw both the *cis* and *trans* isomers.

a 2-methyl-3-hexene
b 2-methyl-2-hexene
c 2-methyl-1-butene
d 3-ethyl-3-methyl-1-pentene
e 2,3-dimethyl-2-butene
f 1-pentene
g 2-pentene
h 1-chloropropene
i 2-chloropropene
j 3-methyl-6-isopropylcyclohexene
k 4-methyl-1-isopropylcyclohexene

3.2 Draw structural formulas for all compounds of molecular formula C_5H_{10} that are:

a alkenes that will not show geometric isomerism
b alkenes that will show geometric isomerism

For each answer to part **b** draw both *cis* and *trans* isomers.

3.3 Draw structural formulas for the four isomeric chloropropenes (C_3H_5Cl).

3.4 There are four geometric isomers of 2,4-heptadiene. Name and draw a structural formula for each.

3.5 PHYSICAL PROPERTIES OF ALKENES

The physical properties of alkenes are much like those of the corresponding alkanes. The first four members of the alkene family are gases at room temperature. The pentanes and higher homologs are colorless liquids, all less dense than water.

Table 3.3 *Physical properties of some alkenes.*

Name	Structural Formula	mp (°C)	bp (°C)	Density at 20°C
ethene	$CH_2{=}CH_2$	−169	−102	
propene	$CH_3CH{=}CH_2$	−185	−48	
1-butene	$CH_3CH_2CH{=}CH_2$	−185	−6	
2-methylpropene	$CH_3C(CH_3){=}CH_2$	−140	−7	
1-pentene	$CH_3(CH_2)_2CH{=}CH_2$	−165	30	0.641
1-hexene	$CH_3(CH_2)_3CH{=}CH_2$	−141	64	0.673
cyclohexene	(cyclohexene ring)	−104	83	0.811

3.6 NATURALLY OCCURRING ALKENES—THE TERPENE HYDROCARBONS

A wide variety of substances in the plant and animal world contain one or more carbon-carbon double bonds. In many of these compounds, the double bond is but one of several functional groups and we shall see a variety of such compounds in later chapters. At this point, however, we shall focus our attention on one group of compounds—the terpene hydrocarbons—in which many individual members of the group contain one or more double bonds as the only functional group.

Aside from the fact that these terpene hydrocarbons are examples of naturally occurring alkenes, there are other perhaps more important reasons for looking at this group of organic compounds. First, they are among the most widely distributed compounds in nature. Second, they will provide a glimpse at some of the wondrous diversity of structure that nature can generate from even a relatively simple carbon skeleton. Third, terpenes illustrate an important principle of the molecular logic of organic chemistry, namely that in the building of what might seem to be rather complex molecules, nature begins by piecing together small, readily available subunits to produce an intriguingly complex but logically designed skeletal framework. At this point we will be concerned only with identifying this framework. Later we shall study reactions by which the framework can be modified to include other functional groups. Finally, a discussion of terpenes is a good point at which to introduce the almost limitless variety of substances known collectively as secondary metabolites.

To put secondary metabolites in perspective in the biological world, recall that the primary synthetic process of nature is photosynthesis, by which plants use the energy of sunlight to power the synthesis of carbohydrates from carbon dioxide and water. (The chemistry of carbohydrates is discussed in Chapter 10.) Through the reactions of intermediary metabolism, plants and animals in turn convert carbohydrates into a pool of small molecules which can be used for chemical energy to perform work, as for example mechanical work of muscle contraction or electrical work during the generation of the nerve impulse. Further, these metabolites can be used as raw materials for the synthesis of amino acids, proteins, lipids, nucleic acids, and the multitude of other molecules necessary for the maintenance of life itself. In addition plants and animals use these same, rather universally distributed intermediates to synthesize compounds which are often unique from one plant or animal family to another. These substances are the so-called secondary metabolites. In fact it has been possible in some instances to develop a chemical form of taxonomy based on the occurrence and distribution of these secondary metabolites in plants.

Before we examine structural formulas of particular terpene hydrocarbons, we should point out that virtually all of these substances have one structural feature in common: their carbon skeletons can be divided into two or more units of isoprene.

$$\begin{array}{c} \quad CH_3 \\ \quad | \\ CH_2{=}C{-}CH{=}CH_2 \end{array}$$

isoprene

C_5H_8

This does not mean that in nature terpenes are built from isoprene (actually they are built from units of acetate). It does mean, however, that the carbon skeleton of essentially every terpene can be divided into two or more subunits that are identical to the carbon skeleton of isoprene. This generalization is known as the isoprene rule.

With this background let us now look at some particular terpene hydrocarbons of plant origin. Probably the terpenes most familiar to you, at least by odor, are components of the so-called "essential oils" which can be obtained by either steam distillation or ether extraction of various parts of plants. These essential oils contain relatively low-molecular-weight, volatile substances which are in large part responsible for the characteristic odor of the plant. Among typical essential oils are those from mint and eucalyptus leaves, pine needles, rose petals, sandalwood, cedar, and oil of cloves. Myrcene, $C_{10}H_{16}$, can be obtained from bayberry wax and from the oils of bay and verbena.

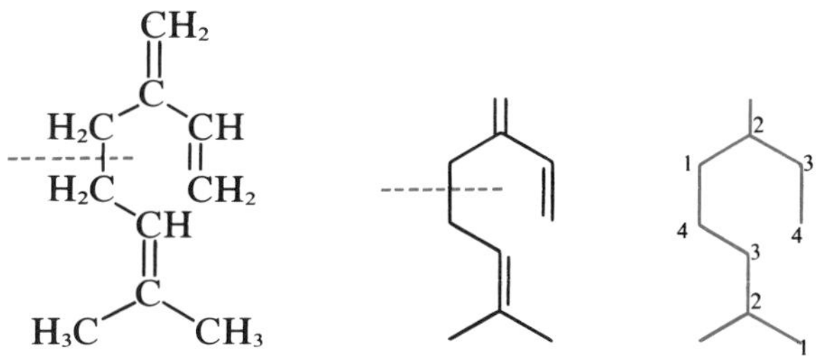

(a) (b) (c)

As you can see from structural formula (a), myrcene contains 10 carbon atoms and 3 carbon-carbon double bonds. For convenience in drawing and in order to show structural features more clearly, the organic chemist sometimes represents structures such as this by using a shorthand notation. Carbon-carbon bonds are represented as lines and carbon atoms are not shown but are understood to be at junctions of lines and where lines end. Structures (a) and (b) are equivalent representations of the structural formula for myrcene. As you can easily see from the position of the dashed lines in (a) or (b) or from the skeletal framework shown in (c), myrcene is divisible into two isoprene units joined from carbon 4 (head) of one unit to carbon 1 (tail) of a second unit. This head-to-tail linkage of isoprene units is vastly more common in nature than the alternative head-to-head(4,4) or tail-to-tail(1,1) linkages. Figure 3.1 shows further examples of acyclic, monocyclic, and bicyclic terpene hydrocarbons. These structural formulas are drawn to show clearly the isoprene-derived skeleton and, except for the one *trans* ring double bond in caryophyllene, do not necessarily show the correct *cis-trans* geometry about double bonds.

Figure 3.1 *Representative terpene hydrocarbons.*

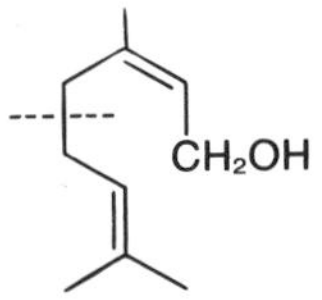

geraniol, $C_{10}H_{18}O$

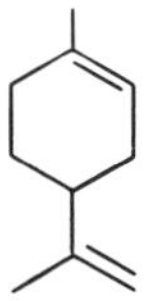

limonene, $C_{10}H_{16}$
oil of orange

α-pinene, $C_{10}H_{16}$
oil of turpentine

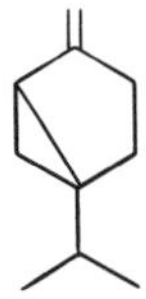

sabinene, $C_{10}H_{16}$
oil of savin

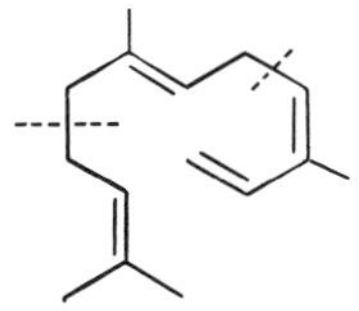

α-farnesene, $C_{15}H_{24}$
oil of citronella

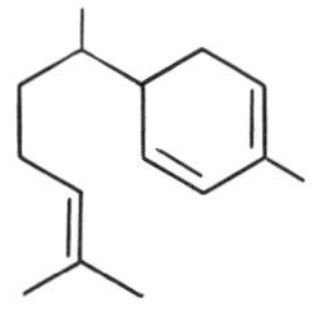

zingiberene, $C_{15}H_{24}$
oil of ginger

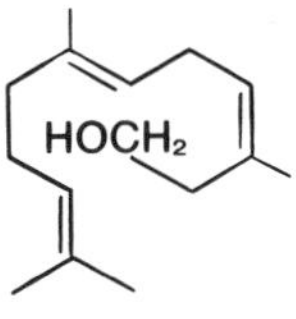

farnesol, $C_{15}H_{26}O$
lily-of-the valley

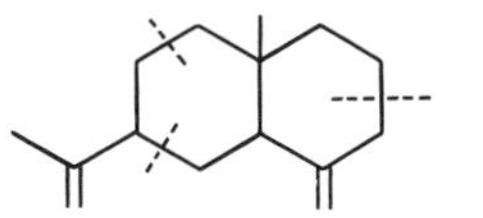

β-selinene, $C_{15}H_{24}$
oil of celery

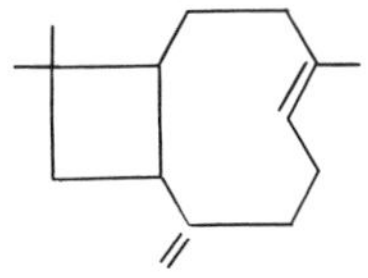

caryophyllene, $C_{15}H_{24}$
oil of cloves

Actually terpenes themselves are the simplest members of a group of compounds more properly called terpenoids. Depending on the number of isoprene units, the various subgroups are known as:

terpenes	C_{10}	two isoprene units
sesquiterpenes	C_{15}	three isoprene units
diterpenes	C_{20}	four isoprene units
triterpenes	C_{30}	six isoprene units
polyterpenes		many isoprene units

The chemistry and function of one important diterpene, vitamin A, is discussed in the mini-essay "The Chemistry of Vision."

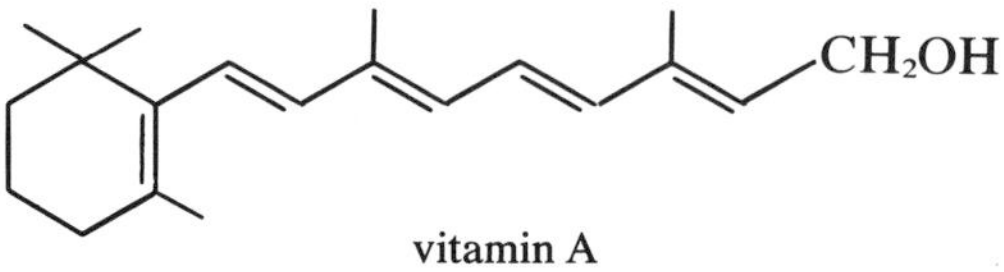

CH_2OH

vitamin A

The plant pigment β-carotene, $C_{40}H_{56}$, has an orange-red color and is commonly used as food coloring. It also has some vitamin A activity because it is apparently cleaved in the body at the central carbon-carbon double bond to give a vitamin A-derived molecule. Notice that β-carotene can be divided into two regular diterpenes with each of the units then joined together in a head-to-head manner.

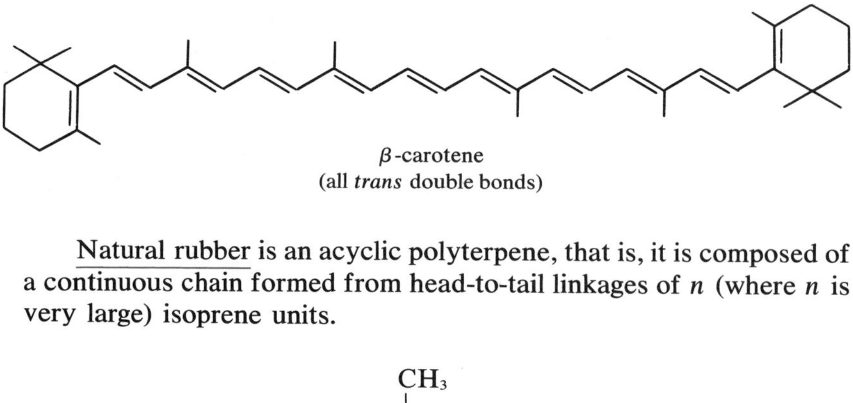

β-carotene
(all *trans* double bonds)

Natural rubber is an acyclic polyterpene, that is, it is composed of a continuous chain formed from head-to-tail linkages of *n* (where *n* is very large) isoprene units.

$$-(CH_2-\overset{\displaystyle CH_3}{\overset{|}{C}}=CH-CH_2)_n$$

Notice that there is a possibility for *cis-trans* isomerism in rubber. The difference in physical properties between natural rubber (the all *cis* configuration) and its isomer gutta-percha (the all *trans* configuration) is dramatic. The outstanding feature of natural rubber is its elasticity. In contrast, gutta-percha is hard and horny. One of its many uses is as a covering for golf balls.

These are but a few of the terpene hydrocarbons that abound in nature, but they should be enough to at least suggest to you their widespread distribution in the biological world, the biochemical individuality that plants are able to achieve through their synthesis, and the structural similarity (the isoprene rule) underlying this apparent structural diversity.

Before closing this discussion we should mention another group of secondary metabolites, this one from the animal world, with which you are undoubtedly familiar on a first-name basis. These are the antibiotics that have been isolated from fungal and related organisms. These antibiotics bear no structural relationship to the terpenes and in fact their biological origins are quite different from that of the terpenes. Examples of some of the important therapeutic agents that have been obtained in this manner are Chloromycetin, Streptomycin, Terramycin, Aureomycin, Erythromycin, and of course the most effective antibiotic of them all, the penicillins. The fascinating story of the development of penicillin from a chance laboratory discovery to a major weapon in the arsenal of chemotherapeutic agents, all within the short span of ten years, is described in the mini-essay "The Penicillins."

Problems

3.5 Draw the structural formula of a terpene hydrocarbon of two isoprene units joined head-to-head; of two isoprene units joined tail-to-tail. Such structures do appear in nature but are very rare.

3.6 Examine the structural formulas of geraniol, α-farnesene, farnesol, zingiberene, and vitamin A. In each, predict the total number of possible *cis-trans* isomers.

3.7 The three isoprene units of α-farnesene and β-selinene are shown by dashed lines on the structural formulas. In the same manner show the isoprene units of the other terpenes listed in Figure 3.1 and verify for yourself that with the exception of one head-to-head linkage in β-carotene they are all head-to-tail linkages.

3.8 Note that α-farnesene, zingiberene, β-selinene, and caryophyllene all have the same molecular formula, $C_{15}H_{24}$, and all are divisible into three isoprene units hooked head-to-tail. Show how the carbon chain of α-farnesene can be coiled and then cross-linked by one or more carbon-carbon bonds to give each of these structures. In doing this do not be concerned with the location of the double bonds.

3.9 Abietic acid, $C_{20}H_{30}O_2$, is the most abundant constituent of pine rosin. It is readily obtained by steam distillation of pine rosin or even shredded pine stumps. On a pound for pound basis, it is one of the cheapest organic acids and is used extensively (as vinyl or glyceryl esters) in lacquers and varnishes and also as the sodium salt in some laundry soaps. Abietic acid is a diterpene and it can be divided into four isoprene units. However, it is what might be called an

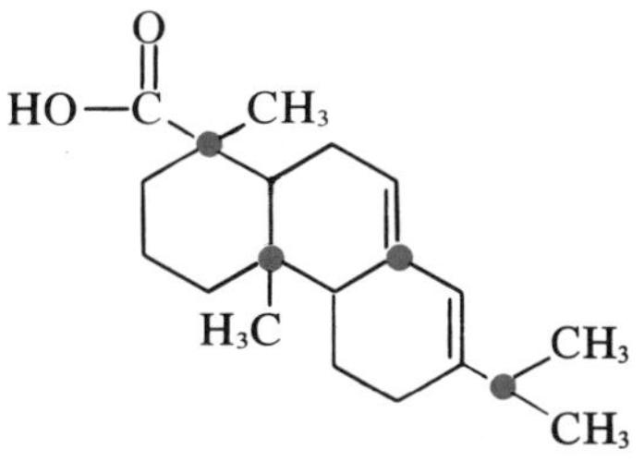

abietic acid

aberrant or irregular terpene in that not all (in this case only three) of the isoprene units are hooked head-to-tail. Show the location of the four isoprene units in abietic acid. (To help you find them, carbon atom 2 of each isoprene unit is indicated by a heavy dot.) Which of the four isoprene units is the aberrant one?

3.10 Santonin, $C_{15}H_{18}O_3$, can be isolated from the flower heads of certain species of *Artemisia*. Work on the structure of santonin was carried out over a fifty-year period by groups of Italian, German, Swiss, and English chemists and the correct structural formula was first proposed in 1930. The first successful laboratory synthesis of santonin was achieved in 1954. Santonin is an anthelmintic, i.e., a drug used to rid the body of worms (helminths). It has been estimated that over one-third of the world's population is infested with these parasites. Santonin is used as an anthelmintic for roundworm (*Ascaris lumbricoides*) in oral doses of 60–200 mg.

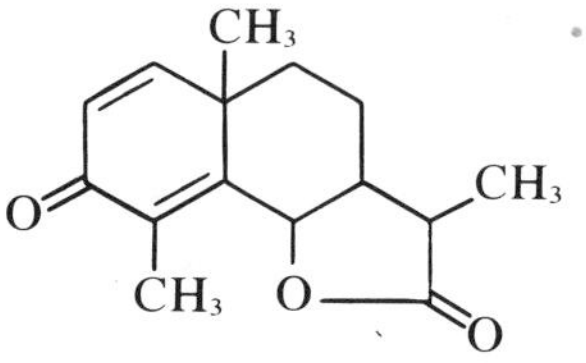

santonin

Santonin is a sesquiterpene. Locate the three isoprene units and show how the carbon skeleton of α-farnesene might be coiled and then cross-linked to give santonin. (There are actually two different coiling patterns that could lead to santonin. See if you can find them both.)

3.7 REACTIONS OF ALKENES

In contrast to alkanes, alkenes react readily with halogens, certain strong acids, a variety of oxidizing and reducing agents, and even water in the presence of concentrated sulfuric acid. Reaction with these and most other reagents is characterized by addition to the double bond.

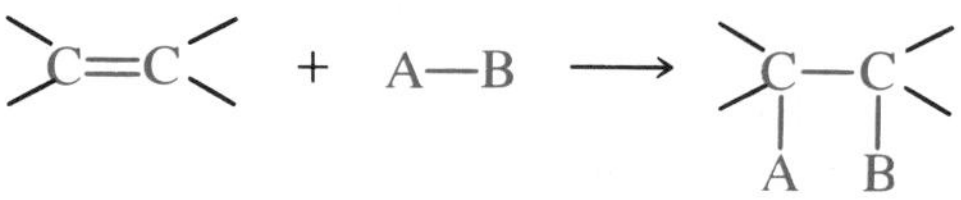

$$\rangle C{=}C\langle \; + \; A{-}B \longrightarrow \rangle \underset{A}{\underset{|}{C}}{-}\underset{B}{\underset{|}{C}}\langle$$

In the above general reaction, there is rupture of one sigma bond (A—B) and one pi bond, and the formation of two sigma bonds. Consequently, addition reactions to the double bond are almost always energetically favorable because there is net conversion of one pi bond into one sigma bond.

In the following sections, we shall look at typical alkene reactions. As we examine these reactions and those of other functional groups in later chapters, we will be asking constantly, "How do these reactions occur?" or in the more common terminology of the chemist, "What is the mechanism of each of these reactions?"

Just exactly what does it mean to specify a reaction mechanism?

In such a description we strive literally to specify the detailed pathway by which the reaction occurs, the exact position of every atom involved during the course of the reaction, the role of the solvent molecules, the interactions and bonds between and within molecules, the role of catalysts, and the rates at which the various changes take place during the reaction. A complete specification of all of these is more than we know about any reaction at the present time, so we will be content in our description of a reaction mechanism if we can specify the important intermediate compounds formed during the course of the reaction and specify in general terms how each step in the overall reaction occurs.

How do we begin to describe a reaction mechanism? First, we assemble all of the available experimental observations and facts about the particular chemical reaction under consideration. Next, through a combination of chemical sophistication, creative insight, and guesswork, we propose several sets of steps, or mechanisms, each of which will account for the overall chemical transformation. Then, we test each of these mechanisms against the experimental observations and exclude those mechanisms that are not consistent with the facts. A mechanism becomes generally established when reasonable alternatives have been excluded and when it has been shown that the mechanism is consistent with every test that the scientist can devise. This of course does not mean that a generally accepted mechanism is in fact a completely true and accurate description of the chemical events, but only that it is consistent with the mass of experimental evidence.

Before we go on to consider reactions and reaction mechanisms, we might also ask, why is it worth the trouble of chemists to establish them and students to learn about them? Certainly one answer lies in the understanding and intellectual satisfaction derived from constructing models that accurately reflect the behavior of chemical systems. Problems of this kind present a particular fascination and exciting challenge to the creative scientist. But there are other, more practical reasons as well. Mechanisms provide us with a theoretical framework within which to organize a great deal of descriptive chemistry. For example, with some insight into how certain reagents add to particular alkenes, we can then generalize about how these same reagents might react with other alkenes and make predictions about how different reagents might also react with the same alkenes. Finally, to the practicing chemist, a knowledge of mechanisms is often helpful in selecting conditions of temperature, pressure, concentration, reaction time, etc. so as to better achieve desired results in a specific reaction.

3.8 ADDITION OF HYDROGEN (HYDROGENATION)

The addition of hydrogen to a carbon-carbon double bond converts an alkene to an alkane. This "saturation" or hydrogenation of the double bond is a very important reaction in the laboratory as well as in nature.

$$CH_2{=}CH_2 + H_2 \xrightarrow[\text{catalyst}]{\text{metal}} CH_3{-}CH_3$$

ETHENE ETHANE

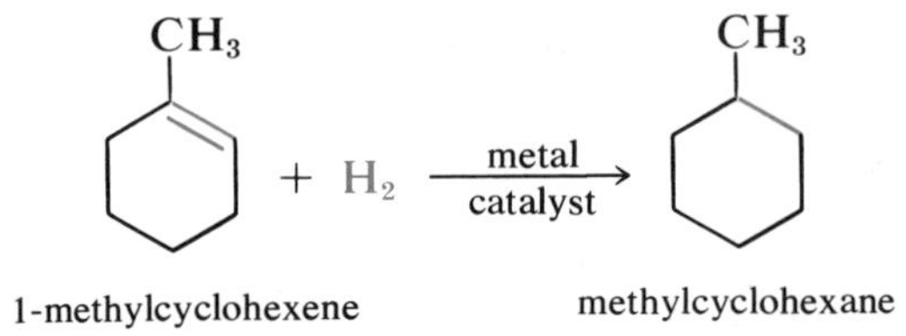

$$\text{1-methylcyclohexene} + H_2 \xrightarrow[\text{catalyst}]{\text{metal}} \text{methylcyclohexane}$$

1-methylcyclohexene methylcyclohexane

Although the addition of hydrogen to an alkene is strongly exothermic, these reactions are immeasureably slow at moderate temperatures in the gas phase but occur readily in the presence of finely divided platinum, palladium, or nickel. Separate experiments have shown that these and other metals near the center of the periodic table are able to absorb large quantities of hydrogen gas by incorporating it into the metallic lattice as hydrogen atoms. These hydrogen atoms are electron-deficient and therefore highly reactive.

Another type of experimental observation that must be incorporated into any acceptable mechanism is the known stereospecificity of the catalytic hydrogenation. For example, addition of hydrogen to 1,2-dimethylcyclopentene yields mostly *cis*-1,2-dimethylcyclopentane.

$$\text{1,2-dimethylcyclopentene} + H_2 \xrightarrow[25°C]{\text{Pt, 1 atm}} \textit{cis}\text{-1,2-dimethylcyclopentane}$$

1,2-dimethylcyclopentene *cis*-1,2-dimethylcyclopentane

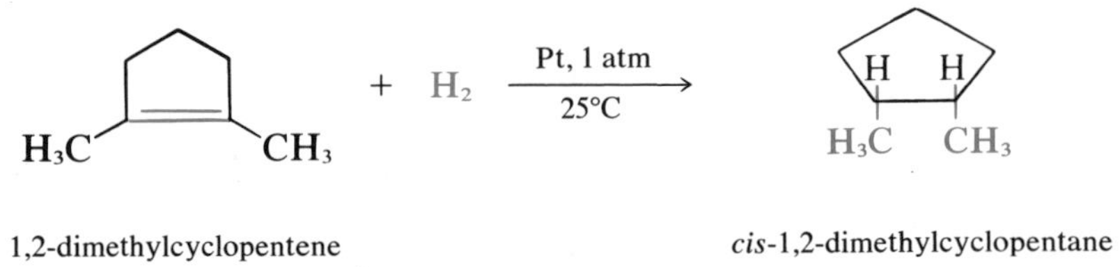

From this and other observations of the stereochemistry of the addition of hydrogen it was concluded that both hydrogen atoms are added simultaneously and from the same side of the alkene molecule.

The generally accepted mechanism for a catalytic hydrogenation reaction postulates that molecular hydrogen is first absorbed on the surface of the metal and the bond between the hydrogen atoms is weakened by interaction with the metal. When the alkene is also absorbed on the surface of the metal, addition occurs—both hydrogen atoms are delivered simultaneously and the resulting alkane is desorbed. A schematic representation of these steps is shown in Figure 3.2.

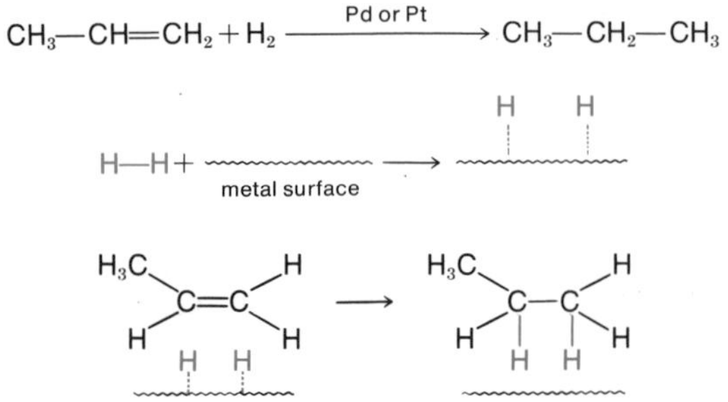

$$CH_3—CH═CH_2 + H_2 \xrightarrow{\text{Pd or Pt}} CH_3—CH_2—CH_3$$

$$H—H + \text{metal surface} \longrightarrow \text{H, H adsorbed on metal surface}$$

$$\text{(}H_3C\text{)(H)C═C(H)(H)} + \text{adsorbed H H} \longrightarrow \text{(}H_3C\text{)(H)(H)C—C(H)(H)(H)}$$

Figure 3.2 *Hydrogenation of propene involving a metal catalyst.*

This concept of the interaction of one or more molecules on a catalytic surface is one that underlies all of our thinking on the catalytic action of enzymes.

3.9 ADDITION OF WATER (MARKOVNIKOV'S RULE)

In the presence of an acid catalyst, often 60% sulfuric acid, water adds to the more reactive alkenes to produce alcohols. The addition of the elements of water to an alkene is called hydration.

$$\underset{\text{ethylene}}{CH_2{=}CH_2} + H_2O \xrightarrow{H^+} \underset{\text{ethanol}}{CH_3CH_2OH}$$

Hydration of ethylene involves the addition of hydrogen to one of the carbon atoms and hydroxide to the other, and only one product is possible. In the hydration of propene we encounter another situation. Without either experimental evidence or a theoretical model to guide us, we would have to expect the formation of two isomeric products, 2-propanol and 1-propanol, depending on the orientation of the addition.

$$\underset{\text{propene}}{CH_3CH{=}CH_2} + H_2O \xrightarrow{H^+} \underset{\substack{\text{2-propanol}\\\text{(major product)}}}{CH_3\underset{\substack{|\\OH}}{C}HCH_3} + \underset{\substack{\text{1-propanol}\\\text{(minor product)}}}{CH_3CH_2CH_2OH}$$

True to our unsophisticated expectation, each of these alcohols is formed, but the fact that 2-propanol predominates in the product mixture clearly indicates that the addition is by no means random.

Acid-catalyzed hydration of 2-methylpropene also produces a mixture of two isomeric alcohols. Again, the addition does not appear to be random for one alcohol predominates over the other.

$$\underset{\text{2-methylpropene}}{CH_3\overset{\substack{CH_3\\|}}{C}{=}CH_2} + H_2O \xrightarrow{H^+} \underset{\substack{\text{2-methyl-2-propanol}\\\text{(major product)}}}{CH_3\overset{\substack{CH_3\\|}}{\underset{\substack{|\\OH}}{C}}CH_3} + \underset{\substack{\text{2-methyl-1-propanol}\\\text{(minor product)}}}{CH_3\overset{\substack{CH_3\\|}}{C}HCH_2OH}$$

After studying a large number of reactions of water and acids such as HCl, HBr, and HI with alkenes, the Russian chemist Vladimir Markovnikov in 1871 made the following empirical generalization: in the addition of an unsymmetrical reagent to an unsymmetrical alkene, the positive part of the reagent adds to the carbon of the double bond that contains the greater number of hydrogen atoms. We should remember that while this rule systematizes many of the observed reactions of addition to alkenes, it is not an explanation for the observations.

3.10 ELECTROPHILIC ATTACK ON THE CARBON-CARBON DOUBLE BOND

In the hydration reaction, the presence of sulfuric acid or other strong acid is required. Weak acids or bases such as sodium hydroxide will not serve to catalyze the addition of water. It became apparent to chemists studying these reactions that the initial attack on the carbon-carbon double bond is by H^+, or to put it in more general terms, the initial attack on the double bond is by an electrophilic (electron-loving or -seeking) reagent. This susceptibility of the double bond to attack by an electrophile is certainly consistent with our electronic formulation, which pictures the double bond as a center of high electron (negative) density.

To account for the hydration of alkenes, organic chemists have proposed an ionic mechanism:

Step 1 $$H^+ + CH_2{=}CH_2 \longrightarrow CH_3{-}CH_2^+$$

Step 2 $$CH_3{-}CH_2^+ + :\ddot{O}(H){-}H \longrightarrow CH_3{-}CH_2{-}\ddot{O}^+(H){-}H$$

Step 3 $$CH_3{-}CH_2{-}\ddot{O}^+(H){-}H \longrightarrow CH_3{-}CH_2{-}\ddot{\underset{..}{O}}{-}H + H^+$$

The first step in the generally accepted mechanism involves the reaction of a pair of electrons of the double bond with H^+ and the formation of a new covalent bond between carbon and hydrogen. Step 1 results in the formation of a carbonium ion (an ion containing a positively charged carbon). The carbon bearing the positive charge has only six valence electrons in its valence shell and is highly reactive. Only in recent years have we had direct experimental evidence for the existence of any carbonium ion. Prior to this the chemist could only infer its existence. Considering the evidence, the inference was very soundly based. In step 2, the carbonium ion completes its valence octet by forming a new covalent bond with an unshared electron pair on the oxygen of a water molecule. Finally, loss of a proton results in the formation of ethyl alcohol and regeneration of a proton.

This mechanism certainly accounts for the observation that hydration of ethylene gives ethyl alcohol and it accounts for the catalysis by strong acid. But how does the carbonium ion mechanism account for the observation that the hydration of propene gives mainly 2-propanol and very little 1-propanol? How does this theory account for the type of observation generalized in Markovnikov's rule? The answer is that the mechanism developed thus far does not account for the rule. At this point, we have two alternatives: either we discard the idea of a carbonium ion and start again, or we can refine the mechanism by introducing a new concept. In this case it is more convenient and fruitful to refine the mechanism and introduce a new concept, namely the relative ease of formation of carbonium ions.

The carbonium ion mechanism proposes that the addition of a proton to an asymmetrical alkene such as propene can give either of two isomeric carbonium ions:

$$CH_3CH{=}CH_2 + H^+ \longrightarrow \begin{cases} CH_3CH_2\overset{+}{C}H_2 & n\text{-propyl carbonium ion (a primary carbonium ion)} \\ CH_3\underset{+}{C}HCH_3 & \text{isopropyl carbonium ion (a secondary carbonium ion)} \end{cases}$$

In the same manner the reaction of 2-methylpropene with a proton can give isomeric carbonium ions:

$$CH_3\overset{CH_3}{\overset{|}{C}}{=}CH_2 + H^+ \longrightarrow \begin{cases} CH_3\overset{CH_3}{\overset{|}{C}}H{-}CH_2^+ & \text{isobutyl carbonium ion (a primary carbonium ion)} \\ CH_3\overset{CH_3}{\overset{|}{C^+}}{-}CH_3 & tert\text{-butyl carbonium ion (a tertiary carbonium ion)} \end{cases}$$

The *n*-propyl carbonium ion will react with water to give 1-propanol, and the isopropyl carbonium ion will react with water to give 2-propanol. Hence, the alcohol formed depends on the carbonium ion formed. It is clear from a detailed study of the addition of H_2O to propene and other alkenes that the slow step in the addition reaction is the formation of the carbonium ion. Once the carbonium ion is formed, it reacts with water to give the observed product.

Since 2-propanol is the major product of the addition of water to propene, the isopropyl carbonium ion must be formed more easily than the *n*-propyl carbonium ion. Since 2-methyl-2-propanol (*tert*-butyl alcohol) predominates over 2-methyl-1-propanol (isobutyl alcohol), the *tert*-butyl carbonium ion must be formed more easily than the isobutyl carbonium ion. From a study of these and many other reactions which proceed via carbonium ions, we infer that the ease of formation of carbonium ions is of the following order:

$$CH_3{-}\overset{CH_3}{\underset{CH_3}{\overset{|}{\underset{|}{C^+}}}} > CH_3{-}\overset{H}{\underset{CH_3}{\overset{|}{\underset{|}{C^+}}}} > CH_3{-}\overset{H}{\underset{H}{\overset{|}{\underset{|}{C^+}}}} > H{-}\overset{H}{\underset{H}{\overset{|}{\underset{|}{C^+}}}}$$

The ease of formation of a carbonium ion increases as the ion becomes more highly substituted at the positive carbon. To generalize, we can write the following order for the ease of formation of carbonium ions:

$$3° > 2° > 1° > \text{methyl}$$

Now let us again ask whether our proposed mechanism is in agreement with the experimental observations? First, combination of steps 1–3 does give the observed stoichiometry for the hydration of one molecule of alkene to yield one molecule of alcohol. Second, the mechanism does account for the observation that, although H^+ is not a part of the stoichiometry of the reaction, it is necessary as a catalyst; acid is involved in the initiation of the reaction and is regenerated in the final step. Finally, given the concept of the relative ease of formation of primary, secondary, and tertiary carbonium ions, this mechanism is consistent with the observations that hydration of propene gives mostly 2-propanol and only very little 1-propanol; it is consistent with the type of observations generalized in Markovnikov's rule.

Problems

3.11 Predict the product(s) of the hydration of the following alkenes:

a 2-butene
b 2-methyl-2-butene
c 2-methyl-1-butene
d cyclohexene
e 1-methylcyclohexene
f *cis*-2-pentene
g *trans*-2-pentene
h 2,3-dimethyl-2-butene

3.12 Terpin hydrate is prepared commercially by the addition of two moles of water to limonene in the presence of dilute sulfuric acid. Limonene is one of the main components of lemon, orange, caraway, dill, bergamot, and some other oils. Terpin hydrate is used medicinally as an expectorant for coughs. It may be given as terpin hydrate and codeine. Propose a structure for terpin hydrate and a reasonable mechanism to account for the formation of the product you have predicted.

$$\text{limonene} + 2H_2O \xrightarrow[H_2SO_4]{\text{dilute}} C_{10}H_{20}O_2 \ (\text{terpin hydrate})$$

limonene: ring with CH_3 on the double-bonded carbon and $C(CH_3)=CH_2$ (H_3C, CH_2) at the 4-position

terpin hydrate

3.11 HALOGENATION OF ALKENES

Bromine and chlorine add readily to alkenes to form single covalent bonds on adjacent carbons. While iodine is generally too unreactive to add, the more reactive iodine monochloride (ICl) and iodine monobromide (IBr) do add readily. (See Problem 3.13 for a quantitative application of iodine monobromide addition). Halogenation with bromine or chlorine is carried out either with the pure reagents (neat) or by mixing the reagents in CCl_4 or some other inert solvent.

$$CH_3CH{=}CHCH_3 + Br_2 \longrightarrow CH_3\underset{Br}{CH}-\underset{Br}{CH}CH_3$$

$$\text{cyclopentene} + Br_2 \longrightarrow \text{cyclopentane with Br, H / H, Br (trans)}$$

trans-1,2-dibromocyclopentane

As illustrated by the second example, the addition of halogen to a simple alkene is quite specific in producing only one dihalide. Where geometric isomerism is possible, it is the *trans* dihalide that is formed as the major product.

Addition of bromine is a particularly useful qualitative test for the detection of alkenes. A solution of bromine in carbon tetrachloride is red. Alkenes and dibromoalkanes are usually colorless. The rapid discharge of red color of a solution of bromine in carbon tetrachloride is a characteristic property of alkenes.

We might pause a moment to compare and contrast the halogenation of alkenes with that of alkanes (Section 2.11). Recall that chlorine and bromine do not react with alkanes unless the halogen–alkane mixture is exposed to strong light or heated to temperatures of 250–400°C. The reaction which then occurs is one of substitution of halogen for hydrogen and the formation of an equivalent amount of HCl or HBr. The halogenation of most higher alkanes inevitably gives a complex mixture of monosubstitution products. In contrast, chlorine and bromine react readily with alkenes at room temperature in the dark. Further, the reaction is the addition of halogen to the two carbon atoms of the double bond with the formation of two new carbon–halogen bonds. If we know the position of the double bond, we can predict with certainty the location of the halogen atoms in the product.

Problem **3.13** Before the recent development of sensitive instrumental techniques, a number of methods were developed to measure the degree of unsaturation of fats and oils. One such method was to experimentally determine an "iodine number." In this procedure, equivalent amounts of I_2 and Br_2 are mixed in acetic acid to produce the highly reactive iodine monobromide, IBr. This reagent adds to alkenes as shown below.

$$R{-}CH{=}CH{-}R + IBr \rightarrow R{-}\underset{\displaystyle I}{\underset{|}{CH}}{-}\underset{\displaystyle Br}{\underset{|}{CH}}{-}R$$

a Propose a mechanism to account for the addition of IBr to an alkene. Based on your mechanism, would you expect the addition of IBr to propene to give

$$CH_3{-}\underset{\displaystyle I}{\underset{|}{CH}}{-}\underset{\displaystyle Br}{\underset{|}{CH_2}} \quad \text{or} \quad CH_3{-}\underset{\displaystyle Br}{\underset{|}{CH}}{-}\underset{\displaystyle I}{\underset{|}{CH_2}}?$$

b The iodine number is defined as the number of grams of iodine that adds to 100 grams of a fat or oil. (For definition of the terms fat and oil see Section 11.2.) Given on the following page are average molecular weights of three oils and one fat and their corresponding iodine numbers.

Fat or Oil	Average Mol Wt (g/mole)	Iodine Number
corn oil	870–900	115–130
soybean oil	860–890	125–140
linseed oil	855–895	175–205
butter fat	700–720	25–40

Which of these substances is the most highly unsaturated; the least highly unsaturated? Explain your reasoning.

3.12 OXIDATION OF ALKENES

The electrons of the carbon-carbon double bond are readily attacked by oxidizing agents. In this section we shall consider two of these agents: potassium permanganate ($KMnO_4$) and ozone (O_3).

Treatment of an alkene with a dilute basic solution of potassium permanganate in water or an aqueous organic solvent results in conversion of the alkene into a 1,2-dialcohol (a glycol).

$$3CH_3CH{=}CH_2 + 2KMnO_4 + 4H_2O \rightarrow 3CH_3\underset{OH}{\underset{|}{CH}}{-}\underset{OH}{\underset{|}{CH_2}} + 2MnO_2 + 2KOH$$

propylene glycol; manganese dioxide (brown ppt)

$$\text{cyclopentene} + KMnO_4 \xrightarrow[CH_3CH_2OH]{H_2O} \text{cis-1,2-cyclopentanediol (HO, OH)} + MnO_2$$

cis-1,2-cyclopentanediol

The *cis* geometry of the product is accounted for by the formation of a cyclic intermediate.

$$\text{cyclopentene} + MnO_4^- \longrightarrow \text{cyclic manganate ester (O–Mn(=O)(O}^-\text{)–O)} \xrightarrow[H_2O]{OH^-} \text{HO, OH diol} + MnO_2$$

This reaction is the basis of a qualitative test for the presence of an alkene because in the course of the reaction, the purple color of the permanganate ion is discharged and a brown precipitate of MnO_2 forms. Unfortunately this test is not completely specific for alkenes since certain other easily oxidized functional groups will also discharge the permanganate color.

Even at low temperatures most alkenes react with ozone, O_3, to cleave the carbon-carbon double bond. This reaction is useful in the preparation of aldehydes and ketones (Chapter 6) and as a means of locating the position of the double bond within an alkene.

In practice, the alkene is dissolved in an inert solvent such as CCl_4 and a stream of ozone gas is bubbled through the solution. Since the ozonide intermediates are usually unstable and explosive in pure form, they are generally not isolated. The reaction mixture is poured into water in the presence of a reducing agent, usually powdered metallic zinc. The resulting aldehydes or ketones are isolated and purified. Through determination of the structures of the products of ozonolysis, it is possible to work back and deduce the structure of an unknown alkene.

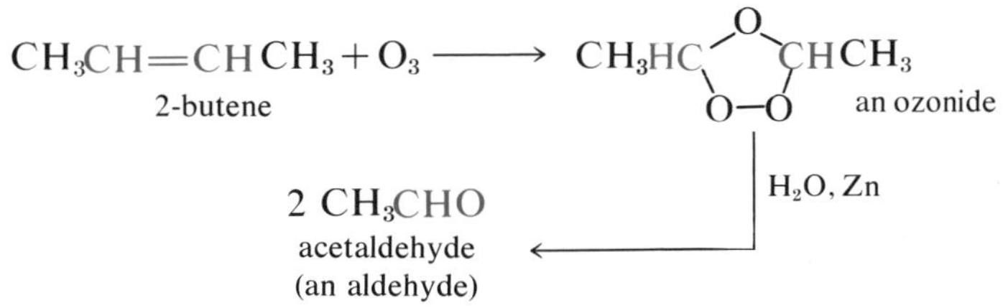

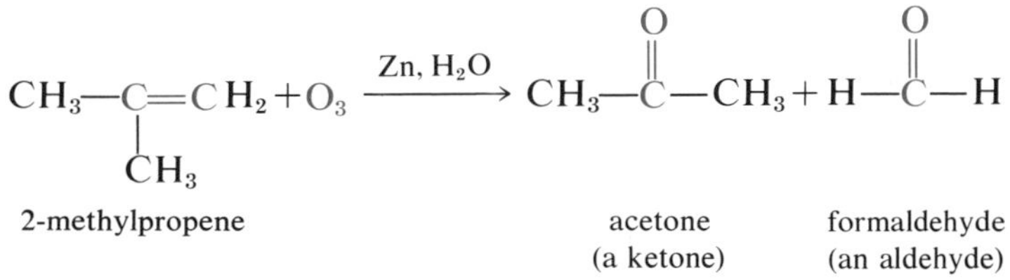

$$\underset{\text{2-butene}}{CH_3CH{=}CHCH_3} + O_3 \longrightarrow \underset{\text{an ozonide}}{CH_3HC\langle O, O{-}O \rangle CHCH_3} \xrightarrow{H_2O,\ Zn} \underset{\substack{\text{acetaldehyde} \\ \text{(an aldehyde)}}}{2\ CH_3CHO}$$

$$\underset{\text{2-methylpropene}}{CH_3{-}\underset{\displaystyle CH_3}{\underset{|}{C}}{=}CH_2} + O_3 \xrightarrow{Zn,\ H_2O} \underset{\substack{\text{acetone} \\ \text{(a ketone)}}}{CH_3{-}\overset{\displaystyle O}{\overset{\|}{C}}{-}CH_3} + \underset{\substack{\text{formaldehyde} \\ \text{(an aldehyde)}}}{H{-}\overset{\displaystyle O}{\overset{\|}{C}}{-}H}$$

Problems

3.14 Write an equation to illustrate the reaction of 2-methyl-2-butene with each reagent:

a H_2/Pt **b** Br_2
c O_3 followed by Zn/H_2O **d** H_2O/H_2SO_4
e $KMnO_4$ (dilute basic solution)

3.15 Repeat Problem 3.14 using cyclohexene.

3.16 Reaction of 2-methylpropene with methanol in the presence of H_2SO_4 yields a compound of formula $C_5H_{12}O$.

$$CH_3{-}\overset{\displaystyle CH_3}{\overset{|}{C}}{=}CH_2 + CH_3OH \xrightarrow{H_2SO_4} C_5H_{12}O$$

Propose a structural formula for this compound. Also propose a mechanism to account for its formation.

3.17 Draw the structural formula of the alkene which will give the indicated product(s) on ozonolysis.

a $C_6H_{12} \rightarrow CH_3CH_2CHO$ as the only product
b $C_6H_{12} \rightarrow CH_3CHO + CH_3COCH_2CH_3$ in equal amounts
c $C_6H_{12} \rightarrow CH_3COCH_3$ as the only product
d $C_7H_{12} \rightarrow CH_3\overset{\displaystyle O}{\overset{\|}{C}}CH_2CH_2CH_2\overset{\displaystyle O}{\overset{\|}{C}}CH_3$

e $C_{10}H_{16} \rightarrow$ (cyclobutane bearing CH_3 and CH_3 on one carbon, $CH_2-\overset{O}{\overset{\|}{C}}H$ on the adjacent carbon, and $\overset{O}{\overset{\|}{C}}-CH_3$ on the carbon across from it)

$$CH_3,\ CH_3 \quad CH_2-\overset{O}{\overset{\|}{C}}H \quad O{=}C{-}CH_3$$

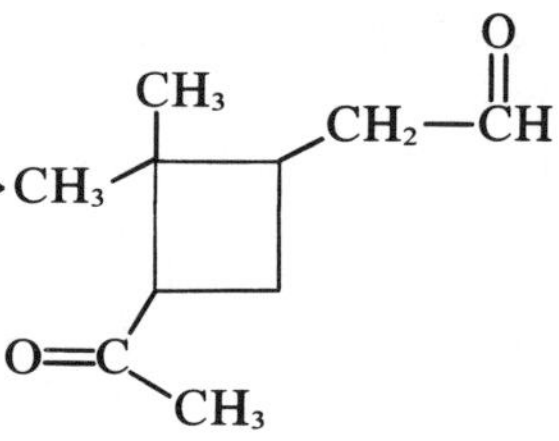

3.18 What product would you expect to isolate after ozonolysis of natural rubber? (Hint: draw a section of the carbon chain including several double bonds. You should then be able to see what small molecule will be produced.)

3.19 Show how the following pairs of compounds can be distinguished by the use of either bromine in carbon tetrachloride or potassium permanganate. In each case tell what you would do experimentally, what you would expect to observe, and write a chemical equation for all reactions observed.

a cyclohexene and cyclohexane (by qualitative observation)
b cyclohexane and hexene (by qualitative observation)
c 2,4-hexadiene and 1-methylcyclohexene (by quantitative observation)

3.13 POLYMERIZATION OF SUBSTITUTED ETHYLENES

From the perspective of the chemical industry, the single most important reaction of alkenes is that of polymerization, the building together of many small units known as monomers (Greek: *mono* + *meros*, single part) into very large, high-molecular-weight polymers (Greek: *poly* + *meros*, many parts). In this section we will be concerned with addition polymerization in which monomer units are joined together without loss of atoms in the process. The alternative process, known as condensation polymerization, involves the loss of one or more atoms during the polymerization process. Condensation polymerization is illustrated in the mini-essay "Nylon and Dacron."

We shall not discuss any mechanism of polymerization, but will show, at least in principle, how addition polymers might be formed. Drawn below are two molecules of ethylene and another molecule we shall call an initiator. The initiator (In) is shown interacting with one molecule of ethylene and causing a rearrangement of bonding electrons, with the result that the carbon skeletons of the two ethylene molecules are now joined together.

$$In \curvearrowright CH_2{=}CH_2 \curvearrowright CH_2{=}CH_2 \curvearrowright \longrightarrow In{-}CH_2{-}CH_2{-}CH_2{-}CH_2{-}$$

The monomer in this case is ethylene. Two monomers attached to one another form a dimer, three monomers attached together form a trimer, and many monomers attached together form a polymer. These are illustrated below.

$-(CH_2CH_2)_2-$	$-(CH_2CH_2)_3-$	$-(CH_2CH_2)_n-$
a dimer	a trimer	a polymer

The subscript "*n*" indicates that the monomer-derived unit repeats *n* times in the polymer chain. Molecular weights of polyethylene range from 50,000 to 500,000 g/mole.

Much of the early interest in polymers arose from the desire to make synthetic rubber and relieve the near-total dependence on natural rubber. In the mid-1920s, several companies, most notably du Pont in the United States and I.G. Farbenindustrie in Germany, began research programs in this area. By the mid-1930s, du Pont was producing the first synthetic rubber, Neoprene, on a commercial basis. Neoprene synthetic rubber is polychloroprene.

$$n\,CH_2{=}\underset{\displaystyle\text{chloropropene}\atop\text{2-chlorobutadiene}}{\overset{Cl}{\overset{|}{C}}{-}CH{=}CH_2} \xrightarrow{\text{polymerization}} -(CH_2{-}\overset{Cl}{\overset{|}{C}}{=}CH{-}CH_2)_n-$$

chloropropene / 2-chlorobutadiene → Neoprene / polychloroprene

The years since have seen extensive research and development in polymer chemistry and physics, and an almost explosive growth in plastics, coatings, and rubber technology has created a world-wide multibillion dollar industry. A few basic factors account for this phenomenal growth. First, the raw materials for plastics, etc. are derived mainly from petroleum; with the development of efficient thermal and catalytic cracking processes, the raw materials became generally cheap and plentiful. Second, within broad limits, scientists have learned how to tailor-make polymers to the requirements of the end use. Third, many plastics can be fabricated more cheaply than the competing materials. For example, plastic technology created the water-based (latex) paints that have revolutionized the coatings industry; plastic films and foams have done the same for the packaging industry. The list could go on and on as we think of the manufactured items that surround us in our daily lives.

Table 3.4 lists several important polymers of substituted ethylenes along with their common names and uses. The polyethylenes, mainly polyethylene and polypropylene, are the largest tonnage plastics in the world.

The tetrafluoroethylene polymers were discovered accidentally in 1938 by du Pont chemists during the search for new refrigerants (the Freons described in Section 2.12). One morning a cylinder of tetrafluoroethylene appeared to be empty (no gas escaped when the valve was opened) and yet the weight of the cylinder indicated it was full. The cylinder was opened and inside was found a waxy solid, the forerunner of Teflon. The solid proved to have very unusual properties: extraordinary chemical inertness, outstanding heat resistance, very high melting point and unusual frictional properties. Du Pont began limited production of Teflon in 1941. The small amount of polymer was preempted at once by the Manhattan Project where it was used in equipment to contain the highly corrosive UF_6 during the separation of the isotopes of uranium. Du Pont built the first commercial Teflon plant in 1948 and the product was used to make gaskets, bearings for automobiles, nonstick equipment for candy

Table 3.4 *Polymers derived from substituted ethylene monomers.*

Monomer	Monomer Name	Polymer Name or Trade Name
$CH_2{=}CH_2$	ethylene	polyethylene, Polythene, for unbreakable containers and tubing
$CH_2{=}CHCH_3$	propylene	polypropylene, Herculon, fibers for carpeting and clothes
$CH_2{=}CHCl$	vinyl chloride	polyvinyl chloride, PVC, Koroseal
$CH_2{=}CCl_2$	1,1-dichloroethylene	Saran, food wrappings
$CH_2{=}CHCN$	acrylonitrile	polyacrylonitrile, Orlon, Acrylics
$CF_2{=}CF_2$	tetrafluoroethylene	polytetrafluoroethylene, Teflon
$CH_2{=}CHC_6H_5$	styrene	polystyrene, Styrofoam, for insulation
$CH_2{=}C(CH_3){-}CO_2CH_3$	methyl methacrylate	polymethyl methacrylate, Lucite, Plexiglas, for glass substitutes
$CH_2{=}CHCO_2CH_3$	methyl acrylate	polymethyl acrylate, Acrylics, for latex paints

manufacturers and commercial bakers, seals for rotating equipment, and a number of other items. Teflon became a household word in 1961 with the introduction of nonstick frying pans in the U.S. market.

3.14 ALKYNES

The alkynes contain one or more carbon-carbon triple bonds. Acetylene is the first member of this class, which is often referred to by the common name of acetylenes.

Simple alkynes often are named as derivatives of acetylene. In the IUPAC nomenclature the ending -yne indicates the presence of a triple bond. The following examples illustrate both systems of nomenclature.

$H{-}C{\equiv}C{-}H$	$CH_3{-}C{\equiv}C{-}H$	$CH_3{-}C{\equiv}C{-}CH_3$
ethyne or acetylene	propyne or methylacetylene	2-butyne or dimethylacetylene

The acetylenes are not widely distributed in nature. Their chief value lies in the ready availability of acetylene itself and its great reactivity under a wide variety of experimental conditions. It is of value as an inexpensive and readily available starting material for a number of commercially important synthetic processes.

3.15 AROMATIC HYDROCARBONS

In 1825 Michael Faraday isolated a compound, later to be called benzene, from the oily liquid which collected in the illuminating gas lines of London. Faraday reported that the substance had an empirical formula of C_2H and named it "bicarburet of hydrogen." This formula was based on the then current (and incorrect) assumption that the atomic weight of carbon was 6. Eilhard Mitscherlich reported in 1834 that this same substance could be obtained by heating benzoic acid (from gum benzoin) and calcium hydroxide. With more accurate values for atomic weights Mitscherlich was able to establish the molecular formula of this substance as C_6H_6 and proposed the name "benzin." This name was criticized on various grounds and soon the name "benzol" became established (and is still used) in the German literature and the name "benzene" became established in the French and English literature.

Benzene is a liquid with a boiling point of 80°C. Its formula suggests a high degree of unsaturation—remember that a saturated alkane would have the formula C_6H_{14}, a saturated cycloalkane the formula C_6H_{12}. With this high degree of unsaturation you might expect that benzene would be highly reactive and would show reactions characteristic of alkenes and alkynes. Surprisingly, benzene does not undergo characteristic alkene reactions. For example, it does not react readily with bromine. Benzene does not undergo addition reactions with hydrogen chloride, hydrogen bromide, and other reagents that usually add to double and triple bonds. Further, it is quite unaffected by the usual oxidizing agents. When it does react, benzene typically does so by substitution. In the presence of bromine and ferric bromide, benzene forms bromobenzene and hydrogen bromide. With fuming sulfuric acid (sulfuric acid containing dissolved SO_3) at room temperature or with concentrated sulfuric acid at elevated temperature, benzene undergoes a substitution reaction producing benzenesulfonic acid. The reaction is known as sulfonation.

$$C_6H_6 + Br_2 \xrightarrow{FeBr_3} \underset{\text{bromobenzene}}{C_6H_5Br} + HBr$$

$$C_6H_6 + SO_3 \xrightarrow[\text{room temp.}]{H_2SO_4\text{(conc)}} \underset{\text{benzenesulfonic acid}}{C_6H_5SO_3H}$$

The terms "aromatic" and "aromatic compounds" were used by Kekulé to classify benzene and a number of its derivatives because many of them have rather pleasant odors. However, after a time it became clear that a sounder classification for these compounds should be based not on aroma but rather on chemical reactivities. Currently the term aromatic is used to refer to the unusual chemical stability of benzene and its derivatives. This chemical inertness of benzene and other aromatic hydrocarbons—a resistance to uncatalyzed halogenation, a resistance to oxidation, and a tendency to react by substitution rather than addition—was both strikingly evident and truly puzzling to Kekulé and his contemporaries.

Let us put ourselves in the mid-19th century and examine the evidence on which chemists attempted to build an adequate model for the structure of benzene. First, it was clear that the molecular formula of benzene is C_6H_6. In the light of Kekulé's theory of a characteristic valence of four for carbon, it seemed evident that the benzene molecule should be highly unsaturated. Yet as we have already demonstrated, benzene does not show the chemical reactivity of the unsaturated compounds known at that time, namely the alkenes. Benzene does undergo reactions, but substitution rather than addition as is characteristic of alkenes. When the monosubstituted benzenes were examined, e.g. bromobenzene, it was found that no isomers were formed. There is one and only one bromobenzene. All efforts to isolate and identify additional isomers were unsuccessful. From this type of evidence, chemists concluded that all six of the hydrogens of benzene are equivalent. Finally, when a monosubstituted benzene was made to undergo further substitution, three different isomers of the disubstituted benzene were isolated. For example, in the case of the further bromination of bromobenzene, three isomers of dibromobenzene can be isolated. Two of the isomers are produced in major amounts, and the other one in only minor amount, but the fact remains that three, and only three, dibromoisomers can be isolated.

For Kekulé and his contemporaries, the problem was to incorporate these observations, along with the accepted tetravalence of carbon, into a structural formulation of the benzene molecule. Before we examine the structural formula for benzene proposed by Kekulé, we should note that the problem of an adequate description of the structure of benzene and other aromatic hydrocarbons has occupied the efforts of chemists for over a century. Only since the 1930s has a general understanding of this problem been realized, and this has required the best efforts of chemists and mathematicians to develop new mathematical tools and physical concepts equal to the problem.

In 1865 Kekulé proposed for benzene the formulas Ia and Ib (cyclohexatriene) as being most consistent with the experimental evidence. He further postulated that the double bonds shift back and forth so rapidly in a kind of mobile equilibrium that the two forms cannot be separated.

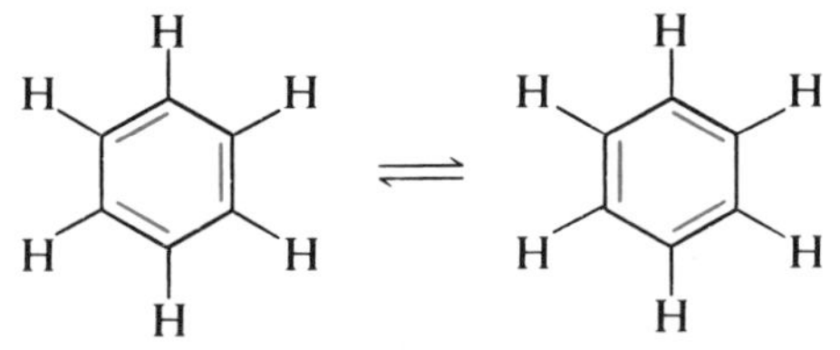

Ia ⇌ Ib

In this formulation, all six hydrogens are equivalent, and the structure of a monosubstituted benzene, e.g. bromobenzene, is illustrated by II. The three possible disubstitution products, the dibromobenzenes, are illustrated by IIIa, IIIb, and IIIc.

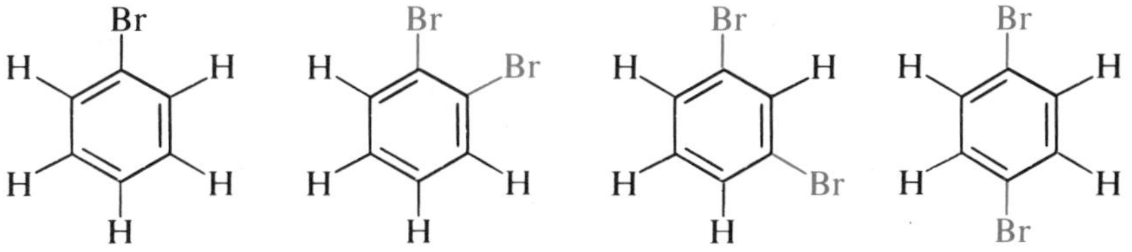

II IIIa IIIb IIIc

Although Kekulé's formulation was consistent with many of the experimental observations, it did not totally solve the problem and was contested for many years because chemists could not reconcile a formula that indicated the presence of three double bonds with the fact that benzene is inert in comparison to alkenes.

We have indicated that substitution is the characteristic reaction of benzene. However, benzene will undergo certain addition reactions which in themselves indicate that it is unsaturated in the same sense that alkenes are unsaturated. For example, in the presence of a platinum catalyst and at high pressure, benzene undergoes addition of three moles of hydrogen to form cyclohexane.

$$C_6H_6 + 3H_2 \xrightarrow[\text{high pressure}]{\text{Pt}}$$

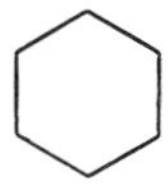

cyclohexane

Further, when a mixture of benzene and chlorine is irradiated with light, benzene adds three moles of chlorine to produce 1,2,3,4,5,6-hexachlorocyclohexane, which is more commonly known as benzene hexachloride (BHC). (See Problem 2.10 for the structural formula of the γ-isomer of BHC, a potent insecticide.)

$$C_6H_6 + 3Cl_2 \xrightarrow{\text{light}}$$

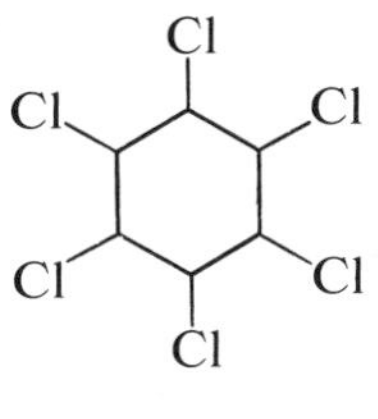

1,2,3,4,5,6-hexachlorocyclohexane

Obviously benzene does show certain of the reactions that we might expect of an alkene. However, it does so under conditions quite different from those typical of alkenes.

The resonance theory of Linus Pauling provided the first adequate description of the structure and unusual reactivities of benzene. According to the resonance theory, when a substance can have two or more equivalent or nearly equivalent structures that are interconvertible simply by the redistribution of valence electrons, the actual molecule does not conform to any one of the contributing structures, but exists as a resonance hybrid of them all. In this formulation, resonance is indicated by the double-headed arrow. Present-day organic chemists abbreviate these contributing structures by leaving out the hydrogen atoms.

Resonance seems an unfortunate term since it suggests a shift back and forth between two or more structures. There is no such shift. The molecule has one and only one real structure. For the remainder of this text, we shall use a Kekulé structure to represent benzene. We shall, of course, understand that it is the resonance hybrid that is intended.

A consequence of resonance is a marked increase in stability of the hybrid over that of any contributing structure. If the contributing structures are equivalent or nearly so, then resonance will be very important. Resonance stabilization is particularly large in the case of benzene and other aromatic hydrocarbons. A benzene ring is so inert in comparison to alkenes that compounds like allylbenzene, which contain both a benzene ring and an alkene, can be put through reactions such as addition of bromine or oxidation or hydrogenation without alteration of the benzene ring.

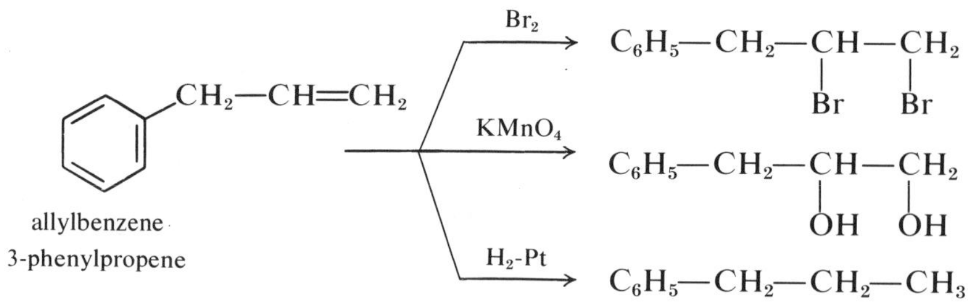

$CH_2—CH{=}CH_2$

allylbenzene
3-phenylpropene

Br_2 → $C_6H_5—CH_2—CH(Br)—CH_2(Br)$

$KMnO_4$ → $C_6H_5—CH_2—CH(OH)—CH_2(OH)$

H_2-Pt → $C_6H_5—CH_2—CH_2—CH_3$

3.16 NOMENCLATURE OF AROMATIC HYDROCARBONS

In the naming of benzene derivatives, as with other classes of organic compounds, we encounter a blend of systematic and common names. As a matter of fact, the use of common names is particularly prevalent and there are many with which you should be familiar. For the IUPAC names, no new rules are necessary.

Some monosubstituted benzenes are named as derivatives of benzene itself.

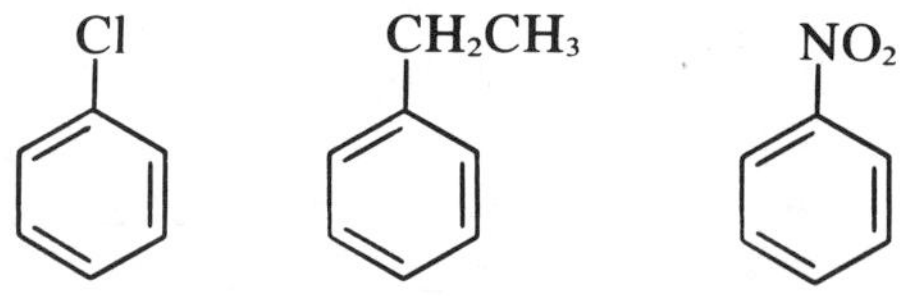

Cl — chlorobenzene

CH_2CH_3 — ethylbenzene

NO_2 — nitrobenzene

Others are known by common rather than IUPAC names, for example, toluene and styrene rather than methylbenzene and phenylethylene.

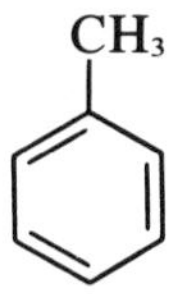

CH_3 | $CH{=}CH_2$ | COOH

toluene (methylbenzene) | styrene (phenylethylene) | benzoic acid

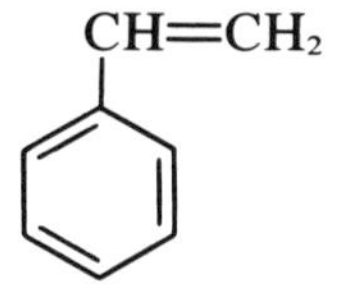

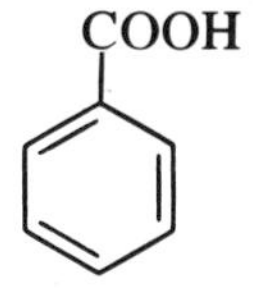

When there are two substituents on the ring, three structural isomers are possible. The substituents may be located by the numbering system 1,2-, 1,3- or 1,4- as used with the cycloalkanes, or alternatively they may be located by using an earlier system of prefixes, *ortho*- (*o*-), *meta*- (*m*-), or *para*- (*p*-). As shown below, the parent name for dimethylbenzene is xylene.

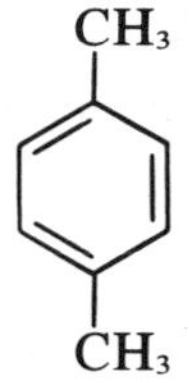

p-xylene 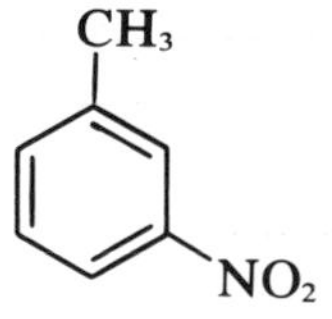 | *m*-nitrotoluene 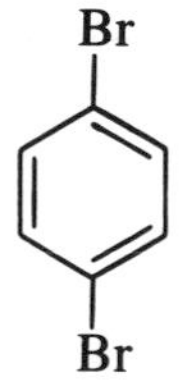 | *p*-dibromobenzene

With three or more substituents, a numbering system is used.

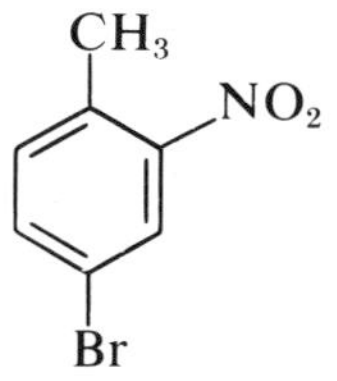

4-bromo-2-nitrotoluene | 2,4,6-trinitrotoluene (TNT)

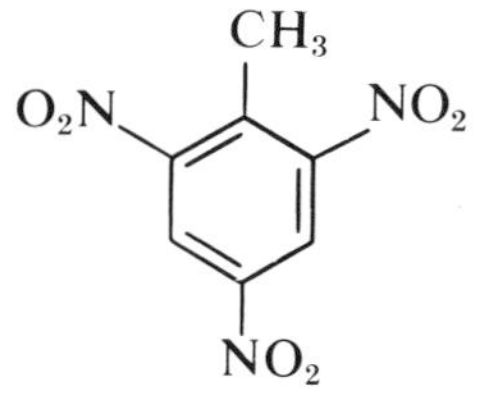

Alkyl benzenes can be named in two ways, either as derivatives of benzene itself or as substituted alkanes. When benzene derivatives are named as substituted alkanes, two names are particularly common.

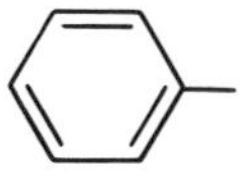

phenyl group 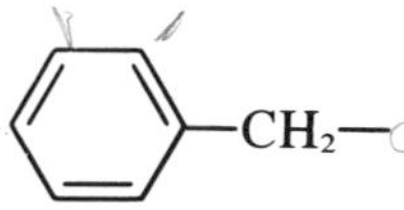 | benzyl group ($-CH_2-$)

The hydrocarbon group C_6H_5— is called a phenyl group and is sometimes abbreviated by the symbol ϕ. The hydrocarbon group $C_6H_5CH_2$— is called a benzyl group. While it might seem logical to call C_6H_5—a benzyl group (by the same reasoning that an alkane becomes an alkyl group), the name benzyl had already been applied to the

$C_6H_5CH_2$— group. Rather than change this already well-established practice and thereby risk obvious confusion, the IUPAC meeting decided to let the established common names stand.

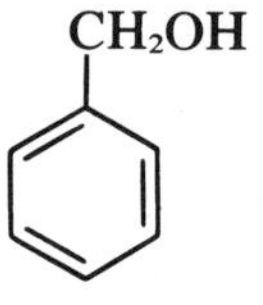

CH_2OH CH_2CH_2OH CH_2CO_2H

benzyl alcohol 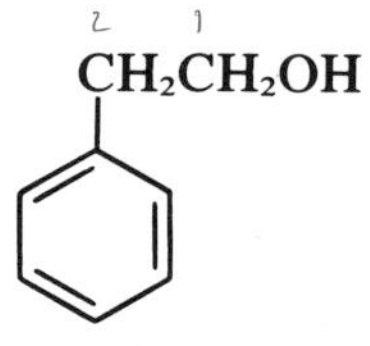 | 2-phenylethanol 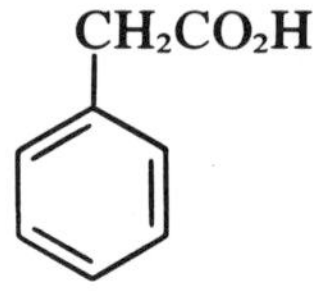 | phenylacetic acid

Closely related to benzene are numerous polynuclear aromatic hydrocarbons having one or more six-membered rings fused together. For each an IUPAC numbering system is used to locate substituents.

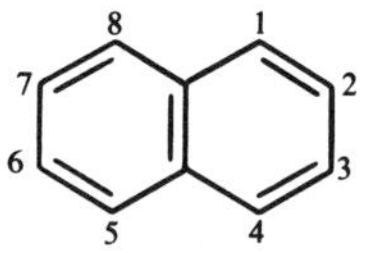

naphthalene, mp 80°C | anthracene, mp 217°C | phenanthrene, mp 99°C

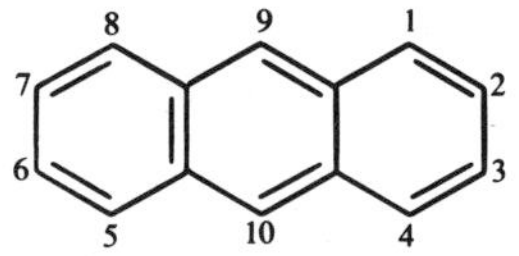

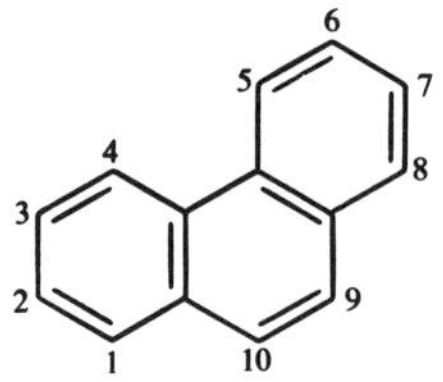

Compounds that contain an hydroxyl group on a benzene ring are known as phenols. The structure of phenol itself is shown below. Other phenols are named either as derivatives of the parent hydrocarbon or by common names.

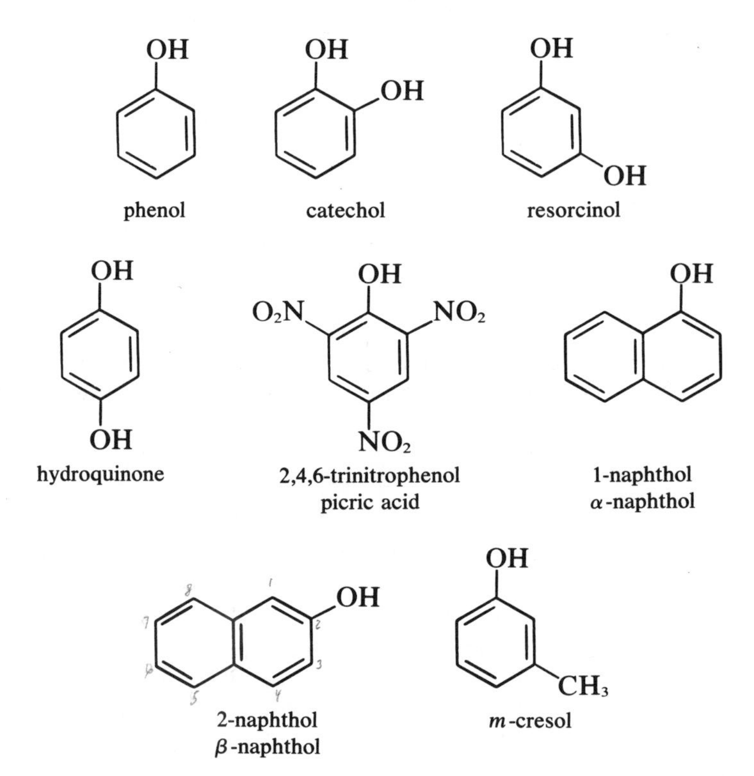

phenol | catechol | resorcinol

hydroquinone | 2,4,6-trinitrophenol picric acid | 1-naphthol α-naphthol

2-naphthol β-naphthol | *m*-cresol

Compounds that contain the —NH_2 group on a benzene ring are known as aromatic amines and are usually named as derivatives of aniline. Many of the simple substitution derivatives are also known by common names, as for example anisidine (methoxyaniline) and toluidine (methylaniline).

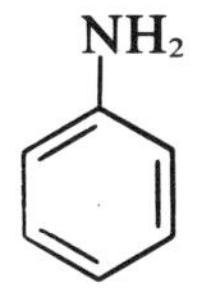

NH_2 — aniline

NH_2, OCH_3 — *p*-methoxyaniline, *p*-anisidine

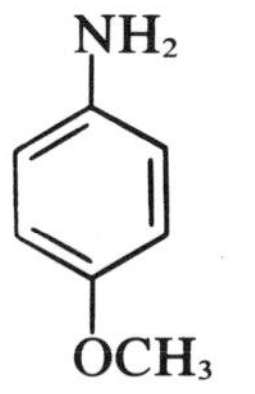

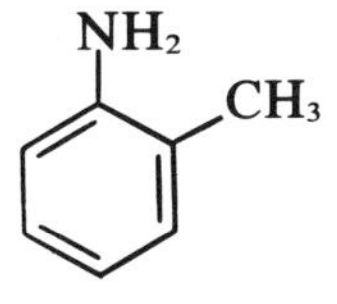

NH_2, CH_3 — *o*-methylaniline, *o*-toluidine

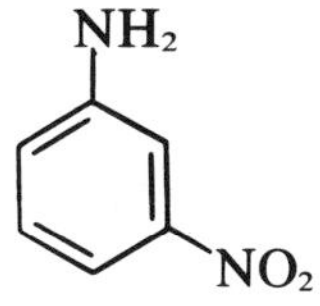

NH_2, NO_2 — *m*-nitroaniline

Substituents on the amino nitrogen are denoted as prefixes and are preceded by the letter N.

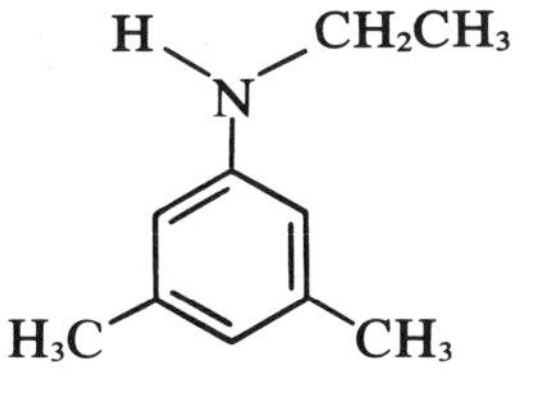

H, CH_2CH_3, N, H_3C, CH_3 — N-ethyl-3,5-dimethylaniline

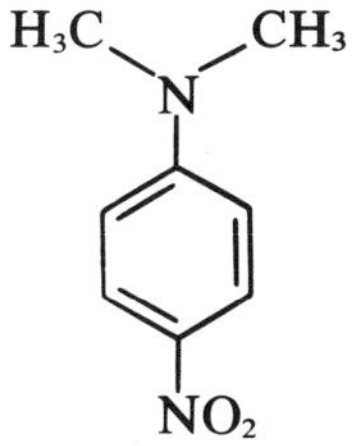

H_3C, CH_3, N, NO_2 — N,N-dimethyl-*p*-nitroaniline

Problems **3.20** Name the following compounds by using the IUPAC system.

a. NO_2, Cl

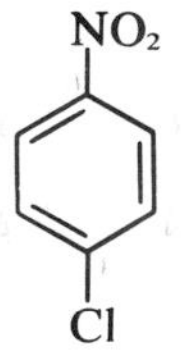

b. CH_3, Br

c. CO_2H, F

d. NH_2, CH_3

e. $NHCH_2CH_2CH_3$, CH_3

f. OH, CH_2CHCH_3 (CH_3), CH_2CH_3

g. NO_2, O_2N, NO_2

h. $CH_3CHCH{=}CHCH_3$

i. CH_3CHCH_3, $CH_3—C—CH_3$, CH_3

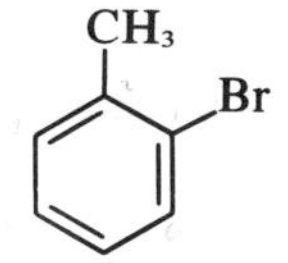

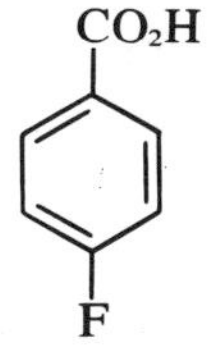

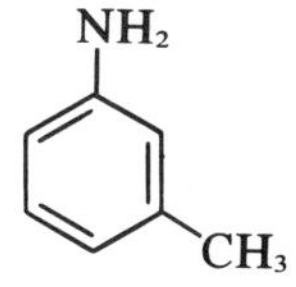

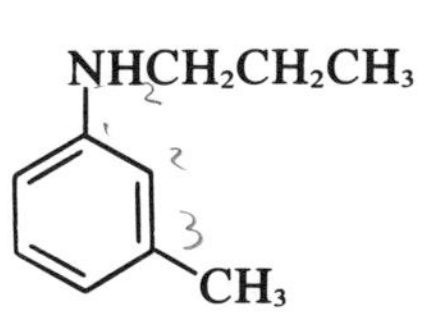

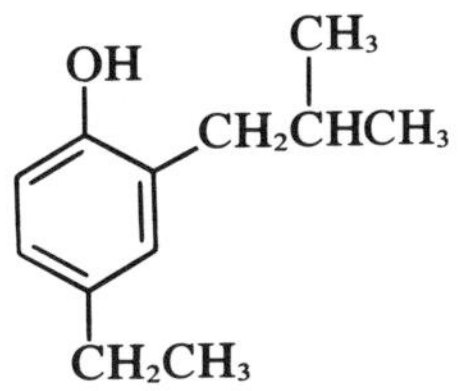

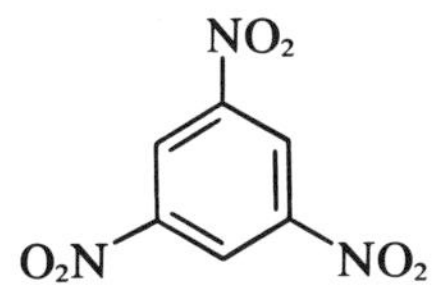

j. OH

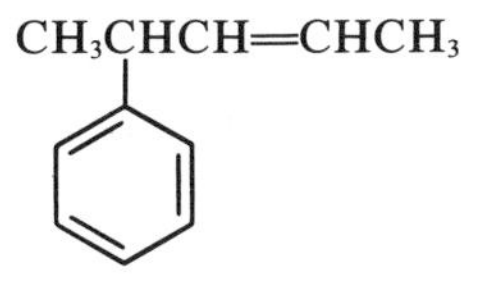

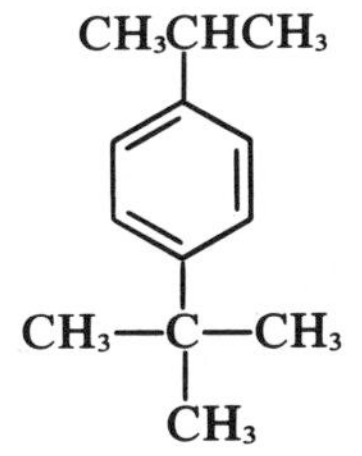

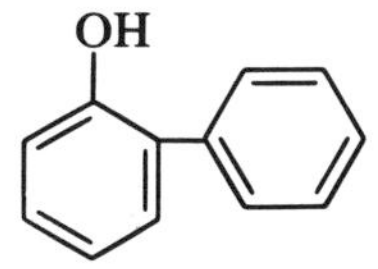

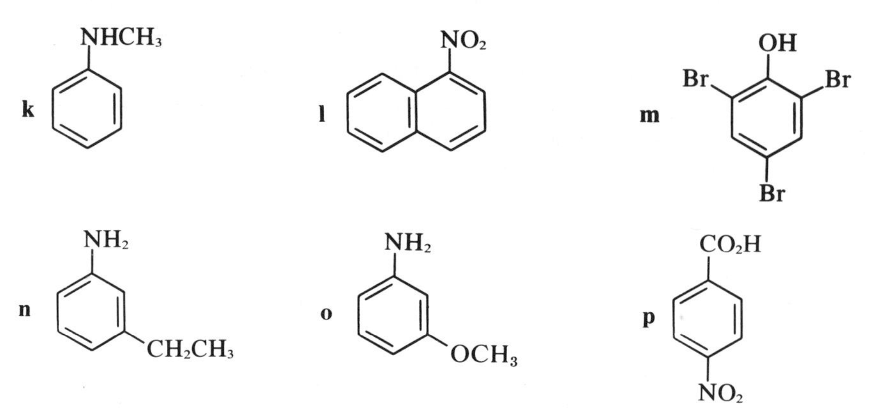

3.21 Draw structural formulas for the following compounds.

a *m*-dibromobenzene
b *p*-aminophenol
c *p*-chloroiodobenzene
d 2-hydroxy-4-isopropyltoluene
e phenol
f benzyl alcohol
g *m*-xylene
h *p*-cresol
i 2-ethylnaphthalene
j *o*-chlorophenol
k *p*-diiodobenzene
l 2-phenyl-2-pentene
m N-methylaniline
n N,N-dimethylaniline
o anthracene
p isopropylbenzene (cumene)

3.22 Name and draw structural formulas for all derivatives of benzene having the following molecular formulas.

a $C_6H_3Br_3$
b C_8H_{10}
c C_8H_9Cl
d C_9H_{12}

3.23 Draw structural formulas for the principal product(s) of the following reactions. Where you predict no reaction, write N.R.

a benzene + $Br_2 \xrightarrow{\text{room temp.}}$

b benzene + $Br_2 \xrightarrow{FeBr_3}$

c cyclopentene + $Br_2 \xrightarrow{\text{dark, room temp.}}$

d cyclopentane + $Br_2 \xrightarrow{\text{dark, room temp.}}$

e cyclopentane + $Br_2 \xrightarrow{250\text{–}400°C}$

f benzene + HBr $\longrightarrow$

g cyclohexene + HBr $\longrightarrow$

h 2-phenyl-2-pentene + $KMnO_4 \xrightarrow{H_2O,\ OH^-}$

i styrene + $H_2O \xrightarrow{H_2SO_4}$

3.24 Below is written a structural formula for a section of polypropylene derived from three units of propylene monomer.

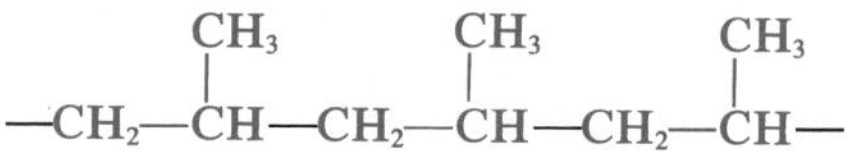

$$\text{—}CH_2\text{—}\underset{|}{\overset{CH_3}{CH}}\text{—}CH_2\text{—}\overset{CH_3}{CH}\text{—}CH_2\text{—}\overset{CH_3}{CH}\text{—}$$

polypropylene

Draw structural formulas for comparable sections of:

a polyvinyl chloride
b Saran
c Teflon
d Orlon
e Styrofoam
f Plexiglas

MINI-ESSAY

THE CHEMISTRY OF VISION

In order for us to see, light must interact with some part of our eye. This interaction involves absorption of light by molecules called visual pigments, located in the retina of the eye. The first recorded observations on visual pigments date from 1877 when the German physiologist Franz Boll demonstrated that the red color of a frog's eye is bleached to yellow by strong light and, if the frog is kept in the dark, the red color slowly returns. This red pigment, found in the retina of many animals including man, was named rhodopsin.

The primary process in vision is the absorption of light by rhodopsin, and in the process rhodopsin changes state. If enough rhodopsin molecules change state, these changes are somehow signaled to the rest of our visual system and we say that we see light. After visual pigment has been in a changed state for a very short time, it is almost incapable of absorbing more light. Therefore, if a flash of light is so intense that it changes the states of almost all rhodopsin molecules, we are virtually insensitive to a second flash. Yet the eye does regain sensitivity and, as it does so, the rhodopsin color returns. Maximum sensitivity is regained after about 40 minutes in the dark. The eye in this state of maximum sensitivity is said to be dark-adapted.

We look at only four aspects of the chemistry of vision: (1) what is the structure of rhodopsin; (2) how does it interact with light and change state; (3) how many rhodopsin molecules must undergo changes in state for us to see a flash of short duration; and (4) how does an insensitive (bleached) visual pigment molecule return to the sensitive state?

The eye contains two types of photoreceptors: rods, the agents of vision in dim light, and cones, the agents of color vision in bright light. In the following discussion we deal only with vision mediated by rods. Although the same principles apply to vision mediated by the cones, the details are not as well understood.

The chemistry underlying visual perception was largely unknown until the problem was taken up by George Wald in 1933, first as a young postdoctoral student in Germany and later as professor of biology at Harvard University. Our present understanding of the chemistry of vision stems directly from his brilliant and pioneering research.

Each of the several million rhodopsin molecules per rod consists of two parts: a colorless protein called opsin; and 11-*cis* retinal, a molecule derived biochemically from vitamin A by isomerization of the double bond between carbons 11 and 12 to the *cis* configuration, followed by oxidation of the primary alcohol to an aldehyde. The structures of vitamin A and 11-*cis* retinal are shown in Figure 1. Unfortunately we know almost nothing of the structure of the protein opsin or how the much smaller molecule of 11-*cis* retinal associates with it.

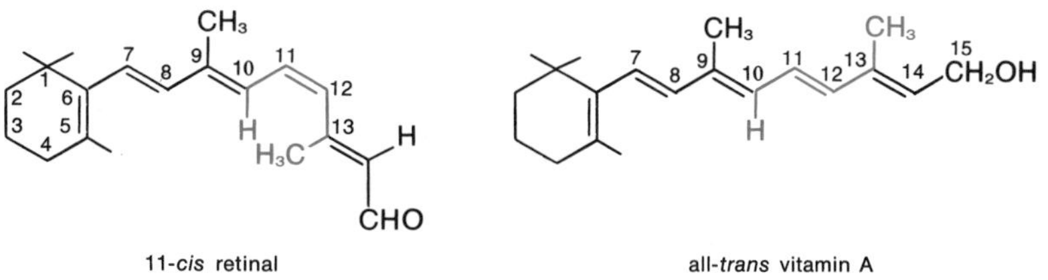

11-*cis* retinal

all-*trans* vitamin A

Figure 1 *11-cis retinal, one component of rhodopsin, and vitamin A, its biochemical precursor.*

How does rhodopsin interact with light and what changes does it undergo? The primary photochemical event itself is quite simple—the isomerization of 11-*cis* retinal to all-*trans* retinal. When a quantum of light is absorbed by rhodopsin, some of the energy causes a partial rupture of the double bond between carbons 11 and 12 by unpairing the two electrons of the pi bond. With the pi bond broken, the bonding has more the character of a single bond and the carbon atoms can rotate more freely. Rotation of one carbon by 180° followed by re-formation of the pi bond gives the more stable all-*trans* retinal. The only function of light in the whole of the visual process is in the catalysis of this isomerization. All other stages—further chemical changes, nerve excitation, etc.—follow from this first step.

According to the model developed by Wald, the isomerization of 11-*cis* retinal to all-*trans* retinal is followed by dissociation of rhodopsin into opsin and all-*trans* retinal. An examination of models suggests why this might happen. All-*trans* retinal is essentially a planar molecule, that is, carbon atoms 7 through 15 all lie in one plane (Figure 2). 11-*Cis* retinal cannot assume a planar conformation without a considerable degree of steric crowding between the $—CH_3$ group on carbon 13 and the —H atom on carbon 10 (Figure 1). This steric crowding is avoided by rotation about the single bond between carbons 10–11. Consequently 11-*cis* retinal has a side chain consisting of two planar sections, one between carbons 7 through 10, the second between carbons 11 through 15.

Due to differences in molecular shapes, all-*trans* retinal interacts with opsin less strongly than does 11-*cis* retinal and the complex dissociates. This isomerization and dissociation somehow produces changes in activity of retinal neurons and visual information is transmitted to the brain. Unfortunately, we do not understand the process by which this nerve excitation occurs. The all-*trans* retinal is enzymatically isomerized to 11-*cis* retinal to complete a sequence of reactions called the visual cycle.

The eye, or more precisely the rhodopsin molecule, is remarkably sensitive compared to most light-sensing devices. A single quantum of light absorbed by one of the several million rhodopsin molecules per rod will activate the rod. Activation of about 10 rods near each other is summed by the visual system so that a person with normal vision will report seeing a flash of short duration. This is an extraordinarily small amount of light. A typical lighted flashlight bulb radiates about 2×10^{15} quanta per millisecond.

The sensitivity of rods depends on a continuing supply of retinal,

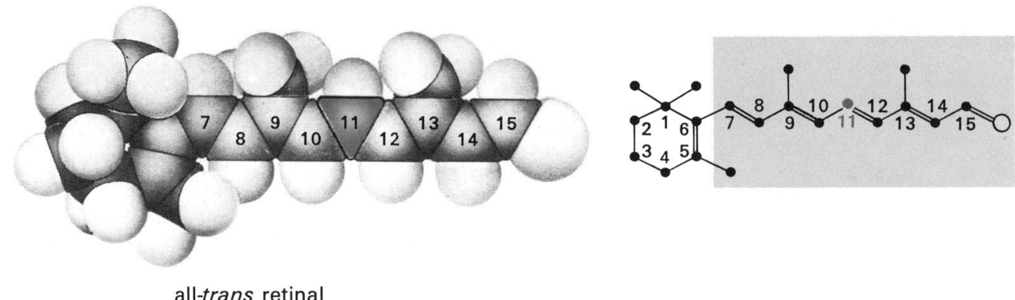

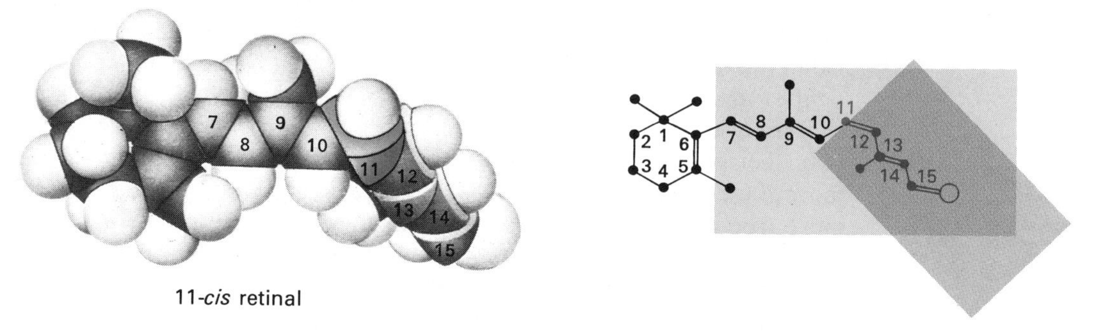

Figure 2 *Space-filling and line representations of all-trans retinal and 11-cis retinal. In all-trans retinal the atoms of the side chain (carbons 7–15) lie in one planar section. In 11-cis retinal the atoms of the side chain lie in two planar sections (the first between carbons 7–10, the second between carbons 11–15). The planar sections are indicated by the shaded background panels (Adapted from "Molecular Isomers in Vision" by Ruth Hubbard and Allen Kropf. Copyright © 1967 by Scientific American, Inc. All rights reserved.)*

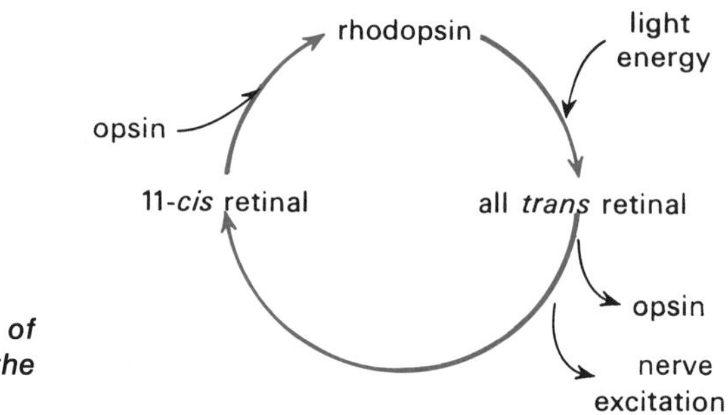

Figure 3 *The visual cycle of rods. Similar cycles occur in the cones.*

which in turn depends on a supply of vitamin A in the diet. In man an early symptom of vitamin A deficiency is "night blindness," a condition in which rods, the agents of vision in dim light, lose their sensitivity.

References

Case J., *Sensory Mechanisms* (The MacMillan Company, New York, 1966), Chap. 6.

Hubbard, R., and Kropf, A., *Bio-Organic Chemistry*, M. Calvin and M. Jorgenson, Editors (W. H. Freeman and Company, San Francisco, 1968).

MINI-
ESSAY

DDT

The middle decades of the 20th century have witnessed a revolution in the field of pest control, a revolution triggered during World War II by the discovery of the insecticidal properties of DDT. This synthetic organic material is undoubtedly the best known, the cheapest, and the most astonishingly effective of all of the so-called second-generation pesticides.

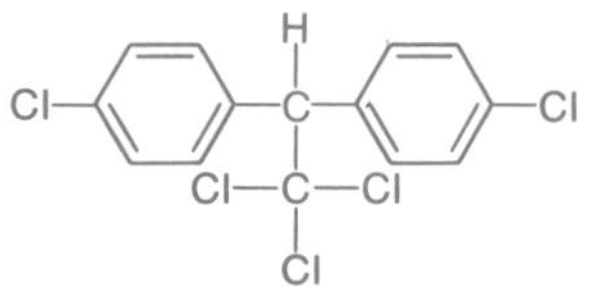

Figure 1 *DDT (dichlorodiphenyltrichloroethane).*

DDT is a polychlorinated hydrocarbon which derives its name from an early and not altogether correct name of *d*ichloro*d*iphenyl*t*richloroethane. DDT is essentially insoluble in water. Further, it has an extremely low vapor pressure and is very stable under a variety of conditions. It is estimated that in moist, fertile soil, 80–90% of DDT residue is lost within ten years. However in dry, nonfertile soils, up to 50% of DDT residues may be present for ten or more years.

Although DDT was first made in 1874 by Othmar Zeidler, a German chemist working on his doctoral dissertation, its effectiveness as an insecticide was not recognized until the mid-1930s. In search of a moth preventative, Paul Mueller, a chemist for the Swiss firm of J. R. Geigy, A.G., synthesized a series of polychlorinated hydrocarbons. Zeidler's DDT proved to be extremely effective and it was patented in 1942. Almost at once it became the "miracle insecticide." Its first large-scale success was in combating a massive typhus epidemic that broke out in Naples, Italy, after the city was liberated by the Allied armed forces. Within a few days after dusting the inhabitants themselves and spraying the area liberally with DDT, the epidemic was broken. This was a spectacular accomplishment for the new and virtually unknown insecticide. Spraying of DDT was used later in the South Pacific to control malaria. In 1948 Paul Mueller was awarded the Nobel Prize for his pioneering work on the insecticidal properties of DDT.

We now know that DDT is effective against a wide variety of agricultural pests and insects that transmit disease. These include the *Anopheles* mosquito, which transmits malaria; the body louse, which transmits typhus; and the tsetse fly, which transmits sleeping sickness. Along with DDT the most widely known and used polychlorinated hydrocarbon insecticides are Aldrin, Dieldrin, Lindane (benzene hexachloride), and Chlordane.

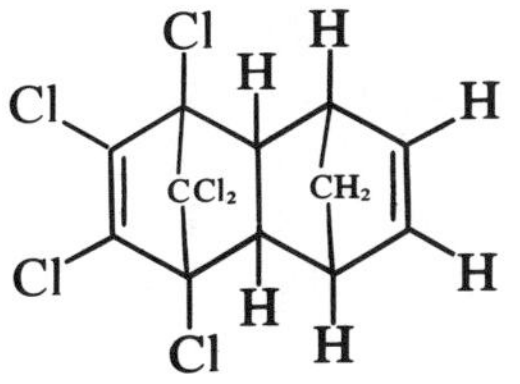

Dieldrin ($C_{12}H_8Cl_6$)

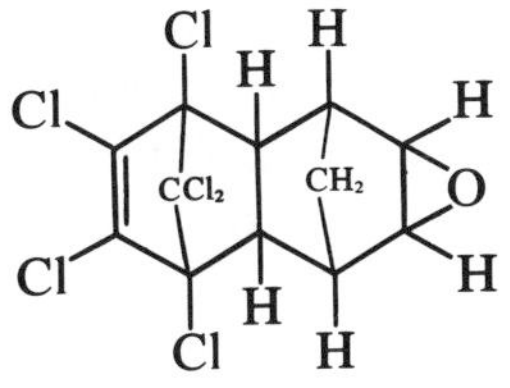

Aldrin ($C_{12}H_8Cl_6O$)

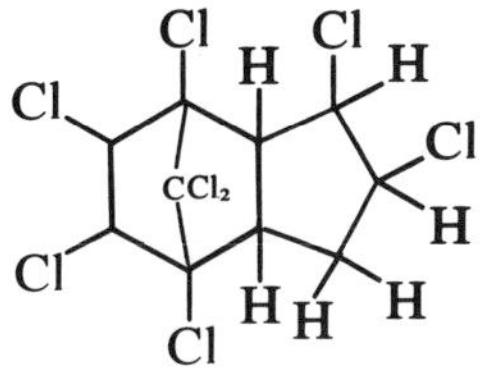

Chlordane ($C_{10}H_6Cl_8$)

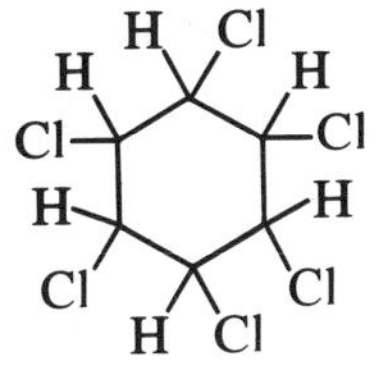

Lindane ($C_6H_6Cl_6$)

The symptoms of DDT poisoning are essentially the same in most insects. First there are tremors and jitters followed by a slow loss of motion, and then paralysis and death. Although the mechanism by which DDT accomplishes this is not totally clear, the evidence so far suggests that the primary effects are almost entirely on the nervous system.

By 1947 the first reports came in, from Italy and Sweden, that DDT was becoming less effective in the control of the housefly. By 1948, the list of resistant insects had grown to twelve. In 1965 tests at Texas A & M University showed that cotton bollworms were 30,000 times more resistant to DDT than they were in 1960. Of all the cases of resistance to insecticides, that of resistance to DDT is probably the best known. It was the first to be demonstrated on a large scale and also the first to be at least partially understood. Resistant insects synthesize an enzyme that catalyzes the dehydrochlorination of DDT and produces dichlorodiphenyldichloroethylene (DDE), a compound with no insecticide activity. The enzyme that catalyzes this reaction is appropriately named DDT-dehydrochlorinase. It has been isolated and purified and has been shown to be a globular protein. It should be pointed out that resistant as well as sensitive insects are able to synthesize this enzyme, but in greatly different amounts.

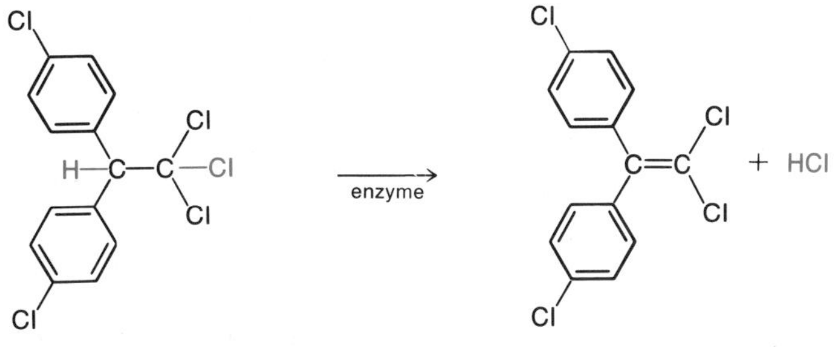

DDT $\xrightarrow{\text{enzyme}}$ DDE + HCl

Figure 2 *Deactivation of DDT by dehydrochlorination catalyzed by DDT-dehydrochlorinase.*

Soon after the appearance of insect resistance, another problem with DDT became apparent. As already noted, DDT is essentially insoluble in water, a polar solvent. However, because DDT has structural characteristics like those of the hydrocarbons, it is very soluble in nonpolar media. For this reason DDT tends to concentrate in fats and oils and in adipose tissue, the site of fat storage in the body. There is now clear evidence of increased DDT concentrations in fat and adipose tissue along food chains—the higher the link on the chain, the greater the concentration. Two well documented examples of concentration in food chains are:

DDT ⟶ lake ⟶ plankton ⟶ fish ⟶ grebe

DDT ⟶ leaf ⟶ earthworm ⟶ robin

In 1957 California's Clear Lake was sprayed with DDT for gnat control. Lake water, after spraying, contained about 0.02 ppm DDT. After a time, lake plankton were found to contain 5 ppm DDT and fish feeding on these microorganisms contained up to 2000 ppm. This is a biological magnification or concentration from water to lake fish by a factor of 100,000. Grebes, diving birds that feed on the lake fish, subsequently died in large numbers.

Oysters seem particularly effective in concentrating DDT from their environment. After only seven days in water containing 10 ppb (parts per billion), East Coast oysters were found to contain 151 ppm DDT, a biological magnification of 15,100. In birds, the presence of DDT seems to interfere with calcium metabolism, which in turn causes egg shells to be thinner and weaker to the point where they are too thin to survive incubation.

In a short essay such as this, it is possible to outline only a bit of the history of DDT. There is now a great deal of very careful research being done on the biological metabolism of DDT; its persistance in plants, animals, and soils; its toxicity and effects on metabolism, reproductive cycles, etc. Irma West, writing in *Organic Pesticides in the Environment,* presents the following perspective on the development and use of DDT.

> There is no question of the great need for effective pest control. There is no question of the immediate efficiency and economic value of modern pesticides in producing food and fiber and in controlling vector-borne disease. There is also no question about the ease of using hindsight to comment on the pesticide problems compared with the difficulties of exercising foresight a decade or more ago in predicting the problems. The problems which have arisen with modern pesticides stem from their ability to do much more than is expected or desired of them. The fact that stable pesticides contaminate, accumulate, and move about in the environment has taken time to be realized. Neither the chemical nor the environment in which it is applied is a simple arrangement. Even the successful immediate control of a target pest can be diminished by a chain of events where the pesticide eventually causes an increase in the pest. The pace of application has long ago exceeded our ability to investigate and comprehend the ultimate results. Regardless of whether

we do or do not escape various small and large disasters potential to this kind of adventure, proceeding so far ahead of understanding is not in the best tradition of science.

References *Cleaning Our Environment—The Chemical Basis for Action*, Section 4 (American Chemical Society, Washington, D.C., 1969).

Giddings, J. C., and Monroe, M. B., Editors, *Our Chemical Environment* (Harper & Row, Publishers, New York, 1972).

Organic Pesticides in the Environment, Advances in Chemistry Series, 60 (American Chemical Society, Washington, D.C., 1966).

4

STEREOISOMERISM AND OPTICAL ACTIVITY

4.1 INTRODUCTION

All compounds that have the same structural formula but different orientations of the atoms in space are grouped collectively under the classification stereoisomers. Geometric isomerism is one type of stereoisomerism; conformational isomerism is a second type; and optical isomerism is still a third type of stereoisomerism. In this chapter we will deal with optical isomers known as enantiomers, diastereomers, and meso compounds.

We shall first examine how optical isomerism can be detected in the laboratory, the structural features giving rise to this type of isomerism, the intimate relationship between optical and geometric isomerism, and finally the significance of stereoisomerism in the biological world.

4.2 POLARIZATION OF LIGHT

The study of optical isomerism began in the early 1800s with the observation that most compounds isolated from natural sources are able to rotate the plane of polarized light, or to use the more common but rather arbitrary term, they are optically active. Ordinary light may be considered as made up of waves vibrating in all planes perpendicular to the direction of propagation. Light in which vibrations are occurring in only parallel planes is said to be plane polarized (Figure 4.1).

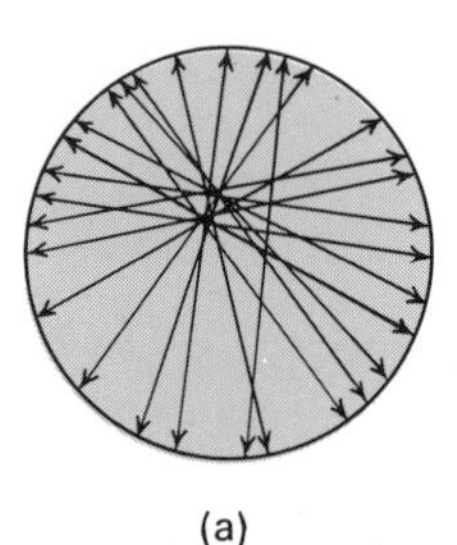

(a)

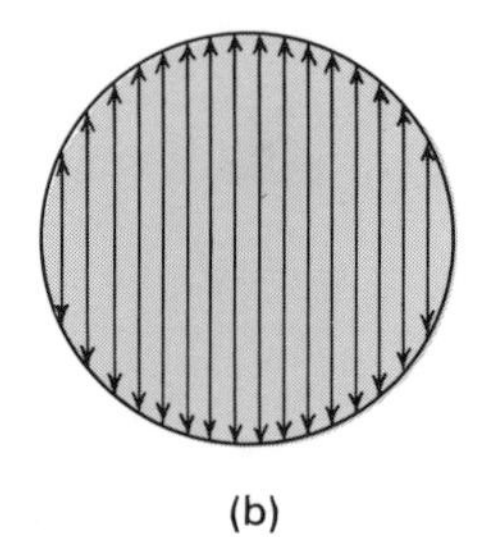

(b)

Figure 4.1 *Polarization of light. (a) Ordinary light vibrating in all planes coming toward the reader. (b) Plane-polarized light.*

There are several ways in which plane-polarized light may be obtained. One such separation is accomplished by means of a prism invented by William Nicol, who first observed that a properly prepared and oriented crystal of Iceland spar (a form of calcite, $CaCO_3$) will transmit waves vibrating in parallel planes. A second method makes use of Polaroid, invented by the American engineer, E. H. Land. Polaroid is made of a crystalline organic compound properly oriented and embedded in plastic.

Materials such as Iceland spar or Polaroid can be used not only to form plane-polarized light, but also to measure quantitatively the effects of optically active substances on the plane of polarized light. The instrument used for such quantitative measurements is the polarimeter.

4.3 THE POLARIMETER

A polarimeter consists essentially of two polarizing prisms such as the Nicol prisms or sheets of Polaroid. One is designated as the polarizer and the other is designated as the analyzer (Figure 4.2).

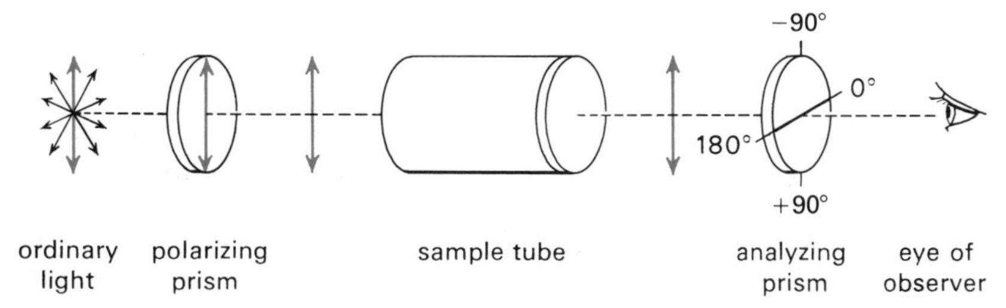

Figure 4.2 *Schematic diagram of a polarimeter, sample tube empty.*

If the sample tube in the polarimeter is empty, the intensity of the light reaching the observer will be a maximum when the polarizer and the analyzer are aligned and are passing light vibrating in the same plane. If the analyzer is turned, less light will be transmitted to the observer. A minimum of light will reach the observer when the analyzer is at right angles to the polarizer. At this point the field of view in the instrument will be dark. We take this as the zero point, or 0° on the scale. Now, if an optically active substance is placed in the sample tube, the plane of polarized light will be rotated by the substance, and a certain amount of light will now pass through the analyzer to the observer. Turning the analyzer clockwise or counterclockwise a few degrees will restore the dark field of view. The number of degrees, α, that the analyzer is turned will be equal to the number of degrees that the optically active substance has rotated the plane of the polarized light. If the analyzer must be turned to the right or clockwise to restore the dark field, the substance is said to be dextrorotatory. If the analyzer must be turned counterclockwise to restore the dark field, then the substance is said to be levorotatory. In either case the substance is optically active. (See Figure 4.3.)

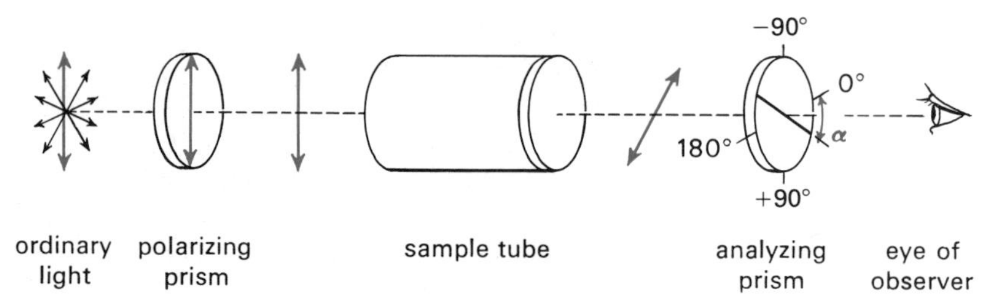

Figure 4.3 *Schematic diagram of a polarimeter with sample tube containing an optically active substance. To restore the dark field of view the analyzer has been rotated clockwise by α degrees.*

The term α is known as the observed rotation and is dependent on the structure and concentration of the compound, the length of the light path through the sample tube, the temperature, the wavelength of the light source, and the nature of the solvent. To facilitate tabulation and communication of data on optically active compounds, chemists have established a set of arbitrarily chosen standard conditions. This commonly reported rotation value, the specific rotation, is defined as the rotation caused by a substance at a concentration of 1 gram per cubic centimeter and in a sample tube of 1 decimeter length. Of course, the specific rotation still depends on the temperature and the wavelength of the light and these values must be reported.

$$\text{specific rotation} = [\alpha]^{\text{temperature}}_{\text{wavelength}} = \frac{\text{observed rotation (degrees)}}{\text{length (decimeters)} \times \text{concentration (g/ml)}}$$

The most commonly used wavelength is the D line of sodium, which is responsible for the yellow color of excited sodium vapor. In reporting either the observed or the specific rotation, it is common practice to indicate dextrorotatory by a positive sign (+) and levorotatory by a negative sign (−). Using these conventions, the specific rotation of sucrose (a common sugar) at 25°C using the D line of sodium as the light source is reported as

$$[\alpha]^{25°\text{C}}_{D} = +66.5\ (H_2O)$$

which indicates that under these standard conditions, an aqueous solution of sucrose rotates plane-polarized light by 66.5° in the clockwise direction (Figure 4.3).

Problem **4.1** A sample of an optically active sugar is prepared by dissolving 0.120 g in 100 ml of water. When this solution is placed in a 1-decimeter tube, the observed rotation is −0.49°. Calculate the specific rotation of this sugar. If the concentration of the sugar were doubled, what would be the observed rotation? the specific rotation?

4.4 STRUCTURE AND OPTICAL ACTIVITY

Now that we are able to detect and measure optical activity, the question naturally arises as to why certain compounds are optically active while others are not. Stated more specifically, what structural features will give rise to optical activity?

By the middle of the 19th century a number of optically active compounds had been isolated from natural sources, and their structures had been determined. Among the compounds known at that time were:

$$CH_3\overset{OH}{\overset{|}{C}}HCO_2H \qquad CH_3CH_2\overset{CH_3}{\overset{|}{C}}HCH_2OH \qquad HO_2C{-}CHOH{-}CH_2{-}CO_2H$$

lactic acid — "active" amyl alcohol (2-methyl-1-butanol) — malic acid

Lactic acid was one of the compounds most intensively investigated. Scheele (1780) discovered lactic acid in sour milk. Subsequently the same lactic acid was found to arise from the bacterial fermentation of milk sugar (lactose) and other naturally occurring sugars. The lactic acid made by fermentation, at least as originally obtained, was optically inactive. Berzelius (1807) discovered a similar acid substance could be extracted from muscle. It was soon established that this too was lactic acid, except for the very important difference that the muscle lactic acid was dextrorotatory, for which reason it is designated (+)-lactic acid. Both acids have the same structure and yet are different. Wislicenus in 1873 concluded that "If molecules can be structurally identical and yet possess dissimilar properties, this difference can be explained only on the ground that it is due to a different arrangement of the atoms in space." By 1850 a great many optically active compounds had been isolated and their structures determined, and yet up to that time no satisfactory explanation for this phenomenon existed.

A solution to this problem was proposed in 1874, when van't Hoff and Le Bel simultaneously but quite independently put forward a bold hypothesis about the geometry of organic molecules. van't Hoff and Le Bel realized that if the four valances of a carbon atom were directed toward the corners of a regular tetrahedron, then four different groups attached to the carbon can assume two and only two different spatial arrangements. These two possible arrangements are related as a person's left and right hands are related, by reflection. One arrangement is the mirror image of the other. Molecules that are related in this way are said to be chiral (from the Greek *cheir,* hand). In such a molecule, the carbon that has four different groups attached to it is called an asymmetric carbon.

Consider for example the arrangement of the four different groups around the carbon 2 of lactic acid (Figure 4.4). Examination of the perspective formulas of the two lactic acids shows that they are different; (a) is the mirror image of (b) and is not superimposable on

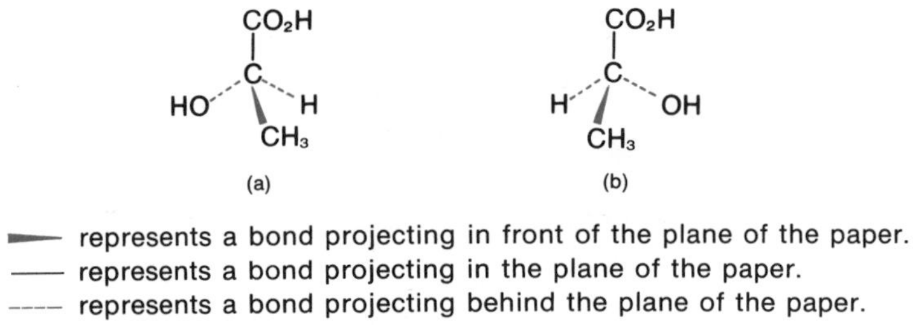

CO_2H / HO–C–H / CH_3 (a) CO_2H / H–C–OH / CH_3 (b)

▬ represents a bond projecting in front of the plane of the paper.
—— represents a bond projecting in the plane of the paper.
---- represents a bond projecting behind the plane of the paper.

Figure 4.4 *Perspective formulas of lactic acid.*

(b). Compounds that are related in this way are termed enantiomers (Greek *enantio*, opposite + *meros*, part). In general terms enantiomers are mirror images that are not superimposable.

Many of the properties of enantiomers are identical: they have the same melting points, the same boiling points, the same solubilities in various solvents. Yet they are isomers and we can expect them to show some differences in their properties. One important difference is optical activity. One member of a pair of enantiomers will be dextrorotatory and the other member will be levorotatory. For each, the absolute magnitude of the optical activity will be the same, but the rotations will differ in sign. We shall see other differences in enantiomeric pairs later. For example, one enantiomer may taste different from the other, or smell different. One enantiomer may be physiologically active whereas the other is physiologically inactive. We will suggest why this might be later in the chapter.

It is a necessary and sufficient condition for enantiomerism that a compound and its mirror image be nonsuperimposable. Conversely, if a compound and its mirror image are superimposable, then the two are identical and the compound will not show enantiomerism. Consider for example the amino acid, glycine (Figure 4.5).

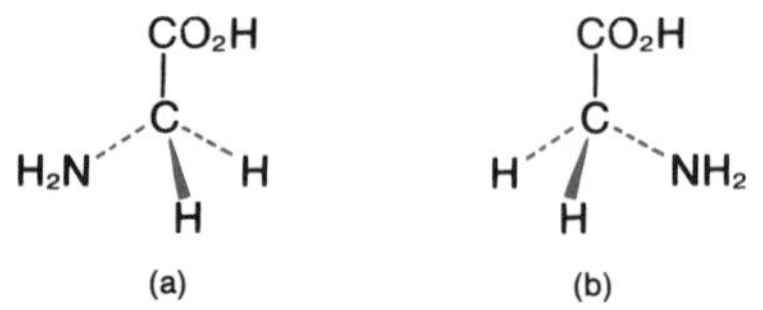

CO_2H / H_2N–C–H / H (a) CO_2H / H–C–NH_2 / H (b)

Figure 4.5 *Glycine. (a) and (b) are mirror images.*

In the perspective representations of glycine, (a) and (b) are mirror images. Are they superimposable? The answer is yes. To see this, simply rotate (b) by 120° around the C—CO_2H bond axis. It is now possible to place (b) directly on (a) and have all groups correspond. Therefore the mirror images are identical; (a) and (b) do not constitute a pair of enantiomers and glycine cannot show optical activity.

Let us emphasize again that every compound has a mirror image. **The key question is whether the mirror images are superimposable.**

Problems

4.2 Draw a stereorepresentation as in Figure 4.4 for each of the following compounds. Also draw its mirror image. Which mirror images are superimposable?

a $CH_3CH(OH)CH_2OH$ **b** $CH_3CH(OH)CH_3$ **c** $CH_3CH(OH)CH_2CH_3$

d $CH_3CH(NH_2)CO_2H$ **e** $CH_2(NH_2)CH_2CO_2H$ **f** $C_6H_5CH(NH_2)CH_3$

4.3 Drawn here are several stereorepresentations of lactic acid. Taking (a) as a reference structure, note that the other structures are stereorepresentations of lactic acid viewed from other perspectives. Which of the alternative representations are identical to (a) and which are mirror images of (a)?

(a) C with CO_2H (top), H, OH, CH_3 (bottom)
(b) C with CH_3 (top), HO, H, CO_2H (bottom)
(c) C with CO_2H (top), HO, CH_3, H (bottom)
(d) C with CH_3 (top), H, CO_2H, OH (bottom)
(e) C with CH_3 (top), H, OH, CO_2H (bottom)

4.5 MOLECULAR SYMMETRY

Any molecule that is superimposable on its mirror image is symmetrical in some way. There are three recognized types of molecular symmetry, but for our purposes we shall consider only one, the existence of a plane of symmetry. (The other types of molecular symmetry are a center of symmetry and a $4n$-fold alternating axis of symmetry.) A plane of symmetry is defined as a plane (often visualized as a mirror) cleaving a molecule in such a way that one side of the molecule is the mirror image of the other.

Inspection of the projection formula of glycine (Figure 4.6a) shows that it possesses a plane of symmetry running through the molecule on the axis of the C—C—N bonds. Alternatively we might rotate the molecule in space into a different projection formula (b), and see the plane of symmetry again running through the axis of the C—C—N bonds, this time oriented in a plane perpendicular to the plane of the paper.

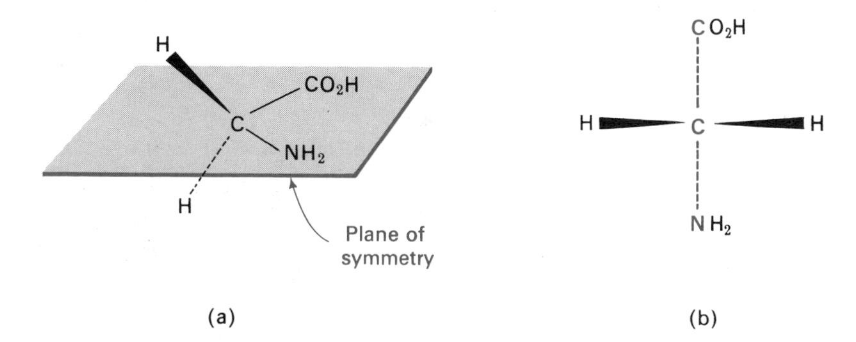

Figure 4.6 *Glycine. A plane of symmetry runs through the axis of the C—C—N bond.*

If you are interested in deciding whether a given structure is symmetrical (and therefore identical to its mirror image and optically inactive), it is often easier to look for a plane of symmetry than to construct a mirror image and compare it with the original.

Molecules that are superimposable on their mirror images are symmetric; molecules that are not superimposable on their mirror images are dissymmetric and can exist as enantiomers. Thus when we ask whether or not a given substance will show enantiomerism, we are in effect searching for the symmetry or dissymmetry of the molecule.

4.6 RACEMIC MIXTURES

Let us return to the earlier discussion of the isolation of two lactic acids, designated as (+)-lactic acid and optically inactive lactic acid. We have already demonstrated that lactic acid and its mirror image are not superimposable, and that lactic acid will show optical isomerism. How is it that one of the lactic acids is optically inactive while the other will show optical isomerism? The answer is straightforward and can easily be demonstrated in the laboratory (Section 4.9). The inactive form of lactic acid is an equal mixture of (+)-lactic acid and (−)-lactic acid. Because the mixture contains equal numbers of molecules that rotate the plane of polarized light to the right and to the left, the mixture does not rotate the plane of polarized light and is optically inactive. Such a mixture containing equal amounts of a pair of enantiomers is called a racemate, a racemic mixture, or simply a ±-mixture.

Problems

4.4 Which will show optical isomerism?

a $CH_2{=}CH{-}CH(OH){-}CH_3$

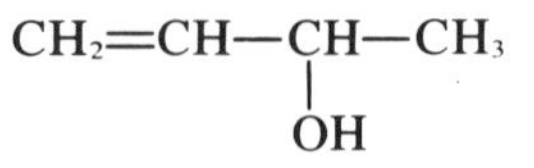

b $CH_2{=}CH{-}CH(OH){-}CH_2{-}CH_3$

c $CH_3{-}CH_2{-}CH(OH){-}CH_2{-}CH_3$

d $CH_2(OH){-}CH_2{-}CH(OH){-}CH_2{-}CH_3$

e $HO{-}C(CH_2CO_2H)_2{-}CO_2H$ (central C bearing CH_2CO_2H above and below, HO and CO_2H to the sides)

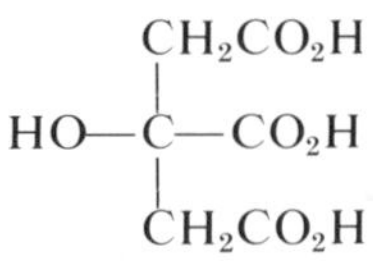

citric acid

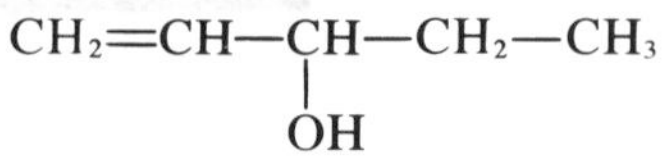

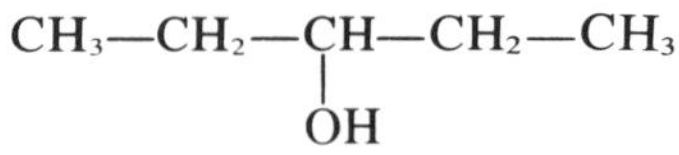

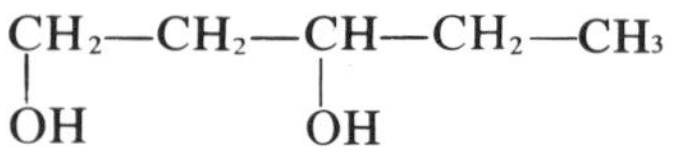

f $H{-}C(CHO)(CH_2OH){-}OH$ (CHO above, CH_2OH below)

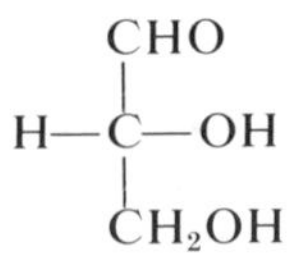

glyceraldehyde

4.5 Draw the structural formula of at least one alkene of molecular formula C_5H_9Br that will show:

a neither geometric nor optical isomerism
b geometric but not optical isomerism
c optical but not geometric isomerism
d both geometric and optical isomerism.

4.7 MULTIPLE CENTERS OF ASYMMETRY

Compounds that contain two or more centers of asymmetry can exist in more than two stereoisomeric modifications. Each additional asymmetric carbon doubles the number of possible stereoisomers. In fact, it can easily be shown that the total number of optical isomers is 2^n, where n is the number of different asymmetric carbon atoms in the structure. Actually this number is the maximum number of optical isomers for any molecule of n centers of asymmetry. Many molecules with n asymmetric carbons have fewer than 2^n optical isomers if two or more of the asymmetric carbons are similar, or if ring formation reduces the number of configurations capable of existence.

As an example of a molecule with two different centers of asymmetry, consider 2,3,4-trihydroxybutanal.

$$\underset{\text{OH}}{\text{CH}_2}-\underset{\text{OH}}{\overset{*}{\text{C}}\text{H}}-\underset{\text{OH}}{\overset{*}{\text{C}}\text{H}}-\text{CHO}$$

The two asymmetric carbons are marked by asterisks. Thus four stereoisomers can be written as given in Figure 4.7, each with a unique configuration. Inspection of the formulas shows that structures I and II are mirror images and therefore a pair of enantiomers. Likewise III and IV are mirror images and represent a second pair of enantiomers. Taken in equal amounts, I and II form one racemate, and III and IV form another racemate.

I	II	III	IV
CHO	CHO	CHO	CHO
H—C—OH	HO—C—H	HO—C—H	H—C—OH
H—C—OH	HO—C—H	H—C—OH	HO—C—H
CH_2OH	CH_2OH	CH_2OH	CH_2OH
erythrose		threose	

Figure 4.7 Stereoisomers of a substance having two dissimilar asymmetric carbons. Trihydroxybutanal.

Neither I nor II is a mirror image of III or IV, nor for that matter are either I or II superimposable on III or IV. Stereoisomer I is a diastereomer of III and IV. Diastereomers are stereoisomers that are not mirror images of each other. Diastereomers have different chemical and physical properties. The diastereomers of trihydroxybutanal are designated by the common names erythrose and threose.

As we indicated earlier, the 2^n isomer number represents the maximum number of stereoisomers. Often molecules have special symmetry properties and the isomer number is reduced. Tartaric acid is a classic example of a molecule possessing two similar asymmetric carbon atoms. The 2^n rule would predict four optical isomers, while in fact only three are known (Figure 4.8).

Structures V and VI are nonsuperimposable mirror images and constitute a pair of enantiomers. VII and VIII are also mirror images but they are superimposable; they represent the same compound. Note

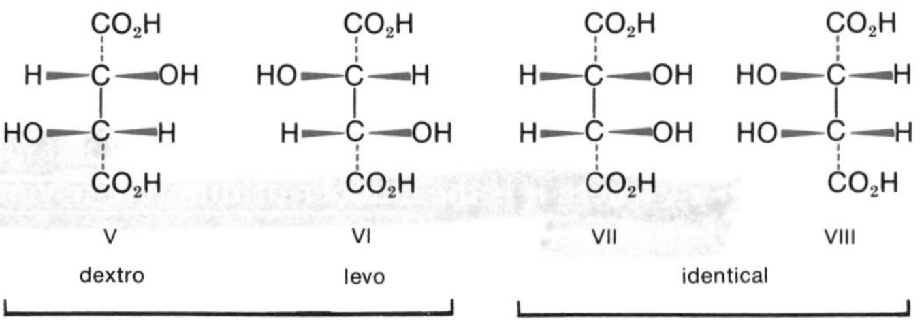

	V	VI	VII	VIII
top	CO_2H	CO_2H	CO_2H	CO_2H
C-2	H—C—OH	HO—C—H	H—C—OH	HO—C—H
C-3	HO—C—H	H—C—OH	H—C—OH	HO—C—H
bottom	CO_2H	CO_2H	CO_2H	CO_2H
	dextro	levo	identical	
	racemate		meso	

***Figure 4.8** Stereoisomers of a substance having two similar asymmetric carbons. Tartaric acid.*

that by rotating VII by 180° in the plane of the paper, it can be superimposed on VIII. In accord with our earlier statement (p. 88) that any molecule superimposable on its mirror image must be symmetrical in some way, we see that there is a plane of symmetry perpendicularly bisecting the central carbon-carbon bond. Symmetrical substances that contain asymmetric carbon atoms are called meso compounds. Meso tartaric acid is a diastereomer of (−)-tartaric acid and of (+)-tartaric acid. See Table 4.1.

***Table 4.1** Physical properties of the tartaric acids.*

Acid	mp (°C)	$[\alpha]_D^{25°C}$
dextro	170	+12°
levo	170	−12°
racemic	206	inactive
meso	146	inactive

4.8 PREDICTING OPTICAL ISOMERISM

There are three methods that you can use when looking at a molecule to determine whether it will show optical isomerism. Since it is a necessary and sufficient condition for optical isomerism that a compound and its mirror image be nonsuperimposable, the most direct test is to build a model of the molecule and one of its mirror image. If these two models are superimposable, then the substance is symmetric and will not show optical isomerism. If they are nonsuperimposable, then the compound will show optical isomerism.

A second method is to look for a plane of symmetry. If the molecule has a plane of symmetry, then the mirror images are identical and the compound will not show optical isomerism.

Third, you can look for the presence or absence of an asymmetric carbon atom. Remember that an asymmetric carbon is one with four different groups attached to it. For each asymmetric carbon atom there will be a maximum of 2^n optical isomers. While looking for asymmetric carbon atoms is perhaps the easiest way to predict optical isomerism, you must remember that a compound may have two or more asymmetric carbon atoms and still have superimposable mirror images—these are called meso compounds; they have a plane of symmetry and are optically inactive.

Problems

4.6 Examine the following structures and mark each asymmetric carbon with an asterisk. Construct the mirror image of each molecule and decide which of the mirror images are superimposable. Verify for yourself that each molecule which is superimposable on its mirror image has a plane of symmetry, and that each molecule which is not superimposable on its mirror image has no plane of symmetry.

a CH_3 on the first CH; CH_3—CH—CH—CO_2H; NH_2 on the second CH

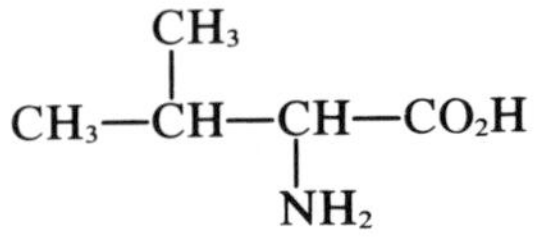

2-amino-3-methylbutanoic acid

b HO—CH_2—CH—CO_2H; NH_2 on CH

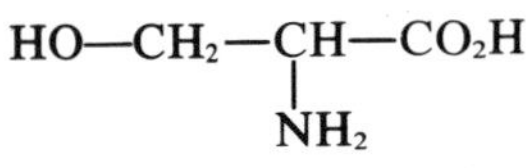

serine

c CO_2H (top); H—C—OH; CH_2OH (bottom)

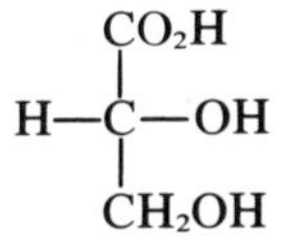

glyceric acid

d CH_2OH (top); H—C—OH; CH_2OH (bottom)

glycerol

e CH_3—C(=O)—CO_2H

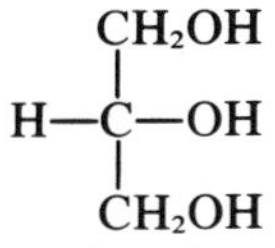

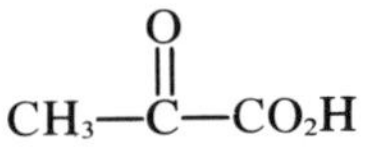

pyruvic acid

f H_3C and H on one carbon, CO_2H and CH_3 on the other: C=C

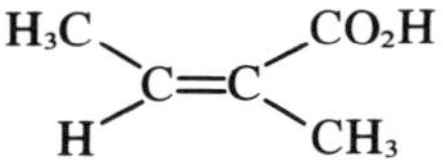

cis-2-methyl-2-butenoic acid

g CH_2—CO_2H (top); HO—C—CO_2H; CH_2—CO_2H (bottom)

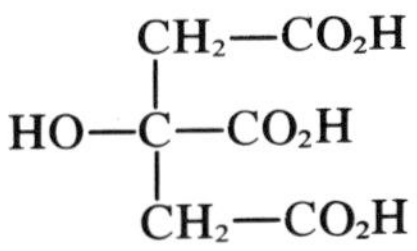

citric acid

h CH_3 on the first CH; CH_3—CH—CH—CH_3; OH on the second CH

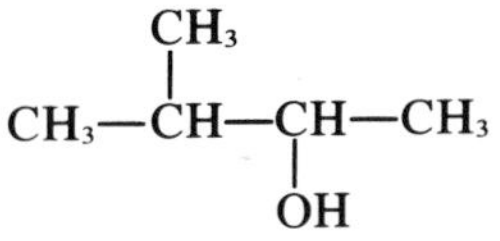

3-methyl-2-butanol

4.7 Inositol is widely distributed in plants and animals and is a growth factor for animals and microorganisms. It has nine possible optical isomers, seven of which are meso and two of which are of enantiomers. The most prevalent natural form is *cis*-1,2,3,5-*trans*-4,6-cyclohexanehexol. Inositol is used medically for treatment of cirrhosis, hepatitis, and fatty infiltration of the liver.

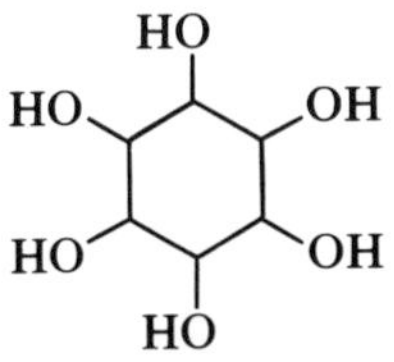

(cyclohexane ring with six OH groups: HO, HO, OH, HO, OH, HO)

inositol

a Draw a chair conformation of the prevalent natural isomer and determine whether it is optically active or meso.

b Draw chair conformations for the single pair of enantiomers.

4.8 Using the molecular formula $C_6H_{12}O$, draw structural formulas for:

a four skeletal isomers

b four functional group isomers

c a pair of acyclic geometric isomers
d a pair of cyclic geometric isomers
e a pair of conformational isomers
f a pair of enantiomers
g a pair of diastereomers

4.9 α-Pinene (p. 49) has two asymmetric carbon atoms and the 2^n rule would predict a maximum of four stereoisomers. However, because of the geometric requirements in fusing two rings together, only two stereoisomers are capable of existence. By building models, show that only one pair of enantiomers can exist. (As an interesting aside, the α-pinene from North American oils is dextrorotatory, while that from most European oils is levorotatory.)

4.10 Draw the four stereoisomers of grandisol. Grandisol, secreted by the hind gut of the male boll weevil (***Anthonomus grandis***), acts as a sex pheromone. A pheromone is a substance secreted by an animal to influence the behavior of another animal of the same species. These sex attractants can be detected by insects in minute amounts and are undoubtedly the most potent physiologically active substances known. For example the male gypsy moth, a serious pest of lumber trees, responds to a quantity of sex pheromone of less than 10^{-13} gram. Sex pheromone-baited traps have major potential in insect control as replacements for DDT and other pesticides which damage the biosphere as a whole.

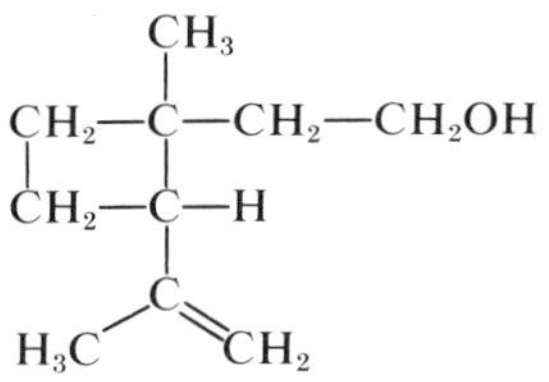

```
       CH3
       |
CH2—C—CH2—CH2OH
|      |
CH2—C—H
       |
       C
     /   \\
 H3C      CH2
```

grandisol

4.11 Chloramphenicol was one of the first broad-spectrum antibiotics to be discovered. Draw the four stereoisomers of chloramphenicol and label those that are mirror images; diastereomers. (Only one of these stereoisomers possesses antibacterial activity.)

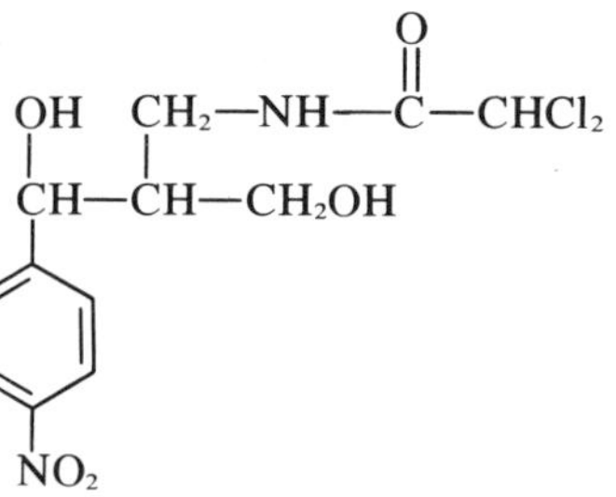

```
                       O
                       ||
OH   CH2—NH—C—CHCl2
|      |
CH—CH—CH2OH
|
(p-phenylene ring)
|
NO2
```

chloramphenicol

4.9 RESOLUTION

The process of separating a pair of enantiomers into (+)- and (−)-isomers is called resolution. The first demonstration of this process was the historic resolution of racemic tartaric acid by Louis Pasteur in 1848. Below 25°C the (+)-enantiomer of sodium ammonium tartrate crystallizes in one type of crystal while the (−)-enantiomer crystallizes in a mirror-image crystal. By carefully hand-picking the crystals, Pasteur separated the mixture into two piles and examined their solutions separately in a polarimeter. He made the exciting observation that a solution of one form rotated the plane of polarized light to the right and the other rotated it to the left. When equal weights of the two kinds of crystals were dissolved in water, the solution of the mixture, like the starting material, had no effect on the plane of polarized light. This initial resolution by Pasteur is particularly remarkable, for since that time very few additional examples have been encountered in which crystallization produces enantiomorphic crystals large enough that manual separation is possible. Fortunately other methods of resolution are now known.

A second and more generally useful method of resolution is chemical, and based on the fact that diastereomers have different physical properties. Pasteur observed that the salts of (+)-tartaric and of (−)-tartaric acid with metals or ammonia were identical in their solubilities. However, salts of (+)- and (−)-tartaric acid derived from certain naturally occurring amines such as strychnine or quinine no longer had the same solubilities. (The amines named are optically active because they contain one or more asymmetric carbons.) Reaction of a racemic acid with an optically active base forms a pair of diastereomeric salts. These salts can be separated by fractional crystallization. Treatment of the separated salts with mineral acid then liberates the original acid in optically pure form.

$$\underset{\text{enantiomers}}{\left\{\begin{matrix}\text{dextro-acid}\\ \text{levo-acid}\end{matrix}\right\}} + \text{levo-base} \rightarrow \underset{\text{diastereomers}}{\left\{\begin{matrix}\text{dextro-acid: levo-base}\\ \text{levo-acid: levo-base}\end{matrix}\right\}}$$

If the starting racemate is an amine, then an optically active acid can be used for the process. This method of resolution is both practical and capable of wide application.

4.10 THE SIGNIFICANCE OF ASYMMETRY IN THE BIOLOGICAL WORLD

Thus far we have looked at ways of detecting and measuring optical activity in the laboratory, predicting optical isomerism by examining molecular structure, and resolving a mixture of enantiomers in the laboratory.

Why is it so important that we be able to describe stereoisomerism (optical as well as geometric) and recognize it in molecules? The reason stems from the fact that we are interested in the reactions between molecules, and particularly between organic substances in the biological world. Living organisms, plant or animal, consist largely of asymmetric and therefore optically active substances. Except for molecules like water, inorganic salts, and a relatively few low-molecular-weight organic molecules, most of the compounds of nature are asymmetric. While these molecules can in principle exist as a mixture of stereoisomers, almost invariably only one stereoisomer is found in nature. There are of course instances where both enantiomers can be found in nature, but they do not seem to exist together in the same biological system. For example, both dextrorotatory and levorotatory lactic acids are found in nature; the (+)-lactic acid is found in living muscle, while the (−)-lactic acid is found in sour milk. We can make the further generalization that not only is just one enantiomer found in nature but also only one enantiomer can be used or assimilated. In fact this latter observation is a basis for a sometimes-used biological resolution technique.

Pasteur himself discovered in 1858–1860 that when the microorganism *Penicillium glaucum*, the green mold found in aging cheese and rotting fruit, is grown in a medium containing racemic tartaric acid, the solution slowly becomes levorotatory. The microorganism preferentially consumes or metabolizes the (+)-tartaric acid. If the process is interrupted at the right time, the (−)-tartaric acid can be crystallized from solution in pure form. If the process is allowed to continue, the microorganism will eventually consume the (−)-tartaric acid as well. Thus while both enantiomers of tartaric acid are metabolized by the microorganism, the (+)-form is metabolized at a much greater rate. As another example, when racemic mevalonic acid is fed to rats, one enantiomer is totally absorbed while almost all of the other enantiomer is excreted in the urine.

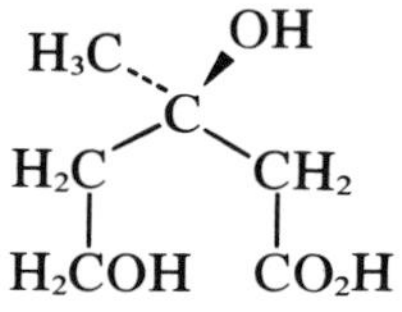

H_3C, OH, C, H_2C, CH_2, H_2COH, CO_2H

mevalonic acid
(metabolized)

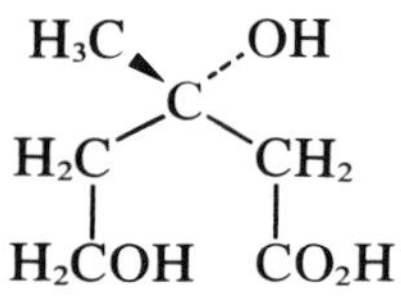

H_3C, OH, C, H_2C, CH_2, H_2COH, CO_2H

mevalonic acid
(excreted)

Many other examples are known in which a mold or other microorganism will use one enantiomer of a racemic mixture. Resolution by this technique is generally referred to as microbiological resolution or separation. This method of resolution, just as with the hand-picking of crystals, is seldom of preparative value. One of the enantiomers, usually the more interesting natural or biologically active form, is sacrificed. The nonmetabolized form may be toxic. Furthermore, since the process must be conducted in dilute solution and in a nutrient medium conducive to the growth of the organism, recovery of the nonmetabolized enantiomer is often difficult. But we have not discussed this method only to conclude that it is seldom of preparative value; we have discussed it for the insight it gives into the operation of biological systems on enantiomeric compounds.

The observations that only one enantiomer is found in a given biological system and that only one enantiomer can be metabolized should be enough to convince us that it is an asymmetric world in which we live. At least it is asymmetric at the molecular level. Essentially all chemical reactions in the biological world take place in an asymmetric environment. Let us develop this last point a bit further. Perhaps the most conspicuous examples of asymmetry among biological molecules are the enzymes. Just to illustrate this point, consider the enzyme chymotrypsin, which functions so efficiently in the intestine at pH 7–8 to catalyze the digestion of proteins. This enzyme is made up of 241 amino acids. Of these, 218 have one asymmetric carbon atom. The number of potential stereoisomers is then 2^{218}, a number that surely seems beyond comprehension. Fortunately nature does not squander its precious resources and energies unnecessarily; only one of these stereoisomers is made in any given organism.

Enzymes catalyze biological reactions by first absorbing on their surface the small molecule or molecules about to undergo reaction. Thus, whether these smaller molecules are asymmetric or not, they are now held for reaction in an asymmetric environment. Let us look in more detail at just two examples, one to illustrate how an enzyme might distinguish between a pair of enantiomers, the second to illustrate how an enzyme might catalyze the conversion of a symmetrical molecule into one pure enantiomer uncontaminated by its mirror image.

Consider first glyceraldehyde (Figure 4.9). This example is chosen because glyceraldehyde is a key intermediate in the metabolism of carbohydrates and is either oxidized to glyceric acid or reduced to glycerol depending on the needs of the cell. Further, it is a small molecule with just one asymmetric carbon atom and it illustrates in a most simple way the interaction of an asymmetric molecule with an asymmetric environment.

How might an enzyme discriminate between one enantiomer of glyceraldehyde and the other? It is generally agreed that an enzyme with specific receptor or binding sites for three of the four substituents on the asymmetric carbon can distinguish readily between two enantiomers. Assume for example that the enzyme involved in the catalysis of a glyceraldehyde reaction has three receptor sites, one specific for —H, another specific for —OH, and a third specific for —CH_2OH, and that these three sites are arranged on the enzyme surface as shown in Figure 4.9.

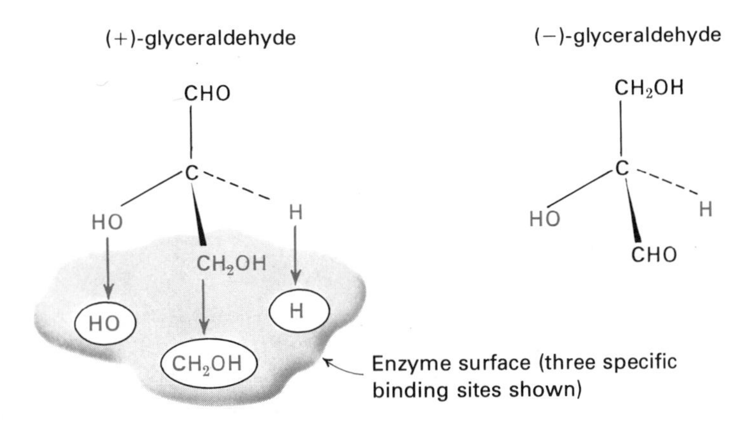

Figure 4.9 *Enzyme stereospecificity. A schematic diagram of an enzyme surface capable of interacting with (+)-glyceraldehyde at three binding sites but with (−)-glyceraldehyde at only two of the three potential binding sites.*

The enzyme can "recognize" (+)-glyceraldehyde (the natural or biologically active form) in the presence of (−)-glyceraldehyde since the correct enantiomer can be absorbed with three groups attached to the appropriate receptor or binding sites while the other enantiomer can bind, at best, to only two of these sites.

The stereospecific course of an enzyme-catalyzed reaction which converts a symmetrical starting material into one pure enantiomer can also be understood with the same type of model. Consider the enzyme-catalyzed reaction of pyruvic acid to (+)-lactic acid.

$$CH_3-\overset{\overset{\displaystyle O}{\|}}{C}-CO_2H \xrightarrow{\text{enzyme}} CH_3-\overset{\overset{\displaystyle OH}{|}}{C}H-CO_2H$$

pyruvic acid (+)-lactic acid

Pyruvic acid is superimposable on its mirror image and therefore will not show optical isomerism. However, because of the requirements of very specific and precise interactions between it and the enzyme surface, it may be possible that the chemical reducing agent can approach the enzyme-bound pyruvic acid molecule from only one direction. Figure 4.10 is a schematic diagram that illustrates how a pyruvic acid molecule might be held on an enzyme surface and the approach of a reducing agent from the "top" of the molecule to form (+)-lactic acid. Of course if the reducing agent were to approach pyruvic acid from the other side, it would lead to the formation of (−)-lactic acid.

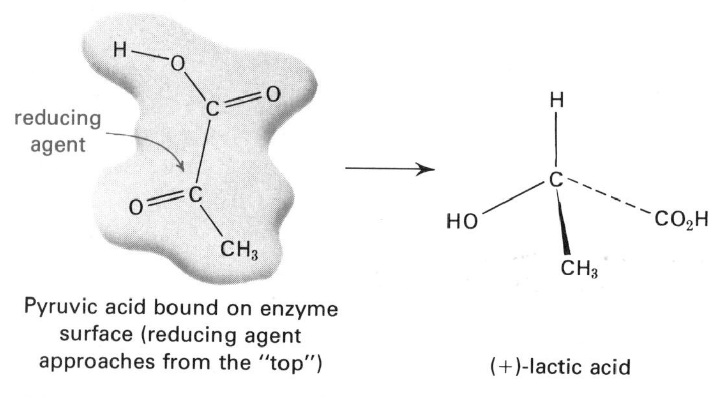

Figure 4.10 *Enzyme stereospecificity. A schematic diagram for the enzyme-catalyzed reduction of pyruvic acid to (+)-lactic acid.*

Quite understandably we do not know the intimate details of how these three molecules (pyruvic acid, enzyme, and reducing agent) interact, but by applying this type of thinking we can at least appreciate the high degree of stereospecificity of the process.

With this insight into the interactions between molecules taking place in a highly asymmetric environment, such things as the microbiological resolution of tartaric acid or the selective excretion of one enantiomer of mevalonic acid should not be surprising to us. This of course does not presume that we have any means to make accurate predictions of which enantiomer might be metabolized or which might be synthesized, but the fact is that the biological world operates with demanding and precise stereospecificity. Further, it should not surprise us (again, though we might not have predicted it) that a molecule might function as a potent antibiotic while its enantiomer either has no potency at all or perhaps is somewhat toxic. It should not surprise us to learn that the physiological and psychological effects of LSD are not also shared by its enantiomer. Neither should it surprise us that a compound may be sweet to the taste while its enantiomer has no taste at all. The fact that the interactions of molecules in the biological world are so very specific in stereochemistry is not surprising—but just how these interactions are accomplished with such high precision and efficiency is one of the great challenges that modern science is only now beginning to unravel.

Problems

4.12 Define and illustrate the following terms:

a stereoisomerism
b optical isomerism
c plane-polarized light
d polarimeter
e dextrorotatory
f levorotatory
g observed rotation
h specific rotation
i asymmetric carbon atom
j enantiomer
k diastereomer
l plane of symmetry
m meso compound
n racemic mixture
o resolution

4.13 Explain the difference in molecular structure between meso tartaric acid and racemic tartaric acid.

4.14 Calculate the observed rotation of a solution prepared by dissolving 5.0 g of (+)-tartaric acid in 100 ml of water and placing a 1-decimeter sample tube filled with this solution in a polarimeter.

4.15 List three experimental methods for the resolution of racemic mixtures and explain each briefly. Which of these methods is the most widely used?

4.16 How might you explain the following?

a An enzyme is able to distinguish between a pair of enantiomers and catalyze a biochemical reaction of one enantiomer but not of its mirror image.

b An enzyme is able to catalyze the conversion of a symmetrical, optically inactive molecule into one pure enantiomer uncontaminated by its mirror image.

c The microorganism *Penicillium glaucum* preferentially metabolizes (+)-tartaric acid rather than (−)-tartaric acid.

5

ALCOHOLS, PHENOLS, AND ETHERS

5.1 INTRODUCTION

Alcohols and phenols are good starting points for our discussion of compounds other than hydrocarbons because their reactions are relatively simple. The reactions of alcohols and phenols are those of the —OH functional group, known as the hydroxyl group.

We shall describe some of the important physical properties of alcohols and the most important types of reactions they undergo in biological chemistry; these are dehydration to form alkenes, and oxidation to form aldehydes, ketones, or carboxylic acids. We shall use the oxidation of alcohols to introduce oxidations and reductions in general and we will discuss the importance of these processes in the biological world. Another important reaction of alcohols, namely esterification with carboxylic acids, will be taken up in Chapter 8.

5.2 STRUCTURE

Alcohols may be viewed as alkyl derivatives of water and are indicated by the general formula R—OH, where the symbol R— is used to indicate any alkyl group. When the hydroxyl group is attached to an aromatic ring, the compounds are known not as aromatic alcohols but rather as phenols, because their properties are significantly different from those of alcohols. Phenols are indicated by the general formula Ar—OH. An aryl group, abbreviated Ar—, is any aromatic group, e.g., a derivative of benzene, naphthalene, etc.

Figure 5.1 shows the structural formulas of a number of the

Figure 5.1 *Structural formulas of some alcohols.*

CH_3OH	methanol methyl alcohol
CH_3CH_2OH	ethanol ethyl alcohol
$CH_3CH_2CH_2OH$	1-propanol *n*-propyl alcohol
$CH_3CH(OH)CH_3$	2-propanol isopropyl alcohol
$CH_3CH_2CH_2CH_2OH$	1-butanol *n*-butyl alcohol
$CH_3CH_2CH(OH)CH_3$	2-butanol *sec*-butyl alcohol
$CH_3CH(CH_3)CH_2OH$	2-methyl-1-propanol isobutyl alcohol
$(CH_3)_3COH$	2-methyl-2-propanol *tert*-butyl alcohol

simpler alcohols. Under each is given the IUPAC name and also a common name. These alcohols are very often referred to by common rather than IUPAC names.

5.3 NOMENCLATURE OF ALCOHOLS

The IUPAC nomenclature of alcohols selects the longest continuous carbon chain that contains the —OH group and considers the alcohol to be derived from that structure. The name is derived by replacing the -e of the parent alkane with -ol, and by indicating the position of the hydroxyl group with a number. Generally the carbon chain of the parent alkane is numbered in such a way as to give the —OH group the lowest possible number. Many alcohols are known by their common or trivial names, e.g., methyl alcohol, ethyl alcohol, etc.

We often refer to alcohols as being primary (1°), secondary (2°), or tertiary (3°). This classification depends on the number of alkyl groups on the carbon bearing the —OH function. (Compare this classification with that of carbonium ions, p. 57).

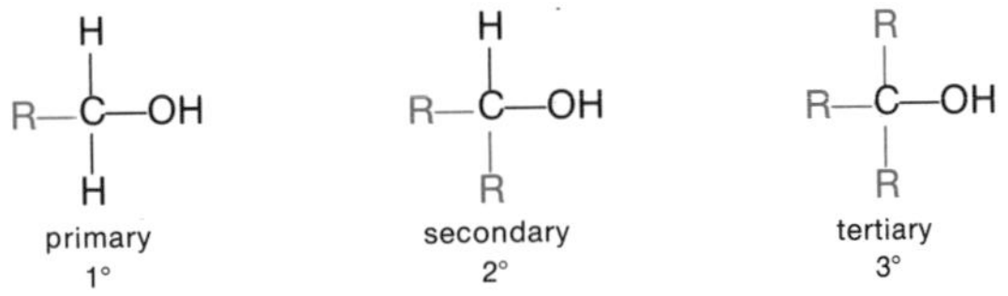

Figure 5.2 *Classification of alcohols: primary, secondary and tertiary.*

Compounds containing two hydroxyl groups are called diols or glycols, those containing three hydroxyl groups are called triols, and those containing several hydroxyl groups are generally referred to as polyols. Several examples are shown below. Under each is given the IUPAC name and the common name

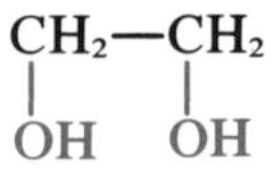

$$\underset{\text{OH}}{\text{CH}_2}-\underset{\text{OH}}{\text{CH}_2}$$

1,2-ethanediol
ethylene glycol

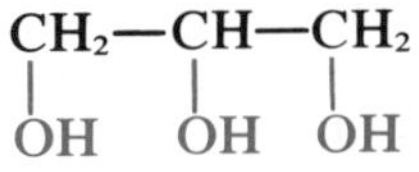

$$\underset{\text{OH}}{\text{CH}_2}-\underset{\text{OH}}{\text{CH}}-\underset{\text{OH}}{\text{CH}_2}$$

1,2,3-propanetriol
glycerol, glycerine

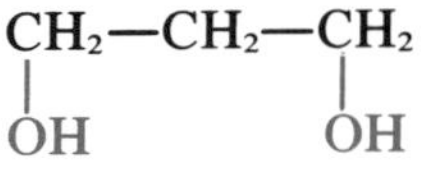

$$\underset{\text{OH}}{\text{CH}_2}-\text{CH}_2-\underset{\text{OH}}{\text{CH}_2}$$

1,3-propanediol
trimethylene glycol

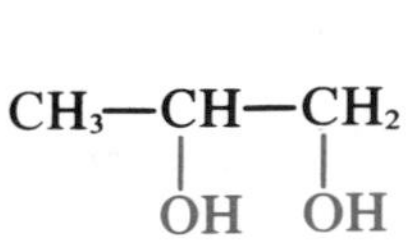

$$\text{CH}_3-\underset{\text{OH}}{\text{CH}}-\underset{\text{OH}}{\text{CH}_2}$$

1,2 propanediol
propylene glycol

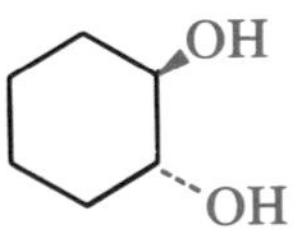

trans-1,2-cyclohexanediol

All of these diols and triols can of course be named as derivatives of the parent alkane. Yet as with so many organic compounds, common names have persisted. Both ethylene glycol and propylene glycol can be prepared by controlled oxidation of ethylene and propylene—hence their common names. Glycerol is a major by-product of the manufacture of soaps and is related in structure to glyceraldehyde and glyceric acid, both key intermediates in the metabolism of carbohydrates. The common name of trimethylene glycol indicates the presence of three methylene groups, $—CH_2—$, between the two hydroxyl groups.

Structures that contain two hydroxyl groups on the same carbon, 1,1-glycols, are almost never isolated.

$$>C(OH)_2 \rightleftharpoons >C{=}O + H_2O$$

a 1,1-glycol

As we shall see in Chapter 6, compounds of this type are in equilibrium with the corresponding aldehyde or ketone and equilibrium generally lies very far to the right.

One last comment on the nomenclature of alcohols. The suffix -ol is generic to alcohols and although names such as glycerol, menthol, and cholesterol contain no clues to their carbon skeletons, the names do indicate that each compound contains one or more hydroxyl groups.

Problem **5.1** Name and draw structural formulas for the eight isomeric alcohols of molecular formula $C_5H_{12}O$. Which of these are primary; secondary; tertiary? How many of these alcohols will show optical isomerism?

5.4 PHYSICAL PROPERTIES OF ALCOHOLS AND PHENOLS

Methanol and ethanol are soluble in water in all proportions, and their chemical and physical properties are more like those of water than the hydrocarbons from which they are derived. The alcohols of higher molecular weight are either partly soluble or insoluble in water. As the length of the alkane chain increases, the physical properties of alcohols resemble more closely those of the alkanes from which they are derived. However, the chemical properties of even the higher alcohols are determined by the —OH group.

Table 5.1 *Physical properties of some simple alcohols.*

Name	mp (°C)	bp (°C)	Solubility (g/100 g H_2O)
methanol	−97	64.5	∞
ethanol	−115	78.3	∞
1-propanol	−126	97	∞
1-pentanol	−78	138	2.3
1-heptanol	−34	176	0.2
1-butanol	−90	118	7.9
2-butanol	−114	99.5	12.5
2-methyl-1-propanol	−108	108	10.5
2-methyl-2-propanol	25	83	∞

Phenol, or carbolic acid as it was once called, is a low-melting solid only slightly soluble in water. In sufficiently high concentration it is corrosive to animal tissue. In dilute solutions it has some antiseptic properties and was used for the first time in the 19th century by Joseph Lister for antiseptic surgery. Its medical use is now limited as it has been supplanted by other antiseptics which are more powerful and have fewer undesirable side effects. *o*-Phenylphenol (Lysol) and *n*-hexylresorcinol (Sucrets and mouthwashes) are used in household preparations for their germicidal properties.

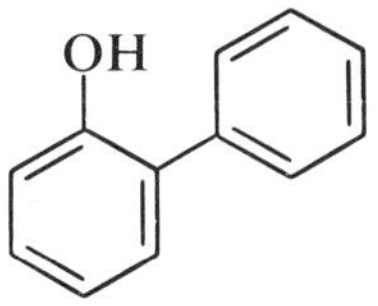

OH

$CH_2CH_2CH_2CH_2CH_2CH_3$

o-phenylphenol 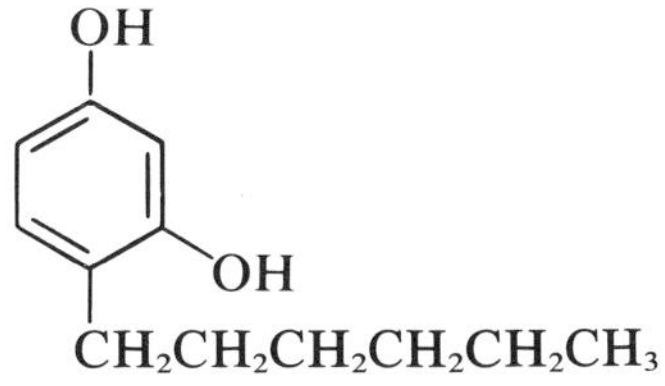 *n*-hexylresorcinol

Phenols are in a different class from alcohols because they have very different chemical properties. Perhaps the most distinctive property of phenols compared to alcohols is that they are fairly acidic compounds, whereas alcohols are even less acidic than water. Because of this higher acidity, phenols react with aqueous solutions of sodium hydroxide to form sodium salts, e.g., sodium phenoxide in the case of phenol.

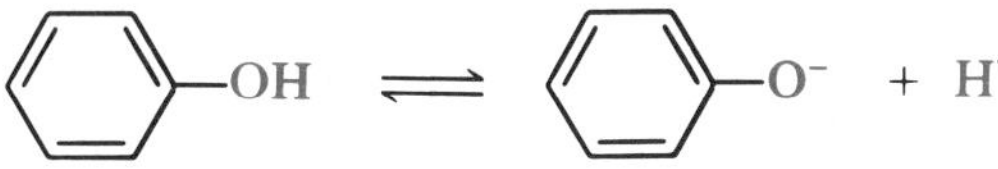

$$C_6H_5\text{—OH} \rightleftharpoons C_6H_5\text{—O}^- + H^+$$

Alcohols do not react in this manner. Mixtures of phenol and water-insoluble alcohols may be separated by shaking with dilute aqueous sodium hydroxide. The phenol enters the water layer as the sodium salt while the alcohols remain as a separate layer. Separation of the aqueous layer followed by acidification with mineral acid regenerates the phenol. Phenols do not dissolve in aqueous sodium bicarbonate because the bicarbonate ion is too weakly basic to remove a proton from a phenol molecule. This fact permits the ready separation of phenols from the more acidic carboxylic acids, as we shall see in Chapter 7.

Let us compare the boiling points and solubilities in water of alcohols with those of other compounds of similar molecular weight but containing different functional groups. As we do this, we will look for relationships between structure and physical properties. To understand anything of these relationships, we must begin by first inquiring into the nature of the forces that hold neutral molecules together. These forces are of two major kinds: dipole-dipole interactions and hydrophobic interactions. In this chapter we shall discuss dipole-dipole interactions as they affect the solubility of polar organic molecules in water and other polar solvents. In later chapters on lipids, amino acids and proteins, and nucleic acids, we shall discuss hydrophobic interactions and the properties of nonpolar organic molecules.

Dipole-dipole interaction is the attraction of a positive part of one polar molecule by the negative part of another. You have already encountered this phenomenon in general chemistry in the study of the physical properties of water and hydrogen sulfide. The formula weight of hydrogen sulfide is 34, nearly twice that of water. Yet its boiling point is 160° below that of water. This is due to the high degree of association of water molecules in the liquid state, by dipole-dipole interaction.

Since the proton is very small, an adjacent water molecule can approach to within a short distance of it. The electrostatic force of attraction between the oppositely charged hydrogen and oxygen is appreciable at this short distance, and the molecules tend to form aggregates in solution. Each water molecule interacts directly with several other water molecules.

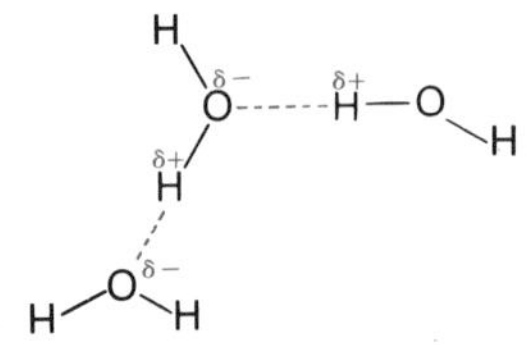

Figure 5.3 *Hydrogen bonding in water. The formation of molecular aggregates.*

This type of attractive force is given the special name hydrogen bonding and is indicated by a dashed line connecting the interacting oxygen and hydrogen. Hydrogen bonds are about 5% as strong as the average C—C, C—N, or C—O single covalent bonds. The higher boiling point of water compared to hydrogen sulfide is due to the extra energy required to break up the aggregates and allow the individual water molecules to enter the vapor state.

This type of association is also possible with alcohols, as illustrated by the association of methyl alcohol, CH_3OH, in Figure 5.4.

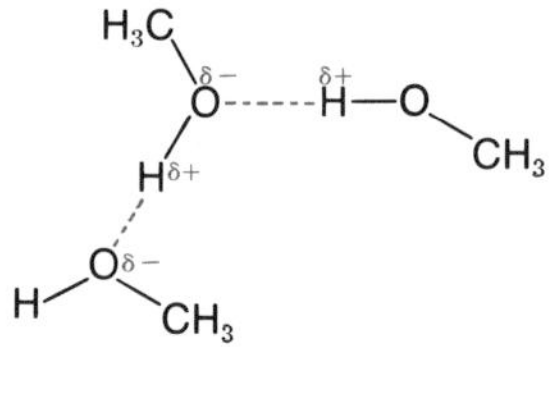

Figure 5.4 *The association of methyl alcohol in the liquid state.*

The effect of hydrogen bonding on boiling point is dramatically illustrated by comparing ethyl alcohol, CH_3CH_2OH, bp 78°C, and its isomer dimethyl ether, CH_3OCH_3, bp −24°C. Each of these compounds has the same empirical formula, C_2H_6O, and the same molecular weight. Therefore differences in boiling point must be due to differences in the degree of association between molecules in the pure liquid. The remarkably high association of CH_3CH_2OH compared to CH_3OCH_3 is due to the fact that both the shape and the structure of the ethyl alcohol molecule allow close approach of the centers of positive and negative charge (compare Figure 5.4). While there is certainly bond and molecular polarity in dimethyl ether, the centers of positive and negative charge are buried more deeply within the molecule than in ethyl alcohol, with the result that the positively charged carbon atom and the negatively charged oxygen atom cannot approach one another closely

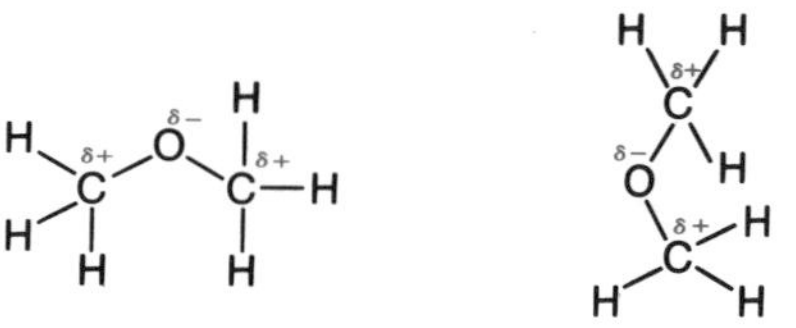

Figure 5.5 Hydrogen bonding is not possible in dimethyl ether for there are no highly polarized hydrogens. Dipole-dipole interaction is not appreciable because the centers of partial negative and positive charge cannot come close enough together.

and therefore the resulting electrostatic interaction is very small (Figure 5.5).

By way of further examples of the consequence of hydrogen bonding, we note that the boiling points of water, alcohols, and carboxylic acids are abnormally high relative to the boiling points of hydrocarbons, ethers, and related compounds of similar molecular weight (Table 5.2).

Table 5.2 *Boiling points of compounds of similar molecular weight.*

Compound	Molecular Weight	Boiling Point (°C)
CH_4	16	−164
H_2O	18	100
CH_3CH_3	30	−89
CH_3OH	32	65
CH_3CH_2OH	46	78
CH_3OCH_3	46	−24
CH_3CO_2H	60	118
$CH_3CH_2CH_2OH$	60	97
$CH_3CH_2OCH_3$	60	11

Problems

5.2 The boiling point of dimethyl sulfide is 38°C, and that of its isomer ethanethiol is 35°C.

$CH_3—S—CH_3$ — dimethyl sulfide, bp 38°C

$CH_3CH_2—SH$ — ethanethiol, bp 35°C

a By considering these structural formulas and relative boiling points, what do you conclude about the importance of hydrogen bonding in compounds containing the thiol (—SH) group?

b By considering the relative molecular weights and boiling points of ethyl alcohol and ethanethiol, what do you conclude about the relative importance of hydrogen bonding in compounds containing the hydroxyl group (—OH) compared to those containing the thiol group (—SH)?

5.3 Compounds that contain the N—H bond also show considerable evidence of association. Would you expect this association to be stronger or weaker than that in compounds containing O—H groups? (Hint: nitrogen is less electronegative than carbon.)

Now let us turn to solubility. When a solid or liquid dissolves in a solvent, the molecules become separated from each other and become surrounded by solvent molecules. In the dissolving process, as in boiling, energy must be supplied to overcome the attractive forces between molecules of the solid or liquid compound. In the case of ionic solids a great deal of energy is required to disrupt the ionic attractive forces. Only water and a few other solvents of high polarity are capable of dissolving appreciable quantities of ionic solids. The energy required to break the attractive forces is supplied through the formation of new attractive forces: between the positive ion and the negative ends of the water molecules, and between the negative ion and the positive ends of water molecules. Such forces are called ion-dipole interactions. In solution, each ion is surrounded by a cluster of water molecules and is said to be hydrated or solvated.

Figure 5.6 *Hydration of sodium and chloride ions in water.*

The process of solution of an ionic solid involves a balance between the energy holding the ions in the solid state and the energy of solvation.

The solubility characteristics of covalent compounds are determined by their polarity or lack of polarity. Highly polar covalent compounds dissolve in highly polar solvents, and do not dissolve in nonpolar solvents. For example, methyl alcohol is soluble in water in all proportions (Figure 5.7), whereas ethane, CH_3CH_3, is insoluble in water. Nonpolar covalent compounds dissolve in nonpolar solvents and are essentially insoluble in polar solvents. We can now appreciate what is involved in the oft-stated solubility generalization of "like dissolves like."

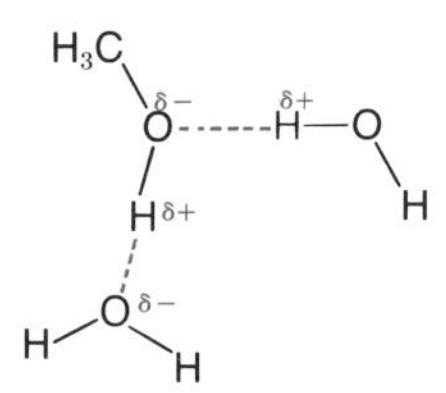

Figure 5.7 *Hydrogen bonding between water and methyl alcohol.*

Problems **5.4** Diethyl ether has a much lower boiling point than *n*-butyl alcohol, yet each of these compounds shows about the same solubility in water, 8 grams per 100 ml of water. How do you account for this observation?

$CH_3CH_2CH_2CH_2OH$	$CH_3CH_2OCH_2CH_3$
n-butyl alcohol	diethyl ether
bp 117°C	bp 35°C

5.5 How would you account for the fact that ethane CH_3CH_3 is a gas (bp − 88°C) whereas hydrazine N_2H_4 is a liquid (bp 113.5°C)?

5.6 Both propanoic acid and methyl acetate have the same molecular formula, $C_3H_6O_2$. One is a liquid, boiling point 141°C; the other is a liquid, boiling point 57°C.

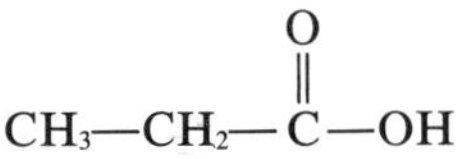

$$\underset{\text{propanoic acid}}{CH_3{-}CH_2{-}\overset{\overset{\displaystyle O}{\|}}{C}{-}OH} \qquad \underset{\text{methyl acetate}}{CH_3{-}\overset{\overset{\displaystyle O}{\|}}{C}{-}O{-}CH_3}$$

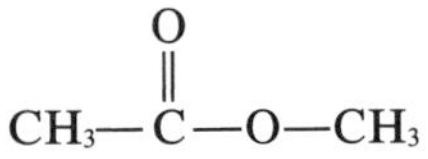

a Which of these two compounds would you predict to have the boiling point of 141°C; the boiling point of 57°C? Explain the basis for your prediction.
b Which of these two compounds would you predict to be more soluble in water? Explain.

5.7 Both acetic acid and methyl formate have the same molecular formula, $C_2H_4O_2$. One is a liquid, boiling point 32°C; the other is a liquid, boiling point 118°C.

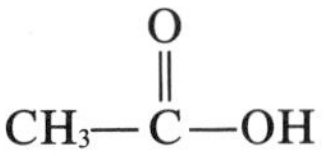

$$\underset{\text{acetic acid}}{CH_3{-}\overset{\overset{\displaystyle O}{\|}}{C}{-}OH} \qquad \underset{\text{methyl formate}}{H{-}\overset{\overset{\displaystyle O}{\|}}{C}{-}O{-}CH_3}$$

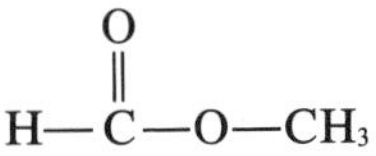

Which of these two compounds would you predict to have the boiling point of 118°C; the boiling point of 32°C? Explain your reasoning.

5.8 Arrange the following sets of compounds in order of decreasing boiling points (highest boiling point to lowest boiling point).

a $CH_3CH_2CH_3$ $\quad$ $CH_3CH_2CH_2CH_2CH_2CH_2CH_3$ $\quad$ $CH_3CH_2CH_2CH_2CH_3$
b N_2H_4 $\quad$ H_2O_2 $\quad$ CH_3CH_3
c CH_3CO_2H $\quad$ CH_3CH_2OH $\quad$ CH_3OCH_3
d $CH_3\underset{\displaystyle OH}{\underset{|}{C}}HCH_3$ $\quad$ $CH_3{-}\underset{\displaystyle OH}{\underset{|}{CH}}{-}\underset{\displaystyle OH}{\underset{|}{CH_2}}$ $\quad$ $\underset{\displaystyle OH}{\underset{|}{CH_2}}{-}\underset{\displaystyle OH}{\underset{|}{CH}}{-}\underset{\displaystyle OH}{\underset{|}{CH_2}}$
e $CH_3CH_2CH_2OH$ $\quad$ $CH_3CH_2OCH_3$ $\quad$ $CH_3CH_2CH_2NH_2$

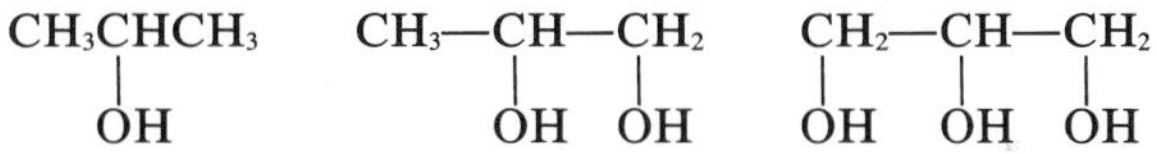

5.9 Arrange the following compounds in order of decreasing solubility in water and explain the principles on which you base your answer.

a ethanol; butane; ethyl ether
b 1-hexanol; 1,2-hexanediol; $\underset{\displaystyle OH}{\underset{|}{CH_2}}{-}\underset{\displaystyle OH}{\underset{|}{CH}}{-}\underset{\displaystyle OH}{\underset{|}{CH}}{-}\underset{\displaystyle OH}{\underset{|}{CH}}{-}\underset{\displaystyle OH}{\underset{|}{CH}}{-}\underset{\displaystyle OH}{\underset{|}{CH_2}}$

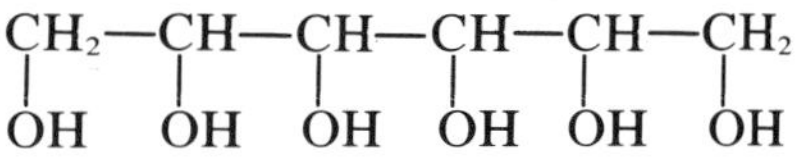

5.5 PREPARATION OF ALCOHOLS

Methyl alcohol (methanol), commonly called wood alcohol, was prepared by the destructive distillation of wood, at least until 1923. When wood is heated to temperatures above 250°C without access to air it decomposes to charcoal and a volatile fraction which partially condenses on cooling. This condensate contains methyl alcohol, acetic acid, and traces of acetone. At the present time methanol is made on a large scale synthetically by the high-pressure hydrogenation of carbon monoxide over a chromic oxide–zinc oxide catalyst.

$$CO + 2H_2 \xrightarrow[350°C,\ 200\ atm]{ZnO\text{-}Cr_2O_3} CH_3OH$$

Methanol is a major commodity worldwide; nearly half of the methanol produced is used to make formaldehyde.

Ethyl alcohol (ethanol), or simply "alcohol" in nonscientific language, has been prepared since antiquity by the fermentation of sugars by yeast, and is the basis for the preparation of alcoholic "spirits." Extensive biochemical investigations have established that the fermentation of sugars, for example glucose, proceeds through an elaborate series of steps, each catalyzed by a particular and specific enzyme.

$$\underset{\text{glucose}}{C_6H_{12}O_6} \xrightarrow{\text{enzymes}} 2CH_3CH_2OH + 2CO_2$$

The sugars for the fermentation process may come from a variety of sources: blackstrap molasses, a residue from the refining of cane sugar; various grains, hence the name "grain alcohol"; grape juice; various vegetables. The immediate product of the fermentation process is an aqueous solution containing up to about 15% alcohol. This alcohol may be concentrated by distillation. Beverage alcohol may contain traces of flavor derived from the source (brandy from grapes, whiskeys from grains) or may be essentially flavor free (vodka). The most important synthetic method for the preparation of ethyl alcohol is the hydration of ethylene (see p. 56 for the mechanism of hydration of alkenes.)

Whatever the method of preparation of ethyl alcohol, be it the hydration of ethylene or fermentation of grains, it is first obtained mixed with water. This mixture must be concentrated and the alcohol separated from the other materials in the reaction mixture by fractional distillation. If we examine the boiling points of pure ethyl alcohol (78.5°C) and of water (100°C), we would predict that the first material to distill would be ethyl alcohol, the component with the lower boiling point. However, in a mixture of ethyl alcohol and water, the material of lowest boiling point is neither alcohol nor water but rather a mixture composed of 95% alcohol and 5% water. Although the boiling point of 95% alcohol is 78.2°C, only 0.3°C below the boiling point of pure alcohol, this small difference makes it impossible to produce absolute alcohol by any kind of fractional distillation of mixtures of alcohol containing more than 5% water. As a result, commercial ethyl alcohol as used for extracts, medicines, beverages, etc., never contains over 95% of the essential component and usually slightly less. This 95% alcohol is said to be a constant-boiling mixture, an azeotrope, or an azeotropic mixture. An azeotrope is a mixture of liquids with a certain definite composition that distills at a constant temperature without change in composition. The boiling point of an azeotropic mixture is usually lower than that of the lowest boiling component, but in some cases it is higher than the boiling point of the highest boiling component. The 95% alcohol azeotrope contains two components and is known as a binary azeotrope.

Since 95% alcohol behaves as if it were a pure compound on distillation, pure or "absolute" alcohol cannot be obtained from it by this technique and other methods must be employed to remove the water from the mixture. In one chemical method calcium oxide, CaO, is used. Calcium oxide reacts with the water to form calcium hydroxide, $Ca(OH)_2$, but it does not react with the alcohol. After all of the water has been removed, the absolute ethyl alcohol is then separated by distillation. A second and more widely used method takes advantage of the fact that a mixture of benzene–water–ethyl alcohol forms a ternary azeotrope that boils at 64.6°C, below the boiling point of any of the three pure components of the azeotrope. The mixture contains 74.1% benzene, 18.5% ethyl alcohol, and 7.4% water. Absolute ethanol is made by adding benzene to the 95% alcohol and removing the water as the volatile azeotrope.

Alcohols can also be prepared by the hydration of alkenes.

$$\underset{\text{propylene}}{CH_3CH{=}CH_2} + H_2O \xrightarrow{H_2SO_4} \underset{\text{isopropyl alcohol}}{CH_3\underset{\displaystyle OH}{\underset{|}{C}}HCH_3}$$

$$\underset{\text{isobutylene}}{CH_3\overset{\displaystyle CH_3}{\overset{|}{C}}{=}CH_2} + H_2O \xrightarrow{H_2SO_4} \underset{\textit{tert}\text{-butyl alcohol}}{CH_3{-}\overset{\displaystyle CH_3}{\overset{|}{\underset{\displaystyle CH_3}{\underset{|}{C}}}}{-}OH}$$

In 1920 Standard Oil (New Jersey) began making isopropyl alcohol from propylene. Isopropyl was the first commercial synthetic alcohol. In general, hydration of alkenes is not an important laboratory method of preparation of alcohols. In fact, alcohols are more generally available than alkenes and often the higher alkenes are prepared either directly or indirectly from alcohols.

Ethylene glycol is prepared commercially from air oxidation of ethylene at high temperature over a silver catalyst to give ethylene oxide, followed by acid-catalyzed hydrolysis.

$$\underset{\text{ethylene}}{CH_2{=}CH_2} + \frac{1}{2}O_2 \xrightarrow[300°C]{Ag} \underset{\text{ethylene oxide}}{\overbrace{CH_2{-}CH_2}^{}\ (\text{bridged by } O)} \xrightarrow[H^+]{H_2O} \underset{\text{ethylene glycol}}{\underset{\displaystyle OH}{\underset{|}{C}}H_2{-}\underset{\displaystyle OH}{\underset{|}{C}}H_2}$$

Ethylene glycol is used as a permanent antifreeze in automotive cooling systems because it is completely miscible with water in all proportions, and a 50% solution with water freezes at −34°C (−29°F). Among its many other important commercial applications is its use as a monomer in the copolymer Dacron (see the mini-essay "Nylon and Dacron").

Glycerol (p. 99) is a major by-product of the manufacture of soaps (Section 7.7); roughly 60 pounds of it are produced per ton of soap. Consumption of glycerol in the United States, both synthetic and by-product from the soap industry, is currently around 300 million pounds per year. Glycerol is a sweet-tasting syrupy liquid. Because of the presence of the three hydroxyl groups on such a small carbon skeleton, glycerol is very high boiling (290°C) and is miscible with water and alcohol in all proportions. It is widely used in the manufacture of resins and nitroglycerine, in the pharmaceutical and cosmetic industries, and as a moistening agent. Glycerol is a trihydroxy compound; through hydrogen bonding it is able to hold water molecules by rather loose intermolecular association and thereby help prevent drying. It is widely used for this purpose in tobacco to help retain a properly moist condition.

Alkyl halides may be converted into the corresponding alcohols by heating in water or in aqueous alkali, as illustrated by the conversion of methyl bromide into methyl alcohol.

$$CH_3\text{—}Br + OH^- \rightarrow CH_3\text{—}OH + Br^-$$

In general the substitution of hydroxyl for halogen in an alkyl halide is not a good preparative method for the synthesis of alcohols. Consequently there would seem to be little need for us to examine this reaction in any detail. Yet in another sense there is very good reason to study this reaction carefully, for it is a specific example of a general reaction class known as nucleophilic substitution at a saturated carbon atom. This is one of the simplest and most important classes of organic reactions, and a variety of industrial and laboratory transformations can be performed through the use of specific reactions belonging to this class.

5.6 NUCLEOPHILIC SUBSTITUTION AT SATURATED CARBON

Chemists have uncovered two fundamental mechanisms for nucleophilic substitution at a saturated carbon atom. In this search, two general tools have been of great usefulness: kinetic studies and stereochemical studies. Kinetic studies, i.e., the measurement of reaction rates, can often tell us something about the timing of the steps in the reaction sequence. Stereochemical studies can often tell us something about the relationship between the configuration of the starting material and the products of a reaction. Let us examine some experimental evidence and then formulate these two general reaction mechanisms.

Methyl, ethyl, isopropyl, and *tert*-butyl bromides can be converted to the corresponding alcohols by reaction with sodium hydroxide in aqueous ethyl alcohol. Further, the rates of these reactions can be measured by analyzing the reaction mixture at known time intervals. When such studies were first carried out in the early 1930s, the results were remarkable in that certain of the findings were totally unexpected.

Let us consider first the reaction of *tert*-butyl bromide with hydroxide ion.

$$(CH_3)_3CBr + OH^- \xrightarrow[\text{ethyl alcohol}]{\text{aqueous}} (CH_3)_3COH + Br^-$$

It was found that the rate of conversion of *tert*-butyl bromide to *tert*-butyl alcohol depended only on the concentration of the tertiary butyl halide; it was completely independent of the concentration of the hydroxide ion. Doubling or halving the concentration of sodium hydroxide had no effect on the rate of conversion to the alcohol. The rate-determining step therefore cannot involve hydroxide ion; it must involve only one reactant, the *tert*-butyl bromide.

This reaction is clearly a nucleophilic substitution in that hydroxide seeks a nucleus to which it can donate a pair of electrons and form a new covalent bond. In the process of donating an electron pair, hydroxide ion displaces bromide ion. Finally, this reaction is unimolecular (or first order) in that only one molecular species is involved in the rate-determining step. We summarize all of these statements by referring to such a reaction as S_N1 (*S*ubstitution; *N*ucleophilic; *Uni*-molecular). This designation of course does not tell us how the reaction takes place, only that it is a unimolecular nucleophilic substitution at a saturated carbon atom.

Next let us look at the reactions of methyl and ethyl bromides with hydroxide ion.

$$CH_3—Br + OH^- \rightarrow CH_3—OH + Br^-$$

$$CH_3CH_2—Br + OH^- \rightarrow CH_3CH_2—OH + Br^-$$

In contrast to the reaction of *tert*-butyl bromide, the rates of reaction of these two compounds were found to be directly proportional to the concentration of both alkyl halide and hydroxide ion. Doubling the concentration of either the alkyl halide or the hydroxide ion doubles the rate of reaction; decreasing the concentration of either by one-half results in a decrease in the rate of reaction by one-half. This reaction is a nucleophilic substitution at a saturated carbon atom and the rate-determining step is bimolecular (or second order) in that it involves two reacting species. We summarize these statements by referring to such reactions as S_N2 (*S*ubstitution; *N*ucleophilic; *Bi*molecular). Again, the designation S_N2 does not tell us how the reaction takes place, only that it is a bimolecular nucleophilic substitution at saturated carbon.

The rate of reaction for isopropyl bromide was found to be directly proportional to the concentrations of isopropyl bromide and hydroxide ion when hydroxide ion was about 1 molar, but independent of hydroxide when hydroxide ion concentration was low. We would then describe this reaction as either S_N1 or S_N2 depending on the experimental conditions.

Studies of the stereochemical course of nucleophilic displacement were first reported around 1900 by Paul Walden, who observed that nucleophilic displacement sometimes, but not always, is accompanied by inversion of configuration. This phenomenon is known as the Walden inversion. In an S_N2 reaction of an optically active alkyl halide, if the carbon undergoing substitution is the center of asymmetry, the product will be optically active and opposite in configuration from the starting material. The stereochemical results of first-order substitution reactions are completely different. In an S_N1 reaction of an optically active alkyl halide, if the carbon undergoing the substitution is the center of asymmetry, the product is most often optically inactive; that is, it is a racemic mixture.

With the experimental observations from kinetic and stereochemical studies, we can now examine the two different mechanisms that have been proposed for nucleophilic substitution reactions at saturated carbon. To account for the S_N2 reaction, chemists propose a one-step mechanism in which the new covalent bond is formed at the same time the old one is broken. In the displacement the nucleophile attacks the saturated carbon from the back side, that is, from the side opposite the group being displaced.

Consider the reaction of hydroxide ion with optically active 2-bromobutane (Figure 5.8). In this reaction, the attack of the hydroxide ion is facilitated by the polarity of the carbon-bromine bond and the reaction begins by the attraction of unlike charge. Hydroxide ion approaches and begins to supply electrons to the carbon. As the new carbon-oxygen bond is formed, the carbon-bromine bond is broken. We speak of this as a concerted or simultaneous process. The final result is the formation of a new carbon-oxygen bond and the loss of a bromine as bromide ion.

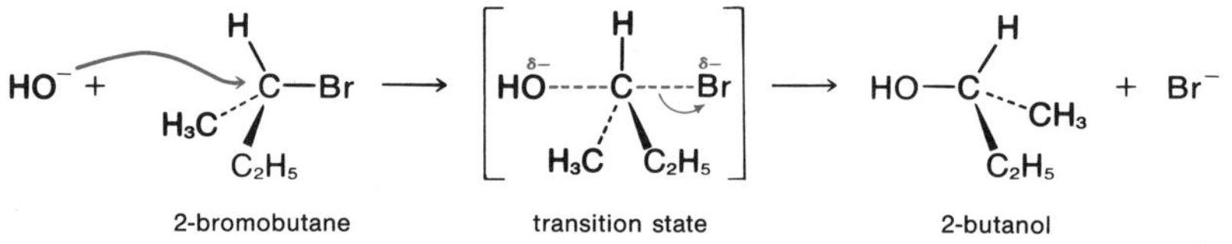

2-bromobutane — transition state — 2-butanol

Figure 5.8 *Inversion of configuration during the* S_N2 *reaction.*

In this formulation of the S_N2 reaction at a saturated carbon atom, both hydroxide ion and alkyl halide must collide. However, not all collisions between these reactants will be effective; only those which meet two stringent requirements will lead to reaction. First, the molecules must collide with sufficient energy to attain a state in which a chemical bond may be made or broken. Second, the molecules colliding must be suitably oriented with respect to each other. Since, in this displacement reaction, hydroxide becomes bonded to the carbon bearing the bromine atom, it is not likely that any collision of the nucleophile with the bromine atom itself or with either of the alkyl substituents will lead to substitution. Furthermore, even collision with proper orientation will not lead to substitution if the energy of collision is insufficient. A useful model for discussing reaction mechanisms is that of the transition state. The transition state is defined as the state of a reacting system that meets both the energy and orientation requirements for effective collision. The transition state for the reaction of hydroxide ion and 2-bromobutane is shown in Figure 5.8.

Note that in the transition state three groups are fully bonded to carbon by single covalent bonds and are planar. This is exactly what we saw earlier for an sp^2 hybridized carbon. This leaves the $2p_z$ orbital to form partial bonds with both the leaving group and the entering group. Because of the geometry of the $2p_z$ orbital, the entering group must approach the carbon atom from the side opposite the leaving group and the result is inversion of configuration. This is represented schematically in Figure 5.9.

To account for the S_N1 reaction, chemists have proposed a different type of mechanism. Consider the conversion of *tert*-butyl bromide to *tert*-butyl alcohol and remember that although the reaction does involve substitution of hydroxide for bromide, the rate of the

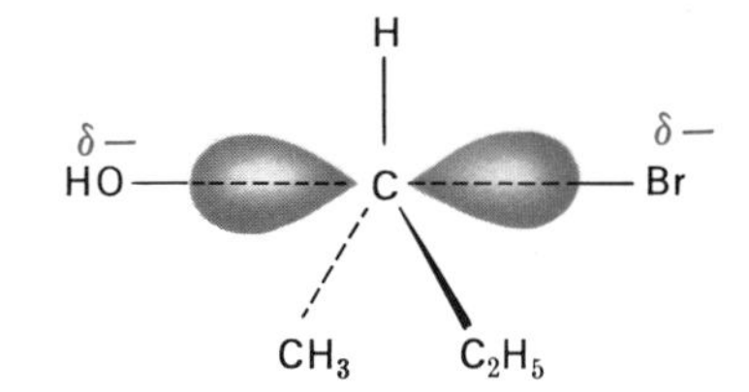

Figure 5.9 *Transition state for an S_N2 reaction showing a $2p_z$ orbital interacting with the leaving group and the entering group.*

reaction does not depend at all on the concentration of the hydroxide ion.

For this type of nucleophilic substitution the first step is proposed to be slow ionization of the C—Br bond to give a carbonium ion, followed by a second and rapid step, the reaction of the carbonium ion with the nucleophile to give the alcohol. In this two-step process, the rate of formation of product is determined by the rate of the slower step, namely the formation of the carbonium ion. Once the carbonium ion is formed, it reacts rapidly with a nucleophile to give the final product.

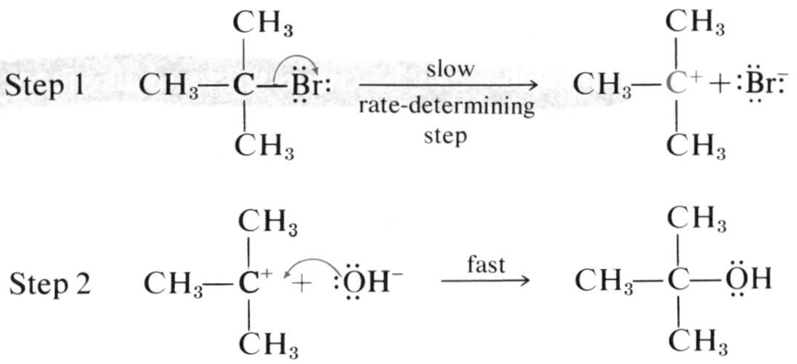

Step 1 $$(CH_3)_3C{-}\ddot{\underset{\cdot\cdot}{Br}}: \xrightarrow[\text{rate-determining step}]{\text{slow}} (CH_3)_3C^+ + :\ddot{\underset{\cdot\cdot}{Br}}:^-$$

Step 2 $$(CH_3)_3C^+ + :\ddot{\underset{\cdot\cdot}{O}}H^- \xrightarrow{\text{fast}} (CH_3)_3C{-}\ddot{\underset{\cdot\cdot}{O}}H$$

What about the stereochemistry of the S_N1 reactions? Is our mechanism consistent with the observation that S_N1 reactions of optically active alkyl halides yield racemic products? The answer is yes. To see this, think about the geometry of the carbonium ion. It is planar with bond angles of 120° about the carbonium carbon, and therefore not a center of asymmetry. The nucleophile can attack the carbonium ion with equal probability from either face of the plane of the molecule and therefore produce a racemic product (Figure 5.10).

Of course, if the carbonium ion is bound in some manner in an asymmetric environment, as for example on the surface of an enzyme, then it is very likely that the product will be entirely one enantiomer. However, in the usual laboratory reactions of carbonium ions, racemization is the rule.

One final comment on these nucleophilic substitutions. We have

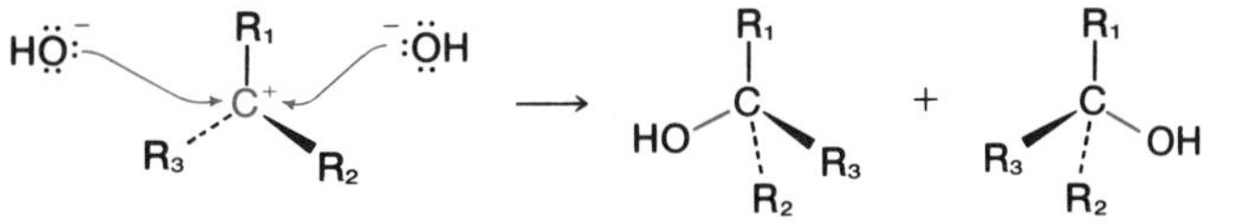

$$HO:^- \;\; R_1R_2R_3C^+ \;\; ^-:OH \longrightarrow HO{-}CR_1R_2R_3 + R_1R_2R_3C{-}OH$$

Figure 5.10 *Attack of a nucleophile from either side of the planar carbonium ion to produce equal amounts of two enantiomers.*

seen that the conversion of 2-bromobutane to 2-butanol proceeds by an S_N2 reaction mechanism, whereas the conversion of *tert*-butyl bromide proceeds by an S_N1 process. What about other substitution reactions? By what mechanisms do they proceed? The answer is that the reaction path depends largely on the ease of forming the carbonium ion. If the carbonium ion is relatively easy to form, as in the case of the tertiary carbonium ion, then the reaction will proceed by the two-step process. If the carbonium ion is formed with difficulty, as in the case of the primary halides, then the reaction will proceed by the one-step process, the S_N2 mechanism. Further, it is possible to define experimental conditions that will favor one or the other of the two mechanisms. A high concentration of nucleophile will favor the S_N2 path; a low concentration will favor the S_N1 path. A solvent of high polarity favors the formation of the carbonium ion by the S_N1 path; a solvent of low polarity favors the S_N2 path. In addition, the nature of the nucleophile will affect the reaction.

5.7 DEHYDRATION OF ALCOHOLS

An alcohol may be converted to an alkene by dehydration—that is, by elimination of a molecule of water. This dehydration requires the presence of an acid catalyst. In practice, dehydration reactions are usually carried out by heating the alcohol and sulfuric or phosphoric acid at temperatures of from 100° to 200°C. Often the water is removed from the reaction mixture by distillation. For example, the dehydration of ethanol yields ethylene.

$$CH_3CH_2OH \xrightarrow{H_2SO_4,\ heat} CH_2{=}CH_2 \quad + \quad H_2O$$

In the first step of such a dehydration, the hydroxyl group is protonated to form an oxonium ion. In the second step, the C—O bond is broken and a carbonium ion is formed. Note that this bond cleavage is made possible through the prior protonation of the hydroxyl group. Finally the carbonium ion loses a proton to form the alkene.

Step 1 $$H{-}\underset{H}{\overset{H}{C}}{-}\underset{H}{\overset{H}{C}}{-}\ddot{\underset{..}{O}}{-}H \quad + \quad H^+ \longrightarrow H{-}\underset{H}{\overset{H}{C}}{-}\underset{H}{\overset{H}{C}}{-}\overset{H}{\underset{..}{O}}{\overset{+}{-}}H$$

Step 2 $$H{-}\underset{H}{\overset{H}{C}}{-}\underset{H}{\overset{H}{C}}{-}\overset{H}{\underset{..}{O}}{\overset{+}{-}}H \longrightarrow H{-}\underset{H}{\overset{H}{C}}{-}\overset{+}{C}\begin{matrix}H\\H\end{matrix} \quad + \quad H{-}\ddot{\underset{..}{O}}{-}H$$

Step 3 $$H{-}\underset{H}{\overset{H}{C}}{-}\overset{+}{C}\begin{matrix}H\\H\end{matrix} \longrightarrow \begin{matrix}H\\H\end{matrix}C{=}C\begin{matrix}H\\H\end{matrix} \quad + \quad H^+$$

Since the dehydration of an alcohol involves an intermediate carbonium ion, we would predict that the relative ease of dehydration of alcohols should be the same as the ease of formation of carbonium ions (p. 57). This prediction is borne out in experimental observations. The ease of dehydration of alcohols is of the order

$$3^\circ > 2^\circ > 1^\circ$$

In dehydrations where more than one alkene is possible, it is generally the more substituted alkene that is the major product.

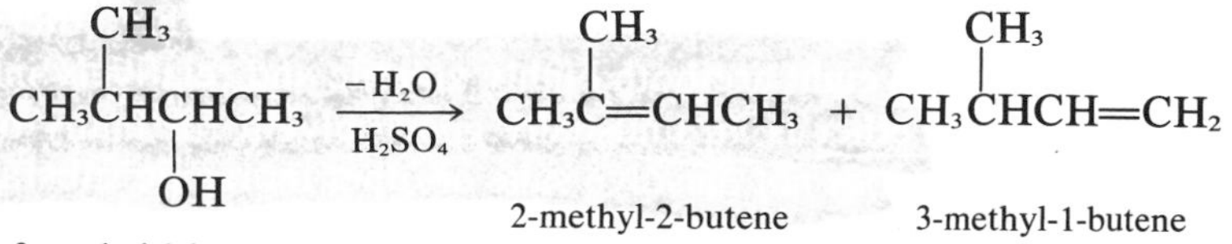

$$CH_3CH(CH_3)CH(OH)CH_3 \xrightarrow[H_2SO_4]{-H_2O} CH_3C(CH_3){=}CHCH_3 + CH_3CH(CH_3)CH{=}CH_2$$

3-methyl-2-butanol → 2-methyl-2-butene (major product) + 3-methyl-1-butene (minor product)

In Section 3.9 we discussed the hydration of alkenes to yield alcohols. In this section we have discussed the dehydration of alcohols to yield alkenes. In fact, the hydration–dehydration reactions are reversible and we must consider the amounts of alkene and alcohol present at equilibrium.

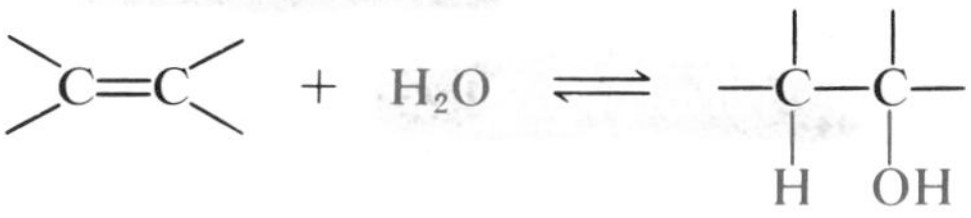

$$>C{=}C< + H_2O \rightleftharpoons -\underset{H}{C}-\underset{OH}{C}-$$

Large amounts of water favor alcohol formation, whereas operation under experimental conditions where water is removed favors formation of alkene. Depending on the conditions, it is possible to use the hydration–dehydration equilibrium to prepare either alcohols or alkenes, each in quite high yields.

5.8 STRUCTURE AND PROPERTIES OF ETHERS

Ethers are derivatives of water in which both hydrogen atoms have been replaced by either alkyl or aryl groups. They are generally named by attaching the names of the alkyl or aryl groups to the generic name ether. For simple ethers, the prefix di- is often not used. Many ethers are known by their common names.

CH_3OCH_3	$CH_3CH_2OCH_2CH_3$	H_2C—CH_2, H_2C—CH_2, ring O	H_2C—O—CH_2, H_2C—O—CH_2 (ring)	C_6H_5—OCH_3
methyl ether bp −24°C	ethyl ether bp 35°C	tetrahydrofuran bp 66°C	1,4-dioxane bp 101°C	anisole or methylphenyl ether bp 154°C

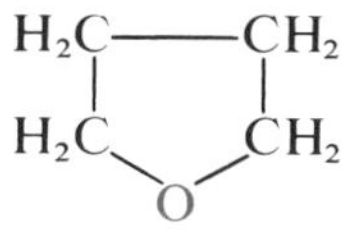

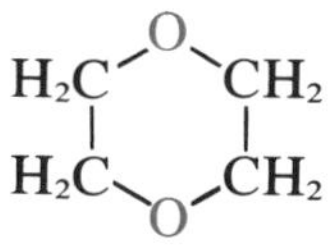

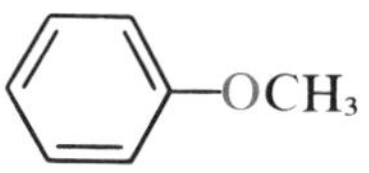

In ethers of more complex structure where one of the attached groups has no simple name, it may be necessary to indicate the —OR group as an alkoxy group.

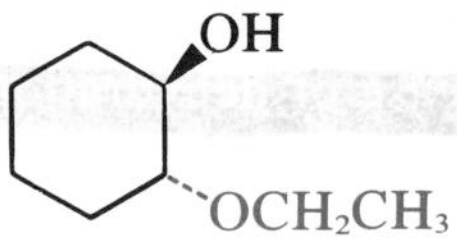

OH

OCH_2CH_3

trans-2-ethoxycyclohexanol

Ethers generally have boiling points about the same as those of alkanes of comparable molecular weight because ethers exhibit no intermolecular hydrogen bonding. The ethers of lower molecular weight are both highly volatile and relatively insoluble in water. Ethyl ether is an unusually good solvent for many organic substances, but does not dissolve salts or other inorganic substances. Because of these good solvent characteristics and because it is not miscible with water, ethyl ether is frequently used to extract organic materials from plant and animal sources. It has a very low boiling point and can easily be removed from the extract by distillation.

Two hazards must be avoided when working with ethers. First, ethers of low molecular weight are highly flammable. Consequently sparks or open flames must be avoided in any area where such an ether is being used. In addition, ethyl ether cannot be extinguished with water because it floats on top of water. Fortunately, carbon dioxide extinguishers are effective. Second, ethers react with oxygen of the air to form peroxides. These peroxides are of higher molecular weight than the parent ethers and accordingly are less volatile. They tend to become concentrated as ether is distilled or evaporated. In concentrated or solid form these peroxides are dangerous because they are highly explosive. Commonly used ethers such as diethyl ether and tetrahydrofuran often become contaminated with peroxides on prolonged storage and exposure to air and light. For this reason purification of ethers by treatment with a reducing agent such as alkaline ferrous sulfate is frequently necessary before use.

Chemically, ethers are far less important than alcohols. In fact ethers resemble the hydrocarbons in their resistance to other types of chemical reaction. They do not react with oxidizing agents such as potassium dichromate, potassium permanganate, or ozone. They are not affected by either strong acids or bases at moderate temperatures. It is precisely this general inertness to chemical reaction on the one hand and good solvent characteristics on the other which make ethers excellent solvents in which to carry out many organic reactions.

5.9 ETHER AND ANESTHESIA*

Prior to the middle of the 19th century, surgery was performed only when absolutely necessary because there were no truly effective general anesthetics. More often than not, patients were drugged with

*Adapted from *Organic Chemistry, A Science and An Art*, by Lloyd N. Ferguson, Willard Grant Press, Boston, 1972.

certain alkaloids, hypnotized, or simply tied down. In 1772 Joseph Priestly isolated nitrous oxide. In 1799 Sir Humphrey Davey demonstrated its anesthetic effect and named it "laughing gas." In 1844 an American dentist, Horace Wells, administered nitrous oxide while extracting teeth. His first anesthetic trials were successful and he soon introduced nitrous oxide into general dental practice. However, one patient awakened prematurely, screaming with pain, and another died. Wells was forced to withdraw from practice, became embittered, depressed, and finally insane. He committed suicide at the age of 33. In the same period, a Boston chemist, Charles Jackson, etherized himself and persuaded a dentist, William Morton, to use it. Subsequently they persuaded a surgeon, John Warren, to give a public demonstration of surgery under anesthesia. The operation was completed painlessly and soon general ether anesthesia was widely adopted for surgical operations. Morton patented his preparation, but soon a great "ether controversy" arose. A Georgia physician, Crawford Long, had used ether anesthesia in surgery and obstetrics several years prior to the Morton-Warren experiment. However, he did not publish his observations or try to bring ether into general anesthesia until 1848 when he gave a report to the Georgia Medical Society. Long, Morton, Jackson, and the descendents of Wells all made claims for the $100,000 prize being offered by the U.S. Congress for development of a safe, effective anesthetic. The controversy continued for years and the award was never made. Embittered by the dispute, Morton died of apoplexy in 1868 and Jackson became insane.

Ethyl ether is still the most widely used anesthetic because it is easy to administer and causes excellent muscle relaxation. If any other agent had been used with the careless methods of earlier times, anesthetic mortality would have been tremendous. Blood pressure and rate of pulse and respiration are usually only slightly affected by ethyl ether. Ether's chief drawbacks are its irritating effect on the respiratory passages and its aftereffect of nausea. Nitrous oxide is pleasant to inhale and rapid in its action, and it is used frequently to induce unconsciousness and pave the way for more potent drugs such as ether.

Vinyl ether does not have a nauseous aftereffect and is a rapid-acting, short-term anesthetic. It is favored by doctors for home and office use, but must be used with caution to prevent too deep a level of unconsciousness. Methylpropyl ether is claimed to be less irritating and more potent than ethyl ether. Cyclopropane, which produces deep anesthesia, would be about the best anesthetic if it were not explosive and did not require so much skill in its administration to avoid deleterious effects on the heart and lungs. Another recent, widely used anesthetic is Halothane, $C_2HBrClF_3$. It is nonflammable, nonexplosive, and causes minimum discomfort to the patient.

The mechanism of action of anesthetics is unknown. Correlations have been observed between anesthetic potency and certain physical properties such as oil/water distribution coefficients, water solubility, vapor pressure, adsorbability, and stability of clathrate hydrates. These generalizations have led to various theories about the mode of action of anesthetics but correlations do not prove a theory, so that no one knows yet how these agents produce the state of unconsciousness.

5.10 OXIDATION OF ALCOHOLS

Alcohols can be oxidized to aldehydes, ketones, or carboxylic acids depending on the number of hydrogen atoms attached to the carbon bearing the —OH and on the experimental conditions.

The oxidation of a primary alcohol under vigorous conditions yields a carboxylic acid. In this oxidation the aldehyde is an intermediate.

$$CH_3CH_2OH + MnO_4^- \xrightarrow{H_2O} \left[CH_3{-}\overset{\overset{\displaystyle O}{\|}}{C}{-}H \right] \longrightarrow CH_3{-}\overset{\overset{\displaystyle O}{\|}}{C}{-}OH + Mn^{2+}$$

acetaldehyde — acetic acid

In general an aldehyde is more easily oxidized than an alcohol. Hence conditions that will suffice to bring about the oxidation of a given primary alcohol will more than suffice to oxidize the intermediate aldehyde to the carboxylic acid. However, by the proper choice of experimental conditions it often is possible to control the oxidation process and isolate aldehydes in reasonable yields. For example, in the case of the oxidation of 1-propanol by aqueous potassium permanganate, the temperature of the reaction mixture is maintained at about 70°C, below the boiling point of the 1-propanol but above that of propanal. Consequently, the aldehyde may be distilled from the reaction mixture as it is formed. It thus escapes further oxidation.

$$CH_3CH_2CH_2OH \xrightarrow[\text{distill } 70°C]{KMnO_4,\ H_2SO_4} CH_3CH_2\overset{\overset{\displaystyle O}{\|}}{C}{-}H$$

1-propanol, bp 97°C — propanal, bp 49°C

The oxidation of secondary alcohols yields ketones. Ketones are resistant to further oxidation, and no special precautions are needed in their preparation.

$$CH_3{-}\overset{\overset{\displaystyle OH}{|}}{C}H{-}CH_3 \xrightarrow[\text{heat}]{KMnO_4,\ H_2SO_4} CH_3{-}\overset{\overset{\displaystyle O}{\|}}{C}{-}CH_3$$

2-propanol (a secondary alcohol) — acetone

$$\text{menthol} \xrightarrow[\text{heat}]{K_2Cr_2O_7} \text{menthone}$$

menthol (a secondary alcohol) — menthone

Tertiary alcohols are stable to oxidation except that if the oxidizing agent is acidic, the alcohol may be dehydrated (Section 5.7); the resulting alkene will be susceptible to oxidative attack.

5.11 THE IMPORTANCE OF OXIDATION IN THE BIOLOGICAL WORLD

We are all aware that there are three major stages in the flow of energy in the biological world. The first stage is photosynthesis, the capture of radiant energy by the pigment chlorophyll in the green plant and its transformation into chemical energy, which is then stored in the form of carbohydrates and other foodstuff molecules. During the process of photosynthesis, carbon dioxide molecules are reduced and carbohydrate molecules formed, and water is oxidized to oxygen.

The second stage in the energy flow is respiration. Through a series of carefully controlled, enzyme-catalyzed oxidations, the chemical energy of carbohydrates and other foodstuff molecules is transformed into a more directly useful form of chemical energy in adenosine triphosphate and other "high-energy" phosphate compounds. In the third stage, the chemical energy recovered from the oxidation of foodstuff molecules is used to do biological work, e.g., the chemical work of synthesis of macromolecules, the mechanical work of muscular contraction, the electrical work of functioning nerve cells, or the osmotic work of moving molecules across cell membranes.

In the following section we shall review how to balance oxidation–reduction reactions. Actually this is a very simple skill and equations of this sort can be balanced easily by following a few simple rules. The reason for developing this skill is so that you can look at a chemical reaction, either a laboratory or a biological reaction, and decide whether it is an oxidation, a reduction, or neither of these. For example, pyruvic acid is converted by one enzyme-catalyzed reaction into acetaldehyde and carbon dioxide; by another enzyme-catalyzed reaction it is converted into acetic acid and carbon dioxide.

$$CH_3-\overset{\overset{\displaystyle O}{\|}}{C}-\overset{\overset{\displaystyle O}{\|}}{C}-OH \xrightarrow{\text{enzyme}} CH_3CHO + CO_2$$

pyruvic acid → acetaldehyde

$$CH_3-\overset{\overset{\displaystyle O}{\|}}{C}-\overset{\overset{\displaystyle O}{\|}}{C}-OH \xrightarrow{\text{enzyme}} CH_3CO_2H + CO_2$$

pyruvic acid → acetic acid

One of these reactions involves an oxidation, the other does not. Which is which?

5.12 BALANCING OXIDATION–REDUCTION EQUATIONS

Oxidation is most conveniently defined as the loss of electrons, and the opposite process of reduction is defined as the gain of electrons. In this sense, oxidation–reduction equations are properly viewed as electron transfer reactions. Every loss of electrons from one atom or molecule (oxidation) must be exactly offset by a compensating gain of electrons in another atom or molecule (reduction). Oxidation and reduction are inextricably linked—one always accompanies the other. Reactions that are described in terms of oxidation and reduction are commonly called redox reactions.

Redox equations are most conveniently balanced by the ion-electron or half-reaction method. Basically this method considers a redox reaction to be the sum of an oxidation half-reaction and a reduction half-reaction. Adding the two half-reactions in the appropriate manner gives the balanced redox equation. The steps in this method of balancing equations are:

1. Write down one equation for the oxidation reaction and another for the reduction reaction.
2. Arrive at a material balance for each by balancing the number of atoms on each side of the equation. To balance the number of oxygen and hydrogen atoms, use H^+ and H_2O if the reaction is done in acid solution; use OH^- and H_2O if the reaction is done in alkaline solution.
3. Arrive at an electrical or charge balance for each equation by adding an appropriate number of electrons to one side or the other.
4. Multiply each half-reaction by an appropriate number so that the number of electrons in the oxidation half-reaction is the same as the number of electrons in the reduction half-reaction.
5. Add the two half-reactions and simplify where possible.

Let us apply this method to balancing the equation for the oxidation of ethanol to acetic acid by potassium permanganate in aqueous acid.

$$CH_3CH_2OH + MnO_4^- \longrightarrow CH_3CO_2H + Mn^{2+} \text{ (in aqueous acid)}$$

This equation is a statement of the reactants and the products, and of the fact that the reaction is carried out in aqueous acid. It is not a balanced chemical equation. Following the rules, we write for the oxidation reaction:

$$(1) \quad CH_3CH_2OH \longrightarrow CH_3CO_2H$$

The number of carbons is balanced as the equation stands. We add H^+ and H_2O to achieve a material balance:

$$(2) \quad CH_3CH_2OH + H_2O \longrightarrow CH_3CO_2H + 4H^+$$

To achieve an electrical or charge balance, we add $4e^-$ to the right-hand side of the equation:

$$(3) \quad CH_3CH_2OH + H_2O \longrightarrow CH_3CO_2H + 4H^+ + 4e^-$$

Equation (3) is a balanced half-reaction. We can see from this half-reaction that the oxidation of ethanol to acetic acid is a four-electron oxidation. Following the same steps for the reduction of MnO_4^- we arrive at (6):

$$(4) \quad MnO_4^- \longrightarrow Mn^{2+}$$

$$(5) \quad MnO_4^- + 8H^+ \longrightarrow Mn^{2+} + 4H_2O$$

$$(6) \quad MnO_4^- + 8H^+ + 5e^- \longrightarrow Mn^{2+} + 4H_2O$$

We see from (6) that the reduction of MnO_4^- to Mn^{2+} is a five-electron reduction.

To add the two equations we must first multiply step (3) by 5, and step (6) by 4. This gives 20 electrons each in the oxidation and the reduction half-reactions. Adding and simplifying gives (7):

$$(7)\quad 5CH_3CH_2OH + 4MnO_4^- + 12H^+ \longrightarrow 5CH_3CO_2H + 4Mn^{2+} + 11H_2O$$

There is a large number of reagents that may be used for the laboratory oxidation of alcohols. We have seen only one—potassium permanganate. Potassium dichromate, $K_2Cr_2O_7$, is another commonly used oxidizing agent. It should come as no surprise that neither of these reagents is involved in the very important oxidation reactions carried out by biological systems. Here, one of the most important agents for enzyme-catalyzed oxidation is nicotinamide adenine dinucleotide, abbreviated NAD$^+$.

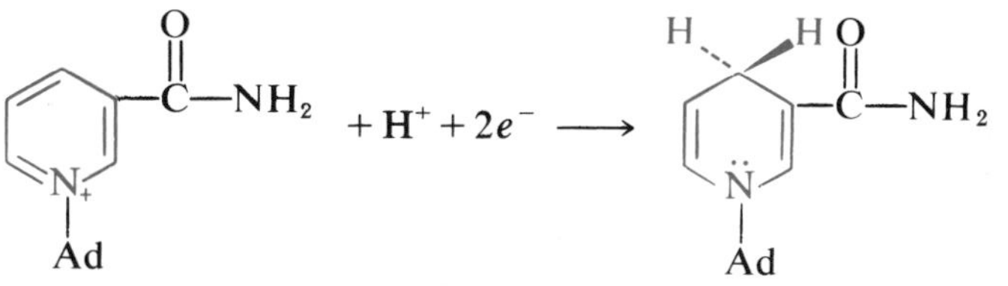

$$\text{NAD}^+ + H^+ + 2e^- \longrightarrow \text{NADH}$$

(Structures: pyridinium ring bearing $-\overset{O}{\overset{\|}{C}}-NH_2$, with N^+–Ad, labeled NAD$^+$; reduced dihydropyridine ring with H, H at C-4 and $-\overset{O}{\overset{\|}{C}}-NH_2$, with $\ddot{N}$–Ad, labeled NADH)

The reactive group of NAD$^+$ in biological oxidation is the pyridine ring (Section 9.4). This ring can accept two electrons and one proton to form the reduced structure NADH.

NAD$^+$ is involved in a great many biological oxidation–reduction reactions. Three examples will show how it is involved in the oxidation of a primary alcohol, a secondary alcohol, and an aldehyde. In each of these oxidations, the reaction is catalyzed by a specific enzyme.

$$\underset{\text{ethanol}}{CH_3CH_2OH} + NAD^+ \xrightarrow[\text{enzyme}]{} \underset{\text{acetaldehyde}}{CH_3\overset{O}{\overset{\|}{C}}-H} + NADH$$

$$\underset{\text{lactic acid}}{CH_3-\overset{OH}{\overset{|}{C}}H-CO_2H} + NAD^+ \xrightarrow[\text{enzyme}]{} \underset{\text{pyruvic acid}}{CH_3-\overset{O}{\overset{\|}{C}}-CO_2H} + NADH$$

$$\underset{\text{3-phosphoglyceraldehyde}}{\begin{matrix} O \\ \| \\ C-H \\ | \\ H-C-OH \quad O \\ | \qquad\quad \| \\ CH_2-O-P-OH \\ | \\ OH \end{matrix}} + NAD^+ \xrightarrow[\text{enzyme}]{} \underset{\text{3-phosphoglyceric acid}}{\begin{matrix} O \\ \| \\ C-OH \\ | \\ H-C-OH \quad O \\ | \qquad\quad \| \\ CH_2-O-P-OH \\ | \\ OH \end{matrix}} + NADH$$

Problems

5.10 Look again at the two reactions of pyruvic acid shown on p. 117. By completing the half-reaction, show that the conversion of pyruvic acid to acetic acid and carbon dioxide is a two-electron oxidation.

5.11 By completing the half-reaction, show that the conversion of pyruvic acid to acetaldehyde and carbon dioxide is neither oxidation nor reduction.

5.12 Complete and balance the NAD^+ oxidations of ethanol, lactic acid, and 3-phosphoglyceraldehyde shown at the end of Section 5.12.

5.13 Each of the following reactions is important in the metabolism of either fats or carbohydrates. State which are oxidations, which are reductions, and which are neither oxidation nor reduction.

a $\underset{\text{glucose}}{C_6H_{12}O_6} \longrightarrow \underset{\text{lactic acid}}{2CH_3\overset{\overset{OH}{|}}{C}HCO_2H}$

b $CH_3(CH_2)_{12}CH_2CH_2CO_2H \longrightarrow CH_3(CH_2)_{12}CH{=}CHCO_2H$

c $CH_3(CH_2)_{12}CH{=}CHCO_2H \longrightarrow CH_3(CH_2)_{12}\underset{\underset{OH}{|}}{C}HCH_2CO_2H$

d $CH_3(CH_2)_{12}\underset{\underset{OH}{|}}{C}HCH_2CO_2H \longrightarrow CH_3(CH_2)_{12}\underset{\underset{O}{\|}}{C}CH_2CO_2H$

e $CH_3(CH_2)_{12}\overset{\overset{O}{\|}}{C}CH_2CO_2H \longrightarrow CH_3(CH_2)_{12}CO_2H + CH_3CO_2H$

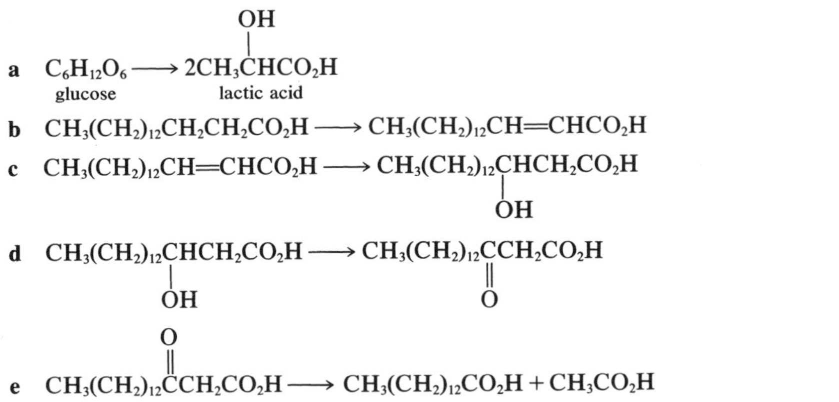

5.13 THIOLS

Sulfur analogs of alcohols are known as thiols, thioalcohols, or mercaptans. The —SH group is called a sulfhydryl group. Mercaptans may be prepared by heating an alkyl halide with sodium hydrosulfide.

$$CH_3CH_2Cl + NaSH \longrightarrow \underset{\substack{\text{ethanethiol}\\ \text{ethyl mercaptan}\\ \text{bp 35°C}}}{CH_3CH_2SH} + NaCl$$

Problem

5.14 How do you account for the fact that ethyl alcohol (molecular weight 46) has a boiling point over 43° higher than that of ethyl mercaptan (molecular weight 62)?

These compounds may be considered as alkyl derivatives of hydrogen sulfide. Like H_2S, they have atrocious odors, they are weakly acidic, and they are easily oxidized. Since they are weakly acidic, thiols yield salts with bases and form insoluble precipitates with lead, mercury, and other heavy metal cations.

$$2RSH + HgCl_2 \longrightarrow (RS)_2Hg + 2HCl$$

Many enzymes contain sulfhydryl groups and can be precipitated or inactivated through the mercury salt.

When treated with mild oxidizing agents such as dilute hydrogen peroxide or cupric salts, thiols are converted into disulfides.

$$2CH_3CH_2SH + H_2O_2 \longrightarrow \underset{\text{ethyl disulfide}}{CH_3CH_2S{-}SCH_2CH_3} + 2H_2O$$

Disulfides are analogous to peroxides but are more stable. They are less volatile than thiols and have less offensive odors.

Problems

5.15 Write balanced half-reactions for the oxidation of ethanethiol by hydrogen peroxide.

5.16 Define oxidation; reduction.

5.17 Oxidation involves loss of electrons. In some, but by no means all oxidations, the loss of electrons is accompanied by the removal of hydrogen and this type of oxidation is known alternatively as dehydrogenation. Which of the following oxidations may also be called dehydrogenations? The oxidation of:

a CH_3CH_2OH to CH_3CHO

b CH_3CHO to CH_3CO_2H

c $CH_3CH{=}CH_2$ to $\begin{array}{ccc} CH_3CH & - & CH_2 \\ | & & | \\ OH & & OH \end{array}$

d $\begin{array}{c} CO_2H \\ | \\ CH_2 \\ | \\ CH_2 \\ | \\ CO_2H \end{array}$ to $\begin{array}{c} CO_2H \\ | \\ CH \\ \| \\ CH \\ | \\ CO_2H \end{array}$

5.18 Write structural formulas for each of the following:

a methylisopropyl ether

b propylene glycol

c 1-chloro-2-hexanol

d 2-methyl-2-*n*-propylpropane-1,3-diol

e 5-methyl-2-hexanol

f 2-isopropyl-5-methyl-1-cyclohexanol

g 2,2-dimethyl-1-propanol

h *tert*-butyl alcohol

i methylcyclopropyl ether

j ethylene glycol

k wood alcohol

l grain alcohol

5.19 Name the following compounds.

a $\begin{array}{c} CH_3CH_2CHCH_2CH_2CH_3 \\ | \\ OH \end{array}$

b (cyclohexane ring bearing CH_2CH_3 and OH on the same carbon)

c (cyclohexene ring bearing OH)

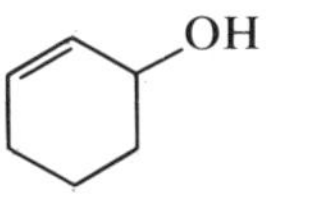

d $CH_3OCH_2CH_2OH$

e $\begin{array}{c} CH_3CH_2CH_2OCHCH_3 \\ | \\ CH_3 \end{array}$

f $C_6H_5CH_2CH_2OH$

g (cyclopentane ring with H, OH and OH, H on adjacent carbons)

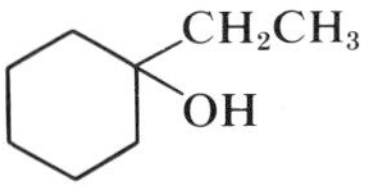

h (benzene ring with OH, NO_2 ortho, and NO_2 para)

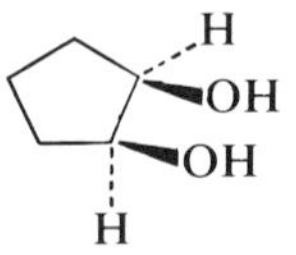

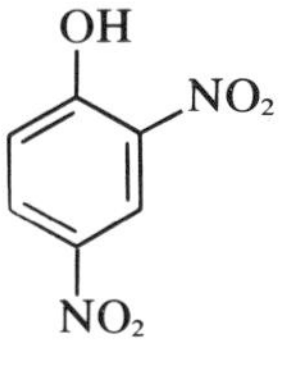

i (benzene ring with OH and CH_3 on adjacent carbons)

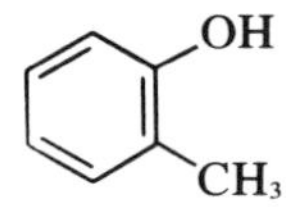

j $CH_3CH_2CH_2CH_2SH$

5.20 Write structural formulas for the six isomeric ethers of molecular formula $C_5H_{12}O$. Name each.

5.21 What is "absolute" ethyl alcohol? How is absolute ethyl alcohol prepared from the more readily available 95% ethyl alcohol?

5.22 Show in a stepwise manner the acid-catalyzed dehydration of cyclohexanol to give cyclohexene and water.

5.23a Write three equations to illustrate nucleophilic substitution at saturated carbon.

b What is meant by the terms S_N1 and S_N2?

c Compare and contrast S_N1 and S_N2 reactions in terms of the rate-determining step, and the stereochemistry of the product as related to that of the starting material.

5.24 Hydrolysis of 2-bromooctane yields 2-octanol.

$$\underset{\text{2-bromooctane}}{CH_3(CH_2)_5\underset{\substack{|\\Br}}{C}HCH_3} + H_2O \longrightarrow \underset{\text{2-octanol}}{CH_3(CH_2)_5\underset{\substack{|\\OH}}{C}HCH_3} + HBr$$

Assume that the starting material is one enantiomer of 2-bromooctane, i.e., that it is optically active. By using suitable stereorepresentations, show the course of this reaction and the stereochemistry of the product if the reaction takes place by an S_N1 mechanism; by an S_N2 mechanism.

5.25 Complete the following substitution reactions showing the major organic products.

a $CH_3\underset{\substack{|\\Cl}}{C}HCH_3 + NaOH \rightarrow$

b $CH_3CH_2\underset{\substack{|\\Br}}{C}HCH_3 + NaSH \rightarrow$

c $CH_3CH_2CH_2OH + HBr \rightarrow$

d $CH_3CH_2\underset{\substack{|\\I}}{C}HCH_3 + CH_3CH_2O^-Na^+ \rightarrow$

5.26 Show how you might convert cyclohexanol into the following:

a cyclohexene
b cyclohexane
c *trans*-1,2-dibromocyclohexane
d *cis*-1,2-cyclohexanediol
e cyclohexanethiol

5.27 Hydration of fumaric acid, catalyzed by aqueous sulfuric acid, produces malic acid.

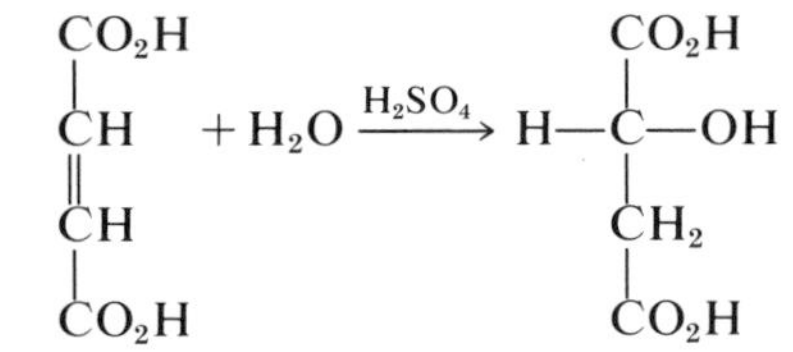

$$\underset{\text{fumaric acid}}{\begin{matrix} CO_2H \\ | \\ CH \\ \| \\ CH \\ | \\ CO_2H \end{matrix}} + H_2O \xrightarrow{H_2SO_4} \underset{\text{malic acid}}{\begin{matrix} CO_2H \\ | \\ H{-}C{-}OH \\ | \\ CH_2 \\ | \\ CO_2H \end{matrix}}$$

a Propose a reasonable mechanism for this hydration.
b Based on your mechanism, would you predict the malic acid so formed to be optically active or racemic? Explain your reasoning.
c The hydration of fumaric acid is one of the steps of the Krebs or tricarboxylic acid cycle. The biological hydration is catalyzed by the enzyme fumarase and produces optically active malic acid. How might you account for the fact that the hydration of fumaric acid catalyzed by aqueous sulfuric acid produces racemic malic acid while the same reaction, catalyzed by fumarase, produces only one enantiomer?

5.28 One of the reactions in the metabolism of glucose is the isomerization of citric acid to isocitric acid. The isomerization is catalyzed by the enzyme aconitase. Propose a reasonable mechanism to account for this isomerization. (Hint: aconitase has within its structure groups that can function as acids.)

$$\underset{\text{citric acid}}{\begin{array}{c} CH_2{-}CO_2H \\ | \\ HO{-}C{-}CO_2H \\ | \\ CH_2{-}CO_2H \end{array}} \underset{}{\overset{\text{aconitase}}{\rightleftharpoons}} \underset{\text{isocitric acid}}{\begin{array}{c} CH_2{-}CO_2H \\ | \\ H{-}C{-}CO_2H \\ | \\ HO{-}C{-}CO_2H \\ | \\ H \end{array}}$$

5.29 Compound A ($C_5H_{10}O$) is optically active, decolorizes a solution of bromine in carbon tetrachloride, and also decolorizes dilute potassium permanganate. Treatment of A with hydrogen gas over a platinum catalyst yields compound B ($C_5H_{12}O$). Treatment of B with warm phosphoric acid forms compound C (C_5H_{10}). Ozonolysis of C yields two compounds, CH_3CHO and CH_3CH_2CHO, in equal amounts. Only compound A is optically active. Propose structural formulas for compounds A, B and C consistent with these observations.

6

ALDEHYDES AND KETONES

6.1 INTRODUCTION

Without doubt the carbonyl group (C=O) is one of the most important functional groups in all of organic chemistry. It is the central feature of aldehydes and ketones, carboxylic acids, esters, amides, and acid anhydrides. Reactions of the carbonyl group are essentially very simple and an understanding of its few basic reaction themes will lead quickly to an understanding of a wide variety of synthetic and biochemical transformations. We shall discuss the chemistry of the carbonyl group in two separate chapters, first in this chapter on aldehydes and ketones, and then again in Chapter 8 on the functional derivatives of carboxylic acids.

6.2 STRUCTURE AND NOMENCLATURE

Aldehydes and ketones contain the carbonyl group, and are referred to as carbonyl compounds. Aldehydes are compounds of the general formula RCHO (or ArCHO); ketones have the general formula R—CO—R′ (either or both R or R′ may be aliphatic or aromatic).

The IUPAC nomenclature of aldehydes follows the familiar pattern of selecting as the name the longest continuous chain that contains the functional group and changing the ending of that hydrocarbon from -e to -al. Common names of aldehydes are very often derived from the common names of the corresponding carboxylic acids by dropping the suffix -ic and adding -aldehyde. Since we will not study the nomenclature of carboxylic acids until the next chapter (Section 7.3), at this point you will not be able to propose common names for aldehydes. However, you should learn the few common names given in this chapter.

We can illustrate how the common system of naming aldehydes works by referring to a carboxylic acid that you are familiar with, namely acetic acid, CH_3CO_2H. The two-carbon aldehyde, CH_3CHO, is related to acetic acid and is therefore named acetaldehyde.

Formaldehyde, the simplest of the aldehydes, derives its name from that of an oxidation product, formic acid. Formaldehyde is a gas (bp −21°C), but is not usually handled in this form. It is commonly produced as a 37% aqueous solution (formalin) which is used for the preservation of biological specimens. Alternatively formaldehyde is produced in solid form as paraformaldehyde (a polymer) or as

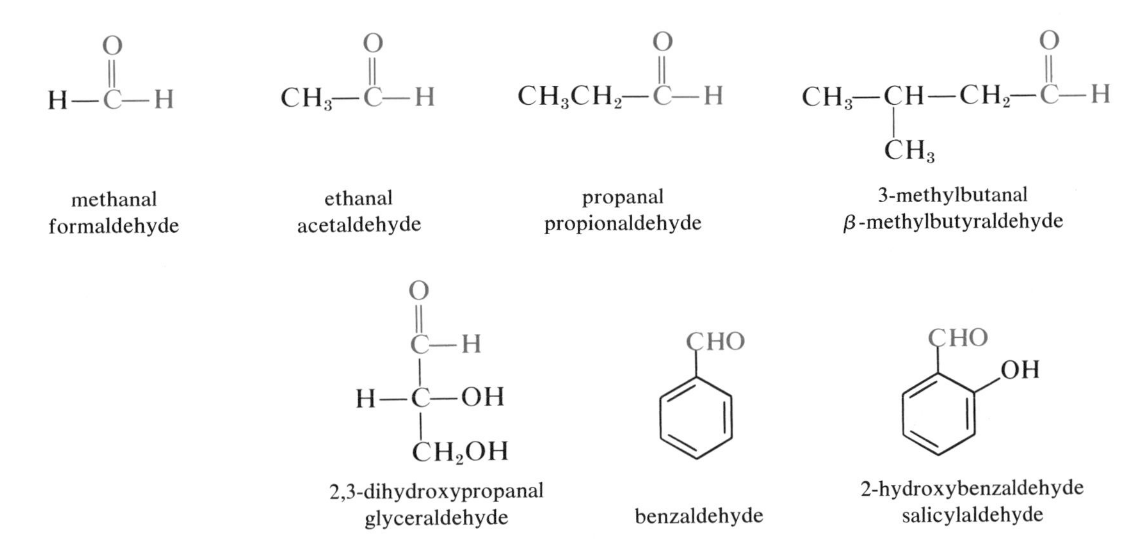

trioxymethylene (a trimer). Formaldehyde is regenerated from either solid form on heating. Candles of paraformaldehyde are sometimes used as fumigating agents.

$HO(CH_2OCH_2OCH_2O)_xH$

paraformaldehyde

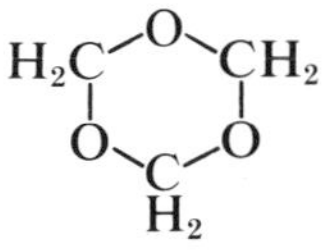

trioxymethylene

The IUPAC system names ketones by using the ending -one to indicate the carbonyl group. As usual the compound is named as a derivative of the longest continuous carbon chain that contains the functional group. In the case of ketones, a number must be used to designate the position of the carbonyl group.

Simple ketones are often named by indicating the names of the two hydrocarbon groups attached directly to the carbonyl.

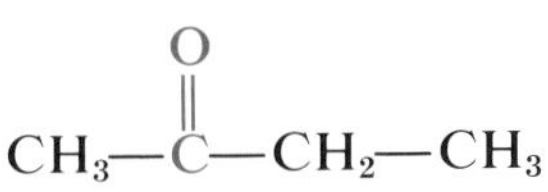

$CH_3-\overset{O}{\overset{\|}{C}}-CH_2-CH_3$

butanone
methyl ethyl ketone

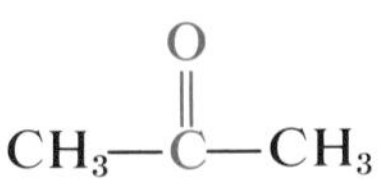

$CH_3-\overset{O}{\overset{\|}{C}}-CH_3$

propanone
acetone

$CH_3-\overset{O}{\overset{\|}{C}}-\underset{}{\overset{CH_3}{\overset{|}{C}H}}-CH_3$

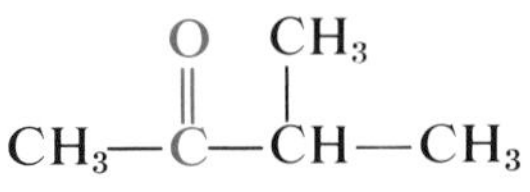

3-methyl-2-butanone
methyl isopropyl ketone

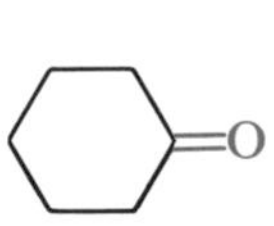

cyclohexanone

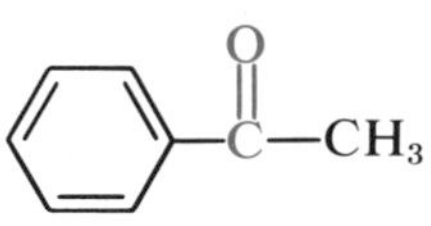

acetophenone
methyl phenyl ketone

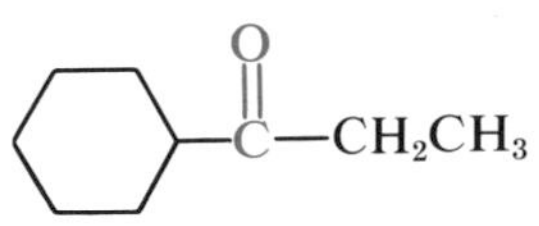

ethyl cyclohexyl ketone

Acetone is an example of a compound whose name is part common or trivial and part derived. Acetone can be prepared by the pyrolysis of the barium salt and certain other heavy metal salts of acetic acid. The name acetone is derived from acetic acid, with the ending -one used to designate that the compound is a ketone.

$$(CH_3COO)_2Ba \xrightarrow[\text{pyrolysis}]{} CH_3-\overset{\overset{\displaystyle O}{\|}}{C}-CH_3 + BaCO_3$$

acetone

Problems

6.1 Draw the structures for all aldehydes of formula C_4H_8O; of formula $C_5H_{10}O$. Give the IUPAC name for each.

6.2 Draw the structures of all ketones of formula C_4H_8O; of formula $C_5H_{10}O$. Give the IUPAC name for each.

6.3 What is the structure of the simplest aldehyde that can show optical isomerism? Name the aldehyde.

6.4 What is the structure of the simplest ketone that can show optical isomerism? Name it.

In the naming of more complicated aldehydes and ketones the carbonyl group generally takes precedence over alkene, hydroxyl, and most other functional groups.

$$CH_3\overset{\overset{\displaystyle OH}{|}}{C}HCH_2\overset{\overset{\displaystyle O}{\|}}{C}CH_3$$

4-hydroxy-2-pentanone

$$CH_3CH_2\overset{\overset{\displaystyle O}{\|}}{C}CH{=}CHCH_2CH_3$$

4-hepten-3-one

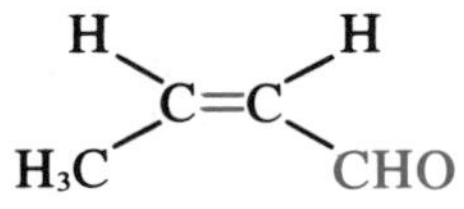

$$\begin{matrix} H & & H \\ & C{=}C & \\ H_3C & & CHO \end{matrix}$$

cis-2-butenal
cis-crotonaldehyde

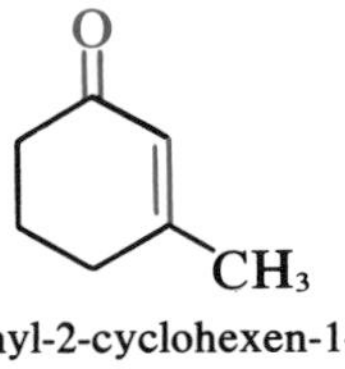

3-methyl-2-cyclohexen-1-one

Note that in the first example the —OH group is indicated by -hydroxy. Also note that in the name 4-hepten-3-one, the heptane root is changed to 4-heptene to indicate the presence of an alkene between carbons 4 and 5 and that the terminal e is converted to -3-one to indicate the presence of a carbonyl group at carbon 3.

A great variety of aldehydes and ketones have been isolated from natural sources, and are best known by their common or trivial names. Such names are usually derived from a source from which the compound can be isolated, or they may refer to a characteristic property of the compound. Figure 6.1 shows structural formulas of several aldehydes and ketones of natural occurrence together with their common names.

Natural or dextrorotatory camphor is characterized by a fragrant

camphor (camphor tree) | irone (violet) | citral (lemon grass oil) | benzaldehyde (bitter almond)

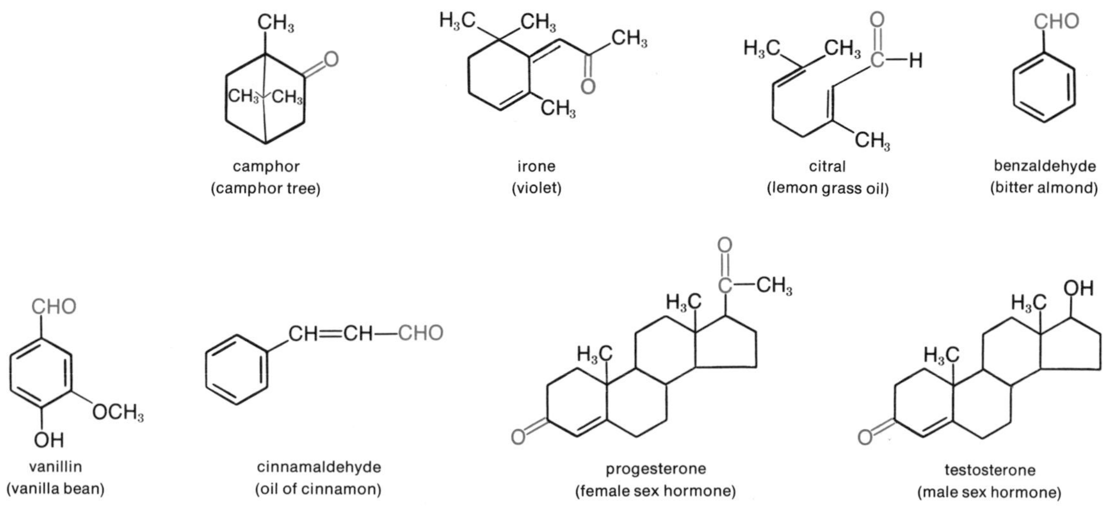

vanillin (vanilla bean) | cinnamaldehyde (oil of cinnamon) | progesterone (female sex hormone) | testosterone (male sex hormone)

Figure 6.1 *Common names of some aldehydes and ketones of natural occurrence.*

penetrating odor. It is obtained from the bark of the camphor tree (*Cinnamomum camphora*) indigenous to the island of Formosa. Camphor has some medicinal uses as a weak antiseptic and as an analgesic and is found in certain liniments. Aldehydes and ketones are particularly abundant in plant essential oils. Irone, the odoriferous principal of the violet, is isolated readily from the rhizomes of iris. Citral is the major constituent (70%) of lemon grass oil. Because of its presence in almond seed, benzaldehyde has been known as oil of bitter almond. Benzaldehyde is also the chief constituent of oils expressed from kernels of the peach, cherry laurel, and other fruits. Vanillin is the fragrant constituent of the vanilla bean. Cinnamaldehyde is the chief constituent of the oil of cinnamon.

Testosterone (Section 11.10), one of the most prominent members of the male sex hormone group, is produced in the testes and is involved in the development of accessory sex functions in the male. Progesterone (Section 11.10), secreted by ovarian tissue (corpus luteum), is one of several sex hormones involved in stimulating growth of the uterine mucosa in preparation for implantation of the fertilized ovum.

In terms of use, acetone is by far the most important ketone. Because of its complete miscibility with water (Table 6.1) and many nonpolar organic solvents, it is used in large volumes as a solvent. Further it is used as an intermediate in the synthesis of many more complex organic molecules. Acetone is formed in the body during fat metabolism. Normally its concentration in the blood is very low, but under conditions where fat metabolism is accelerated, as in the disease diabetes mellitus, there is an accumulation of acetone in the blood. The odor of acetone on the breath of a patient is evidence that he may be a diabetic.

Formaldehyde is used for the preservation of biological materials and in the manufacture of silvered mirrors (Section 6.5). Probably the largest commercial use of formaldehyde is in the base-catalyzed condensation with phenol to yield resins known as Bakelites.

6.3 PHYSICAL PROPERTIES

The carbonyl group makes aldehydes and ketones polar molecules. Hence they have higher boiling points than nonpolar compounds of comparable molecular weight. Since they do not have a hydrogen atom directly bonded to oxygen, they do not show the same degree of interaction and association as alcohols and carboxylic acids of comparable molecular weight. The lower aldehydes and ketones are appreciably soluble in water because of hydrogen bonding between the carbonyl group and the solvent water.

Table 6.1 Physical properties of aldehydes and ketones.

Structure	Name: Common	Name: IUPAC	bp (°C)	Solubility (g/100 g H_2O)
$HCHO$	formaldehyde	methanal	−21	very
CH_3CHO	acetaldehyde	ethanal	20	∞
CH_3CH_2CHO	propionaldehyde	propanal	49	16
$CH_3(CH_2)_2CHO$	butyraldehyde	butanal	76	7
$CH_3(CH_2)_3CHO$	valeraldehyde	pentanal	103	slightly
$CH_3(CH_2)_4CHO$	caproaldehyde	hexanal	129	slightly
CH_3COCH_3	acetone	propanone	56	∞
$CH_3COCH_2CH_3$	methyl ethyl ketone	butanone	80	26
$CH_3COCH_2CH_2CH_3$	methyl *n*-propyl ketone	2-pentanone	102	6
$CH_3CH_2COCH_2CH_3$	diethyl ketone	3-pentanone	101	5

6.4 PREPARATION OF ALDEHYDES AND KETONES

We have already seen (Section 5.10) that controlled oxidation of a primary alcohol will yield an aldehyde and that oxidation of a secondary alcohol will yield a ketone. Aldehydes and ketones are very important as intermediates in the synthesis of other compounds, and since it is often necessary to create an aldehyde or ketone in a molecule containing other potentially reactive functional groups, the synthetic organic chemist has developed many different reagents and experimental techniques for the introduction of the carbonyl group. Oxidation is but one of these techniques. Our concern is not so much with the methods employed by the chemist in the laboratory as it is with the reactions of organic molecules in biological systems, and accordingly we shall confine our discussion to the methods by which aldehydes and ketones are created in biological systems. As we might anticipate, oxidation of primary and secondary alcohols is the most important biochemical reaction by which aldehydes and ketones are produced. Nicotinamide adenine dinucleotide (NAD^+) is the most important, but not the only, biological oxidizing agent.

6.5 OXIDATION OF ALDEHYDES

As pointed out in Chapter 5, aldehydes are very susceptible to oxidation to give carboxylic acids. Ketones, on the other hand, are oxidized only under the most vigorous conditions. In the oxidation of primary alcohols to aldehydes, special experimental techniques such as removal of the product from the reaction mixture (Section 5.10) must be used to prevent further oxidation.

Aldehydes are readily oxidized to carboxylic acids even by such weak oxidizing agents as silver ion, Ag^+. In Tollens' test, a solution of silver nitrate in ammonium hydroxide is added to the substance suspected of being an aldehyde. Within a few minutes at room temperature, silver ion is reduced to metallic silver and the aldehyde is oxidized to the acid. This test requires an alkaline medium, and to prevent the precipitation of the insoluble silver oxide, the silver ion is converted with ammonia into the soluble silver–ammonia complex ion, $Ag(NH_3)_2^+$.

$$RCHO + 2Ag(NH_3)_2^+ + 3OH^- \longrightarrow RCOO^- + 2Ag + 4NH_3 + 2H_2O$$

If the test is done properly, the metallic silver will deposit as a mirror on a glass surface. It is for this reason that the test is commonly called the silver mirror test. Oxidation by silver ion is a convenient test to distinguish aldehydes from ketones (which do not react). The reduction of silver ion by formaldehyde is used in a commercial process to silver mirrors.

6.6 REACTIONS OF THE CARBONYL GROUP

The typical reactions of aldehydes and ketones may be divided into two classes:

1. nucleophilic addition to the carbonyl group, and
2. reaction at the carbon alpha to the carbonyl group.

These reactions are among the most important in organic and biochemistry for the formation of new carbon-carbon bonds and for this reason they deserve special attention. In this chapter we shall develop the chemistry of nucleophilic addition to the carbonyl group of aldehydes and ketones, and in Chapter 8 we shall discuss nucleophilic addition to the carbonyl groups of carboxylic acids, esters, amides, and anhydrides. In all of these nucleophilic additions, the same generalizations apply—the reaction produces a tetrahedral carbonyl addition compound. Certain of these reactions are acid catalyzed, others are base catalyzed, but in every instance the chemical reactions begin with the formation of a tetrahedral carbonyl addition intermediate.

$$R(H)C{=}O + H{:}Z \longrightarrow R(H)C(OH)(Z)$$

In the case of aldehydes and ketones, the carbonyl addition product is often the final product. With carboxylic acids, esters, amides, and anhydrides, the carbonyl addition derivative subsequently undergoes breakdown to yield another and final product.

6.7 ADDITION OF WATER—HYDRATION

Aldehydes and ketones add water reversibly to form hydrated carbonyl compounds. In pure water at 20°C, formaldehyde is about 99.99% hydrated.

$$\underset{\text{formaldehyde}}{H_2C{=}O} + H_2O \rightleftharpoons \underset{99.99\%}{H_2C(OH)_2}$$

Under the same conditions acetaldehyde is about 58% hydrated. The extent of hydration of ketones at equilibrium is very small. While it is not possible to observe directly acetone hydrate in aqueous solution, it is possible to infer indirectly its existence using $H_2{}^{18}O$ and to demonstrate that there is oxygen exchange between water and acetone. This exchange involves the formation of hydrated acetone.

$$(H_3C)_2C{=}O + H_2{}^{18}O \rightleftharpoons (H_3C)_2C({}^{18}OH)(OH) \rightleftharpoons (H_3C)_2C{=}{}^{18}O + H_2O$$

Exchange is slow in pure water but is accelerated by the addition of small amounts of HCl or other strong acids.

Generally the hydrated forms of aldehydes and ketones cannot be isolated because water is readily lost and the carbonyl group is reformed. In a few special cases where the carbonyl group is heavily substituted with electron-withdrawing groups, the resulting 1,1-diol is stable. For example, chloral hydrate can be isolated as a stable, crystalline solid.

$$\underset{\text{trichloroacetaldehyde}}{CCl_3{-}\overset{O}{\overset{\|}{C}}{-}H} + H_2O \longrightarrow \underset{\text{chloral hydrate}}{CCl_3{-}C(OH)_2{-}H}$$

6.8 ADDITION OF ALCOHOLS—FORMATION OF ACETALS AND KETALS

Alcohols add to aldehydes and ketones in the same manner as described for the addition of water. Addition of one molecule of alcohol to an aldehyde results in the formation of a hemiacetal. The comparable reaction with a ketone forms a hemiketal.

$$CH_3{-}CHO + CH_3OH \rightleftharpoons CH_3{-}C(OH)(OCH_3){-}H$$

a hemiacetal
(acetaldehyde hemiacetal)

Hemiacetals are only minor constituents of an equilibrium mixture of alcohol and aldehyde, except in one very important case. When the hydroxyl group is a part of the same molecule that contains the carbonyl group, and a five- or six-membered ring can form, the cyclic hemiacetal structure is frequently the normal form in which the compound exists at equilibrium.

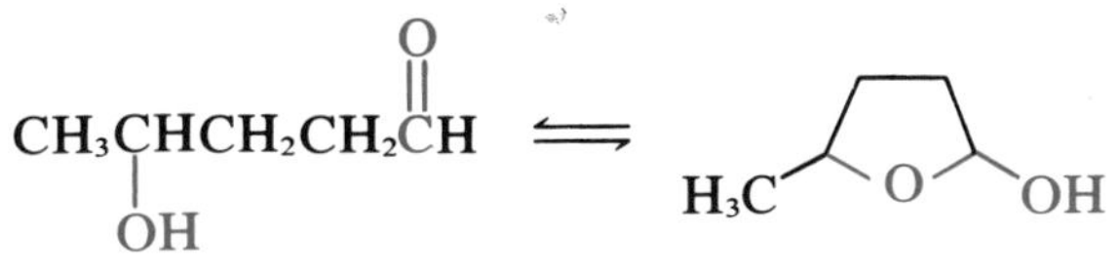

$$CH_3\underset{\substack{|\\OH}}{C}HCH_2CH_2\overset{\substack{O\\||}}{C}H \rightleftharpoons \text{(cyclic hemiacetal: } H_3C\text{–ring(O)–}OH\text{)}$$

4-hydroxypentanal

Problem **6.5** How many stereoisomers are possible for the cyclic hemiacetal shown above? Suppose the starting aldehyde is (+)-4-hydroxypentanal. Now how many stereoisomeric cyclic hemiacetals are possible?

Hemiacetals and hemiketals react further with alcohol, with the formation of compounds known as acetals and ketals. The reaction is catalyzed by acids. The overall reaction of an aldehyde and alcohol may be written as follows:

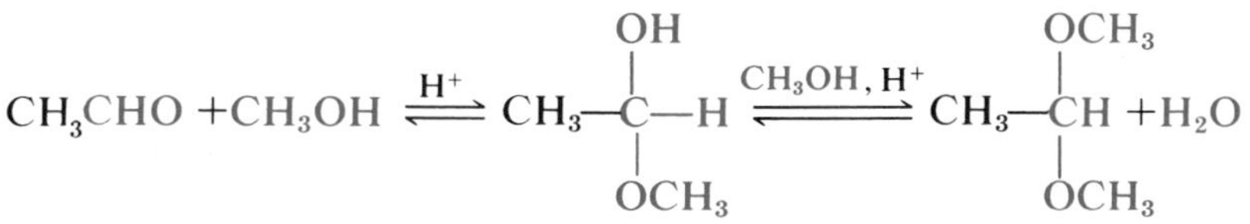

$$CH_3CHO + CH_3OH \underset{}{\overset{H^+}{\rightleftharpoons}} CH_3-\underset{\substack{|\\OCH_3}}{\overset{\substack{OH\\|}}{C}}-H \underset{}{\overset{CH_3OH,\ H^+}{\rightleftharpoons}} CH_3-\underset{\substack{|\\OCH_3}}{\overset{\substack{OCH_3\\|}}{C}}H + H_2O$$

a hemiacetal (acetaldehyde hemiacetal) — an acetal (acetaldehyde dimethyl acetal)

A mechanism for the acid-catalyzed conversion of acetaldehyde hemiacetal into acetaldehyde dimethyl acetal is shown below. It involves the protonation of —OH to form an oxonium ion, loss of a molecule of water to form a resonance-stabilized cation, condensation with a second molecule of alcohol, and finally loss of a proton to give the acetal.

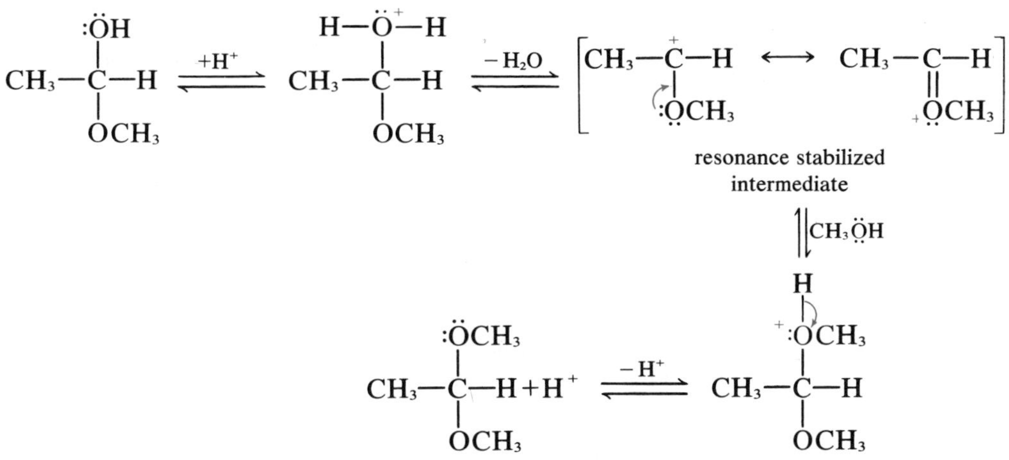

$$CH_3-\underset{\substack{|\\OCH_3}}{\overset{\substack{:\ddot{O}H\\|}}{C}}-H \overset{+H^+}{\rightleftharpoons} CH_3-\underset{\substack{|\\OCH_3}}{\overset{\substack{H-\ddot{O}^+-H\\|}}{C}}-H \overset{-H_2O}{\rightleftharpoons} \left[CH_3-\underset{\substack{|\\:\ddot{O}CH_3}}{\overset{+}{C}}-H \longleftrightarrow CH_3-\underset{\substack{||\\{}^+\ddot{O}CH_3}}{C}-H\right]$$

resonance stabilized intermediate

$$\Big\downarrow\!\!\uparrow\ CH_3\ddot{O}H$$

$$CH_3-\underset{\substack{|\\OCH_3}}{\overset{\substack{:\ddot{O}CH_3\\|}}{C}}-H + H^+ \overset{-H^+}{\rightleftharpoons} CH_3-\underset{\substack{|\\OCH_3}}{\overset{\substack{H\\|\\{}^+:\ddot{O}CH_3\\|}}{C}}-H$$

Acetal formation is an equilibrium reaction and in order to obtain high yields of the acetal, it is necessary to remove water from the reaction mixture and thus favor the formation of the product acetal. Ketones in the same type of reaction form ketals. The addition mechanism is essentially the same as that described for acetal formation but the equilibrium is less favorable to the formation of ketals than it is for acetals.

$$(H_3C)_2C{=}O + 2CH_3OH \underset{}{\overset{H^+}{\rightleftharpoons}} (H_3C)_2C(OCH_3)_2 + H_2O$$

Acetal and ketal formation is catalyzed by acid. Their hydrolysis in water is also catalyzed by acid. However, acetals and ketals are stable and unreactive to aqueous base.

Problems

6.6 Propose a reasonable mechanism for the formation of acetone dimethyl ketal.

$$\underset{\text{acetone}}{(H_3C)_2C{=}O} + 2CH_3OH \overset{H^+}{\rightleftharpoons} \underset{\text{acetone dimethyl ketal}}{(H_3C)_2C(OCH_3)_2} + H_2O$$

6.7 Propose a reasonable mechanism to account for the formation of cyclic acetal from 4-hydroxypentanal and one molecule of methyl alcohol.

6.9 ADDITION OF PRIMARY AMINES—FORMATION OF SCHIFF BASES

Primary amines (R—NH_2) and certain monosubstituted amines such as hydroxylamine (HO—NH_2) are strong enough nucleophiles to add directly to aldehydes and ketones. The amine nitrogen adds to the carbonyl carbon and an amine hydrogen adds to the carbonyl oxygen. These tetrahedral carbonyl addition products undergo ready dehydration to give substances called Schiff bases.

$$>C{=}O + H_2NR \longrightarrow \left[>C\begin{matrix} OH \\ NHR \end{matrix} \right] \longrightarrow \underset{\text{a Schiff base}}{>C{=}NR} + H_2O$$

In the case of certain derivatives of ammonia, the reaction products are often crystalline solids with sharp melting points whose physical properties can be used to identify and characterize unknown aldehydes and ketones.

$$>C{=}O \xrightarrow[\text{hydroxylamine}]{HO{-}NH_2} >C{=}N{-}OH + H_2O \quad \text{(an oxime)}$$

$$>C{=}O \xrightarrow[\text{phenylhydrazine}]{H_2N{-}NH{-}C_6H_5} >C{=}N{-}NH{-}C_6H_5 + H_2O \quad \text{(a phenylhydrazone)}$$

The nucleophilic addition of a primary amine to a carbonyl group is a particularly important part of nitrogen metabolism in biological systems (Chapter 18).

6.10 ADDITION OF HYDROGEN—REDUCTION

Aldehydes can be reduced to primary alcohols, and ketones to secondary alcohols, by an amazing variety of reagents. In this section we shall discuss one laboratory method, namely catalytic hydrogenation. In addition we shall also discuss how aldehydes and ketones are reduced in biological systems by NADH.

Catalytic hydrogenation is completely analogous to the reduction of carbon-carbon double bonds. The carbonyl compound in an inert solvent and hydrogen are shaken under pressure and in the presence of a heavy metal catalyst. The most commonly used catalysts are palladium, platinum, nickel, or copper chromite.

$$\text{cyclohexanone} + H_2 \xrightarrow{Ni} \text{cyclohexanol (OH)}$$

The reduction product is obtained by filtration from the catalyst followed by fractional distillation of the alcohol. Catalytic hydrogenation of an aldehyde or ketone is generally more difficult than the hydrogenation of an alkene. Consequently, if a molecule contains both a carbon-carbon double bond and an aldehyde or ketone, the conditions necessary to reduce the carbonyl group may also saturate the alkene.

In biological systems, the reducing agent most often involved in the conversion of aldehydes or ketones to alcohols is NADH. One such example is the reduction of acetaldehyde to ethanol as the terminal step in alcoholic fermentation of sugar by yeast.

$$\underset{\text{acetaldehyde}}{CH_3CHO} + NADH + H^+ \xrightarrow{\text{enzyme}} \underset{\text{ethanol}}{CH_3CH_2OH} + NAD^+$$

We have already seen the reverse of this reaction in Section 5.12, but there we were concerned with recognizing and balancing oxidation–reduction reactions.

Let us look at this reaction again, and this time examine the steps by which the aldehyde is reduced and NADH is oxidized. Recall that in the discussion of the significance of asymmetry in the biological world (Section 4.10) we developed the concept that reactions in the biological world take place in an asymmetric environment and involve specific interactions between the reactant(s) and an enzyme surface. In the schematic diagram in Figure 6.2, the enzyme is shown binding NADH, acetaldehyde, and the proton needed to complete the formation of the primary alcohol. There is very good evidence that the actual reduction step involves the simultaneous transfer of a pair of electrons and a hydrogen from NADH to acetaldehyde. To complete the reaction, the reduced acetaldehyde reacts with H^+ to form ethanol.

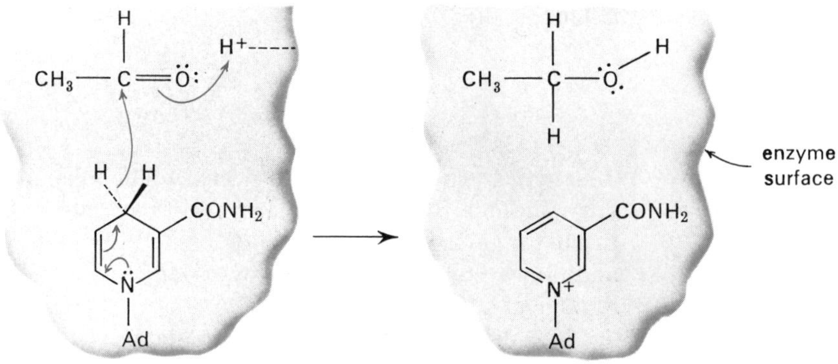

Figure 6.2 *Schematic diagram for the enzyme-catalyzed reduction of acetaldehyde to ethanol by NADH showing the transfer of a hydride ion.*

To appreciate the significance of this scheme, remember that in Section 5.12 we developed a formal method for determining whether a reaction involves oxidation, reduction, or neither. By applying this method, we can show that the conversion of acetaldehyde to ethanol is a two-electron reduction and that the conversion of NADH to NAD^+ is a two-electron oxidation.

$$CH_3CHO + 2H^+ + 2e^- \rightarrow CH_3CH_2OH$$
$$NADH \rightarrow NAD^+ + H^+ + 2e^-$$
$$\overline{CH_3CHO + NADH + H^+ \rightarrow CH_3CH_2OH + NAD^+}$$

Thus we are able to determine what process (oxidation, reduction, or neither) is taking place, but not how it is taking place. The schematic diagram in Figure 6.2 shows the coupling of NADH and acetaldehyde on an enzyme surface and tells us how it happens.

6.11 TAUTOMERISM

Aldehydes, ketones, and other carbonyl compounds (e.g. esters) with an α-hydrogen are in rapid and reversible equilibrium with an isomer formed by the migration of a proton from the α-carbon to oxygen. This new isomer is called an enol, a name derived from the IUPAC designation of it as both an alkene and an alcohol.

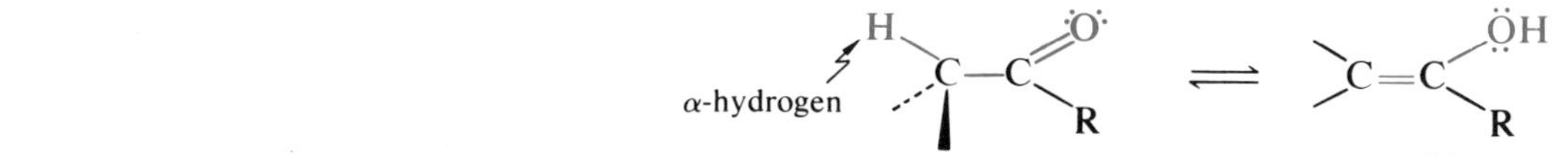

$$\underset{\text{keto form}}{\overset{\alpha\text{-hydrogen}}{H}-\overset{|}{C}-\overset{\overset{\displaystyle O}{\|}}{C}-R} \rightleftharpoons \underset{\text{enol form}}{>C=C(OH)-R}$$

$$CH_3-C(=O)-H \rightleftharpoons CH_2=C(OH)-H$$

$$CH_3-C(=O)-CH_3 \rightleftharpoons CH_2=C(OH)-CH_3$$

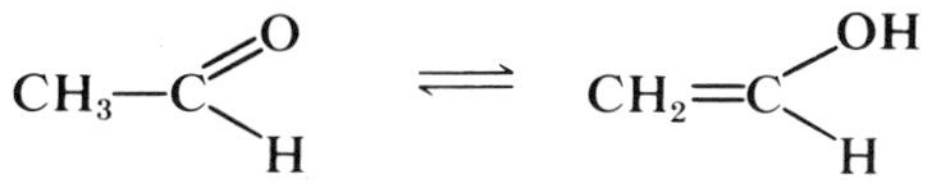

The term tautomerism is used to designate this rearrangement. Most generally tautomeric shifts are observed with migration of a proton between two or more sites within a molecule. In the case of simple aldehydes and ketones, the keto form predominates over the enol form at equilibrium by factors of well over 100 to 1 and no one has yet isolated a pure enol form of a simple aldehyde or ketone because the enol form converts to the more stable keto forms.

However, in molecules that contain an aldehyde or ketone and one or more different functional groups, the enol form may predominate over the keto form. One such example is phenol. Here the enol form is a part of a substituted benzene ring and it is the special stability of the benzene ring that accounts for the stability of this enol form over that of the keto form.

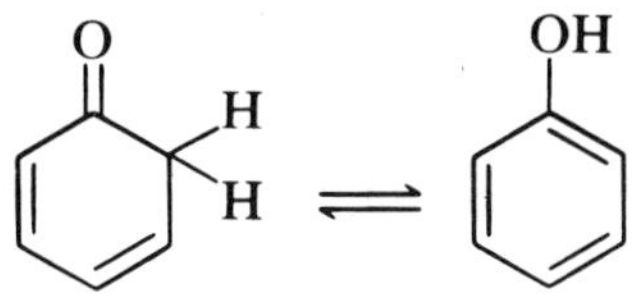

$$\text{cyclohexa-2,4-dienone } (C_6H_6O,\ CH_2) \rightleftharpoons \underset{100\%}{C_6H_5OH}$$

phenol

Another example is ethyl acetoacetate. Liquid ethyl acetoacetate consists of an equilibrium mixture containing approximately 75% of the enol form. Here the enol is stabilized by internal hydrogen bonding between the hydrogen of the —OH group and the carbonyl oxygen of the ester.

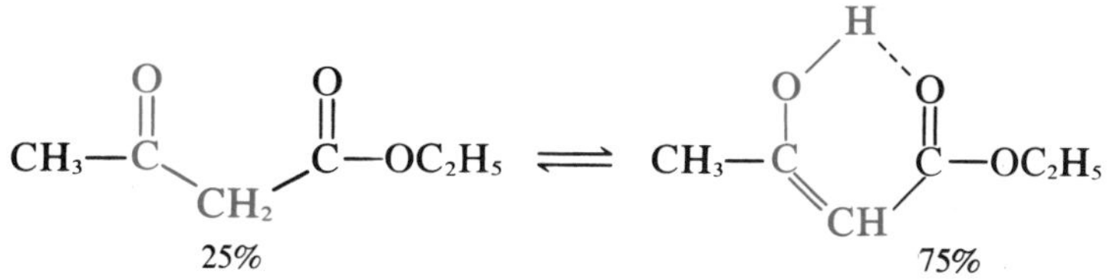

$$\underset{25\%}{CH_3-C(=O)-CH_2-C(=O)-OC_2H_5} \rightleftharpoons \underset{75\%}{CH_3-C(-O-H\cdots O=)=CH-C-OC_2H_5}$$

ethyl acetoacetate

6.12 ACIDITY OF α-HYDROGENS

Because hydrogen and carbon have essentially the same electronegativity, there is normally no appreciable polarity to the C—H bond, and no tendency for the C—H bond to ionize or to show any acidic properties.

In the case of hydrogens α to a carbonyl, the situation is different. Two factors contribute to enhance the acidity of the α-hydrogens. First, the presence of the adjacent carbon-oxygen dipole polarizes the electron pair of the C—H bond to such an extent that the hydrogen may be removed as a proton by a strong base. Ionization of the α-hydrogen results in the formation of an anion. The second and perhaps the more important factor in the acidity of the α-hydrogens is the fact that the resultant anion can be stabilized by resonance. The enolate anion is a hybrid of two important structures, one with the negative charge on carbon (a carbanion) and the other with the negative charge on oxygen.

$$\underset{\text{keto form}}{CH_3{-}\underset{\displaystyle H}{\underset{|}{C}}{=}O} \underset{+H^+}{\overset{-H^+}{\rightleftharpoons}} \left[\,^{-}\ddot{C}H_2{-}\underset{\displaystyle H}{\underset{|}{C}}{=}\ddot{O}: \longleftrightarrow CH_2{=}\underset{\displaystyle H}{\underset{|}{C}}{-}\ddot{O}:^{-} \right]_{\text{enolate anion}} \underset{-H^+}{\overset{+H^+}{\rightleftharpoons}} \underset{\text{enol form}}{CH_2{=}\underset{\displaystyle H}{\underset{|}{C}}{-}OH}$$

The enolate anion may accept a proton on carbon to regenerate the starting material (the keto form) or it may accept a proton on oxygen to form an enol. It should be emphasized that in aqueous solution, most carbonyl compounds show no acidic properties. It is only in the presence of a strong base such as sodium hydroxide, or sodium ethoxide in ethanol, that aldehydes and ketones can be converted into anions.

6.13 ADDITION OF CARBANIONS—THE ALDOL CONDENSATION

The most important type of reaction of an anion derived from an aldehyde or ketone is the addition to the carbonyl of another aldehyde or ketone. For example, the condensation of two molecules of acetaldehyde in aqueous alkali results in the formation of 3-hydroxybutanal, commonly known as aldol. The name aldol is derived from the structural features of the compound; it is an aldehyde and an alcohol.

$$\underset{\text{acetaldehyde}}{2\,CH_3CHO} \xrightleftharpoons{NaOH,\ H_2O} \underset{\substack{\text{3-hydroxybutanal} \\ \text{aldol}}}{CH_3{-}\overset{\displaystyle OH}{\overset{|}{C}}H{-}CH_2{-}CHO}$$

Step 1 in the mechanism for this reaction uses a molecule of base to form the anion of acetaldehyde. Actually the resonance-stabilized enolate anion is produced, but for convenience we focus our attention on the carbanion contributing structure. Nucleophilic addition of the carbanion to another molecule of acetaldehyde in step 2 gives a carbonyl addition intermediate, which in turn reacts with water to give the final product and to regenerate a molecule of base.

Step 1 $\quad CH_3CHO + OH^- \rightleftharpoons :\overset{-}{C}H_2{-}CHO + HOH$

Step 2

$$CH_3{-}\overset{\overset{\displaystyle :\ddot{O}}{\|}}{C}{-}H + \bar{:}CH_2{-}CHO \rightleftharpoons CH_3{-}\underset{\underset{\displaystyle H}{|}}{\overset{\overset{\displaystyle :\ddot{O}:^-}{|}}{C}}{-}CH_2{-}CHO$$

Step 3

$$CH_3{-}\underset{\underset{\displaystyle H}{|}}{\overset{\overset{\displaystyle :\ddot{O}:^-}{|}}{C}}{-}CH_2{-}CHO + HOH \rightleftharpoons CH_3{-}\underset{\underset{\displaystyle H}{|}}{\overset{\overset{\displaystyle :\ddot{O}H}{|}}{C}}{-}CH_2{-}CHO + OH^-$$

Ketones also undergo the aldol condensation, as illustrated by the condensation of acetone in the presence of a barium hydroxide catalyst.

$$2CH_3{-}\overset{\overset{\displaystyle O}{\|}}{C}{-}CH_3 \xrightleftharpoons{Ba(OH)_2} CH_3{-}\underset{\underset{\displaystyle CH_3}{|}}{\overset{\overset{\displaystyle OH}{|}}{C}}{-}CH_2{-}\overset{\overset{\displaystyle O}{\|}}{C}{-}CH_3$$

4-hydroxy-4-methyl-2-pentanone
diacetone alcohol

The ingredients in the key step of the aldol condensation are fundamentally an anion and a carbonyl group. In the self-condensation both roles are played by one kind of molecule, but there is no reason why this should be a necessary condition for the reaction. Many kinds of mixed aldol condensations are possible. Consider the aldol condensation of acetone and formaldehyde. Formaldehyde cannot function as an enolate anion because it contains no α-hydrogen, but it can function as a particularly good enolate anion acceptor because of the freedom from steric crowding around the carbonyl group. Acetone forms an enolate anion easily but its carbonyl group is a relatively poor enolate anion acceptor compared to that of formaldehyde. Consequently the aldol condensation of acetone and formaldehyde occurs readily.

$$CH_3{-}\overset{\overset{\displaystyle O}{\|}}{C}{-}CH_3 + H{-}\overset{\overset{\displaystyle O}{\|}}{C}{-}H \xrightleftharpoons{OH^-} CH_3{-}\overset{\overset{\displaystyle O}{\|}}{C}{-}CH_2{-}\overset{\overset{\displaystyle OH}{|}}{C}H_2$$

4-hydroxy-2-butanone

In the case of mixed aldol condensations where there is not an appreciable difference in reactivity between the two compounds, mixtures of products result. For example, in the condensation of equimolar quantities of propanal and butanal, both α-carbons are alike and the carbonyls are alike. As a consequence, the aldol condensation in which a mixture of these two aldehydes is used results in a mixture of four possible products.

Problems

6.8 Draw structural formulas for the four products resulting from the aldol condensation between propanal and butanal.

6.9 Predict the structures of the principal products to be expected from the following aldol condensations.

a propanal $\xrightarrow{\text{NaOH}}$

b cyclohexanone $\xrightarrow{\text{Ca(OH)}_2}$

c glyceraldehyde $\xrightarrow{\text{NaOH}}$

d acetaldehyde + acetone $\xrightarrow{\text{NaOH}}$

e formaldehyde + cyclohexanone $\xrightarrow{\text{Ca(OH)}_2}$

f acetophenone $\xrightarrow{\text{NaOH}}$

g benzaldehyde + acetaldehyde $\xrightarrow{\text{base}}$

β-Hydroxyaldehydes and β-hydroxyketones are very easily dehydrated. Often the conditions necessary to bring about the condensation itself will suffice to cause dehydration. Alternatively, warming the aldol product in dilute mineral acid leads to dehydration. The major product of the loss of water is one with the carbon-carbon double bond α-β to the carbonyl.

$$\underset{\text{aldol}}{CH_3{-}\overset{\overset{OH}{|}}{C}H{-}CH_2{-}CHO} \xrightarrow[\text{warm}]{\text{dilute HCl}} \underset{\substack{\text{2-butenal} \\ \text{crotonaldehyde}}}{CH_3{-}\overset{\beta}{C}H{=}\overset{\alpha}{C}H{-}CHO} + H_2O$$

The double bonds of alkenes, aldehydes, and ketones are readily reduced by catalytic hydrogenation or by chemical means. Hence the aldol condensation is often used for the preparation of saturated alcohols. For example, acetaldehyde may be converted into butanol by first making 2-butenal.

$$\underset{\text{acetaldehyde}}{2CH_3CHO} \xrightarrow[\text{then dehydration}]{\text{aldol condensation}} \underset{\text{2-butenal}}{CH_3CH{=}CHCHO} \xrightarrow{2H_2/Pt} \underset{\text{1-butanol}}{CH_3CH_2CH_2CH_2OH}$$

Alternatively, if the β-hydroxyaldehyde is isolated, selective oxidation of the aldehyde with silver nitrate in ammonia solution will produce a β-hydroxycarboxylic acid.

$$\underset{\text{aldol}}{CH_3\overset{\overset{OH}{|}}{C}HCH_2CHO} + 2Ag^+ \xrightarrow{NH_4OH} \underset{\substack{\text{3-hydroxybutanoic acid} \\ \beta\text{-hydroxybutyric acid}}}{CH_3\overset{\overset{OH}{|}}{C}HCH_2CO_2H} + 2Ag$$

Problem 6.10 Show how the following compounds can be synthesized from the indicated starting materials by way of aldol condensation products.

a $CH_3—CH_2—CH_2—CH_2OH$ from acetaldehyde

b $CH_3—CH{=}CH—CO_2H$ from acetaldehyde

c $CH_3—CH(OH)—CH_2—CH_2(OH)$ from acetaldehyde

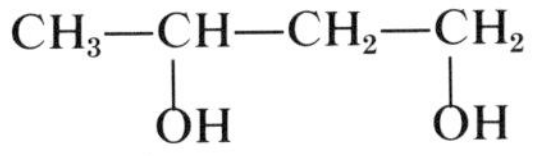

d $CH_3—CH_2—CH_2—CH_2—CH(CH_2—CH_3)—CH_2OH$ from butanal

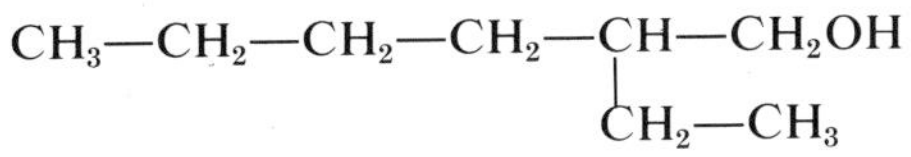

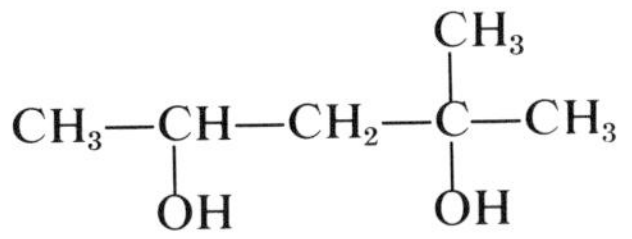

e $CH_3—CH(OH)—CH_2—C(CH_3)(OH)—CH_3$ from acetone

f $CH_3—CH(CH_3)—CH_2—CH_2—CH_3$ from acetone

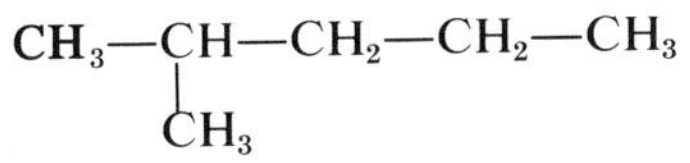

An important example of a biochemical process that may be formulated as an aldol condensation is the reaction, catalyzed by the enzyme aldolase, which forms fructose 1,6-diphosphate from dihydroxyacetone phosphate and glyceraldehyde 3-phosphate. Note that at physiological pH, near 7, the phosphate groups are ionized.

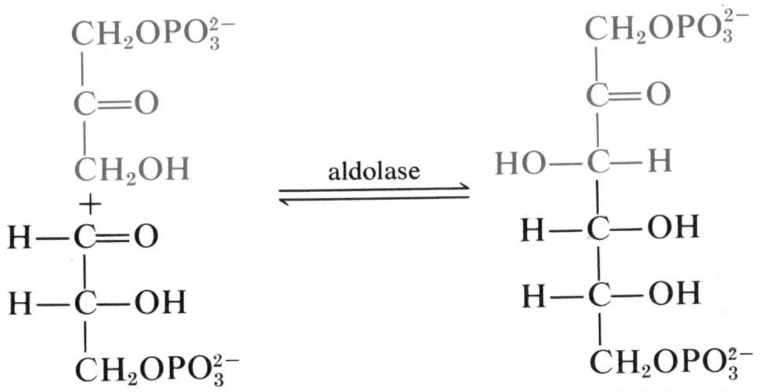

$$\begin{array}{c} CH_2OPO_3^{2-} \\ | \\ C{=}O \\ | \\ CH_2OH \\ + \\ H—C{=}O \\ | \\ H—C—OH \\ | \\ CH_2OPO_3^{2-} \end{array} \xrightleftharpoons{\text{aldolase}} \begin{array}{c} CH_2OPO_3^{2-} \\ | \\ C{=}O \\ | \\ HO—C—H \\ | \\ H—C—OH \\ | \\ H—C—OH \\ | \\ CH_2OPO_3^{2-} \end{array}$$

dihydroxyacetone phosphate + glyceraldehyde 3-phosphate → fructose 1,6-diphosphate

This reaction occurs in the biosynthesis of glucose in green plants. The reverse reaction occurs during the metabolism of glucose to produce energy (glycolysis).

Problems

6.11 Name the following compounds.

a $CH_3—CH(CH_3)—CHO$

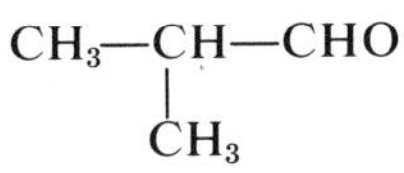

b $CH_3CH_2CH_2\overset{O}{\overset{\|}{C}}CH_2CH_2CH_3$

c (cyclopentanone ring, C=O, with adjacent CH_3)

d (benzene ring)$—\overset{O}{\overset{\|}{C}}—CH_2CH_3$

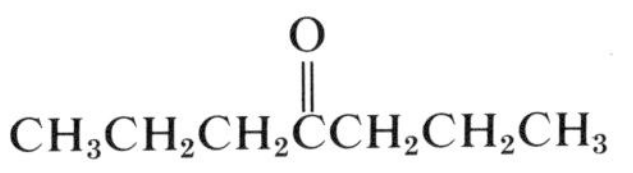

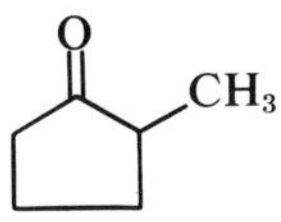

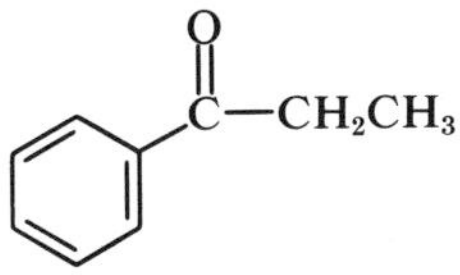

e $CH_3-CH(OH)-C(=O)-CH_2-CH_3$ **f** $CH_3-CH{=}CH-CHO$

g 1,1-dimethoxycyclohexane ($C_6H_{10}(OCH_3)_2$) **h** $C_6H_5-CH(OCH_2CH_3)_2$

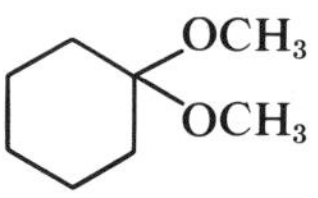

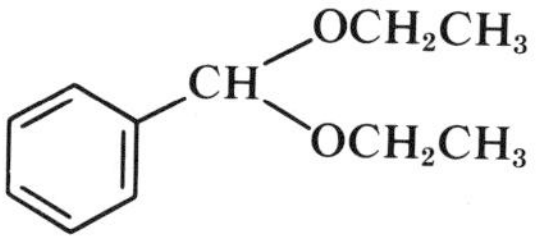

i $CH_3-O-CH_2-CH_2-CHO$ **j** $CH_3O-C_6H_4-CO-CH_3$ (para)

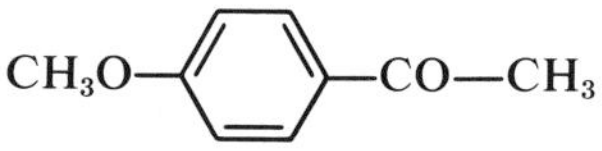

k $CH_3-C(OH)(CH_3)-CH_2-C(=O)-CH_3$ **l** $CH_3CH_2CH(OH)CH_2CH(CH_3)CH_2OH$

m cyclopentane-1,3-dione **n** $CH_3C(=O)CH_2CH{=}CHCH_2CH_2C(=O)-H$

6.12 Write structural formulas for the following compounds.

a cycloheptanone
b propanal
c 2-methylpropanal
d benzaldehyde
e methyl *tert*-butyl ketone
f 1,1-dimethoxycyclohexane
g the phenylhydrazone of acetophenone
h the oxime of cyclopentanone
i propenal
j *p*-bromocinnamaldehyde
k 3-methoxy-4-hydroxybenzaldehyde (vanillin from the vanilla bean)
l 3-phenyl-2-propenal (from oil of cinnamon)
m 3,7-dimethyl-2,6-octadienal (from orange blossom oil)

6.13 Write equations for the reactions (if any) of the following reagents with cyclopentanone and with benzaldehyde.

a silver nitrate in ammonium hydroxide
b methyl alcohol in the presence of a few drops of sulfuric acid
c hydroxylamine
d dilute potassium permanganate in aqueous acid
e formaldehyde in the presence of base
f phenylhydrazine
g H_2, platinum

6.14 Show how you could distinguish between cyclohexene and cyclohexanone using each of the following reagents.

a 5% potassium permanganate in aqueous acid
b bromine in carbon tetrachloride
c phenylhydrazine in aqueous alcohol

6.15 5-Hydroxyhexanal readily forms a six-membered cyclic hemiacetal. Draw a chair conformation for this hemiacetal. Label substituent groups axial or equatorial.

6.16 What is meant by the term tautomerism? Illustrate tautomerism by drawing keto and enol tautomers for acetaldehyde; acetone; butanal; and 2-methylcyclohexanone (two enol forms).

6.17 Describe how resonance and tautomerism differ.

6.18 Compound A ($C_5H_{12}O$) does not give a precipitate with phenylhydrazine. Oxidation of A with potassium dichromate gives B ($C_5H_{10}O$). Compound B reacts readily with phenylhydrazine but does not give a precipitate with silver nitrate in ammonia. The original compound, A, can be dehydrated with sulfuric acid to give a hydrocarbon, C (C_5H_{10}). Ozonolysis of C gives acetone and acetaldehyde. Propose structural formulas for compounds A, B, and C.

6.19 Acetaldehyde reacts with ethylene glycol in the presence of a trace of sulfuric acid to give a cyclic acetal of formula $C_4H_8O_2$. Draw the structural formula of this acetal and propose a mechanism for its formation.

6.20 In dilute aqueous base, the α-hydrogens of butanal show acidity, but neither the β- nor γ-hydrogens show any acidity whatsoever. Account for this difference in acidities.

6.21 How would you account for the fact that in dilute aqueous alkali, glyceraldehyde is converted into an equilibrium mixture of glyceraldehyde and dihydroxyacetone?

CHO
|
CH—OH
|
CH_2OH
glyceraldehyde

dilute base ⟶

CHO
|
CH—OH
|
CH_2OH
glyceraldehyde

+

CH_2OH
|
C=O
|
CH_2OH
dihydroxyacetone

6.22 Cyclohexene can be converted into cyclopentene carboxaldehyde by the following series of steps. Ozonolysis of cyclohexene followed by work up in the presence of zinc and acetic acid forms a compound $C_6H_{10}O_2$. Treatment of this compound with dilute base transforms it into an isomer, also of formula $C_6H_{10}O_2$. Warming this isomer in dilute acid yields cyclopentene carboxaldehyde.

cyclohexene $\xrightarrow[\text{then zinc in acetic acid}]{O_3}$ $C_6H_{10}O_2$ $\xrightarrow{\text{dilute base}}$ $C_6H_{10}O_2$ $\xrightarrow[-H_2O]{\text{warm}}$ cyclopentene carboxaldehyde (CHO)

Propose structural formulas for the isomeric compounds of formula $C_6H_{10}O_2$ and account for the conversion of one isomer into the other in the presence of dilute base.

MINI-
ESSAY

INSECT JUVENILE HORMONE

Present knowledge of insect juvenile hormone stems quite logically from basic research on insect physiology and on the chemical events that control insect development. In the four decades since the existence of insect juvenile hormones was recognized, scientists have isolated and identified the hormone from one insect, discovered an extremely potent and selective hormone-mimicking substance in the plant world, and synthesized hundreds of hormone analogs in the laboratory. As a result we may now be on the verge of realizing a new generation of pesticides, compounds so selective in their action that they will attack only certain insect pests without presenting a hazard to other organisms.

To put these results in perspective we must first look at insect development itself. Insect growth and metamorphosis generally proceed through four stages—egg to larva to pupa to adult. Internal glands secrete certain hormones which control these various stages. One such hormone, known as juvenile hormone, is secreted by the corpora allata, two tiny glands in the head of the insect. At certain stages in development this hormone must be present; at certain other stages it must be absent. For example, juvenile hormone must be present for the immature larva to progress through the usual stages of larval growth. Then, for the mature larva to undergo metamorphosis into a mature adult, the secretion must stop. If juvenile hormone is supplied at this critical time, either by implantation of active corpora allata or by application of the pure hormone itself, then the pupa does not form a viable, mature adult. Juvenile hormone must also be absent from insect eggs for them to undergo normal embryonic development. If the hormone is applied to the eggs, either they fail to hatch or the immature insects die without reproducing.

Although the existence of juvenile hormone was recognized as early as 1939 and its site of production in the corpora allata was established, all efforts to extract and isolate it from living insects failed. Then in 1956 Carroll Williams of Harvard University made the fortuitous discovery that the abdomen of the adult male Cecropia moth contains a rich storage depot of the hormone. In retrospect, the discovery of this depot seems most remarkable, for even today the only other known insect from which juvenile hormone can be extracted is the closely related male Cynthia moth.

The first crude preparations of Cecropia juvenile hormone were obtained by extracting the excised abdomens with ether. Evaporation of the ether extracts leaves a golden colored oil, about 0.2 ml per abdomen. Injection of this crude oil produced all the effects achieved by implanting active corpora allata. In fact, injection of the hormone is not even necessary to produce these effects. Merely placing the oily extract on unbroken skin has the same results—derangement of growth and the formation of nonviable adults. It is just this type of disrupted development coupled with the simplicity of application that first suggested juvenile hormone as a potential insecticide. Although its activity varies from family to family, natural Cecropia hormone affects such diverse insects as representatives of Coleoptera, Lepidoptera, Hemiptera, and Orthoptera.

Before the structure of Cecropia juvenile hormone was established, scientists had observed some slight degree of juvenile hormone activity in farnesol, a naturally occurring alcohol (Figure 1). Following this lead, William Bowers of the United States Department of Agriculture experimental station at Beltsville, Maryland, began to make systematic structural modifications of farnesol, hoping to synthesize new and even more active compounds. In 1965 he prepared methyl 10,11-epoxyfarnesoate from farnesol by forming an epoxide (a cyclic ether derived from an alkene) between carbons 10 and 11 and oxidizing the alcohol on carbon 1 to a carboxylic acid, followed by esterification.

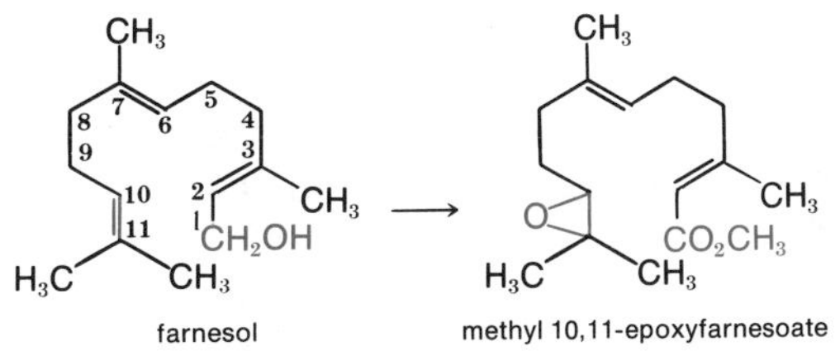

Figure 1 *Farnesol, a naturally occurring alcohol, and methyl 10,11-epoxyfarnesoate, a juvenile hormone analog synthesized in 1965.*

The new substance was 1600 times more active than farnesol, and even though it had only about 0.02% of the Cecropia hormone activity, this suggested that, when juvenile hormone was identified, it would bear at least some structural resemblance to farnesol. Bowers speculated that "... it is quite possible that the juvenile hormone will be synthesized before it is structurally identified from natural sources."

In 1964 another juvenile hormone analog was discovered by what most certainly must be regarded as serendipity smiling and chance favoring the prepared mind. Karel Sláma from Czechoslovakia arrived at Harvard University, bringing with him species of the European bug *Pyrrhocoris apterus*, a species he had reared and studied for many years in Prague. To his considerable surprise and mystification, these bugs, when reared in the Harvard environment, failed to metamorphose into sexually mature adults. Instead they continued to grow as larvae or molted into adultlike forms while retaining many larva-like characteristics. All ultimately died without attaining maturity. Behavior of this type had previously been observed only upon application of juvenile hormone, and it appeared that the bugs had access to some unknown source of the hormone. Finally it was established that the source was none other than the paper towels placed in each rearing jar to give the bug a surface to walk on. Almost all American paper had the same effect, but surprisingly, paper of European or Japanese manufacture had no effect. For want of a better name, the substance was termed "paper factor." The origin of paper factor was traced to balsam fir, a principal source of American paper pulp. Balsam fir synthesizes the active material, which stays with the paper pulp through the entire manufacturing process. "Paper factor" eventually became known as juvabione (Figure 2), suggesting its relation to juvenile hormone.

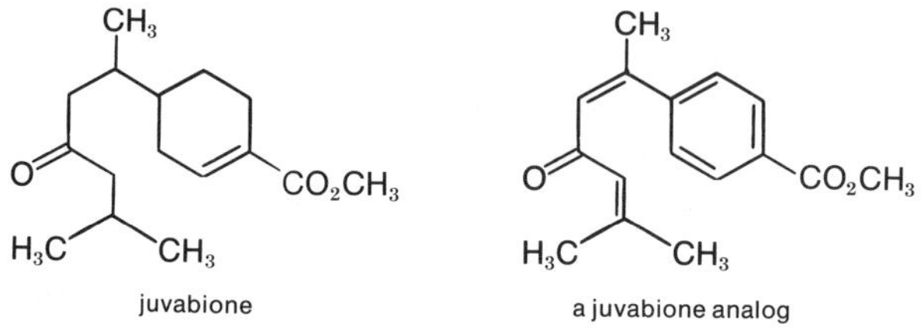

juvabione a juvabione analog

Figure 2 *Juvabione, a juvenile hormone analog of the balsam fir, active only on insects of the family Pyrrhocoridae. A juvabione analog derived from p-(1,5-dimethylhexyl) benzoic acid.*

Unlike the juvenile hormone of the male Cecropia moth, juvabione is active only in the Pyrrhocoridae, an insect family that contains some of the most destructive pests of cotton. Closely related families appear to be totally unaffected. This exciting discovery was the first evidence for the existence of juvenile hormone-like material with highly selective action on a particular family of insects. This type of specificity is essential if we hope to tailor-make insecticides against only certain predetermined pests.

With the structure of juvabione determined, Sláma and his associates in Czechoslovakia prepared a number of compounds structurally related to juvabione but incorporating a benzene rather than a cyclohexene ring. One such analog is shown in Figure 2. Some of these derivatives are about 100 times more active than juvabione itself, and all retain specific action for only Pyrrhocoid bugs.

Research in this area came full cycle when in 1965 Herbert Röller of the University of Wisconsin isolated the male Cecropia hormone itself and, in 1967, using less than 0.3 mg of pure material, determined its structure.

Figure 3 *Juvenile hormone of the male Cecropia moth.*

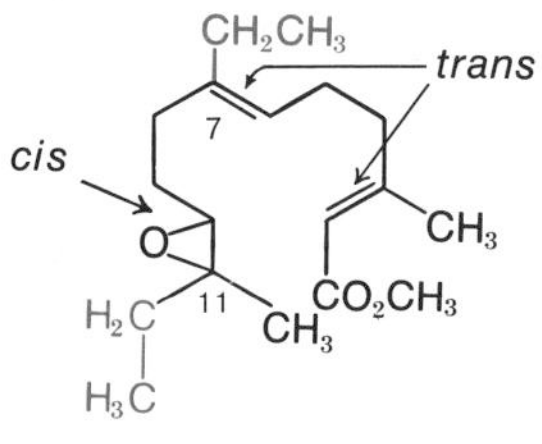

Bower's prediction in 1963 that Cecropia hormone would bear some structural similarity to farnesol is amply confirmed. Furthermore Cecropia hormone (Figure 3) is remarkably similar to the farnesol derivative synthesized by Bowers in 1965. The difference is only two carbon atoms; the alkyl groups at carbons 7 and 11 in Cecropia hormone are —CH_2CH_3 rather than —CH_3. Cecropia hormone contains two *trans* double bonds and a *cis* epoxide. The *trans* configuration of both double bonds appears crucial for biological activity. In contrast, the stereochemistry of the epoxide ring is of secondary importance.

Clearly juvenile hormone analogs offer promise of a new approach in pest control. The strategy is to discover, either from natural sources or through synthesis, compounds that successfully mimic the insect's own hormones and then to use them at critical times during the insect's development. After all, what defense does an insect have against its own hormones? It now appears likely that more compounds will be discovered; they will be relatively simple in structure, easy to synthesize in large quantities, extremely potent, and will be so highly selective in their action that they will scarcely affect the surrounding biosphere. It is on this third generation of pesticides than an ecology-minded public is pinning its hopes.

References

Williams, C., "Third-Generation Pesticides," *Scientific American*, Vol. 217, July 1967, p. 13.

Williams, C., in *Chemical Ecology*, E. Sondheimer and J. Simeone, Editors (Academic Press, New York, 1970).

MINI-ESSAY

PHEROMONES

Chemical communication abounds in nature: the clinging, penetrating odor of the skunk's defensive spray; the hound, nose to the ground in pursuit of prey; the female cat advertising her sexual availability. As biologists and chemists cooperate to extend our knowledge to other animals, it becomes increasingly apparent that chemical communication is a widespread biological phenomenon, perhaps even the primary mode of communication in many species. Prior to 1950, isolation of enough active principal to permit identification seemed an insurmountable task, preventing us from deciphering any of these chemical communications. Rapid advances in instrumental analysis now make it possible to carry out structural determinations on milligrams and even micrograms of material. Structures of an impressive number of chemical code words have been determined. Table 1 lists several of these along with the behavioral responses they evoke. Even with these advances in chemical analysis, the isolation and identification of such chemicals remains a major feat of technical and experimental excellence. For example, obtaining a mere 12 milligrams of gypsy moth sex attractant required the processing of some 500,000 female moths, each yielding about 0.02 micrograms of attractant. In other species, it is not uncommon to process at least 20,000 insects to obtain enough material for identification.

The term pheromone refers to chemicals secreted by an organism which evoke a specific response in another member of the same species. Pheromones are generally divided into two classes—releaser and primer pheromones—depending on their mode of action. Releaser pheromones produce a relatively rapid, reversible change in behavior, such as sexual attraction and stimulation, assembly of elementary aggregations, territorial and home range marking, recruiting for foraging efficiency, etc. Primer pheromones produce more subtle effects. They cause important physiological changes that affect the animal's development and later behavior. The most clearly understood primer pheromones regulate caste systems in social insects (bees, ants, and termites). A typical honey bee colony consists of one queen, several hundred drones (males), and thousands of workers (underdeveloped females). The queen bee, the only fully developed female in the colony, secretes a "queen substance" which prevents construction of royal cells for rearing new queens and which also prevents ovarian development in workers. One of the primer pheromones in the queen substance has been identified as 9-keto-*trans*-2-decenoic acid.

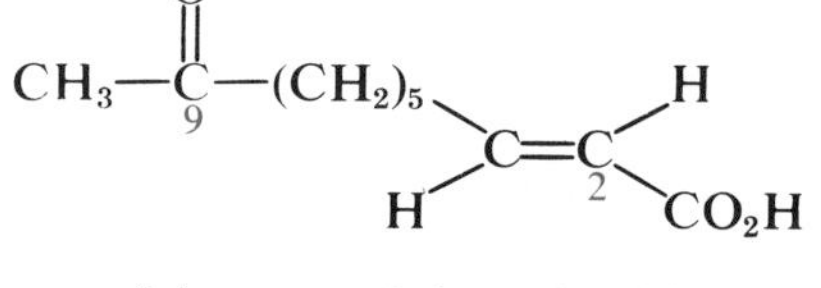

$$CH_3-\overset{\overset{O}{\|}}{\underset{9}{C}}-(CH_2)_5-CH\overset{2}{=}CH-CO_2H$$

9-keto-*trans*-2-decenoic acid

Table 1 *Six insect pheromones that have been isolated and identified, listed according to the behavior elicited.*

Pheromone	Insect
Sex pheromones	
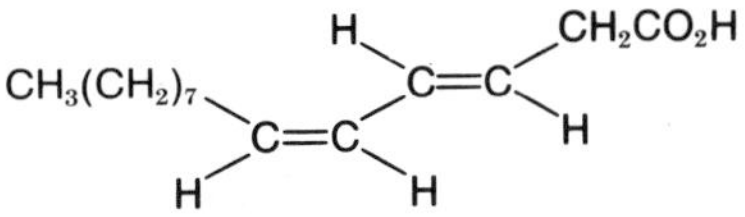 *trans*-3-*cis*-5-tetradecadienoic acid	*Black carpet beetle* (*Attogenus megatoma*)
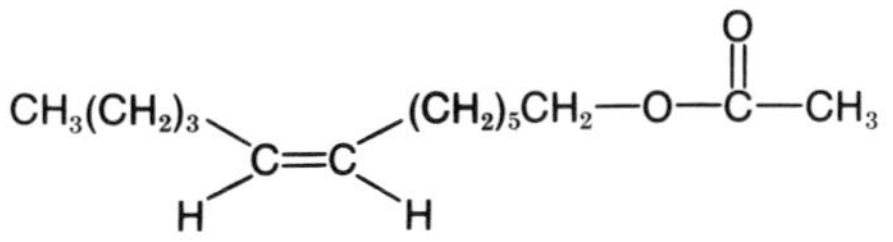 *cis*-7-dodecenyl acetate	Cabbage looper (*Trichoplusa mi*)
Alarm pheromones	
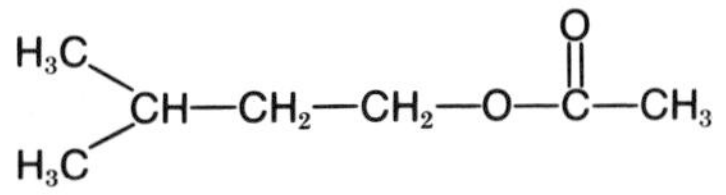 isoamyl acetate	Honey bee (*Hymenoptera*)
Recruiting pheromones	
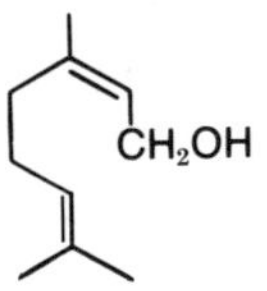 geraniol; 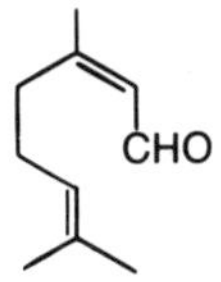 citral	Honey bee (*Hymenoptera*)
Aggregating pheromones	
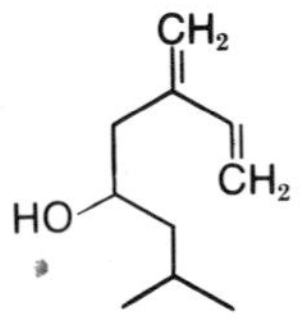 2-methyl-6-methylene-7-octen-4-ol	Bark beetle (*Ips confusus*)

This primer pheromone acts like a birth control pill, regulating the queen–worker relationship. In addition, the same compound serves as a sex pheromone, attracting drones to the queen during her mating flight.

Sex pheromones have received a great deal of attention because the larvae of certain insects which release them, particularly those of moths and beetles, are among the most serious agricultural pests. To date over 100 different insect sex pheromones have been isolated and identified. Sex pheromones are commonly referred to as "sex attractants," but this term is misleading for it implies only attraction. Actually, the behavioral response is considerably more complicated. Low levels of pheromone stimulation elicit orientation and flight of the male toward the female and, if the level of stimulation is high enough, copulation follows. At least in the case of the cabbage looper, none of these behavioral responses requires the presence of the female. A spot of sex pheromone on a piece of filter paper elicits orientation, flight, and even copulatory behavior, all directed toward the spot of the evaporating pheromone.

Some of the earliest observations on alarm pheromones have been recorded on the honey bee. Bee-keepers, and perhaps some of the rest of us too, are well aware that the sting of one bee often causes swarms of angry workers to attack the same spot. When a worker stings an intruder, it discharges, along with venom, an alarm pheromone that evokes the aggressive attack by other bees. One component of this alarm pheromone is isoamyl acetate, a sweet-smelling substance with an odor similar to that of banana oil.

The release of recruiting pheromones near food sources assists honey bees in their foraging activity. Particularly heavy pheromone secretion occurs where copious amounts of sugar syrup are available, and this strongly attracts other foragers. Both geraniol and citral have been identified in this recruiting pheromone.

One of the few aggregating pheromones thus far identified is that of the bark beetle, *Ips confusus.* The bark beetle kills up to five billion board feet of timber each year. Invasion begins with an initial attack by a few beetles, followed by a massive secondary attack that kills the tree. In the case of *Ips confusus*, a few males initially bore into the tree to construct nuptial chambers. During this process they excrete frass, a mixture of fecal pellets and wood fragments. These fecal pellets contain the aggregating pheromone that triggers the massive secondary invasion by both males and females. One male mates with three females, and colonization is on its way. The aggregating pheromone of frass is (−)-2-methyl-6-methylene-7-octen-4-ol.

Impressive though these recent accomplishments are, we are just beginning to understand pheromone communication, and a great deal has yet to be learned. For example, do the same chemicals have different meanings in different contexts (time of day, season, temperature, etc.)? Do pheromones operate singly as well as in mixtures? Can a pheromone mixture carry different information depending on the composition of the mixture? Can animals create new messages by modulating the intensity of emission or the rate of pheromone emission; that is, can they communicate by AM and FM? And another tantalizing question—do human pheromones exist? Hopefully we will begin to discover answers to some of these questions in the decades ahead.

References

Wilson, E. O., *Bio-Organic Chemistry*, M. Calvin and M. Jorgensen, Editors (W. H. Freeman and Company, San Francisco, 1968).

Wood, D. L., Silverstein, R. M., and Nakajima, M., Editors, *Control of Insect Behavior by Natural Products* (Academic Press, New York, 1970).

Sondheimer, E., and Simeone, J. B., Editors, *Chemical Ecology* (Academic Press, New York, 1970).

7

CARBOXYLIC ACIDS

7.1 INTRODUCTION

Of the various classes of organic compounds that show acidity, carboxylic acids are by far the most important. These acids and their functional derivatives are ubiquitous both in the biological world and, thanks to the blend of research and technology, in the world of man-made materials (for example, see the mini-essay "Nylon and Dacron").

While we are presenting some of the chemistry of carboxylic acids, we shall also discuss some of the related chemistry of phosphoric acid, an inorganic acid of special importance in the biological world. Carboxylic acids and phosphoric acid are considered together because they each ionize in water to give acidic solutions and they each form functional derivatives known as esters, amides and anhydrides. The structure and acidity of carboxylic acids are discussed here in Chapter 7. In Chapter 8 we shall discuss the functional derivatives of carboxylic acids and phosphoric acid.

7.2 STRUCTURE

Substances that contain a carbonyl group (C=O) bonded directly to a hydroxyl group (—OH) are called carboxylic acids. We have already described (Problem 1.1) the Lewis electronic structure for the carboxylic acid functional group. Based on our discussions of bonding, we predict bond angles of 120° about the carbonyl carbon, and 109.5° about the hydroxyl oxygen (see Figure 7.1). In addition, the three atoms attached to the carbonyl carbon are in the same plane.

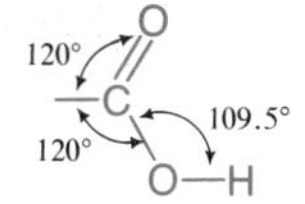

Figure 7.1 *Structure of the COOH group.*

7.3 NOMENCLATURE

Carboxylic acids are easily isolated from natural sources and quite a few have been known since the early days of alchemy. As is often the case with organic nomenclature, the common names refer to the natural sources of these acids and these names have no relation to any systematic nomenclature. For example formic acid adds to the sting of the bite of an ant (Latin, *formica*, ant); butyric acid gives rancid butter its characteristic smell (Latin, *butyrum*, butter); caproic acid is found in goat fat (Latin, *caper*, goat); and lauric acid is derived from laurel.

Carboxylic acids derived from open-chain hydrocarbons are often termed fatty acids because many of them can be obtained from the hydrolysis of naturally occurring fats (Section 11.2). Table 7.1 shows the structural formulas and names of some representative aliphatic monocarboxylic and dicarboxylic acids.

Table 7.1 *Nomenclature of carboxylic acids.*

Formula	Name (IUPAC)	Name (Common)
$HCOOH$	methanoic	formic
CH_3COOH	ethanoic	acetic
CH_3CH_2COOH	propanoic	propionic
$CH_3(CH_2)_2COOH$	butanoic	butyric
$CH_3(CH_2)_3COOH$	pentanoic	valeric
$CH_3(CH_2)_4COOH$	hexanoic	caproic
$CH_3(CH_2)_{10}COOH$	dodecanoic	lauric
$CH_3(CH_2)_{14}COOH$	hexadecanoic	palmitic
$CH_3(CH_2)_{16}COOH$	octadecanoic	stearic
$HOOC{-}COOH$		oxalic
$HOOC{-}CH_2COOH$		malonic
$HOOC{-}CH_2CH_2COOH$		succinic
$HOOC{-}CH_2CH_2CH_2COOH$		glutaric

The IUPAC system of nomenclature employs the name of the longest continuous carbon chain that contains the —COOH group, and indicates the acid function by using the suffix -oic acid. All of the other rules for naming organic compounds apply. Since the —COOH group must be on the terminal carbon of the chain, it is usually not necessary to indicate its position by the number 1-. For common names, Greek letters (α, β, γ, δ, etc.) are often used to locate substituents. Note that the α-carbon is the one next to the carboxyl group and therefore an α-substituent is equivalent to a 2-substituent.

$$\underset{5}{\overset{\delta}{C}}-\underset{4}{\overset{\gamma}{C}}-\underset{3}{\overset{\beta}{C}}-\underset{2}{\overset{\alpha}{C}}-\underset{1}{\overset{\overset{O}{\|}}{C}}-OH$$

The aliphatic dicarboxylic acids are known almost exclusively by their common names. Aromatic acids are usually named by common names or as derivatives of benzoic acid.

In more complex structural formulas, the carboxyl group may be named as a substituent, a carboxy group, as in 2-carboxycyclohexanone shown on p. 151. Alternatively the —COOH group may be named by

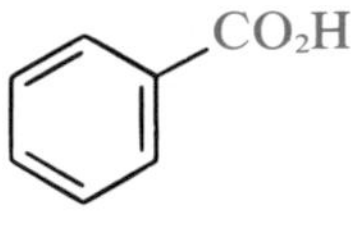

$C_6H_5CO_2H$ — benzoic acid

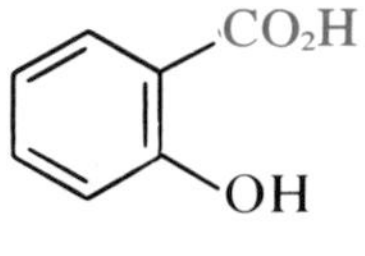

o-$HOC_6H_4CO_2H$ — salicylic acid, o-hydroxybenzoic acid

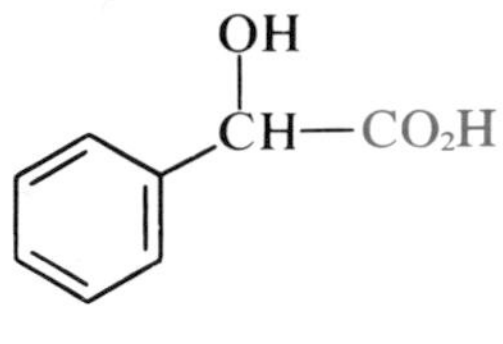

$C_6H_5CH(OH)CO_2H$ — mandelic acid, α-hydroxyphenylacetic acid

adding the words -carboxylic acid to the name of the parent hydrocarbon structure.

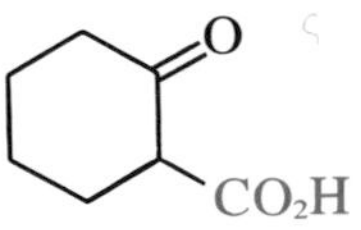

2-carboxycyclohexanone

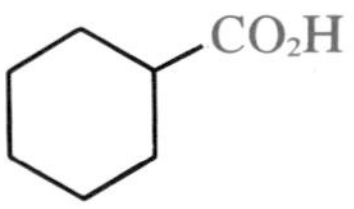

cyclohexanecarboxylic acid

Salts of organic acids are named in much the same manner as inorganic salts; the cation is named first and then the carboxylate anion. The anions are named by dropping the terminal -ic from the name of the acid and adding -ate.

$CH_3-\overset{O}{\overset{\|}{C}}-O^- Na^+$

sodium acetate

$\begin{array}{l} CH_2-\overset{O}{\overset{\|}{C}}-O^- Na^+ \\ | \\ CH_2-\overset{O}{\overset{\|}{C}}-O^- Na^+ \end{array}$

sodium succinate

Problems **7.1** Name the following compounds.

a $CH_3CH(OH)CH_2CH_2CO_2H$

b o-$ClC_6H_4CO_2H$

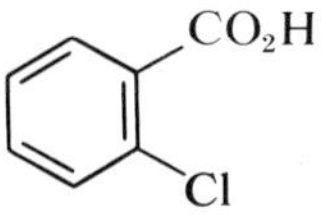

c $ClCH_2CO_2H$

d $CH_3CH(CH_3)CO_2H$

e $C_6H_5CH_2CO_2H$

f cyclopentane–CO_2H

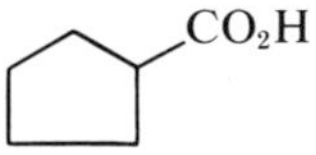

g $C_6H_5CO_2Na$

h $CH_3CH_2CH_2CH_2CO_2NH_4$

i $CH_3CH(CH_2CO_2H)CO_2H$

j CF_3-CO_2H

k $(CH_3CH_2CO_2)_2Ca$

l $CH_3\overset{O}{\overset{\|}{C}}CH_2CH_2CH_2CH_2CH_2CH{=}CHCO_2H$

7.2 Draw structural formulas for the following.

- **a** 3-hydroxybutanoic acid
- **b** sodium oxalate
- **c** trichloroacetic acid
- **d** 4-aminobutanoic acid
- **e** sodium hexadecanoate
- **f** calcium stearate
- **g** potassium phenylacetate
- **h** octanoic acid
- **i** 2-hydroxypropanoic acid (lactic acid)
- **j** 2-aminopropanoic acid (alanine)
- **k** *p*-aminobenzoic acid
- **l** potassium 2,4-hexadienoate (the food preservative potassium sorbate)

7.4 PHYSICAL PROPERTIES

As we would expect from their structure, the carboxylic acids are polar compounds. They show evidence of strong hydrogen bonding. Even the higher members of the aliphatic series show a miscibility with water considerably greater than that of alkanes, alkyl halides, and ethers of comparable molecular weight. Boiling points of carboxylic acids indicate that they are more highly associated in the liquid state than even the alcohols. For example, propanoic acid (mw 74, bp 141°C) boils over 20° higher than 1-butanol (mw 74, bp 117°C).

The first four members of the series (formic–butanoic) are colorless liquids, completely miscible with water, and they have sharp, disagreeable odors. Vinegar is a 4–5% solution of acetic acid in water. It is the acetic acid that gives vinegar its characteristic odor and flavor. Pure acetic acid (mp 16°C, 61°F) solidifies to an icy-looking mass when cooled slightly below normal room temperature and for this reason is often called glacial acetic acid. Table 7.2 gives physical properties of some aliphatic carboxylic acids.

***Table 7.2** Physical properties of carboxylic acids.*

Name	mp (°C)	bp (°C)	Solubility g/100 g H_2O
formic	8	100	∞
acetic	16	118	∞
propanoic	−22	141	∞
butanoic	−6	164	∞
hexanoic	−3	205	1.0
palmitic	63	390	insoluble
oxalic	189	dec.*	9
malonic	136	dec.*	74
succinic	185	dec.*	6

*dec. indicates decomposes before boiling.

Problem

7.3 Both formic acid and acetic acid exist in the gas phase as dimers held together by hydrogen bonding. Propose several geometric arrangements for two molecules of acetic acid held together by hydrogen bonding. (The actual arrangement involves the formation of an eight-membered ring with two hydrogen bonds as part of the ring. Draw this structure also.)

7.5 FATTY ACIDS

Fatty acids are long-chain aliphatic carboxylic acids, so named because they are readily obtained by either acid, alkaline, or enzyme-catalyzed hydrolysis of fats. The common neutral fats are triesters of glycerol and their general formula is shown in Figure 7.2. We shall discuss the structure and properties of fats in more detail in Chapter 11. For the moment we shall look at the fatty acids themselves.

$$\begin{array}{l} CH_2{-}O{-}C({=}O){-}R \\ | \\ CH{-}O{-}C({=}O){-}R' \\ | \\ CH_2{-}O{-}C({=}O){-}R'' \end{array} + 3H_2O \xrightarrow[\text{or enzyme catalysis}]{H^+ \text{ or } OH^-} \begin{array}{l} CH_2{-}OH \\ | \\ CH{-}OH \\ | \\ CH_2{-}OH \end{array} + \begin{array}{l} R{-}CO_2H \\ R'{-}CO_2H \\ R''{-}CO_2H \end{array}$$

a neutral fat — glycerol — fatty acids

Figure 7.2 The hydrolysis of a neutral fat.

Over 70 fatty acids have been isolated from various cells and tissues. Table 7.3 gives the structures of some important fatty acids.

Table 7.3 *Some important naturally occurring fatty acids.*

Carbon Atoms	Structure	Common Name	mp (°C)
Saturated fatty acids			
12	$CH_3(CH_2)_{10}COOH$	lauric	44
14	$CH_3(CH_2)_{12}COOH$	myristic	58
16	$CH_3(CH_2)_{14}COOH$	palmitic	63
18	$CH_3(CH_2)_{16}COOH$	stearic	70
20	$CH_3(CH_2)_{18}COOH$	arachidic	77
Unsaturated fatty acids			
16	$CH_3(CH_2)_5CH{=}CH(CH_2)_7COOH$	palmitoleic	−1
18	$CH_3(CH_2)_7CH{=}CH(CH_2)_7COOH$	oleic	16
18	$CH_3(CH_2)_4CH{=}CHCH_2CH{=}CH(CH_2)_7COOH$	linoleic	−5
18	$CH_3CH_2(CH{=}CHCH_2)_3(CH_2)_6COOH$	linolenic	−11
20	$CH_3(CH_2)_3(CH_2CH{=}CH)_4(CH_2)_3COOH$	arachidonic	−49

We may make certain generalizations about the fatty acid components of higher plants and animals.

1. Nearly all fatty acids have an even number of carbon atoms, usually between 14 and 22 carbons in an unbranched chain.
2. Fatty acids having 16 or 18 carbon atoms are by far the most abundant.
3. In general, unsaturated acids predominate over saturated acids.
4. The unsaturated fatty acids have lower melting points than their saturated counterparts. The physical properties of the particular fatty acid components also affect the fats into which they are

incorporated and, as we shall see in Section 11.2, fats rich in unsaturated fatty acids are lower melting than those rich in the saturated fatty acids.

5. In most of the unsaturated fatty acids of higher organisms there is a double bond between carbon atoms 9 and 10.
6. The double bonds of nearly all naturally occurring fatty acids are in the *cis* configuration. The *trans* configuration is very rare.

Because of their long hydrocarbon chains, fatty acids are essentially insoluble in water. However, they do interact with water in a particular manner. If a droplet of fatty acid is placed on the surface of water, it will spread out to form a thin film one molecule thick (a monomolecular layer) with the polar carboxyl groups dissolved in the water and the nonpolar hydrocarbon chains forming a hydrocarbon layer on the surface of the water. This is illustrated in Figure 7.3.

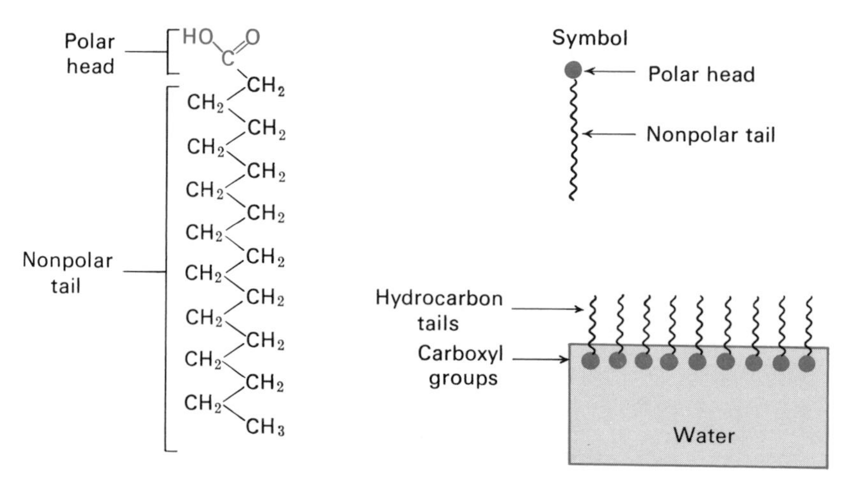

Figure 7.3 *The interaction of a fatty acid droplet with water to form a monomolecular layer.*

7.6 ESSENTIAL FATTY ACIDS

If fatty acids (in the form of fats) are entirely withheld from the diet of rats, the rats soon begin to suffer from retarded growth, scaly skin, kidney damage, and eventually, early death. Addition of the unsaturated fatty acids—linoleic, linolenic, and arachidonic—will cure this condition. Strictly speaking, linoleic is the critical fatty acid for it can be converted within the cell to linolenic and arachidonic acids. Because linoleic acid must be obtained in the diet for normal growth and well-being of higher animals and man, it is classified as an essential fatty acid. Table 7.4 shows some possible dietary sources of linoleic acid.

As you can see from Table 7.4, most animal fats are relatively rich in saturated fatty acids, and though the percentage of unsaturated fatty acids is also high, this unsaturation is due mostly to oleic acid. Vegetable fats, on the other hand, generally have a lower content of

Table 7.4 *Distribution of saturated and unsaturated fatty acids in some foods.**

	% Fat in Edible Portion of Food	% total fat		
		saturated	oleic	linoleic
Animal Fats				
beef	5–37	43–48	43	0.5–3.0
butter	81	57	33	3
eggs	11.5	35	44	8.7
fish (tuna)	4.1	24.4	24.4	0.5
milk (whole, pasturized)	3.7	57	33	3
pork	52	36.5	42	9.6
Vegetable Fats				
coconut oil	100	85	6	0.5
corn oil	100	10	28	53
margarine	81	22.2	58	17.3
peanut oil	100	18	47	29
soybean oil	100	15	20	52
cottonseed oil	100	25	21	50

*Figures adopted from *USDA Handbook 8* (1963).

saturated fatty acids and a relatively higher content of unsaturated fatty acids including linoleic acid. Corn, cottonseed, soybean, and wheat-germ oils are especially rich in linoleic acid.

There is no set minimum requirement for this essential fatty acid, but the Food and Nutrition Board states that for adults a linoleic acid intake of about 6 grams per day in a diet of 2700 calories should be sufficient. For infants and premature babies, the requirements are higher. Human milk and commercially prepared infant formulas provide a generous allowance of linoleic acid.

7.7 SOAPS

Alkaline hydrolysis of naturally occurring fats and oils (esters of long-chain acids and glycerol) is called saponification. The products of saponification are glycerol and the sodium or potassium salts of carboxylic acids. These fatty acids are known as soaps. One of the most ancient organic reactions is the boiling of lard (a fat) with a slight excess of soda (sodium hydroxide) in an open kettle and the eventual isolation of soap. In the present-day industrial manufacture of soap, molten tallow (e.g., the fat of cattle and sheep) is heated with a slight excess of sodium hydroxide. After the saponification is complete, an inorganic salt such as sodium chloride is added to precipitate the soap as thick curds. The water layer is drawn off and the glycerol is recovered from it by vacuum distillation.

The crude soap curds contain salt, alkali, and glycerol as impurities. These are removed by boiling the curds in water and reprecipitating with salt. After several such purifications, the soap may be used without further processing as an inexpensive industrial soap.

Fillers such as sand or pumice may be added to make a scouring soap. Other treatments transform the crude soap into laundry soaps, medicated soaps, cosmetic soaps, liquid soaps, etc.

Soap owes its remarkable cleansing properties to its ability to act as an emulsifying agent. Let us consider the sodium salt of stearic acid as a specific example of a soap. Regarded from one end, sodium stearate is a highly ionic salt and therefore is strongly attracted to water (it is hydrophilic). Regarded from the other end, sodium stearate is a long hydrocarbon chain and is said to be hydrophobic (repelling water) and lyophilic (attracting) toward nonpolar organic solvents. Because its long hydrocarbon chain is intrinsically insoluble in water, there is little tendency for sodium stearate to dissolve in water to form a true solution. However, it readily disperses in water to form micelles in which the charged carboxylate groups form a negatively charged surface and the nonpolar, water-insoluble hydrocarbon chains lie buried within the center (Figure 7.4). Such micelles have a net negative charge and remain suspended or dispersed because of mutual repulsion of one for another.

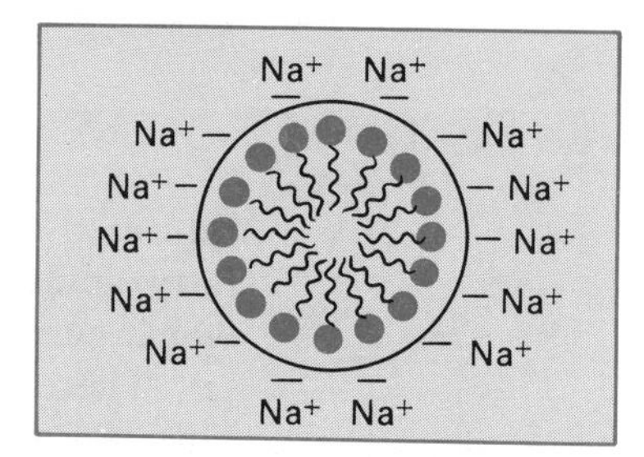

Figure 7.4 *Sodium stearate micelle.*

Soaps also seek out the interface between water and fats, oils, or grease, substances that by themselves are quite insoluble in water. (Most dirt is held to clothes by a thin film of grease or oil.) If the oil is dispersed into tiny droplets throughout the water by shaking, the soap again forms micelles, now with the oil droplet at the center (Figure 7.5). In this way, the oil, grease, or dirt is then emulsified and may be carried away in the wash water.

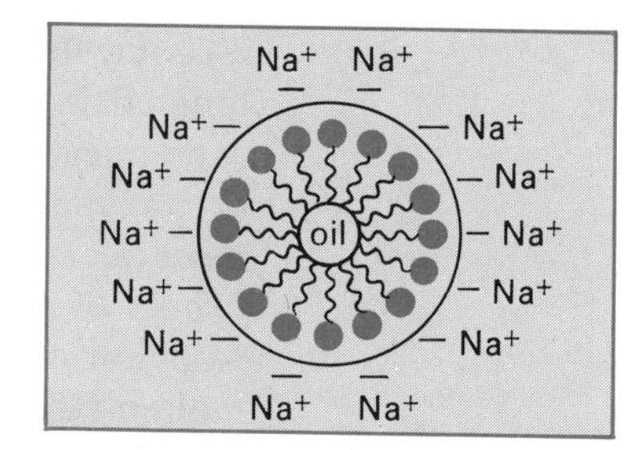

Figure 7.5 *Sodium stearate micelle with a "dissolved" oil droplet.*

Yet soaps are not without their disadvantages. First, soaps are sodium or potassium salts of weak acids and are converted by strong mineral acid into the free fatty acids.

$$CH_3(CH_2)_{16}CO_2^-Na^+ + H^+ \rightarrow CH_3(CH_2)_{16}CO_2H\downarrow + Na^+$$

The free acids are far less soluble than their potassium or sodium salts and they precipitate or form a scum. Therefore soaps cannot be used in acidic solution. Second, soaps form insoluble salts when used in water containing calcium, magnesium, or ferric ions ("hard water").

$$2CH_3(CH_2)_{16}CO_2^- + Ca^{2+} \rightarrow [CH_3(CH_2)_{16}CO_2^-]_2Ca^{2+} \downarrow$$

This precipitate or scum formation creates problems including rings around the bathtub, the films that spoil the luster of hair, and the grayness and harshness of feel that build up on textiles after repeated washing.

Given these limitations on the use of the natural soaps, the problem for the chemist is to create a new type of cleansing agent that will be readily soluble in both acidic and alkaline solutions, and will not form insoluble precipitates when used in hard water. Despite considerable effort, there was no significant progress until late in the 1940s with the introduction of synthetic detergents.

7.8 SYNTHETIC DETERGENTS

One of the most useful innovations in cleansing has been the development in recent years of synthetic detergents (often called syndets). These synthetic products have as good or better cleansing power as ordinary soaps and at the same time they avoid the two major difficulties already listed for soaps. Given an understanding of the mechanism of action of the soaps, the design criteria for a synthetic detergent are as follows: a molecule with a long hydrocarbon chain (preferably 12 to 18 carbon atoms) and a highly polar group or groups at one end of the molecule. It was recognized that the essential characteristics of a soap could be produced in a molecule containing a sulfate group rather than a carboxylate group. Such compounds, known as alkyl acid sulfate esters, are strong acids, comparable in strength to sulfuric acid. Furthermore, the calcium, magnesium, and ferric salts of alkyl acid sulfate esters are soluble in water.

In the earliest method of syndet production, a long-chain alcohol is allowed to react with sulfuric acid and form an alkyl acid sulfate ester. Neutralization with sodium hydroxide forms a synthetic detergent.

$$\underset{\text{lauryl alcohol}}{CH_3(CH_2)_{10}CH_2OH} + H_2SO_4 \rightarrow \underset{\substack{\text{lauryl acid sulfate} \\ \text{(lauryl hydrogen sulfate)}}}{CH_3(CH_2)_{10}CH_2O{-}SO_3H} + H_2O$$

$$\downarrow \text{NaOH}$$

$$\underset{\substack{\text{sodium lauryl sulfate} \\ \text{(a syndet)}}}{CH_3(CH_2)_{10}CH_2O{-}SO_3^-Na^+} + H_2O$$

The physical resemblances between this synthetic detergent and the ordinary soaps are obvious: a long nonpolar hydrocarbon chain and a highly polar end.

Yet the major advance in detergents came in the late 1940s when it became technologically feasible to make the so-called alkylbenzene sulfonate detergents. The essential raw materials for this synthesis, propylene and benzene, had become readily available from the petroleum refining industry. Propylene was polymerized to a tetramer and reacted with benzene. This product was then sulfonated and reacted with sodium hydroxide to yield an alkylbenzene sulfonate salt of the type shown below.

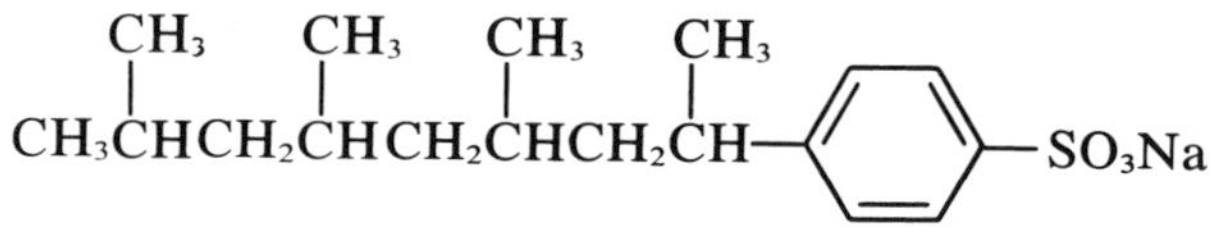

$$\begin{matrix} CH_3 & & CH_3 & & CH_3 & & CH_3 \\ | & & | & & | & & | \\ CH_3CH & CH_2 & CH & CH_2 & CH & CH_2 & CH \end{matrix}\!\!-\!C_6H_4\!-\!SO_3Na$$

a sodium alkylbenzene sulfonate

These alkylbenzene sulfonate detergents were introduced in the 1950s and were accepted very rapidly. Within a decade U.S. production of synthetic detergents increased twenty-fold and today they command close to 90% of the market once held by soaps.

The cleansing power of synthetic detergents of this type can be enhanced enormously by certain additives. Sodium tripolyphosphate ($Na_5P_3O_{10}$), or STPP, is added as a "builder"; it has the ability to coordinate with and suspend calcium, magnesium, copper, iron, and many other ions. Thus STPP is able to break up and suspend certain clays and pigments by forming complexes with the metal ions, thereby facilitating their removal. Phosphates were introduced into cleansing agents in 1948 when Proctor & Gamble Company introduced Tide. By 1953, STPP was used in more than half of the detergents sold in the United States and by 1970 almost all detergents contained phosphates, sometimes as much as 60% by weight. Other common additives are whitening agents (optical brighteners), sudsing enhancers and repressors, and granular salts to create a satisfactory consistency for the commercial product.

Yet as useful as the synthetic detergents proved to be, they have been plagued by two major problems. One of the problems, that of disposability and biodegradability, has been solved. The other, the phosphate additives, is now being brought under control.

The first of these problems began to appear as excessive foaming in natural waters and sewage treatment plants. Most of it was found to be caused by the alkylbenzene sulfonate detergents. Soaps are removed from sewage waters by precipitation or by degradation in the treatment plants by microorganisms that are able to metabolize the linear alkyl hydrocarbon chains of the natural soaps derived from fats and oils. Such soaps are said to be biodegradable. It was discovered that the first alkylbenzene sulfonate detergents marketed could not be removed in either of these ways; they could not be precipitated and they could not be degraded by the microorganisms in sewage treatment plants. Instead they remained in suspension, causing sudsing and foaming, polluting streams, and in some cases finding their way into municipal drinking supplies. The solution to the problem seemed to be in replacing the non-biodegradable branched-chain part of the alkylbenzene sulfonate by a biodegradable linear-chain hydrocarbon. Fortunately just such linear hydrocarbons had become available through advances in petroleum refining, and in particular the use of molecular sieves to separate the desired detergent-range hydrocarbons. In 1965 the detergent industry converted entirely to the new linear alkylbenzene sulfonates of the type shown below.

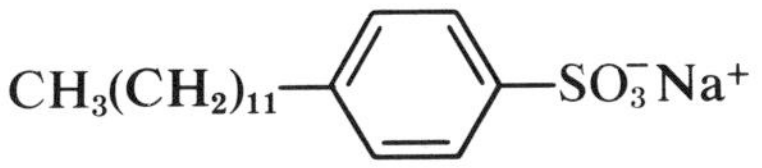

$$CH_3(CH_2)_{11}—C_6H_4—SO_3^- Na^+$$

sodium dodecylbenzenesulfonate
(a syndet)

The second problem is that of the phosphate additives themselves and their contribution to water pollution. Strictly speaking, phosphate is not a pollutant, but rather a fertilizer, and it is as a fertilizer that phosphate has created a problem. The tremendous quantities of phosphates added to lakes, rivers and streams through the use of detergents (and agricultural phosphate-based fertilizers as well) have greatly increased the nutrient quality of the water. According to the House of Representatives' Subcommittee on Conservation and Natural Resources, when a rich stream of fertilizer flows steadily into a lake:

> Overstimulated, the water plants grow to excess. Seasonally they die off and rot. . . . In the process of decay they exhaust the dissolved oxygen of the water and produce the rotten-egg stench of hydrogen sulfide. . . . The game fish die of oxygen deficiency. . . . Intake filters for potable water become clogged, and boat propellers fouled with algae. The lake loses its value as a water supply, as an esthetic and recreational asset, and as an avenue of commerce. Finally the water itself is displaced by the accumulating masses of living and dead vegetation and their decay products and the lake becomes a bog, and, eventually, dry land.

This evolution of a lake is, of course, a natural process, but one that has been greatly accelerated. Lake Erie is the most notorious example of a dying American lake. It is said to have "aged" 15,000 years in the last 50. Since phosphate-containing detergents were introduced in 1948, the phosphate content of the lake's western basin has more than tripled. This situation must not be allowed to continue. Some low-phosphate detergents are already on the market. However, this problem will not be solved until public pressure and intelligent legislation combine to require much lower phosphate levels in all detergents.

7.9 ACIDITY OF CARBOXYLIC ACIDS

All carboxylic acids, whether soluble or insoluble in water, react quantitatively with aqueous solutions of sodium hydroxide to form acid salts. They also react quantitatively with solutions of sodium bicarbonate and sodium carbonate to liberate carbon dioxide. The evolution of CO_2 can be used as a simple qualitative test to distinguish carboxylic acids from most other organic compounds.

In addition to reacting quantitatively with strong inorganic bases,

$$R-\overset{O}{\overset{\|}{C}}-OH + NaOH \longrightarrow R-\overset{O}{\overset{\|}{C}}-O^- \, Na^+ + H_2O$$

$$R-\overset{O}{\overset{\|}{C}}-OH + NaHCO_3 \longrightarrow R-\overset{O}{\overset{\|}{C}}-O^- \, Na^+ + CO_2 + H_2O$$

$$2\,R-\overset{O}{\overset{\|}{C}}-OH + Na_2CO_3 \longrightarrow 2\,R-\overset{O}{\overset{\|}{C}}-O^- \, Na^+ + CO_2 + H_2O$$

carboxylic acids also ionize in water to give acidic solutions, and at least in this operational sense are like the common inorganic acids. However, quantitatively they are quite different from inorganic acids such as HCl, HBr, HNO_3, and H_2SO_4. These inorganic acids are strong acids, that is, they are completely ionized in water.

$$HCl \longrightarrow H^+ + Cl^-$$

Carboxylic acids, on the other hand, are only incompletely ionized in dilute aqueous solution and an equilibrium is established between the carboxylic acid, the carboxylate anion, and H^+.

$$CH_3CO_2H \rightleftharpoons CH_3CO_2^- + H^+$$

The equilibrium constant for this ionization is called K_a, the acid dissociation or ionization constant, and has the form

$$K_a = \frac{[CH_3CO_2^-][H^+]}{[CH_3CO_2H]}$$

Values of K_a for some representative carboxylic acids are given in Table 7.5.

Table 7.5 *K_a and pK_a for some carboxylic acids.*

Name	Structure	K_a	pK_a
formic	HCOOH	1.8×10^{-4}	3.74
acetic	CH_3COOH	1.8×10^{-5}	4.74
propanoic	CH_3CH_2COOH	1.4×10^{-5}	4.85
butanoic	$CH_3CH_2CH_2COOH$	1.6×10^{-5}	4.80
chloroacetic	$ClCH_2COOH$	1.4×10^{-3}	2.85
dichloroacetic	$Cl_2CHCOOH$	3.3×10^{-2}	1.48
trichloroacetic	Cl_3CCOOH	2.3×10^{-1}	0.64
methoxyacetic	CH_3OCH_2COOH	3.3×10^{-4}	3.48
benzoic	C_6H_5COOH	6.5×10^{-5}	4.19

It should be obvious from the expression for K_a that strong (highly dissociated) acids will have large values of K_a and, conversely, that weak (incompletely dissociated) acids will have considerably smaller values of K_a. From the K_a values in Table 7.5 we know that formic acid is a stronger acid than acetic acid. Other than the K_a for formic acid, the values for unsubstituted aliphatic carboxylic acids are essentially the same as that for acetic acid. Therefore the value of K_a for acetic acid is a useful number to remember.

As a matter of convenience, dissociation constants of acids are very often reported as pK_a, a number defined as the negative logarithm of K_a. For acetic acid,

$$K_a = 1.8 \times 10^{-5}$$

$$pK_a = -\log 1.8 \times 10^{-5} = 4.74$$

Obviously pK_a numbers are easier to say, simpler to write, and more convenient to express in tables. Table 7.5 also gives pK_a values for the carboxylic acids.

We might wonder why carboxylic acids are so much more acidic than alcohols, other compounds containing the —OH functional group. The acidity of ethyl alcohol is undectable in water, but the dissociation constant is estimated to be about 10^{-16}. The value of K_a for acetic acid is over 10^{10} times larger than that for ethanol. The question may be stated in another way: comparing acetic acid and ethyl alcohol, why is acetic acid so much more extensively ionized in water?

$$CH_3CH_2OH \rightleftharpoons CH_3CH_2O^- + H^+ \quad \text{ethanol}$$

$$CH_3CO_2H \rightleftharpoons CH_3CO_2^- + H^+ \quad \text{acetic acid}$$

The answer is that there is appreciable resonance stabilization of the carboxylate anion. There is no comparable resonance stabilization for the ethylate anion.

$$CH_3{-}C(=\ddot{O}{:})(\ddot{O}{:}^-) \longleftrightarrow CH_3{-}C(\ddot{O}{:}^-)(=\ddot{O}{:})$$

once the H+ is given up the anion is stable & does not want it back

These two structures differ only in the position of the electrons; they are contributing structures to the resonance hybrid. The two forms are structurally equivalent and the resonance stabilization is large. This resonance effect is largely responsible for the greater stability of the carboxylate anion relative to the ethylate anion, and the greater acid strength of the carboxylic acid.

7.10 CALCULATION OF [H⁺]

If we know the K_a or pK_a of a carboxylic acid, we can then calculate the hydrogen ion concentration of a solution containing the carboxylic acid. As an example, consider a solution of 60.0 grams (one mole) of acetic acid in enough water to make one liter of solution, that is, 1.00*M* acetic acid. If we let x be the concentration of H^+ present at equilibrium, then the concentration of CH_3COO^- must also be x since these two species are formed in equal amounts through the dissociation of acetic acid. The concentration of CH_3COOH remaining after the ionization will be $1.00 - x$.

$$\begin{array}{llll} & CH_3COOH \rightleftharpoons & CH_3COO^- & + \; H^+ \\ \text{at equilibrium} & 1.0 - x & x & x \end{array}$$

By substituting these values in the dissociation constant for acetic acid we have

$$\frac{x^2}{1.00 - x} = 1.8 \times 10^{-5}$$

We can solve this quadratic equation by making a straightforward approximation. If we examine the original mathematical expression, it can be simplified if we recognize that the extent of ionization will be small (a little experience helps to recognize this), and that the concentration of CH_3COOH will be essentially $1.00M$. This is equivalent to saying that since x is small compared to 1.00, then $1.00 - x$ is essentially 1.00. Therefore the original quadratic equation can be simplified to

$$\frac{x^2}{1} = 1.8 \times 10^{-5} \qquad \text{or} \qquad x = [H^+] = 0.0042M$$

The degree of ionization of $1.00M$ acetic acid is 0.0042 parts per 1.00 mole, or 0.42%.

7.11 pH

In dealing with $[H^+]$ we find ourselves dealing with and relating numbers such as 4.2×10^{-3} and so on. Sørensen in 1904 introduced the term pH as a more convenient notation for expressing the concentration of H^+. pH is defined as the negative logarithm of the hydrogen ion concentration:

$$pH = -\log [H^+]$$

In the same manner we may define pOH as the negative logarithm of the hydroxide ion concentration:

$$pOH = -\log [OH^-]$$

In Section 7.10 we determined that the concentration of H^+ in $1.00M$ acetic acid is 0.0042 molar. The pH of this solution is 2.38:

$$[H^+] = 0.0042 = 4.2 \times 10^{-3}$$
$$pH = -\log [4.2 \times 10^{-3}] = 2.38$$

It is important to realize that pH is a logarithmic function. Thus when the pH is decreased by one unit, from 7 to 6, the concentration of H^+ is increased by a factor of 10, from 10^{-7} to 10^{-6}. The more acidic solution has the smaller pH.

Table 7.6 shows typical pH values for some common fluids. In several cases these are average values. For example, the pH of human blood plasma under normal conditions varies within the range pH 7.35–7.45. The normal pH of human gastric juice varies within the range pH 1.0–3.0.

Table 7.6 *pH values of some common fluids.*

Fluid	pH
human gastric juice	1.5
lemon juice	2.8
orange juice	3.7
coffee	5.2
urine	6.0
milk	6.4
pure water	7.0
human blood plasma	7.4
household ammonia	11.0

7.12 THE HENDERSON–HASSELBALCH EQUATION

This useful equation may be obtained directly from the expression for the dissociation constant for a weak acid. Consider the ionization of a generalized weak acid, HA:

$$HA \rightleftharpoons H^+ + A^- \qquad K_a = \frac{[H^+][A^-]}{[HA]}$$

By taking the logarithm of the expression for K_a and rearranging we get

$$-\log [H^+] = -\log K_a + \log \frac{[A^-]}{[HA]}$$

$$pH = pK_a + \log \frac{[A^-]}{[HA]}$$

This expression, known as the Henderson–Hasselbalch equation, is very convenient for calculating the pH of a solution containing a weak acid and the salt of that weak acid. For example, in a solution prepared by mixing 0.05 mole of acetic acid and 0.05 mole of sodium acetate, and diluting with water to one liter:

$$pH = 4.74 + \log \frac{[0.05]}{[0.05]} = 4.74$$

7.13 TITRATION CURVES

The titration curve observed when 50 ml of $0.10M$ acetic acid is titrated with $0.10M$ sodium hydroxide is shown in Figure 7.6. This curve is obtained by measuring the pH of $0.10M$ acetic acid before and after the addition of successive portions of $0.10M$ sodium hydroxide. This curve may also be calculated given the value of K_a for acetic acid. We shall examine the curve in some detail, especially the region around point B, the mid-point of the titration.

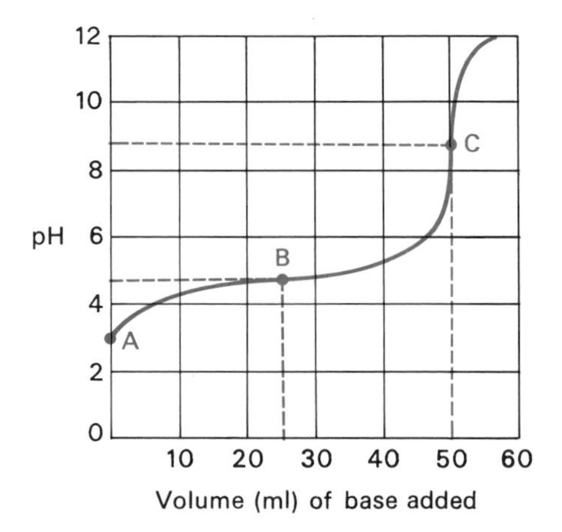

Figure 7.6 *Graph showing the titration of 50 ml of 0.10 M acetic acid with 0.10 M NaOH.*

Point A of the curve represents the pH of $0.10M$ acetic acid and is pH 2.83. Point C represents the end-point or equivalence point of the titration. Exactly enough $0.10M$ NaOH has been added to react with, or neutralize, the 50 ml of acetic acid. The pH of this solution is 8.72.

Point B represents the mid-point of the titration, where enough NaOH has been added to react with, or neutralize, one-half of the acetic acid. At this point, the concentrations of CH_3COOH and CH_3COONa are equal and the pH of the solution is equal to 4.74, the pK_a of acetic acid. Notice that the titration curve is relatively flat on either side of point B, i.e., the change in pH per ml of NaOH added is relatively small. In contrast, the change in pH per ml of NaOH added around point C, the end-point of the titration, is very large. In the region of the graph around point B the solution functions as an acid–base buffer.

7.14 BUFFER SOLUTIONS

A "buffer" is something that resists change. In chemistry, a pH buffer is a substance or mixture of substances that allows solutions to resist large changes in pH upon the addition of small amounts of H^+ or OH^- ions. It should be emphasized that buffered solutions do change in pH on the addition of H^+ or OH^- but the change is much less than if no buffer were present.

Common buffer mixtures contain two substances. An "acidic" buffer contains a weak acid and the salt of the weak acid. A "basic" buffer contains a weak base and the salt of the weak base. Note that point B on the titration curve (Figure 7.6) represents the point of maximum buffering action. This solution contains equal amounts of a weak acid (acetic acid) and the salt of a weak acid (sodium acetate). Together these two species resist changes in pH by partially absorbing H^+ or OH^- ions added to the solution.

If H^+ ions are added to the buffered solution, they react partially with acetate ion to form acetic acid. Thus most of the added H^+ ions are removed from solution through the formation of an equivalent amount of un-ionized acetic acid.

$$H^+ + CH_3CO_2^- \rightarrow CH_3CO_2H$$

When strong base is added, it in effect reacts with un-ionized acetic acid to yield water and more acetate ion.

$$CH_3CO_2H + OH^- \rightarrow CH_3CO_2^- + H_2O$$

Consequently addition of either H^+ or OH^- ions to a buffered solution results in changes in the relative amounts of the weak acid and the salt of the weak acid but only minor changes in the concentrations of H^+ and OH^-.

There are two factors important in determining the effectiveness or the capacity of a buffer. First, the capacity is directly proportional to the molar concentration of the buffer. By convention, the buffer concentration refers to the total concentration of both the weak acid and the salt of the weak acid. For example, $0.1M$ acetate buffer may be made up of 0.025 mole of acetic acid and 0.075 mole of sodium acetate in a liter of solution, or any other amounts of the two totaling 0.1 mole per liter. The other factor affecting the capacity of the buffer is the ratio of the concentration of the salt of the weak acid to the concentration of the acid itself.

For practical purposes, a solution can act as an effective buffer when its pH is within the range $pK_a \pm 1$ of the weak acid. We can see the significance of this range by using the Henderson–Hasselbalch equation. When the concentration of the acid, HA, is 10 times that of the salt, A^-, then

$$\begin{aligned} pH &= pK_a + \log\frac{[A^-]}{[HA]} \\ &= pK_a + \log\frac{1}{10} \\ &= pK_a - 1 \end{aligned}$$

In other words, the pH is one unit less than the pK_a of the weak acid when the concentration of HA is 10 times that of A^-.

When the concentration of the salt, A^-, is 10 times that of the acid, HA, then

$$\begin{aligned} pH &= pK_a + \log\frac{[A^-]}{[HA]} \\ &= pK_a + \log\frac{10}{1} \\ &= pK_a + 1 \end{aligned}$$

In other words, the pH of the solution is one unit greater than the pK_a of the weak acid when the concentration of A^- is 10 times that of HA.

For acetic acid, pK_a is 4.74; therefore a solution of acetic acid and sodium acetate will function as a buffer within the pH range of approximately 3.74 to 5.74.

Of course, the most effective buffer is one with equal concentrations of weak acid and salt, i.e., one in which the pH of the solution is equal to the pK_a of the weak acid. While a solution of acetic acid and sodium acetate will buffer within the pH range 3.74–5.74, it is most effective as a buffer at or very near pH 4.74.

7.15 ACID–BASE BALANCE IN BLOOD PLASMA

As we shall see in Chapters 15–18, the biological oxidation of intermediary metabolites results in the production of a variety of inorganic and organic acids and bases. For example, lactic acid is produced during the degradation of glucose; ammonia is produced during the degradation of amino acids. The ability of the blood to transport these metabolically produced acids and bases from the sites of their production in the cells to the sites of their excretion in the lungs and kidneys without appreciable change in pH depends on the presence in the blood of very effective buffer systems.

In a healthy person the pH of blood and other body fluids remains at a remarkably constant level of 7.35 to 7.45. The principal buffers of blood plasma are bicarbonate, phosphate, and plasma proteins. In addition, the red cells of the blood have another buffer system, namely the protein of the hemoglobin molecules with which they are so tightly packed. We shall discuss the buffering action of proteins in Chapter 12. At this point we will concentrate on the bicarbonate and phosphate buffer systems.

Phosphoric acid is a triprotic acid and successive ionizations liberate one, two, and three protons.

$$H_3PO_4 \rightleftharpoons H_2PO_4^- + H^+ \qquad K_1 = 7.5 \times 10^{-3} \qquad pK_1 = 2.1$$

$$H_2PO_4^- \rightleftharpoons HPO_4^{2-} + H^+ \qquad K_2 = 2.0 \times 10^{-7} \qquad pK_2 = 6.7$$

$$HPO_4^{2-} \rightleftharpoons PO_4^{3-} + H^+ \qquad K_3 = 4.8 \times 10^{-13} \qquad pK_3 = 12.3$$

At pH 7.4, the normal pH of blood plasma, it is the acid–base pair $H_2PO_4^-/HPO_4^{2-}$ that is responsible for buffering action. By using the Henderson–Hasselbalch equation, we can calculate the ratio of HPO_4^{2-} to $H_2PO_4^-$ at this pH.

$$pH = pK_2 + \log\frac{[HPO_4^{2-}]}{[H_2PO_4^-]}$$

$$7.4 = 6.7 + \log\frac{[HPO_4^{2-}]}{[H_2PO_4^-]}$$

$$0.7 = \log\frac{[HPO_4^{2-}]}{[H_2PO_4^-]}$$

and therefore

$$\frac{[HPO_4^{2-}]}{[H_2PO_4^-]} = 5.1$$

In other words, at pH 7.4 the phosphate buffer system is about 1/6 (or 16%) $H_2PO_4^-$ and 5/6 (or 83%) HPO_4^{2-}. This provides good buffering capacity, particularly since normal metabolism produces slightly more organic acids than bases.

While phosphate and blood proteins play a part in controlling the acid–base balance in blood, it is the bicarbonate buffer which is by far the most important in maintaining this pH balance within such narrow limits.

Carbonic acid is a dibasic acid. In terms of buffering action in blood plasma, it is the first dissociation constant which is important.

$$H_2CO_3 \rightleftharpoons HCO_3^- + H^+ \qquad K_1 = 8.0 \times 10^{-7} \qquad pK_1 = 6.1$$

In this equation, carbonic acid is written as H_2CO_3. However, since about 99% of the carbonic acid in blood is in the form of dissolved carbon dioxide, it is more common to refer to the acid form as CO_2. Therefore, the dissociation of carbonic acid can also be written as

$$CO_2 + H_2O \rightleftharpoons HCO_3^- + H^+ \qquad pK_1 = 6.1$$

By using the Henderson–Hasselbalch equation we can calculate the ratio of the acid form, CO_2, to the salt form, HCO_3^-, at pH 7.4.

$$pH = pK_1 + \log\frac{[HCO_3^-]}{[CO_2]}$$

$$7.4 = 6.1 + \log\frac{[HCO_3^-]}{[CO_2]}$$

Solving this equation shows that in blood plasma at pH 7.4,

$$\frac{[HCO_3^-]}{[CO_2]} = \frac{20}{1}$$

In other words, in blood plasma at pH 7.4, the ratio of bicarbonate to carbonic acid is 20 to 1. The normal concentration of bicarbonate in plasma is about 0.025 mole per liter and that of carbonic acid is about 0.0012 mole per liter. Therefore, the concentration of this buffer system is approximately 0.026 mole per liter, mainly as bicarbonate.

Recall from our discussion in Section 7.14 that a weak acid and its salt are most effective as an acid–base buffer in the region $pH = pK_a \pm 1$, i.e., in the concentration range from 10% salt and 90% acid to 90% salt and 10% acid. Yet the pK_1 of carbonic acid is 1.3 units smaller than the pH of the fluid which it is to buffer. At pH 7.4, this buffer is approximately 95% bicarbonate and 5% carbonic acid. How then can bicarbonate act as such an extremely effective buffer? It is because the respiratory system provides a means for making very rapid adjustments in the concentration of carbonic acid in blood. In addition, the kidneys provide a means for making slower, more long-term adjustments in the concentration of bicarbonate. Through the cooperative interaction of these two systems, the bicarbonate to carbonic acid ratio can be kept very close to 20 to 1.

From the clinical standpoint, respiratory acidosis, respiratory alkalosis, metabolic acidosis, and metabolic alkalosis are the four major disturbances in acid–base balance. Acidosis is brought about by any abnormal condition that leads to the accumulation of excess acid in the body or excessive loss of alkali. In acidosis, the pH of blood falls below 7.30. Alkalosis is brought about by any abnormal condition that leads to loss of acid or accumulation in the body of excess alkali. In alkalosis, the pH of blood rises above 7.50.

Any chronic respiratory difficulty or depression of breathing rate can increase the CO_2 concentration in blood plasma.

$$\frac{[HCO_3^-]}{[CO_2]}$$ ← increases in chronic respiratory difficulty or depression of breathing

Because of this increase, the ratio of HCO_3^- to CO_2 decreases to something less than 20 to 1, and accordingly the pH decreases. For example, even hypoventilation due to holding your breath to get rid of the hiccups can temporarily decrease the pH of blood to 7.30 or below.

Hyperventilation for any reason, including rapid, deep breathing, results in a "blow off" or decrease in the concentration of dissolved CO_2.

$$\frac{[HCO_3^-]}{[CO_2]}$$ ← decreases with hyperventilation

Because CO_2 decreases, the ratio of HCO_3^- to CO_2 increases to greater than 20 to 1 and the pH of blood increases. Even mild hyperventilation can increase blood pH to as high as 7.51.

Hyperventilation and hypoventilation bring about what are called respiratory alkalosis and respiratory acidosis. To compensate for these conditions, the kidneys react to either increase or decrease the concentration of HCO_3^- in an effort to restore the 20 to 1 ratio of bicarbonate to carbonic acid. For example, in the case of respiratory acidosis, the kidneys compensate by increasing the reabsorption of bicarbonate. As long as the concentration of carbon dioxide is elevated, the kidneys will stabilize the bicarbonate at an elevated level.

In respiratory alkalosis, the kidneys will decrease the concentration of H^+, thereby allowing for the reaction of H^+ and HCO_3^- to form more carbonic acid. Instead of excreting H^+ in an acidic urine, the kidneys will excrete other positive ions, mainly Na^+ and K^+, and the urine will become less acidic or even slightly alkaline.

Metabolic acidosis may result from a variety of causes including increased biosynthesis of acids such as the ketone bodies (Section 17.5) and decreased excretion of H^+ due to kidney failure. Metabolic alkalosis may result from impaired nitrogen metabolism or any other factor that leads to an increase in the production of bases.

Problems

7.4 Define and give an example of the following:

a fatty acid **b** soap **c** detergent

7.5 Draw structural formulas for:

a sodium palmityl sulfate
b calcium palmitate
c sodium dodecylbenzenesulfonate

7.6 Draw structural formulas for lauric, palmitic, stearic, oleic, linoleic, and arachidonic acids. For each that will show geometric isomerism, state the total number of isomers possible.

7.7 By using structural formulas illustrate how a molecule of fatty acid interacts with water to form a monomolecular layer on the surface of water.

7.8 By using structural formulas, show how a soap (e.g., sodium stearate) "dissolves" fats, oils, and grease.

7.9 What is the function of the phosphate additives in detergents? In what way have these phosphate additives of detergents (as well as phosphates from other sources) contributed to the crisis in ecology?

7.10 Show by balanced equations the reaction of a soap with (a) hard water, and (b) acidic solution.

7.11 Characterize the structural features necessary to make a detergent of good cleansing ability. Illustrate by structural formulas two different classes of synthetic detergents. Name each example.

7.12 Below are given structural formulas for a cationic detergent and a nonionic detergent. How would you account for the detergent properties of each?

$$C_6H_5—CH_2—\overset{\overset{CH_3}{|}}{\underset{\underset{C_8H_{17}}{|}}{N^+}}—CH_3 \quad Cl^-$$

benzyldimethyloctylammonium chloride

$$CH_3(CH_2)_{14}\overset{\overset{O}{\|}}{C}—O—CH_2—\overset{\overset{CH_2OH}{|}}{\underset{\underset{CH_2OH}{|}}{C}}—CH_2OH$$

pentaerythrityl palmitate

7.13 What does it mean to say that linoleic acid is an "essential" fatty acid? Name several dietary sources of linoleic acid.

7.14 Calculate the pH of the following solutions.

a $0.023M$ H^+ **b** $7.6 \times 10^{-7}M$ H^+
c $5.25 \times 10^{-12}M$ H^+ **d** $3.27 \times 10^{-4}M$ OH^-

7.15 Calculate the hydrogen ion concentration of each fluid listed in Table 7.6.

7.16 Calculate the pH for $1.0M$ HCl; for $1.0M$ acetic acid.

7.17 Vinegar is a 4–5% solution of acetic acid in water. For this problem assume you are dealing with a 5% solution and that the density of vinegar is 1.00 g/ml.

a Calculate the concentration of acetic acid in moles per liter of solution.
b Calculate the hydrogen ion concentration and the pH of this solution.
c How many grams of sodium bicarbonate are required to neutralize 100 ml of this vinegar?

7.18 Compare the pH of a 0.1*M* acetic acid solution with that of a solution 0.1*M* in acetic acid and also 0.1*M* in sodium acetate.

7.19 Calculate the pH of a solution prepared by mixing 2.40 g (0.04 mole) of acetic acid and 4.92 g (0.06 mole) of sodium acetate in enough water to make one liter of solution.

7.20 Given pure acetic acid and pure sodium acetate, what weight of each would you mix in water to prepare one liter of 0.1*M* buffer solution of pH 4.74? (The total concentrations of acetic acid and sodium acetate added should equal 0.1 mole.)

7.21 Calculate the pH change on adding 1.0 ml of 1.0*M* HCl to:

a 100 ml of water
b 100 ml of a solution containing 0.05 mole of acetic acid and 0.05 mole of sodium acetate.

7.22 The remarkable buffering capacity of blood is shown by the fact that adding 1 ml of 10*M* HCl to 1 liter of plasma causes the pH to drop only slightly, from 7.4 to about 7.2.

a Calculate the expected change in hydrogen ion concentration if blood had no buffering capacity whatsoever.
b Calculate the actual change in hydrogen ion concentration.

7.23 Calculate the ratio of bicarbonate to carbonic acid at the following pH values.

a pH 7.31 as might occur during diabetic acidosis where there is decreased metabolism of glucose, increased metabolism of fatty acids, and the production of acetoacetic and β-hydroxybutyric acids. These acids cause a decrease in the pH of blood plasma. This is an example of metabolic acidosis.
b pH 7.71 as might occur during anesthesia with the patient on a respirator. This is an example of respiratory alkalosis.

7.24 In the event of acidosis, the kidneys respond to restore the acid–base balance by excreting acidic urine of pH as low as 4.8. Calculate the ratio of HPO_4^{2-} to $H_2PO_4^-$ at this pH.

7.25 In the event of alkalosis, the kidneys respond to restore the acid–base balance by excreting less acidic or even slightly alkaline urine of pH as high as 8.0. Calculate the ratio of HPO_4^{2-} to $H_2PO_4^-$ at this pH.

7.26 Compound A ($C_5H_{10}O_3$) readily dissolves in water to give an acidic solution. Compound A also dissolves in 0.1 *M* $NaHCO_3$ with the evolution of carbon dioxide. In addition, compound A is optically active and contains an alcohol group. Oxidation of compound A by potassium permanganate gives compound B ($C_5H_8O_4$). Compound B is a dicarboxylic acid and is optically inactive. Deduce structures for compounds A and B consistent with these observations.

7.27 There are four isomeric alcohols of molecular formula $C_4H_{10}O$. Name and draw structural formulas for each. Which of these alcohols is indicated by the following experimental observations? Compound D ($C_4H_{10}O$), on oxidation by potassium permanganate in acid solution, gives compound E ($C_4H_8O_2$), a carboxylic acid. Treatment of compound D with warm phosphoric acid brings about dehydration and yields compound F (C_4H_8). Treatment of compound F with warm aqueous sulfuric acid gives G ($C_4H_{10}O$), a new alcohol isomeric with compound D. Compound G is resistant to oxidation. Propose structures for compounds D, E, F, and G consistent with these observations.

MINI-
ESSAY

THE PROSTAGLANDINS

The prostaglandins are ubiquitous biological compounds that have a wide range of physiological activity. Since their discovery less than 50 years ago, scientists have isolated and characterized many different prostaglandins, developed means for synthesizing natural prostaglandins and synthetic analogs in the laboratory, and have begun to unravel the mysteries of the physiological activity of these remarkable compounds.

In 1930 Raphael Kurzrok and Charles Lieb, two gynecologists in New York, noticed the actions of human seminal plasma on the state of contraction of isolated human uteri and reported that "... uteri from patients who had gone through successful pregnancies responded with relaxation, but uteri from women who had a history of complete or long-standing sterility were stimulated on addition of seminal fluid."

During the next few years these studies generated little interest. However in 1933, Maurice Goldblatt in England began investigating the unusual effects of sheep vesicular gland homogenates on tissues of other organisms. He established that the effect of these homogenates was not due to histamines, catecholamines, adenosine, or choline. Independently and unaware of Goldblatt's work, Ulf von Euler in Sweden observed that human semen and extracts of sheep vesicular glands lowered blood pressure upon injection and stimulated isolated preparations of intestinal, uterine, and smooth muscle. He also established that this factor was lipid soluble and proposed the name prostaglandin because the first such factors were discovered in male genital glands and secretions. von Euler also managed to characterize prostaglandins as nitrogen-free carboxylic acids containing at least one carbon-carbon double bond but he was unable to isolate pure prostaglandin. We know now that concentrations of prostaglandins are very low, usually in the nanogram (10^{-9} g) range, so it is not surprising that von Euler was unable to isolate a pure sample.

From 1939 to 1948 the war prevented further research. In 1948 von Euler encouraged Dr. Sune Bergström to begin the purification of concentrates prepared from Iceland sheep seminal vesicle glands. However the supply was limited and the project was halted until 1956 when Bergström established a seminal gland collection system and began the tedious job of isolating and characterizing the structure of prostaglandin.

At about the same time, other scientists showed that the prostate gland was in fact not the principal source of prostaglandins but that the seminal vesicles were the source. Thus the name prostaglandin is a misnomer but since the literature already had used the term prostaglandin so widely, the name was retained.

In 1959 Bergström reported the isolation and structural characterization of two natural prostaglandins. These two compounds exhibited different solubilities in ether and phosphate buffer. The compound in the ether layer was named prostaglandin E. The compound found in the phosphate buffer layer was named prostaglandin F because phosphate is spelled with an F in Swedish. Laboratory work moved rapidly from this point and by 1968 Dr. John Pike and his associates at The Upjohn Company and Professor E. J. Corey at Harvard each announced the laboratory synthesis of several prostaglandins and prostaglandin analogs. At the present time practically every major pharmaceutical company has an active prostaglandin research program under way.

The prostaglandin family of compounds consists of naturally occurring fatty acids which each contain 20 carbon atoms and have the same carbon skeleton as prostanoic acid.

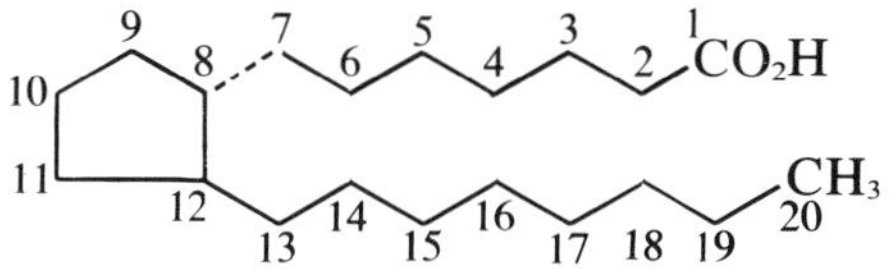

prostanoic acid

Various arrangements of double bonds, hydroxyl groups, and ketones divide the acids into four categories. Structural formulas for the so-called six primary prostaglandins are shown in Figure 1. Also shown are the three polyunsaturated fatty acids from which they are synthesized in living cells.

In these formulas, the letters PG stand for prostaglandin. All E series prostaglandins contain a C-9 ketone and a C-11 hydroxyl group below the plane of the ring. All F series prostaglandins contain C-9 and C-11 hydroxyls *cis* and below the plane of the ring. When referring to the F series the designation α indicates that the C-9 hydroxyl group is below the plane of the ring. Finally, the side chains beginning with C-1 to C-7 and C-13 to C-20 may contain one, two, or three double bonds. The total number of double bonds is denoted by a subscript number following the name of the prostaglandin. For example, $PGF_{2\alpha}$ describes a prostaglandin of the F series containing two carbon-carbon double bonds (location on the side chains unspecified) and a C-9 hydroxyl group below the plane of the five-membered ring.

The six E and F prostaglandins arising from tri-, tetra-, and penta-unsaturated derivatives of eicosanoic acid shown in Figure 1 make up the six primary prostaglandins. These are in turn converted into secondary prostaglandins.

By 1962 scientists had begun to unravel the intricate steps in the biosynthesis of prostaglandins. Certain naturally occurring fatty acids seemed to be the most active and efficient biological precursors, and in 1964 B. Samuelsson in Sweden pointed out the striking similarity in structure between PGE_2 and 5,8,11,14-eicosatetraenoic acid (Figure 1). He showed further that PGE_2 could be synthesized from this polyunsaturated fatty acid by homogenates of sheep seminal vesicles.

The structural requirements for PG formation are now known to be a free, 20-carbon atom, linear chain carboxylic acid with *cis* double bonds at carbons 8, 11, and 14. The specific fatty acid present within the tissue appears to determine the particular prostaglandin produced.

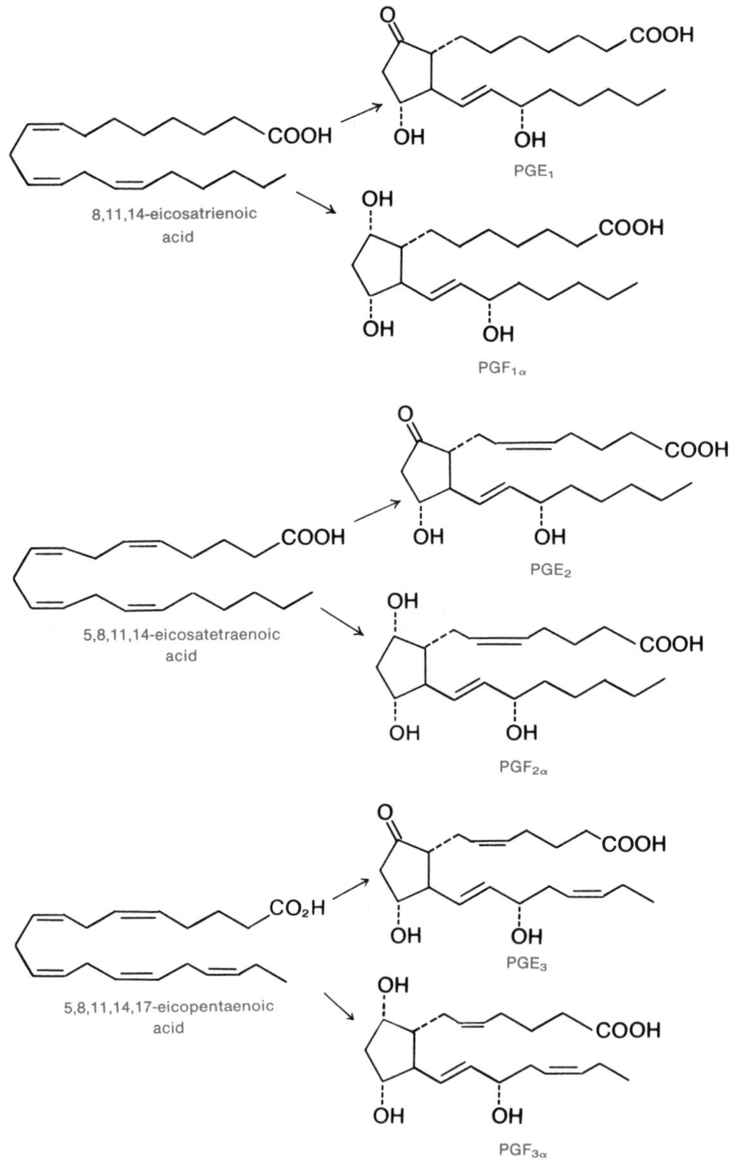

8,11,14-eicosatrienoic acid

PGE_1

$PGF_{1\alpha}$

5,8,11,14-eicosatetraenoic acid

PGE_2

$PGF_{2\alpha}$

5,8,11,14,17-eicopentaenoic acid

PGE_3

$PGF_{3\alpha}$

Figure 1 *Structural formulas of the six primary prostaglandins and the polyunsaturated carboxylic acids from which each is synthesized in living cells.*

For example, the principal polyunsaturated fatty acid found in sheep vesicular glands is 8,11,14-eicosatrienoic acid with smaller amounts of 5,8,11,14-eicosatetraenoic acid. In the lungs the opposite is true. Consequently sheep seminal vesicles produce prostaglandins of the two double bond series while lungs produce prostaglandins of the one double bond series. Furthermore, seminal vesicles produce prostaglandins predominantly of the E series while lungs produce mainly those of the F series.

All six primary prostaglandins contain three oxygen atoms as either ketone or hydroxyl groups. Interestingly all three oxygen atoms originate from molecular oxygen and furthermore, the two oxygen atoms on the five-membered ring (C-9 and C-11) arise from the same molecule of oxygen.

Although their mechanism of action is not well understood even at the present time, it is clear that prostaglandins are involved at the cellular level in regulating many functions, including gastric acid secretion, contraction and relaxation of smooth muscles, inflammation and vascular permeability, body temperature, food intake, and blood platelet aggregation. Further, there seems little doubt that they play a role in almost every stage of reproductive physiology.

The therapeutic potential of prostaglandins and the lack of any abundant natural source of these compounds led chemists to investigate laboratory synthesis as a method for obtaining them. The first totally synthetic prostaglandin became available in 1968. However research was still hampered by the cost of the products. The price of the laboratory-synthesized prostaglandins at one time was as high as $1000 per gram. The price of prostaglandins dropped dramatically with the discovery in 1969 that the gorgonian sea whip or sea fan, *Plexaura homomalla*, which grows on the coral reefs off the coast of Florida and in the Caribbean, is a rich source of prostaglandin-like materials. The concentration of these PG-like substances is about 100 times the normal concentration found in mammalian sources. These PG-like compounds are then converted by laboratory means into prostaglandins and prostaglandin analogs. In effect, chemists take advantage of the ability of *P. homomalla* to perform steps which in the laboratory give very poor yields. At the present time, however, there appears little need to depend on even this natural source since chemists have succeeded in devising a variety of highly effective laboratory syntheses of prostaglandins from readily available chemicals that make production on a commercial scale possible.

As indicated in the introduction to this essay, the first recorded observations on the biological activity of the prostaglandins were those of gynecologists Kurzrok and Lieb, and it now appears that the first widespread clinical application of these substances will also be by gynecologists and obstetricians. The observation that prostaglandins mediate the contraction of smooth muscle, including the uterus, led quite naturally to a widespread interest in their use to induce labor and abortion and to stimulate menstruation. The first published study of the use of prostaglandins for the induction of labor was by Dr. Sultan Karim of Uganda's Makerere University medical school. Between 1968 and 1971 he reported induction of labor at term in over 500 women by the use of PGE_2 and $PGF_{2\alpha}$. These studies coupled with those of M. P. Embrey at Oxford and M. Bygdeman and M. Wiqvist at Stockholm leave no doubt about the ability of PGE_2 and $PGF_{2\alpha}$ to induce labor at term by intravenous infusion. The rate of infusion of PGE_2 (which is 8–10 times more potent than $PGF_{2\alpha}$) varies from patient to patient but a typical value is 0.05 microgram (0.05×10^{-6} g) per kilogram of body weight per minute for several hours with a total dose of several milligrams. More recent studies have described the oral administration of prostaglandins for the induction of labor.

In 1966 Bygdeman reported on the marked and specific stimulation effect of certain pure prostaglandins on the contractibility of the uterus of both pregnant and non-pregnant women. This and other clinical observations suggested that these substances could be used to induce therapeutic abortion. Subsequent studies on experimental animals have shown that PGE_1, PGE_2, and $PGF_{2\alpha}$ are effective in inducing abortions when administered by intravenous infusion.

A final extension of these results would suggest that if small amounts of prostaglandins can induce labor at term, or induce abortion in early pregnancy, these same or related compounds might also be used in contraception. Preliminary findings by Karim and his associates on a small number of patients have suggested that PGE_2 and $PGF_{2\alpha}$ do in fact have potential usefulness as once-a-month contraceptives for inducing menstruation. In reporting on these preliminary findings, the *British Medical Journal* was led to comment, "This remarkable result carries the implication that if prostaglandins become readily available, abortion on demand at any time in pregnancy will become a practical possibility. The social, ethical, and clinical consequences of putting them to such use will need careful evaluation."

But before we can expect to see any of these substances in general clinical use, a great deal of research needs to be done. At this stage even the mechanism of action of prostaglandins on smooth muscle is still largely unknown. And what of their absorption and distribution to specific tissues, their metabolism, the biological activity of their specific metabolites, and the whole question of their role in other biological processes in addition to reproductive physiology? Many if not most of these problems must be probed and solved through collaboration of the basic scientist and the clinical scientist.

References

*Bergström, S., "Prostaglandins: Members of a New Hormonal System," *Science*, **157**, 382 (1967).

*Bergström, S., Carlson, L. A., and Weeks, J. R., "The Prostaglandins: A Family of Biologically Active Lipids," *Pharmacological Review*, **20**, 1 (1968).

Bylinsky, G., "Upjohn Puts the Cell's Own Messengers to Work," *Fortune* (June 1972).

"Clinical Use of Prostaglandins," *British Medical Journal*, No. 5730, 253 (October 1970).

Karim, S. M. M., and Sharma, S. D., "Oral Administration of Prostaglandins for the Induction of Labour," *British Medical Journal*, **1**, 260 (1971).

Ramwell, P. W., Editor, *The Prostaglandins*, Vol. 1 (Plenum Press, New York, 1973).

---

*These two references to S. Bergström provide key references to the basic scientific and historical literature of these fascinating compounds.

8

ESTERS, AMIDES, AND ANHYDRIDES

In addition to undergoing the reactions with bases discussed in Chapter 7, which involve the loss of a proton from the carboxyl group, acids can be transformed into a variety of derivatives in which the structural alteration of the carboxyl group is somewhat greater. In this chapter we examine three such structural alterations.

8.1 STRUCTURE

Organic esters, amides, and anhydrides are functional derivatives of carboxylic acids in which the —OH of the carboxyl group has been replaced by —OR, —NH_2, or —OCOR. These derivatives all contain the acyl group.

$$\underset{\text{acyl group}}{R\text{—}\overset{\overset{\displaystyle O}{\|}}{C}\text{—}} \qquad \underset{\text{ester}}{R\text{—}\overset{\overset{\displaystyle O}{\|}}{C}\text{—}OR} \qquad \underset{\text{amide}}{R\text{—}\overset{\overset{\displaystyle O}{\|}}{C}\text{—}NH_2} \qquad \underset{\text{anhydride}}{R\text{—}\overset{\overset{\displaystyle O}{\|}}{C}\text{—}O\text{—}\overset{\overset{\displaystyle O}{\|}}{C}\text{—}R}$$

Fats, proteins, and many other naturally occurring substances are derivatives of carboxylic acids.

8.2 NOMENCLATURE

Esters are named as derivatives of carboxylic acids by dropping the suffix *-ic* from the IUPAC or the common name of the acid and adding *-ate*. The alkyl or aryl group on oxygen is named first, followed by the name of the acid from which the ester is derived. In the case of esters of thiols the prefix *thio-* is used to indicate the presence of the sulfur atom.

$$\underset{\text{ethyl acetate}}{CH_3\text{—}\overset{\overset{\displaystyle O}{\|}}{C}\text{—}O\text{—}CH_2CH_3} \qquad \underset{\text{methyl pyruvate}}{CH_3\text{—}\overset{\overset{\displaystyle O}{\|}}{C}\text{—}\overset{\overset{\displaystyle O}{\|}}{C}\text{—}OCH_3}$$

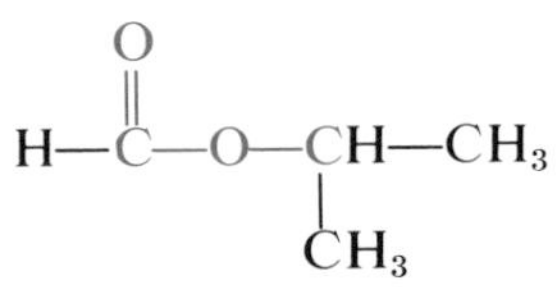

$$H-\overset{\overset{O}{\|}}{C}-O-\underset{\underset{CH_3}{|}}{CH}-CH_3$$

isopropyl formate

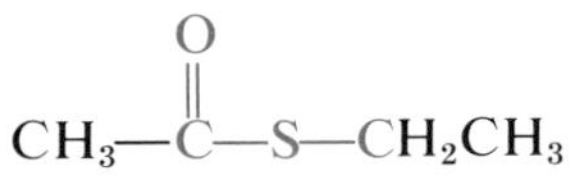

$$CH_3-\overset{\overset{O}{\|}}{C}-S-CH_2CH_3$$

ethyl thioacetate

Several organic esters of nitric acid and nitrous acid have been used as drugs for more than 100 years. Two of these are glyceryl trinitrate, or as it is more commonly known, nitroglycerine, and isoamyl nitrite.

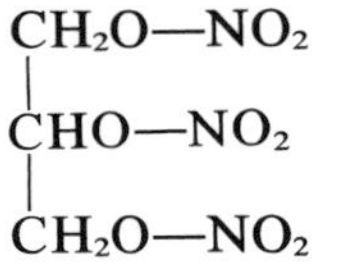

$$\begin{array}{l} CH_2O-NO_2 \\ | \\ CHO-NO_2 \\ | \\ CH_2O-NO_2 \end{array}$$

nitroglycerine

$$\begin{array}{l} CH_3\diagdown \\ \qquad CH-CH_2-CH_2O-NO \\ CH_3\diagup \end{array}$$

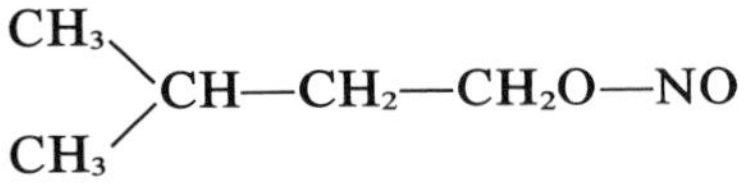

isoamyl nitrite

These organic nitrate and nitrite esters produce rapid relaxation of most smooth muscles of the body. The most important effect medically is relaxation of the smooth muscle of blood vessels and dilation of all large and small arteries including those of the heart. For this reason they are called vasodilators. The organic nitrate esters are used extensively in treating patients with angina pectoris.

Amides are named as derivatives of carboxylic acids by dropping the suffix -oic from the IUPAC name of the acid, or the suffix -ic from the common name of the acid, and adding -amide.

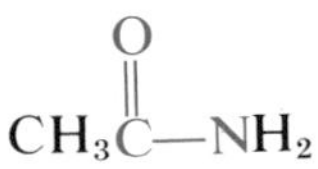

$$CH_3\overset{\overset{O}{\|}}{C}-NH_2$$

acetamide

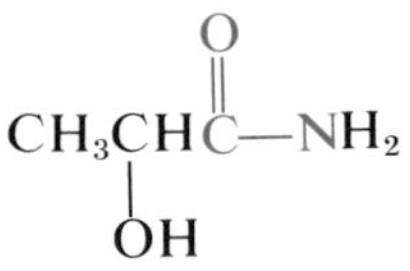

$$CH_3\underset{\underset{OH}{|}}{CH}\overset{\overset{O}{\|}}{C}-NH_2$$

lactamide

$$CH_3\underset{\underset{CH_3}{|}}{CH}CH_2\overset{\overset{O}{\|}}{C}-NH_2$$

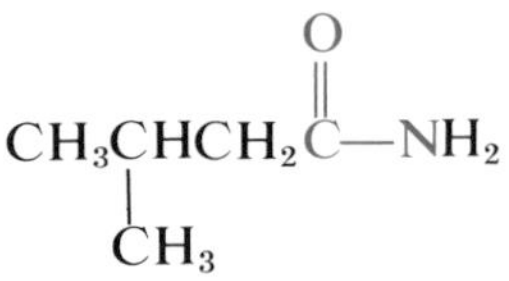

3-methylbutanamide
β-methylbutyramide

If the nitrogen atom is substituted with an alkyl or aryl group, the substituent is named and its location on nitrogen is indicated by a capital N-.

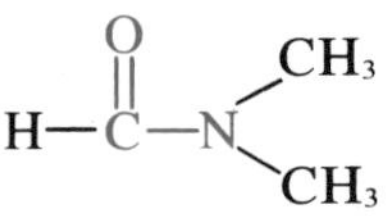

$$H-\overset{\overset{O}{\|}}{C}-N\begin{array}{l} \diagup CH_3 \\ \diagdown CH_3 \end{array}$$

N,N-dimethylformamide

$$CH_3CH_2CH_2\overset{\overset{O}{\|}}{C}-NH-C_6H_5$$

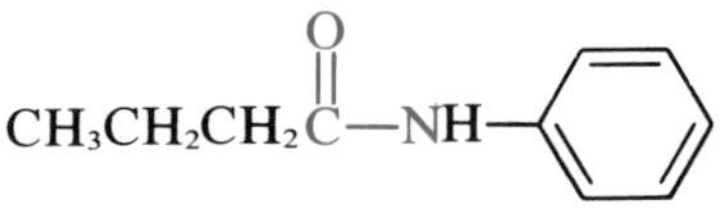

N-phenylbutanamide
N-phenylbutyramide

Anhydrides are named by adding anhydride to the name of the acid. For our purposes, the most important organic anhydride is acetic anhydride.

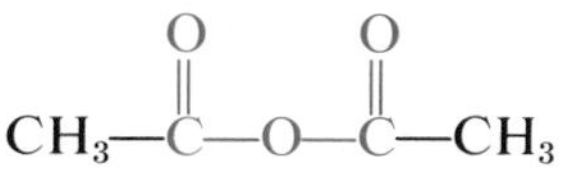

$$CH_3-\overset{\overset{\large O}{\|}}{C}-O-\overset{\overset{\large O}{\|}}{C}-CH_3$$

acetic anhydride

The esters, anhydrides, and amides of phosphoric acid deserve special mention because they are exceedingly important in the whole of biological chemistry. The combination of organic molecules with phosphate (H_3PO_4) is one of the important metabolic reactions because many organic substances can be metabolized only in the phosphorylated state. The esters of phosphoric acid are named in the same manner as esters of carboxylic acids.

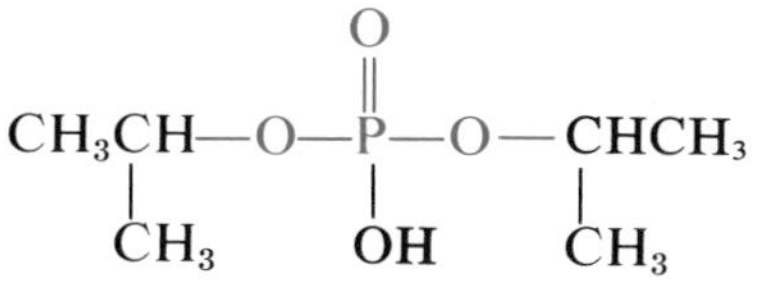

$$\underset{\underset{\large CH_3}{|}}{CH_3CH}-O-\underset{\underset{\large OH}{|}}{\overset{\overset{\large O}{\|}}{P}}-O-\underset{\underset{\large CH_3}{|}}{CHCH_3}$$

diisopropyl hydrogen phosphate

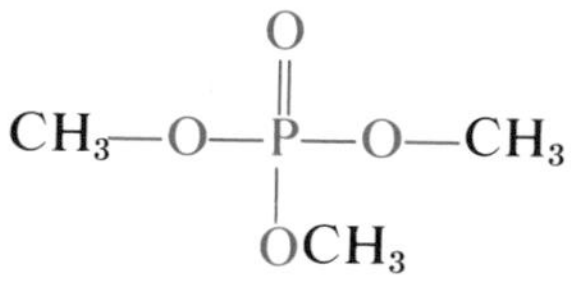

$$CH_3-O-\underset{\underset{\large OCH_3}{|}}{\overset{\overset{\large O}{\|}}{P}}-O-CH_3$$

trimethyl phosphate

Heating of phosphoric acid first gives a monoanhydride, $H_4P_2O_7$, known as pyrophosphoric acid. Further heating will yield higher-molecular-weight polymers of phosphoric acid.

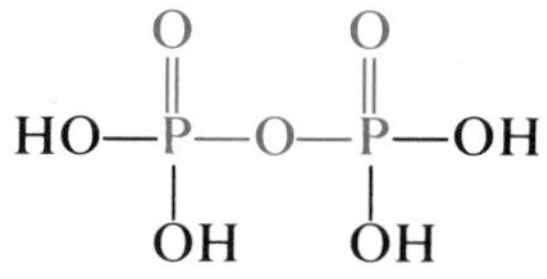

$$HO-\underset{\underset{\large OH}{|}}{\overset{\overset{\large O}{\|}}{P}}-O-\underset{\underset{\large OH}{|}}{\overset{\overset{\large O}{\|}}{P}}-OH$$

pyrophosphoric acid

$$HO-\underset{\underset{\large OH}{|}}{\overset{\overset{\large O}{\|}}{P}}-O-\underset{\underset{\large OH}{|}}{\overset{\overset{\large O}{\|}}{P}}-O-\underset{\underset{\large OH}{|}}{\overset{\overset{\large O}{\|}}{P}}-OH$$

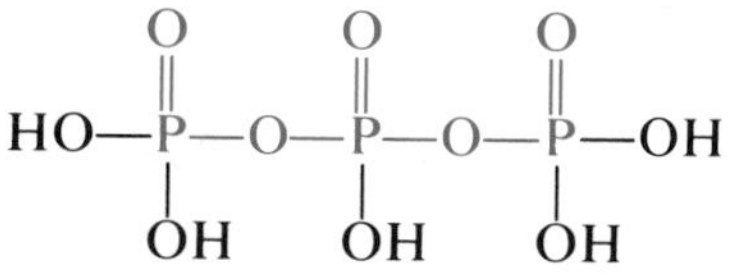

triphosphoric acid

Acetyl phosphate is a mixed anhydride, that is, it is an anhydride derived from two different acids, in this case from one molecule of acetic acid and one molecule of phosphoric acid.

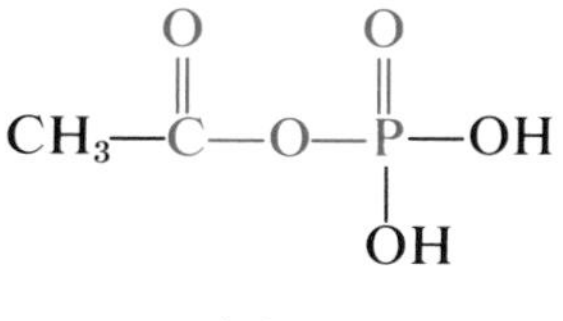

$$CH_3-\overset{\overset{\large O}{\|}}{C}-O-\underset{\underset{\large OH}{|}}{\overset{\overset{\large O}{\|}}{P}}-OH$$

acetyl phosphate

Problems

8.1 Draw structural formulas for the nine isomeric esters of molecular formula $C_5H_{10}O_2$. Name each.

8.2 Draw structural formulas for the following compounds.

a	phenyl acetate	**b**	diethyl carbonate
c	benzamide	**d**	cyclobutyl butanoate
e	methyl 3-butenoate	**f**	isopropyl 3-methylhexanoate

g diethyl oxalate
h ethyl *cis*-2-pentenoate
i acetamide
j *p*-chlorophenyl acetate
k methyldiethyl phosphate
l N-phenylbutanamide
m diethyl malonate
n formamide

8.3 NUCLEOPHILIC SUBSTITUTION AT UNSATURATED CARBON

The basic reaction theme common to the carbonyl group of aldehydes, ketones, carboxylic acids, esters, amides, and anhydrides is addition to the carbonyl group to form a tetrahedral carbonyl addition intermediate. In the case of aldehydes and ketones, this addition product either is isolated as such or it undergoes loss of water to give an unsaturated derivative of the original aldehyde or ketone. For example, recall that aldol condensation products often are isolated as such (Section 6.13),

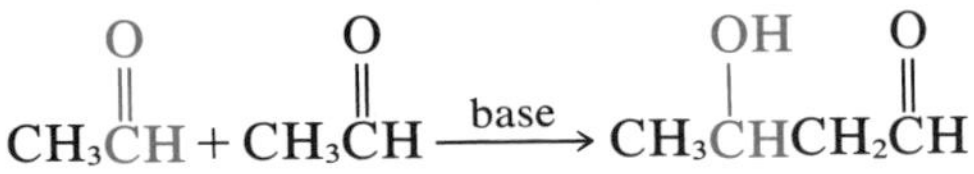

$$CH_3\overset{O}{\overset{\|}{C}}H + CH_3\overset{O}{\overset{\|}{C}}H \xrightarrow{\text{base}} CH_3\overset{OH}{\overset{|}{C}}HCH_2\overset{O}{\overset{\|}{C}}H$$

while the reaction of an aldehyde or ketone with hydroxylamine (Section 6.9) yields an unsaturated derivative of the original aldehyde or ketone.

$$CH_3\overset{O}{\overset{\|}{C}}CH_3 + H_2NOH \longrightarrow \left[CH_3-\underset{CH_3}{\overset{OH}{C}}-N\begin{matrix} H \\ OH \end{matrix} \right] \longrightarrow CH_3\underset{CH_3}{C}=NOH + H_2O$$

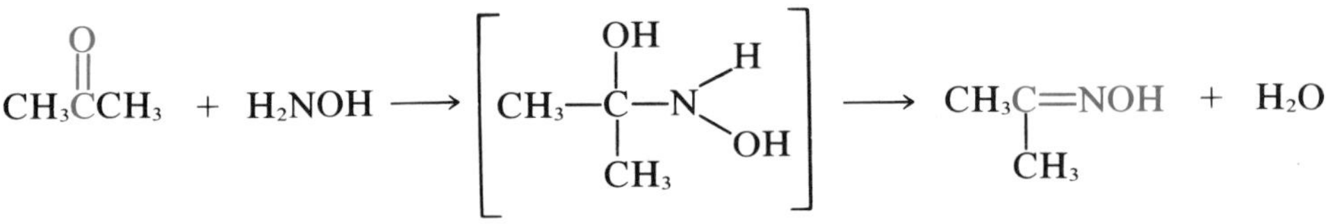

However, with the new functional groups to be studied in this chapter, namely esters, amides, and anhydrides, the addition compounds undergo subsequent collapse to regenerate the carbonyl group.

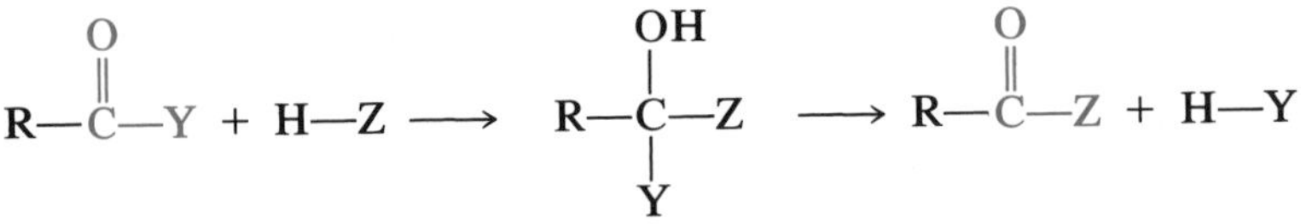

$$R-\overset{O}{\overset{\|}{C}}-Y + H-Z \longrightarrow R-\underset{Y}{\overset{OH}{C}}-Z \longrightarrow R-\overset{O}{\overset{\|}{C}}-Z + H-Y$$

It is for this reason that we characterize these reactions of the carbonyl group as nucleophilic substitution at unsaturated carbon. As an example, shown below is the reaction of methyl acetate, an ester, with ammonia.

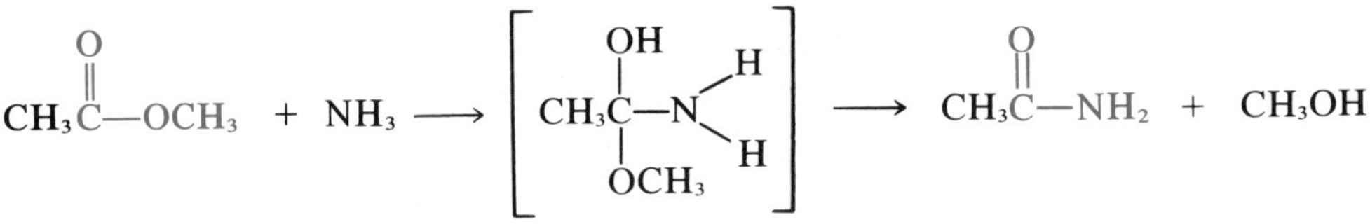

$$CH_3\overset{O}{\overset{\|}{C}}-OCH_3 + NH_3 \longrightarrow \left[CH_3\underset{OCH_3}{\overset{OH}{C}}-N\begin{matrix} H \\ H \end{matrix} \right] \longrightarrow CH_3\overset{O}{\overset{\|}{C}}-NH_2 + CH_3OH$$

The result of this sequence of reactions is substitution of $-NH_2$ for $-OCH_3$.

8.4 PREPARATION OF ESTERS

A carboxylic acid is converted into an ester by heating with an alcohol in the presence of an acid catalyst, usually dry hydrogen chloride, concentrated sulfuric acid, or an ion-exchange resin in the hydrogen ion form. The direct esterification of alcohols and acids in this manner is called Fischer esterification.

$$R-\overset{\overset{\large O}{\|}}{C}-OH + HO-R' \overset{H^+}{\rightleftharpoons} R-\overset{\overset{\large O}{\|}}{C}-O-R' + H_2O$$

The reaction involves condensing a molecule of acid and a molecule of alcohol through the elimination of a molecule of water. This reaction is reversible, and generally at equilibrium there are appreciable quantities of both ester and alcohol present. For example, if 60.0 g (one mole) of acetic acid and 60.0 g (one mole) of *n*-propyl alcohol are refluxed for a short time in the presence of a few drops of concentrated sulfuric acid, the reaction mixture at equilibrium will contain about 68.0 g (0.67 mole) of ester, 12.0 g (0.67 mole) of water, and 20.0 g (0.33 mole) each of acetic acid and *n*-propyl alcohol. In other words, at equilibrium there is about 67% conversion of acid and alcohol into ester.

Direct esterification can be used to prepare esters in high yields. For example, if the alcohol is particularly inexpensive, we may use a large excess of it and achieve a high conversion of the acid into ester. Or we may take advantage of a situation in which the boiling points of the reactants and the ester are higher than that of water. Heating the reaction mixture somewhat above 100°C will remove water as it is formed and shift the equilibrium toward the production of higher yields of ester.

Fischer esterification is but one of the general methods of preparing esters. From the standpoint of the organic chemist interested in the laboratory synthesis of esters, a much more important method is the reaction of hydroxyl compounds (alcohols, phenols) with acid anhydrides. We shall see this method of ester preparation in Section 8.11.

Problem

8.3a Write an equation for the equilibrium established when acetic acid and *n*-propyl alcohol are refluxed in the presence of a few drops of concentrated sulfuric acid.

b Using the data in Section 8.4, calculate the equilibrium constant for this reaction.

8.5 ACID CATALYSIS OF ESTERIFICATION—MECHANISM

When an ester is prepared by the Fischer esterification, the question arises as to whether the oxygen eliminated as water comes from the acid or from the alcohol. A clear demonstration of the source of the water oxygen comes from an experiment in which one of the oxygen atoms is labeled. The use of a mass spectrometer shows that ordinary oxygen in nature is a mixture of three isotopes: 99.7% ^{16}O, 0.04% ^{17}O and 0.20% ^{18}O. Through the use of modern methods of separating isotopes, a variety of compounds with significant enrichment in ^{18}O are commercially available.

When methanol enriched with ^{18}O is allowed to react with acetic acid containing ordinary oxygen, the ester is found to contain the enriched oxygen. None is present in the product water. This demonstrates that esterification involves the rupture of the C—O bond of the acid and the H—O bond of the alcohol.

$$CH_3\overset{\overset{\displaystyle O}{\|}}{C}—OH + H—{}^{18}OCH_3 \xrightleftharpoons{H^+} CH_3\overset{\overset{\displaystyle O}{\|}}{C}—{}^{18}OCH_3 + H_2O$$

A mechanism consistent with this observation and with the fact that Fischer esterification is acid catalyzed postulates initial attack of a proton on the carbonyl oxygen. As a result of this attack, the carbonyl carbon becomes more electron deficient and more susceptible to nucleophilic attack. Addition of an alcohol molecule to the protonated carbonyl group gives a tetrahedral addition intermediate. This intermediate may then either (1) lose a molecule of alcohol to regenerate the starting material, or (2) lose a molecule of water to give an ester. Shown below is a detailed step-by-step mechanism for acid-catalyzed esterification.

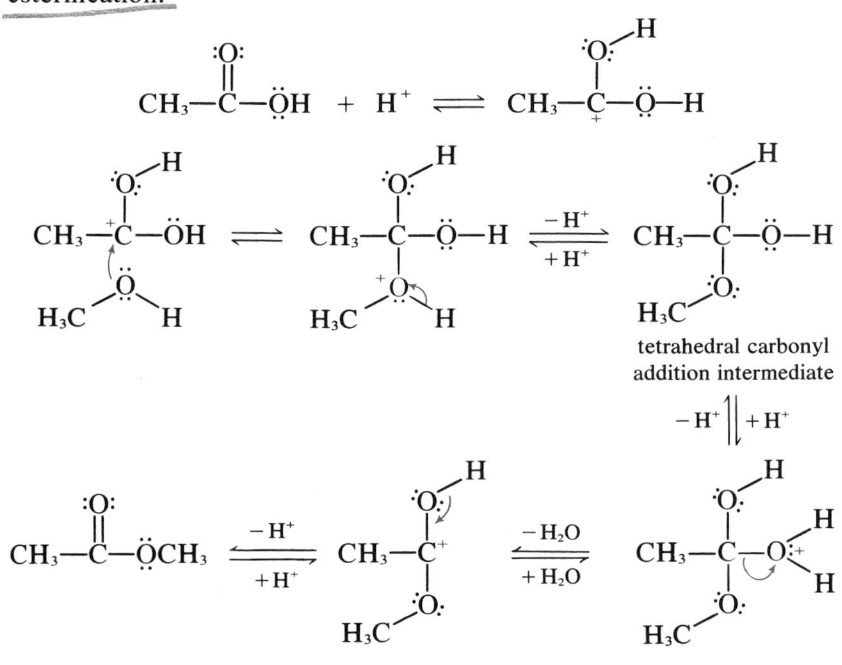

$$CH_3—C(=O)—OH + H^+ \rightleftharpoons CH_3—\overset{+}{C}(—OH)—OH$$

$$CH_3—\overset{+}{C}(OH)—OH + H_3C—O—H \rightleftharpoons CH_3—C(OH)(OH)—\overset{+}{O}(H)—CH_3 \underset{+H^+}{\overset{-H^+}{\rightleftharpoons}} CH_3—C(OH)(OH)—OCH_3$$

tetrahedral carbonyl addition intermediate

$$\underset{+H^+}{\overset{-H^+}{\rightleftharpoons}}$$

$$CH_3—C(OH)(OCH_3)—\overset{+}{O}H_2 \underset{+H_2O}{\overset{-H_2O}{\rightleftharpoons}} CH_3—\overset{+}{C}(OH)—OCH_3 \underset{+H^+}{\overset{-H^+}{\rightleftharpoons}} CH_3—C(=O)—OCH_3$$

This may seem complicated because of the several steps in the pathway. However you should be able to break the mechanism down into the following components:

(i) four proton-transfer reactions;

(ii) formation of a carbon-oxygen single bond by the reaction of a nucleophile and an electrophile;

(iii) rupture of a carbon-oxygen single bond by the separation of a nucleophile and an electrophile.

If we omit all proton-transfer reactions, we see that this mechanism is a specific example of the general mechanism (Section 8.3) of addition to the carbonyl group to form a tetrahedral carbonyl addition intermediate,

$$CH_3-\overset{\overset{O}{\|}}{C}-OH + HOCH_3 \rightleftharpoons \left[CH_3-\underset{\underset{OH}{|}}{\overset{\overset{OH}{|}}{C}}-OCH_3 \right] \rightleftharpoons CH_3-\overset{\overset{O}{\|}}{C}-OCH_3 + H_2O$$

and that Fischer esterification is an example of nucleophilic substitution at unsaturated carbon.

8.6 PHYSICAL PROPERTIES OF ESTERS

Esters are neutral substances, less soluble in water and lower boiling than isomeric carboxylic acids (see Problems 5.6 and 5.7). Unlike the carboxylic acids from which they are derived, the low-molecular-weight esters have rather pleasant odors. The characteristic fragrances of many flowers and fruits are in many instances due to the presence of esters, either singly or in mixtures. Some of the more familiar esters are ethyl formate (artificial rum flavor), methyl butanoate (apples), octyl acetate (oranges), and ethyl butanoate (pineapples). Artificial fruit flavors are made largely from mixtures of lower-molecular-weight esters.

Naturally occurring fats and oils are triesters of glycerol and long-chain carboxylic acids. The structure and physical properties of these ubiquitous biomolecules are discussed in Chapter 11. Dacron polyester is a condensation polymer of ethylene glycol and terephthalic acid. The structure and commercial development of dacron polyester is presented in the mini-essay "Nylon and Dacron."

8.7 HYDROLYSIS OF ESTERS

Esters of carboxylic acids, as well as those of other acids such as phosphoric and sulfuric acid, may be reconverted to the corresponding acids and alcohols by hydrolysis in either aqueous acid or base.

$$CH_3CO_2CH_2CH_3 + H_2O \xrightarrow{H^+} CH_3CO_2H + CH_3CH_2OH$$

In the case of acid-catalyzed hydrolysis, the H^+ functions in the same manner as we encountered in esterification, namely, to make the carbonyl carbon more susceptible to nucleophilic attack. Since each step in the mechanism of acid-catalyzed esterification (Section 8.5) is reversible, this mechanism can equally well account for acid-catalyzed hydrolysis. If the reaction is carried out in a large excess of water, the position of the equilibrium can be shifted to favor the formation of acid and alcohol.

In alkaline hydrolysis of esters, the OH^- adds directly to the carbonyl carbon to form a tetrahedral carbonyl addition intermediate. This intermediate then eliminates a molecule of alcohol and generates the carboxylate anion, $RCOO^-$.

$$R{-}\overset{\overset{\large O}{\|}}{C}{-}OR' + OH^- \rightleftharpoons \left[R{-}\underset{\underset{\large R'}{|}}{\underset{\large O}{|}}{\overset{\overset{\large O^-}{|}}{C}}{-}OH \right] \longrightarrow R{-}\overset{\overset{\large O}{\|}}{C}{-}O^- + HOR'$$

tetrahedral carbonyl
addition intermediate

This mechanism, like that for acid-catalyzed esterification and hydrolysis, involves cleavage of the C—O bond of the carboxylic acid portion of the molecule rather than the C—O bond of the alcohol. Alkaline hydrolysis of an ester is not an equilibrium reaction because the carboxylate anion, the final product, shows no tendency to react with alcohol.

Problem **8.4** Explain why it is preferable to hydrolyze esters in aqueous base rather than in aqueous acid.

8.8 AMMONOLYSIS OF ESTERS

Treatment of esters with ammonia, often in a solvent such as ethanol, yields an amide. This reaction is similar to hydrolysis and is called ammonolysis. Ammonia is a strong nucleophile, and adds directly to the carbonyl carbon without a catalyst being necessary.

$$CH_3{-}\overset{\overset{\large O}{\|}}{C}{-}OCH_2CH_3 + NH_3 \longrightarrow \underset{\text{acetamide}}{CH_3{-}\overset{\overset{\large O}{\|}}{C}{-}NH_2} + CH_3CH_2OH$$

Another example of this ammonolysis reaction is the laboratory synthesis of barbituric acid and barbituates. Heating urea and diethyl malonate at 110°C in the presence of sodium ethoxide yields barbituric acid.

$$\underset{\text{diethyl malonate}}{H_2C(COOCH_2CH_3)_2} + \underset{\text{urea}}{(H_2N)_2C{=}O} \longrightarrow \underset{\text{barbituric acid}}{H_2C\begin{matrix}\diagup C(=O){-}NH \diagdown \\ \diagdown C(=O){-}NH \diagup\end{matrix}C{=}O} + 2CH_3CH_2OH$$

Mono- and disubstituted malonic esters yield substituted barbituric acids known as barbiturates.

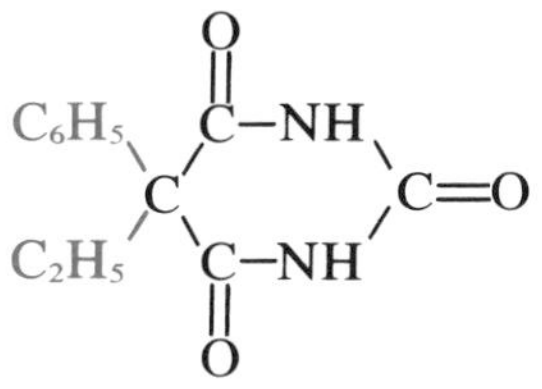

$$(C_6H_5)(C_2H_5)C\left\langle\begin{matrix}\overset{O}{\overset{\|}{C}}-NH\\ \underset{\underset{O}{\|}}{C}-NH\end{matrix}\right\rangle C{=}O \qquad (CH_3CH_2)(CH_3CH_2CH_2\underset{\underset{CH_3}{|}}{C}H)C\left\langle\begin{matrix}\overset{O}{\overset{\|}{C}}-NH\\ \underset{\underset{O}{\|}}{C}-NH\end{matrix}\right\rangle C{=}O \qquad (CH_3CH_2)(CH_3CH_2CH_2\underset{\underset{CH_3}{|}}{C}H)C\left\langle\begin{matrix}\overset{O}{\overset{\|}{C}}-NH\\ \underset{\underset{O}{\|}}{C}-NH\end{matrix}\right\rangle C{=}S$$

phenobarbital (Luminal) 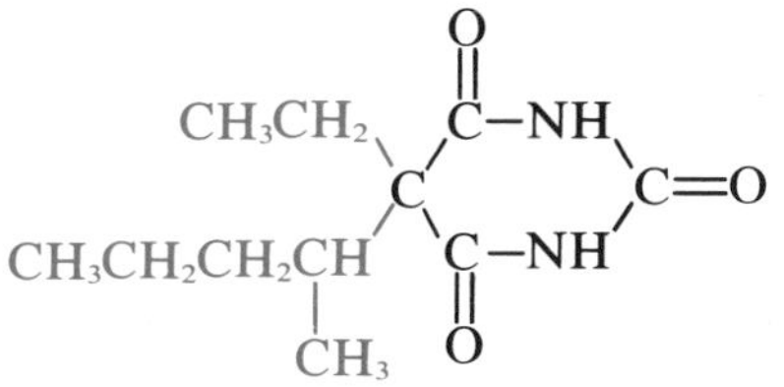 — pentobarbital 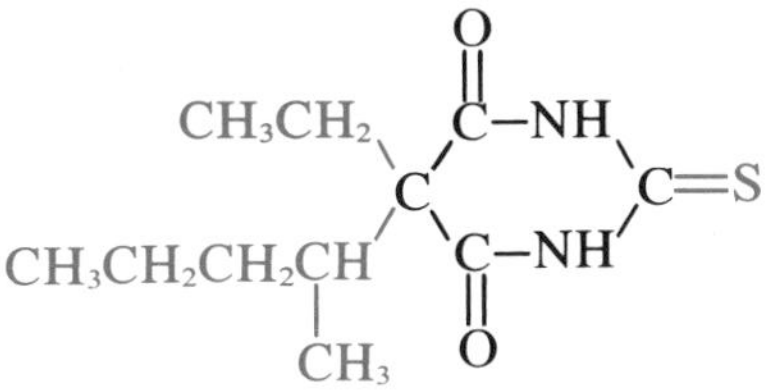 — thiopental

The barbiturates produce effects ranging from mild sedation to deep anesthesia, and even death, depending on the dose and the particular barbiturate. The sedation, long or short acting, depends on the structure of the barbiturate. Phenobarbital (Luminal) is long acting while pentobarbital acts for a shorter time, about three hours. Thiopental is very fast acting and is used as an anesthetic for producing deep sedation quickly. With barbiturates in general, sleep can be produced with as little as 0.1 g (1 capsule) and toxic symptoms and even death can result from 1.5 g (15 capsules). Barbiturates produce both a general tranquilizing effect and a sleep-producing effect. Recently it has been possible to separate these effects with another class of drugs. Reserpine and the chloropromazine derivatives tranquilize without producing sleep at the same time (see the mini-essay "Biogenic Amines and Emotion").

8.9 PREPARATION OF AMIDES

Amides may be prepared by heating the ammonium salt of a carboxylic acid above its melting point.

$$CH_3-\overset{O}{\overset{\|}{C}}-O^-NH_4^+ \xrightarrow{\text{heat}} CH_3-\overset{O}{\overset{\|}{C}}-NH_2 + H_2O$$

They are also prepared by the reaction of an ester with ammonia or a derivative of ammonia as illustrated in Section 8.8. Ammonolysis of ethyl nicotinate yields nicotinamide.

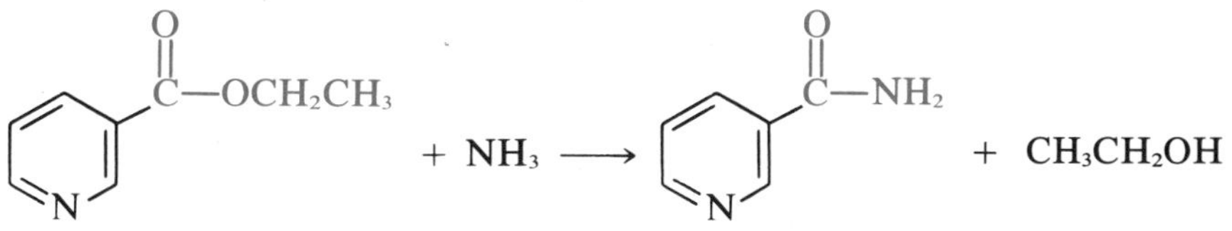

$$C_5H_4N-\overset{O}{\overset{\|}{C}}-OCH_2CH_3 + NH_3 \longrightarrow C_5H_4N-\overset{O}{\overset{\|}{C}}-NH_2 + CH_3CH_2OH$$

ethyl nicotinate — nicotinamide

In Section 8.11 we shall see another method for amide preparation, namely the reaction of anhydrides with ammonia and ammonia derivatives.

8.10 HYDROLYSIS OF AMIDES

Amides are very resistant to hydrolysis even in boiling water. However in the presence of moderately concentrated aqueous acid or base, hydrolysis does occur, though not as rapidly as in the case of esters. It is not unusual to reflux an amide for several hours in concentrated hydrochloric acid to effect hydrolysis.

$$\underset{\text{phenylacetamide}}{C_6H_5CH_2\overset{O}{\overset{\|}{C}}NH_2} + H_2O \xrightarrow[\text{reflux}]{35\%\ HCl} \underset{\text{phenylacetic acid}}{C_6H_5CH_2CO_2H} + NH_4Cl$$

8.11 REACTIONS OF ANHYDRIDES

Anhydrides undergo rapid reaction with water, ammonia, alcohols, thioalcohols, and phenols. In fact, many of the anhydrides of carboxylic acids react so readily with water that they must be protected from atmospheric moisture during storage.

$$CH_3-\overset{O}{\overset{\|}{C}}-O-\overset{O}{\overset{\|}{C}}-CH_3 \xrightarrow{H_2O} CH_3COOH + CH_3COOH$$

$$\xrightarrow{NH_3} CH_3CONH_2 + CH_3COOH$$

$$\xrightarrow{CH_3OH} CH_3CO_2CH_3 + CH_3COOH$$

$$\xrightarrow{CH_3CH_2SH} CH_3COSCH_2CH_3 + CH_3COOH$$

Aspirin is prepared by the reaction of acetic anhydride and salicylic acid.

$$\underset{\text{salicylic acid}}{o\text{-}HOC_6H_4CO_2H} + CH_3-\overset{O}{\overset{\|}{C}}-O-\overset{O}{\overset{\|}{C}}-CH_3 \longrightarrow \underset{\text{acetylsalicylic acid, aspirin}}{o\text{-}(CH_3\overset{O}{\overset{\|}{C}}-O)C_6H_4CO_2H} + CH_3CO_2H$$

salicylic acid

acetylsalicylic acid
aspirin

8.12 RELATIVE REACTIVITIES OF ANHYDRIDES, ESTERS, AND AMIDES

The common carboxylic acid derivatives we have described show quite marked differences in reactivities. Consider for example the ease of hydrolysis of an anhydride, an ester, and an amide. Acetic anhydride reacts so readily with water that it must be protected from atmospheric moisture during storage. Ethyl acetate reacts very slowly with water at room temperature but hydrolyzes readily on heating in the presence of an acid or base catalyst. Acetamide is very resistant to hydrolysis except in the presence of moderately concentrated acid or base. The reactivity decreases in the following order.

$$\underset{\text{anhydride}}{CH_3-\overset{O}{\overset{\|}{C}}-O-\overset{O}{\overset{\|}{C}}-CH_3} > \underset{\text{ester}}{CH_3-\overset{O}{\overset{\|}{C}}-OCH_2CH_3} > \underset{\text{amide}}{CH_3-\overset{O}{\overset{\|}{C}}-NH_2}$$

It follows directly from this order of reactivity that any less reactive derivative of a carboxylic acid may be prepared directly from a more reactive derivative, but not vice versa. Esters and amides may be synthesized from anhydrides,

$$CH_3\overset{O}{\overset{\|}{C}}-O-\overset{O}{\overset{\|}{C}}CH_3 \xrightarrow{ROH} CH_3\overset{O}{\overset{\|}{C}}-OR + CH_3CO_2H$$

$$CH_3\overset{O}{\overset{\|}{C}}-O-\overset{O}{\overset{\|}{C}}CH_3 \xrightarrow{R_2NH} CH_3\overset{O}{\overset{\|}{C}}-NR_2 + CH_3CO_2H$$

but an anhydride cannot be prepared from an ester or an amide plus an acid. Also, an amide may be synthesized from an ester plus a derivative of ammonia,

$$CH_3\overset{O}{\overset{\|}{C}}-OCH_2CH_3 + R_2NH \longrightarrow CH_3\overset{O}{\overset{\|}{C}}-NR_2 + CH_3CH_2OH$$

but an ester cannot be synthesized from an amide plus an alcohol.

$$CH_3\overset{O}{\overset{\|}{C}}-NR_2 + CH_3CH_2OH \not\longrightarrow CH_3\overset{O}{\overset{\|}{C}}-OCH_2CH_3 + R_2NH$$

8.13 THE CLAISEN CONDENSATION. β-KETOESTERS

The Claisen condensation is a carbonyl reaction closely related to the aldol condensation. The reaction involves a carbanion derived from an ester in a condensation with another molecule of ester. As an example, when ethyl acetate is treated with sodium ethoxide in ethanol and the resulting solution is acidified, ethyl 3-ketobutanoate (commonly known as ethyl acetoacetate) is obtained.

$$2CH_3-\overset{O}{\overset{\|}{C}}-OCH_2CH_3 \xrightarrow[CH_3CH_2OH]{CH_3CH_2O^-Na^+} \underset{\substack{\text{ethyl 3-ketobutanoate} \\ \text{ethyl acetoacetate}}}{CH_3-\overset{O}{\overset{\|}{C}}-CH_2-\overset{O}{\overset{\|}{C}}-OCH_2CH_3} + CH_3CH_2OH$$

The net reaction is one of substitution at one carbonyl by the carbanion from a second ester molecule according to the following mechanism:

Step 1 $$CH_3CH_2\ddot{\underset{..}{O}}{:}^- + H-CH_2-\overset{O}{\overset{\|}{C}}-OC_2H_5 \rightleftharpoons CH_3CH_2OH + {:}CH_2-\overset{O}{\overset{\|}{C}}-OC_2H_5$$

Step 2

$$CH_3{-}\overset{\overset{\ddot{O}}{\parallel}}{C}{-}OC_2H_5 + {:}\overset{-}{C}H_2{-}\overset{\overset{O}{\parallel}}{C}OC_2H_5 \rightleftharpoons CH_3{-}\underset{\underset{OC_2H_5}{|}}{\overset{\overset{:\ddot{O}:^-}{|}}{C}}{-}CH_2{-}\overset{\overset{O}{\parallel}}{C}{-}OC_2H_5$$

tetrahedral carbonyl
addition intermediate

Step 3

$$CH_3{-}\underset{\underset{:\ddot{O}C_2H_5}{|}}{\overset{\overset{:\ddot{O}:^-}{|}}{C}}{-}CH_2{-}\overset{\overset{O}{\parallel}}{C}{-}OC_2H_5 \rightleftharpoons CH_3{-}\overset{\overset{:O:}{\parallel}}{C}{-}CH_2{-}\overset{\overset{O}{\parallel}}{C}OC_2H_5 + CH_3CH_2\ddot{\underset{..}{O}}{:}^-$$

In the case of a mixed Claisen condensation, i.e., a condensation between two different esters, a mixture of four possible products will result unless there is an appreciable difference in reactivity between one ester and the other. One such difference is if one of the esters has no α-hydrogens and therefore cannot serve as a carbanion. Examples of such esters are ethyl formate, ethyl benzoate, and diethyl carbonate.

$$\underset{\text{ethyl benzoate}}{C_6H_5\overset{\overset{O}{\parallel}}{C}OCH_2CH_3} + \underset{\text{ethyl propanoate}}{CH_3CH_2\overset{\overset{O}{\parallel}}{C}OCH_2CH_3} \longrightarrow \underset{\text{ethyl 2-methyl-3-keto-3-phenylpropanoate}}{C_6H_5\overset{\overset{O}{\parallel}}{C}\underset{\underset{CH_3}{|}}{C}H\overset{\overset{O}{\parallel}}{C}OCH_2CH_3} + CH_3CH_2OH$$

Ester condensations are of great importance in biosynthesis. One example is the biosynthesis of acetoacetyl coenzyme A, a key intermediate in synthesis of a variety of compounds of plant and animal origin.

$$\underset{\text{acetyl coenzyme A}}{2CH_3{-}\overset{\overset{O}{\parallel}}{C}{-}SCoA} \xrightarrow{\text{enzyme}} \underset{\text{acetoacetyl coenzyme A}}{CH_3{-}\overset{\overset{O}{\parallel}}{C}{-}CH_2{-}\overset{\overset{O}{\parallel}}{C}{-}SCoA} + \underset{\text{coenzyme A}}{CoA{-}SH}$$

The starting material for this reaction is acetyl coenzyme A, an ester of acetic acid and a complex molecule ($C_{21}H_{36}O_{16}N_7P_3S$) called coenzyme A. A major structural feature of coenzyme A is the presence of a sulfhydryl group (—SH). This molecule is generally abbreviated CoA—SH to emphasize the importance of the sulfhydryl group, or often is it written more simply as CoA. An enzyme-catalyzed Claisen-type ester condensation of two molecules of acetyl coenzyme A leads to the formation of acetoacetyl coenzyme A. We shall return to the discussion of acetoacetyl coenzyme A in Chapter 17 and show that it is a key intermediate in the biosynthesis of ketone bodies, steroids, and terpenes.

8.14 HYDROLYSIS AND DECARBOXYLATION OF β-KETOESTERS

Hydrolysis of β-ketoesters in warm acid or base yields the corresponding β-ketoacids.

$$CH_3-\overset{O}{\overset{\|}{C}}-CH_2-\overset{O}{\overset{\|}{C}}-OCH_2CH_3 + H_2O \longrightarrow CH_3-\overset{O}{\overset{\|}{C}}-CH_2-\overset{O}{\overset{\|}{C}}-OH + CH_3CH_2OH$$

Such β-ketoacids readily lose carbon dioxide on heating to yield ketones.

$$CH_3-\overset{O}{\overset{\|}{C}}-CH_2-\overset{O}{\overset{\|}{C}}-OH \xrightarrow[\text{heat}]{} CH_3-\overset{O}{\overset{\|}{C}}-CH_3 + CO_2$$

In the more usual reaction, ester hydrolysis and decarboxylation occur together and only the resulting ketone is isolated.

Such decarboxylation is a unique property of β-ketoacids and is not observed with α, γ or higher ketoacids. The reaction involves a cyclic six-membered ring transition state and, by rearrangement of electrons, leads directly to the enol form of the product ketone and carbon dioxide. Equilibration of the keto and enol forms generates the ketone.

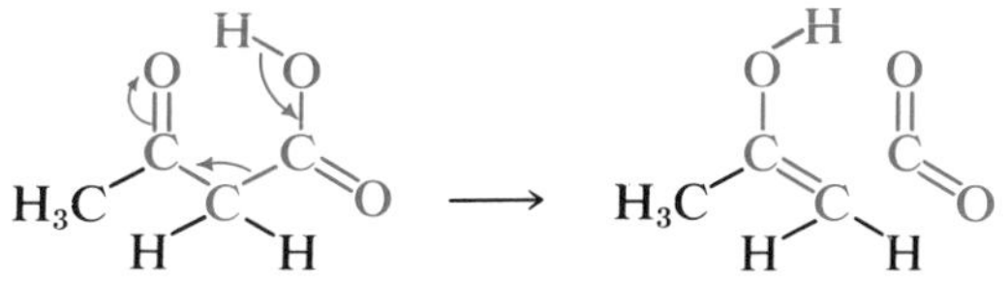

$$H_3C-C(=O\cdots H)-CH_2-C(=O)-O \longrightarrow H_3C-C(OH)=CH_2 + O=C=O$$

Problems **8.5** Write structural formulas for the products of hydrolysis of the following esters, amides, and anhydrides.

a $CH_3\overset{O}{\overset{\|}{C}}OCH_2CH_3 + H_2O \longrightarrow$

b $CH_2OCOCH_3-CHOCOCH_3-CH_2OCOCH_3 + 3H_2O \longrightarrow$

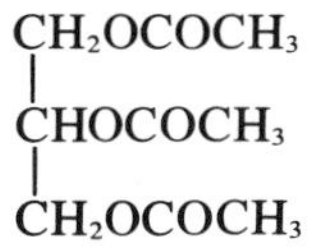

c $H-C(CHO)(CH_2OPO_3^{2-})-OH + H_2O \longrightarrow$

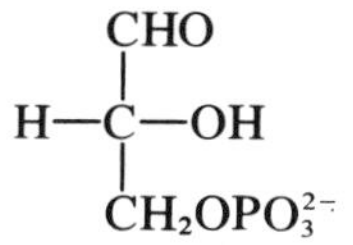

d $CH_3\overset{O}{\overset{\|}{C}}OCH_2CH_2O\overset{O}{\overset{\|}{C}}CH_3 + 2H_2O \longrightarrow$

e $CH_2ONO_2-CHONO_2-CH_2ONO_2 + 3H_2O \longrightarrow$

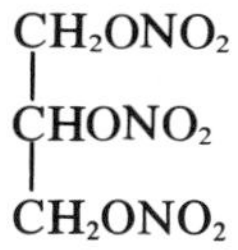

f $(H_3C)_2CHCH_2CH_2ONO + H_2O \longrightarrow$

g $CH_3(CH_2)_{10}CH_2OSO_3H + H_2O \longrightarrow$

h $CH_3\overset{O}{\overset{\|}{C}}SCH_2CH_3 + H_2O \longrightarrow$

i $H_2N\overset{O}{\overset{\|}{C}}NH_2 + 2H_2O \longrightarrow$

j 3-pyridyl-$\overset{O}{\overset{\|}{C}}NH_2$ (pyridine ring with N) $+ H_2O \longrightarrow$

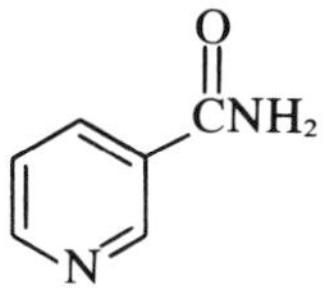

8.6 Propose structural formulas for the tetrahedral carbonyl addition intermediates formed during the following reactions.

a $CH_3CO_2H \; + \; HOCH_3 \xrightarrow{H^+} CH_3CO_2CH_3 \; + \; 3H_2O$

b $CH_3CO_2CH_2CH_3 \; + \; NH_3 \longrightarrow CH_3CONH_2 \; + \; HOCH_2CH_3$

c $CH_3CH_2CONH_2 \; + \; H_2O \xrightarrow{H^+} CH_3CH_2CO_2H \; + \; NH_4^+$

d $CH_3\overset{\overset{O}{\|}}{C}O\overset{\overset{O}{\|}}{C}CH_3 \; + \; H_2O \longrightarrow 2CH_3CO_2H$

e $CH_3\overset{\overset{O}{\|}}{C}O\overset{\overset{O}{\|}}{C}CH_3 \; + \; NH_3 \longrightarrow CH_3CONH_2 \; + \; CH_3CO_2^- \; NH_4^+$

f $CH_3\overset{\overset{O}{\|}}{C}O\overset{\overset{O}{\|}}{C}CH_3 \; + \; CH_3OH \longrightarrow CH_3CO_2CH_3 \; + \; CH_3CO_2H$

g $CH_3\overset{\overset{O}{\|}}{C}O\overset{\overset{O}{\|}}{C}CH_3 \; + \; HSCH_2CH_3 \longrightarrow CH_3\overset{\overset{O}{\|}}{C}SCH_2CH_3 \; + \; CH_3CO_2H$

h $2\,CH_3\overset{\overset{O}{\|}}{C}S{-}CoA \xrightarrow{\text{enzyme}} CH_3\overset{\overset{O}{\|}}{C}CH_2\overset{\overset{O}{\|}}{C}S{-}CoA \; + \; CoA{-}SH$

8.7 Complete the following reactions.

a $CH_3CH_2CO_2CH_3 + CH_3CH_2NH_2 \longrightarrow$

b $C_6H_4(CO_2H)(OH) \; + \; CH_3\overset{\overset{O}{\|}}{C}O\overset{\overset{O}{\|}}{C}CH_3 \longrightarrow$

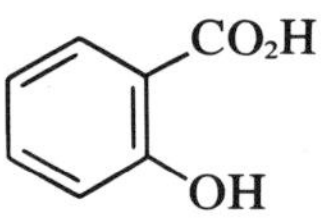

c $CH_3CH_2O{-}C_6H_4{-}NH_2 \; + \; CH_3\overset{\overset{O}{\|}}{C}O\overset{\overset{O}{\|}}{C}CH_3 \longrightarrow$ (phenacetin, a pain reliever)

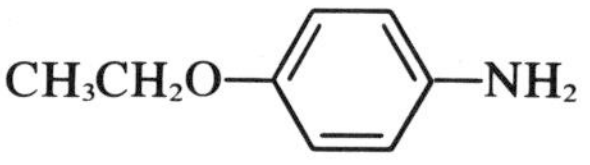

d $C_6H_5CO_2CH_3 + NH_3 \longrightarrow$

e $C_6H_4(CO_2H)(OH) \; + \; CH_3OH \xrightarrow{H^+}$ (oil of wintergreen)

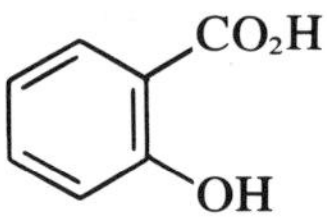

f $CH_3\underset{\underset{CH_3}{|}}{C}HCO_2H + C_6H_5NHCH_3 \xrightarrow{\text{heat}}$

g $2CH_3\overset{\overset{O}{\|}}{C}O\overset{\overset{O}{\|}}{C}CH_3 + HOCH_2CH_2OH \longrightarrow$

8.8 If 15 grams of salicylic acid is reacted with an excess of acetic anhydride, how many grams of aspirin could be formed?

8.9 Hydrolysis of a neutral compound (A), $C_6H_{12}O_2$, yielded an acid (B) and an alcohol (C). Oxidation of the alcohol, C, by potassium dichromate in aqueous sulfuric acid gave an acid identical to B. What are the structural formulas of A, B, and C?

8.10 The following compounds all contain a carbon-oxygen-carbon linkage. Compare the reactivity of each to water.

a diethyl ether
b acetic anhydride
c ethyl acetate

8.11 The following compounds are derivatives of carboxylic acids. Compare the reactivity of each to water.

a ethyl acetate
b acetic anhydride
c acetamide

8.12 Devise a simple chemical test that would enable you to distinguish between the following pairs of compounds. State clearly what you would do, what you would expect to observe, and how you would interpret the observations.

a CH_3CO_2Na and $CH_3CO_2NH_4$
b $CH_3CO_2CH_2CH_3$ and $CH_3CH_2CH_2CO_2H$
c CH_3CONH_2 and $CH_3CO_2NH_4$
d $CH_3CO_2CH_2CH_3$ and $CH_3OCH_2CH_2CO_2H$
e $CH_3CO_2CH_2CH_2CH_3$ and $CH_3CH_2O\overset{O}{\overset{\|}{C}}OCH_2CH_3$

8.13 Carboxylic acids and alcohols may be converted to esters by a variety of chemical methods. When the acid contains more than one COOH group and the alcohol contains more than one —OH group, then under the appropriate experimental conditions, hundreds of molecules may be linked together to give a polyester. Dacron is a polyester of terephthalic acid and ethylene glycol.

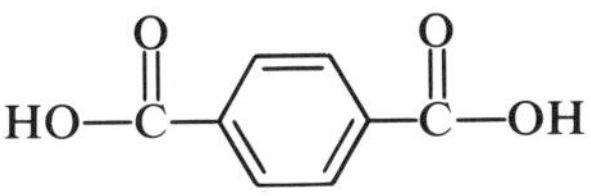

$HO-\overset{O}{\overset{\|}{C}}-C_6H_4-\overset{O}{\overset{\|}{C}}-OH$

terephthalic acid

$HO-CH_2CH_2-OH$

ethylene glycol

a Formulate a structure for Dacron polyester. Be certain to show in principle how several hundred molecules can be hooked together to form the polyester.
b Write an equation for the chemistry involved when a drop of concentrated hydrochloric acid makes a hole in a Dacron polyester shirt (blouse).
c From what starting materials do you think the condensation fiber Kodel Polyester is made?

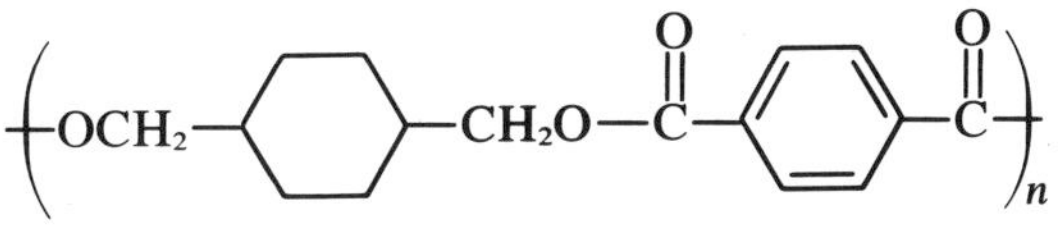

$\left(-OCH_2-C_6H_{10}-CH_2O-\overset{O}{\overset{\|}{C}}-C_6H_4-\overset{O}{\overset{\|}{C}}-\right)_n$

Kodel Polyester

8.14 Draw the structures for each product of the following reactions. In some cases the empirical formula of the product is given. You should approach these problems by first considering how each of the reagents might interact. For example in part **a**, diethyl carbonate is a diester. In what way do esters interact with ammonia? By considering a plausible mechanism for each reaction you should be able to arrive at correct structural formulas for the products.

a $C_2H_5O{-}\overset{\overset{\displaystyle O}{\|}}{C}{-}OC_2H_5 + 2NH_3 \longrightarrow CH_4ON_2 + 2CH_3CH_2OH$

diethyl carbonate — urea

b urea (CH_4ON_2) + $H_2O \xrightarrow[\text{heat}]{\text{aqueous acid}}$

c $(CH_3CH_2)_2C(COOCH_2CH_3)_2$ + urea $\longrightarrow C_8H_{12}O_3N_2 + 2CH_3CH_2OH$

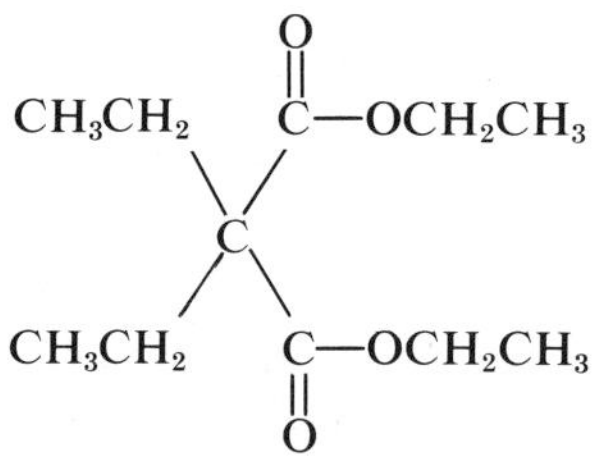

diethyl diethylmalonate — veronal or barbital

d diethyl carbonate + $NH_3 \longrightarrow C_3H_7O_2N + CH_3CH_2OH$

urethane

Note that urethane itself has mild hypnotic properties. The bisurethane of 2-methyl-2-*n*-propylpropane-1,3-diol is widely used as a tranquilizer. Miltown is one of the 22 trade names of this tranquilizer.

8.15 Phenobarbital is a long-acting sedative and hypnotic, and a central nervous system depressant which is used to treat mild hypertension and temporary emotional strain. Predict the products of complete hydrolysis of phenobarbital in aqueous acid.

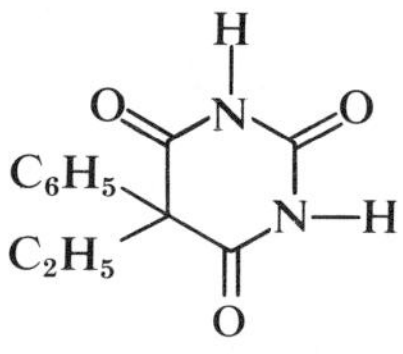

(structure: six-membered ring with two N—H groups, three C=O groups, and a carbon bearing C_6H_5 and C_2H_5)

phenobarbital

8.16 Drawn below are structural formulas for three phosphate-containing compounds. In each case the name given is that in common usage in the biological sciences. Each plays a major role in the flow of energy in the biological world.

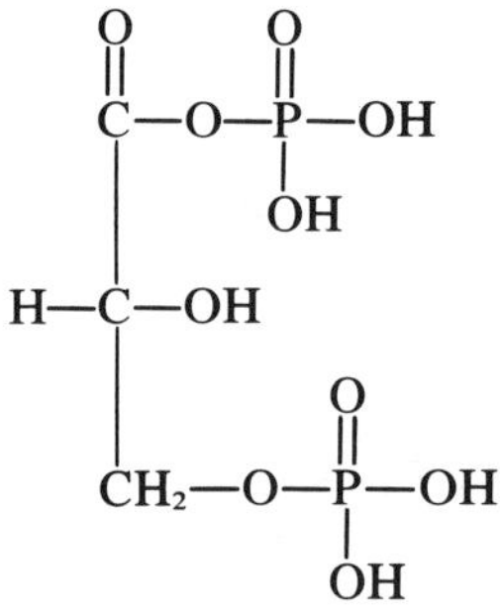

$HO{-}P(O)(OH){-}O{-}C(=O){-}CH(OH){-}CH_2{-}O{-}P(O)(OH){-}OH$

1,3-diphosphoglyceric acid

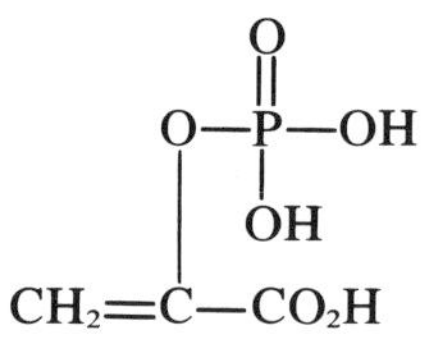

$CH_2{=}C(O{-}P(O)(OH){-}OH){-}CO_2H$

phosphoenolpyruvic acid

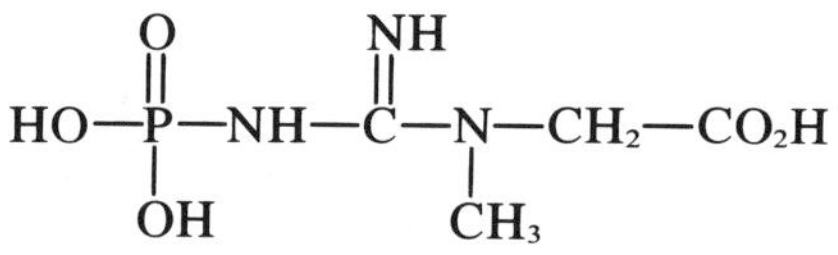

$HO{-}P(O)(OH){-}NH{-}C(=NH){-}N(CH_3){-}CH_2{-}CO_2H$

creatine phosphate

a Identify the four functional derivatives of phosphoric acid in the structural formulas drawn on p. 192.
b In these three structural formulas, the acid functional groups (both carboxylic and phosphoric) are shown in the un-ionized form, i.e., as —CO_2H and —OPO_3H_2. Yet, in biological systems at pH 7.4, both acid functional groups are fully ionized. Rewrite these structural formulas to show the proper ionization of these groups at pH 7.4.
c Write structural formulas for the products of hydrolysis of the phosphate functional groups in each compound. Note that at pH 7.4, the ion HPO_4^{2-} is the major phosphate species.

8.17 Show how you might convert ethyl propanoate into the following:

a ethyl 2-methyl-3-ketopentanoate **b** 3-pentanone

8.18 Starting with any aldehyde, ketone, or ester, synthesize each of the following by using the aldol or Claisen condensation and any necessary subsequent steps.

a $CH_3CH_2CH_2$—CH(OH)—CH(CH_2—CH_3)—CH_2OH

b H—C(=O)—CH(CH_3)—C(=O)—OCH_2CH_3

c C_6H_5—CH(OH)—CH_2—CH_2OH

d C_6H_5—CH=CH—C(=O)—CH_3

e C_6H_5—CH=CH—C(=O)—OH

8.19 One biosynthetic route to the formation of aromatic compounds, including derivatives of benzene, is thought to involve the formation and cyclization of a polycarbonyl chain derived from acetate units. Examples include the biosynthesis of 6-methylsalicylic acid and orsellinic acid. 6-Methylsalicylic acid is formed as a metabolite of the mold *Penicillium urticae*. If the acetic acid supplied to the mold is labeled at the carboxyl group with carbon-14 (indicated here in color), the 6-methylsalicylic acid has the label pattern shown.

4 CH_3—C(=O)—SCoA → [polycarbonyl chain: CH_3—C(=O)—CH_2—C(=O)—CH_2—C(=O)—CH_2—C(=O)—SCoA] → 6-methylsalicylic acid; orsellinic acid

6-methylsalicylic acid

orsellinic acid

Devise a reasonable scheme for the biosynthesis of each of these compounds starting from acetyl coenzyme A. (Hint: use a combination of the aldol condensation, the Claisen condensation, dehydration, reduction and keto-enol tautomerism).

8.20 The same polycarbonyl chain shown in brackets in Problem 8.19 can also cyclize in an alternative manner to give phloroacetophenone.

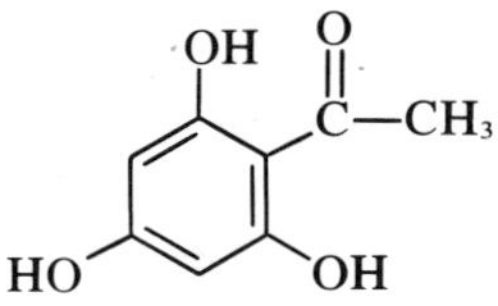

OH O
C—CH_3
HO OH
phloroacetophenone

a Propose a sequence of reactions that might lead from the bracketed polycarbonyl compound to phloroacetophenone.
b If the polycarbonyl compound is labeled with carbon-14 as shown in the previous problem, predict the positions of the carbon-14 labels in phloroacetophenone.

MINI-ESSAY

NYLON AND DACRON

Shortly after World War I, a number of creative and far-sighted chemists recognized the need for developing basic knowledge of polymer chemistry. One of the most productive of these pioneers was Wallace M. Carothers, of E. I. du Pont de Nemours and Company. Some 50 years ago, Carothers began fundamental research into the reactions of aliphatic dibasic acids and glycols (Section 5.3) and obtained polyester products of high molecular weight, over 10,000. These products were tough, opaque solids which melted at moderate temperatures to give clear, viscous liquids which could be drawn into fibers. However, the melting points of these polyester fibers were too low for them to be of use as textile fibers and they were not investigated further (that is, not for another decade). Carothers next turned his attention to reactions between dibasic acids and diamines, and in 1934 synthesized nylon, the first purely synthetic fiber.

The first step in the reaction is that of one molecule of hexamethylenediamine with one molecule of adipic acid to form a dimer with the elimination of water.

$$\underset{\text{adipic acid}}{HO-\overset{\overset{\displaystyle O}{\|}}{C}-(CH_2)_4-\overset{\overset{\displaystyle O}{\|}}{C}-OH} + \underset{\text{hexamethylenediamine}}{H_2N-(CH_2)_6-NH_2} \xrightarrow{275°C}$$

$$HO-\overset{\overset{\displaystyle O}{\|}}{C}-(CH_2)_4-\overset{\overset{\displaystyle O}{\|}}{C}-NH-(CH_2)_6-NH_2 + H_2O$$

The product of this condensation has a functional group at either end of the molecule and can react further. As dimer concentration builds up, it reacts with a monomer to give a trimer, or with another dimer to give a tetramer, etc. The reaction proceeds in a series of condensations, forming very long chains.

$$-\left(\overset{\overset{\displaystyle O}{\|}}{C}-(CH_2)_4-\overset{\overset{\displaystyle O}{\|}}{C}-NH-(CH_2)_6-NH\right)_n-$$

nylon 66

The term nylon has become accepted as the generic name for polyamides of this type. The polyamide resulting from the condensation of hexamethylenediamine and adipic acid is called nylon 66 to indicate its origin from two monomers each having six carbon atoms.

In the commercial synthesis of nylon 66, equimolar amounts of hexamethylenediamine and adipic acid dissolved in aqueous alcohol are heated in an autoclave to about 225°C. Since this is a closed system containing both water and alcohol vapor, the internal pressure rises to about 250 pounds per square inch (about 15 atmospheres). As the condensation proceeds and more water is produced, steam is bled from the autoclave to maintain a constant pressure. The temperature is gradually raised to about 275°C and as steam continues to be removed, the internal pressure falls to that of the atmosphere. After a total reaction time of 2–4 hours, polymer molecules of molecular weights 10,000–20,000 are formed. The crude nylon is extruded, cooled and cut into small chips. This is the raw material from which nylon fibers and plastics are fabricated.

As a first stage in fiber production, crude nylon, in the molten state at 260–270°C, is melt-spun into fibers and carefully cooled to room temperature. At this stage, the polymer is best described as unoriented crystalline. The crystallinity of polymers is not quite the same as the crystallinity of such common substances as copper sulfate and glucose. These substances show obvious crystalline structure. The characteristic property of the crystalline state is a high degree of order—the atoms are arranged regularly in a three-dimensional lattice which is built up of a simple unit repeated over and over again. With a crystal of a low-molecular-weight substance like glucose, there is complete order, i.e., 100% crystallinity. Although polymers do not crystallize in the conventional sense, they often have regions where chains are precisely oriented with respect to one another. Such regions are called crystallites. Melt-spun nylon 66 fibers have many crystallite regions, but they are randomly oriented and the fiber in this state is not suitable for textile use. However, the character of the fibers can be improved dramatically by drawing them at room temperature to about four times their original length. In this cold-drawing process, the fiber elongates and the crystalline regions are drawn together and oriented in the direction of the fiber axis. Hydrogen bonds between the oxygen atoms of one chain and the N—H groups of another chain hold the chains together. In the case of nylon 66, this process does not produce additional crystallites but rather orients existing crystallites parallel to the fiber axis. The effects of orientation of the polymer molecules on the physical properties can be profound; both tensile strength and stiffness increase with increasing orientation. Cold-drawing is an important step in the production of all synthetic fibers (see Figure 1).

Nylon has been widely accepted as a textile fiber because of its very high strength, its good elastic properties, and because fibers of very small denier can be drawn. Denier, a measure of the fineness of a fiber, is arbitrarily defined as the weight in grams of 9000 meters of fiber. Thus a 15 denier stocking is one in which 9000 meters of the fiber used in its manufacture weighs 15 grams. Nylon is also heat-settable, meaning that permanent creases can be introduced at high temperature and the material is then crease-resistant at the moderate temperatures of normal use.

An interesting alternative to the synthesis of nylon is exemplified in the manufacture of nylon 6. The essential feature in the condensation polymerization is amide formation between an amine and a carboxylic acid. The manufacture of nylon 66 uses two different starting materials—a diamine and a dicarboxylic acid. The manufac-

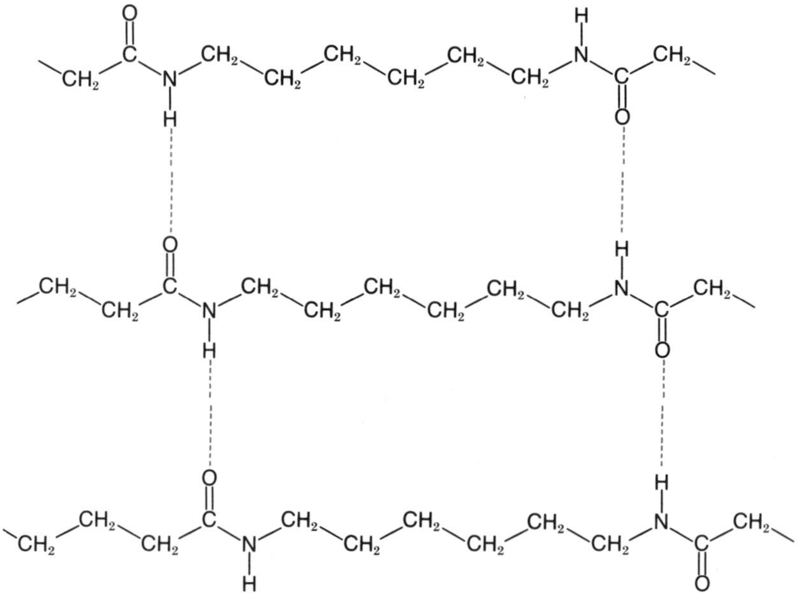

Figure 1 *Structure of cold-drawn nylon 66 fiber. Hydrogen bonds between adjacent chains hold molecules together in a crystalline structure.*

ture of nylon 6 uses one starting material, caprolactam (the cyclic amide of 6-aminohexanoic acid, a substance containing an amino group and a carboxyl group in the same molecule).

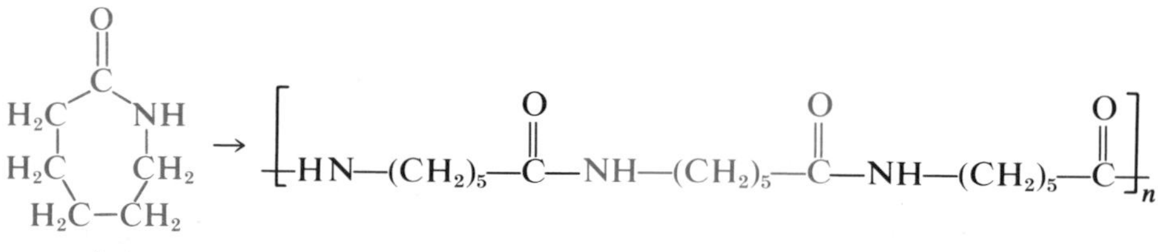

$$\text{caprolactam} \rightarrow \left[HN-(CH_2)_5-\overset{O}{\overset{\|}{C}}-NH-(CH_2)_5-\overset{O}{\overset{\|}{C}}-NH-(CH_2)_5-\overset{O}{\overset{\|}{C}} \right]_n$$

caprolactam nylon 6

Formerly, at least, nylon 6 was the principal variety in Europe while nylon 66 was the principal variety in the United States. The reason for this was the availability of different raw materials.

A variety of other polyamides have been made in the laboratory, but only a few of these have met with commercial success. Table 1 lists the commercially available nylons. Of these nylon 66 and nylon 6 account for most of the total output.

By the 1940s, the relationships between molecular structure and bulk properties of polymers were quite well understood, and the polyester condensations were reexamined. Carothers had already concluded that the polyester fibers from aliphatic dibasic acids and ethylene glycol were not suitable for textile use because they were too low melting. Winfield and Dickson at the Calico Printers Association in England reasoned, quite correctly as it turned out, that a greater resistance to rotation in the polymer backbone would confer a higher degree of crystallinity, stiffen the polymer chain, and thereby lead to a

Table 1 *Types of nylon in commercial production.*

Name	Starting Materials	Applications
66	$H_2N—(CH_2)_6—NH_2$ and $HOOC—(CH_2)_4—COOH$	fibers and moldings
6	$H_2N—(CH_2)_5—COOH$	fibers and moldings
7	$H_2N—(CH_2)_6—COOH$	fibers
610	$H_2N—(CH_2)_6—NH_2$ and $HOOC—(CH_2)_8—COOH$	moldings
11	$H_2N—(CH_2)_{10}—COOH$	moldings
12	$H_2N—(CH_2)_{11}—COOH$	moldings

more acceptable polyester fiber. To create this stiffness in the polymer chain they used terephthalic acid, an aromatic dicarboxylic acid. Whereas poly(ethylene adipate) melts much too low to be a suitable textile fiber, poly(ethylene terephthalate), formed from ethylene glycol and terephthalic acid, has proven tremendously successful and is marketed under the trade names Terylene and Dacron.

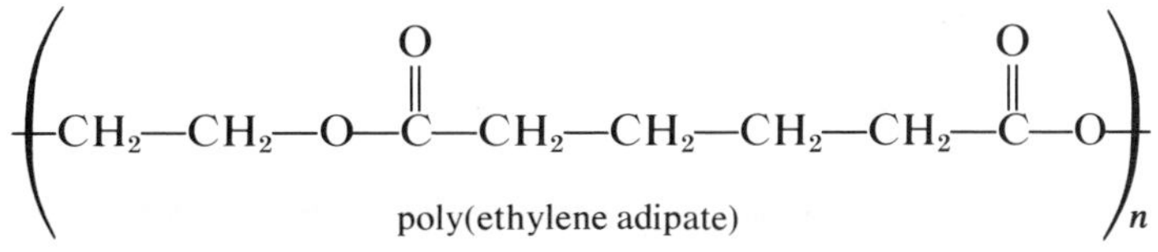

$$\left(CH_2—CH_2—O—\overset{O}{\overset{\|}{C}}—CH_2—CH_2—CH_2—CH_2—\overset{O}{\overset{\|}{C}}—O \right)_n$$

poly(ethylene adipate)

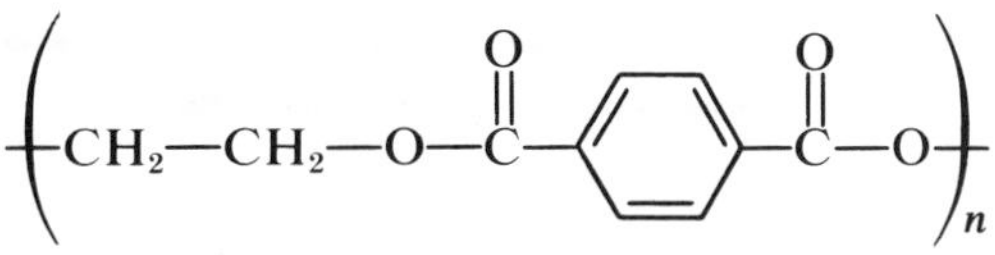

$$\left(CH_2—CH_2—O—\overset{O}{\overset{\|}{C}}—C_6H_4—\overset{O}{\overset{\|}{C}}—O \right)_n$$

poly(ethylene terephthalate)
Dacron, Terylene

Just as with nylon, the crude polyester is first spun into fibers and then is cold-drawn. The outstanding feature of Dacron is its stiffness, about four times that of nylon 66, very high strength, and a remarkable resistance to creasing and wrinkling. Because the Dacron fibers are much harsher (due to their stiffness) to the touch than wool, they are usually blended with wool or cotton. The polyester also is heat-settable, hence permanent creases may be introduced in the manufacturing of garments which then can be used repeatedly without ironing.

References

Billmyer, F. W., *Textbook of Polymer Chemistry* (Interscience Publishers, New York, 1957).

Kaufman, M., *Giant Molecules* (Doubleday and Company, Garden City, N.Y., 1968).

9

AMINES

9.1 INTRODUCTION

The organic derivatives of ammonia are known as amines and are very important in the whole of organic and biochemistry. Most of their chemistry stems directly from the presence of the unshared pair of electrons on nitrogen and from their resulting nucleophilicity and weak basicity.

9.2 STRUCTURE AND CLASSIFICATION

Amines are derivatives of ammonia in which one or more of the hydrogens are replaced by alkyl or aryl groups. Amines are classified as primary, secondary, or tertiary according to the number of hydrocarbon groups attached to the nitrogen atom. A nitrogen with four hydrocarbon groups attached is positively charged and is known as a quaternary ammonium salt.

9.3 ALIPHATIC AND AROMATIC AMINES

The simpler aliphatic amines are named by specifying the alkyl groups attached to the nitrogen atom and adding the ending -amine.

$CH_3—NH_2$
methylamine
(primary)

$CH_3CH_2—N(H)—CH_2CH_3$
diethylamine
(secondary)

$CH_3CH_2CH(CH_3)—NH_2$
sec-butylamine
(primary)

$(CH_3)_3N$
trimethylamine
(tertiary)

$HO—CH_2CH_2—NH_2$
ethanolamine
2-aminoethanol
(primary)

$(CH_3)_4N^+\ Cl^-$
tetramethylammonium
chloride
(quaternary ammonium salt)

Aryl amines are named as derivatives of the parent member of the family, aniline.

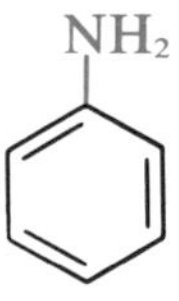

aniline

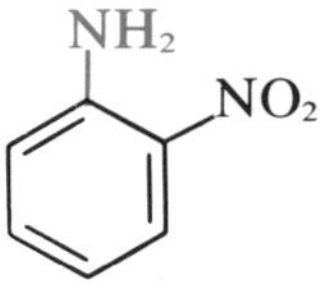

2-nitroaniline
o-nitroaniline

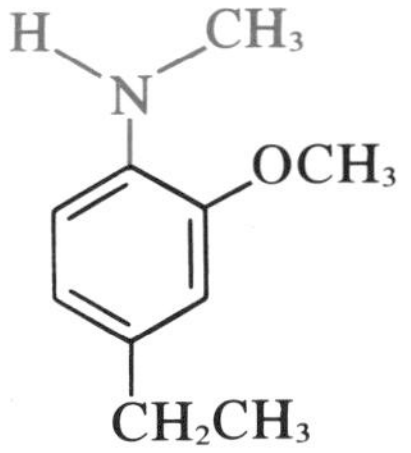

2-methoxy-4-ethyl-N-methylaniline

In the IUPAC system, the $—NH_2$ group may be treated as a substituent and named as an amino group.

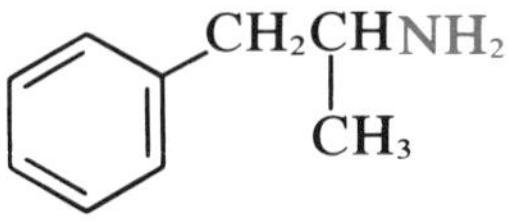

$$C_6H_5CH_2CH(NH_2)CH_3$$

1-phenyl-2-aminopropane
amphetamine

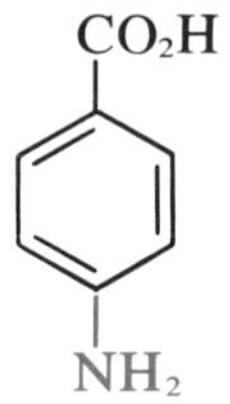

p-aminobenzoic acid

$$H_2NCH_2CH_2CH_2CO_2H$$

4-aminobutanoic acid
γ-aminobutyric acid

One quaternary ammonium salt of particular importance in human physiology is choline and its ester, acetylcholine.

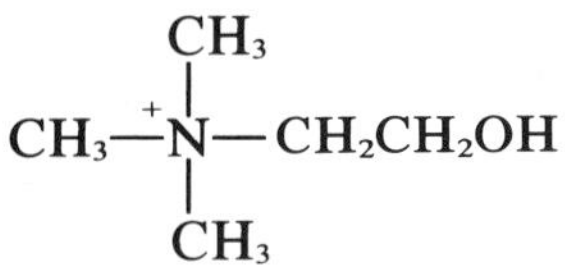

$$(CH_3)_3\overset{+}{N}—CH_2CH_2OH$$

choline

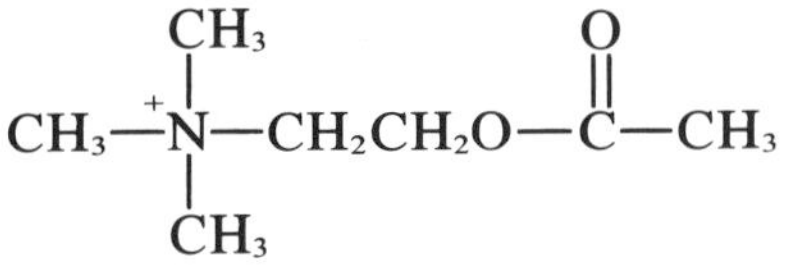

$$(CH_3)_3\overset{+}{N}—CH_2CH_2O—\overset{O}{\overset{\|}{C}}—CH_3$$

acetylcholine

Choline is a building block of a group of phospholipids known as lecithins (Chapter 11) which are important in the construction of cell membranes. Acetylcholine is a transmitter substance in certain motor neurons responsible for causing contraction of voluntary muscles. Acetylcholine is stored in synaptic vesicles and released in response to electrical activity in the neuron; it diffuses across the synapse and interacts with receptor sites on a neighboring neuron to cause transmission of a nerve impulse. After interacting with receptor sites, acetylcholine is deactivated through hydrolysis catalyzed by the enzyme acetylcholinesterase.

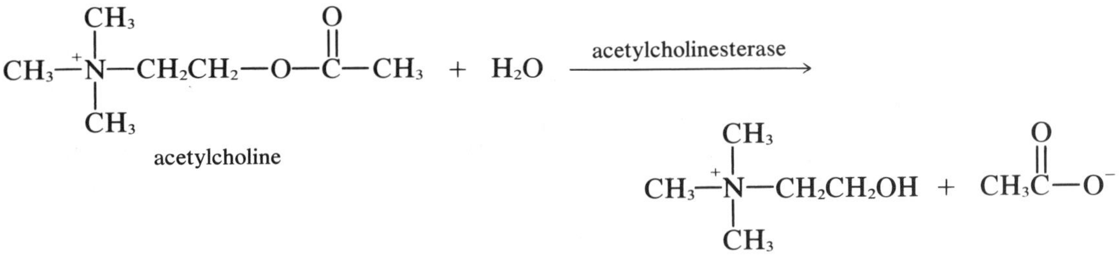

$$(CH_3)_3\overset{+}{N}—CH_2CH_2—O—\overset{O}{\overset{\|}{C}}—CH_3 + H_2O \xrightarrow{\text{acetylcholinesterase}} (CH_3)_3\overset{+}{N}—CH_2CH_2OH + CH_3\overset{O}{\overset{\|}{C}}—O^-$$

acetylcholine → choline + acetate

Inhibition of this enzyme leads to increased concentrations of acetylcholine in the synapse and increased stimulation of the particular neurons. In response to this overload, the neurons cease to fire, leading quickly to loss of response in voluntary muscles.

There are several compounds, natural and synthetic, that affect acetylcholine-mediated nerve transmission. Among the most widely known of these are the nerve gases and related insecticides. Diisopropyl fluorophosphate (DFP) and Tabun are potent inhibitors of acetylcholinesterase and a few milligrams of either one can kill a man in a few minutes through paralysis and respiratory failure.

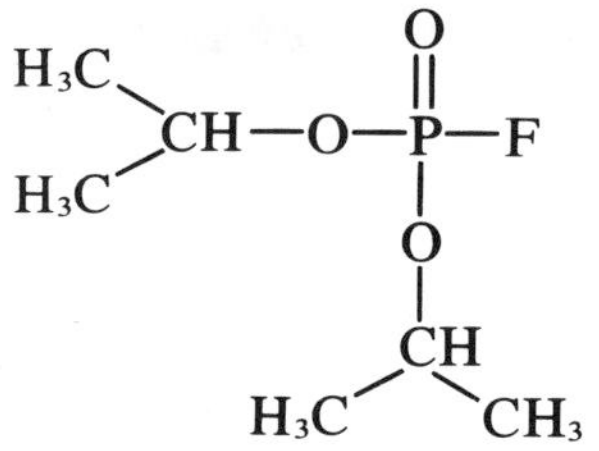

$$(H_3C)_2CH-O-\overset{\overset{O}{\|}}{\underset{\underset{\displaystyle CH(CH_3)_2}{|}}{\underset{O}{\underset{|}{P}}}}-F \qquad CH_3CH_2-O-\overset{\overset{O}{\|}}{\underset{\underset{\displaystyle N(CH_3)_2}{|}}{P}}-CN$$

diisopropyl fluorophosphate (DFP) 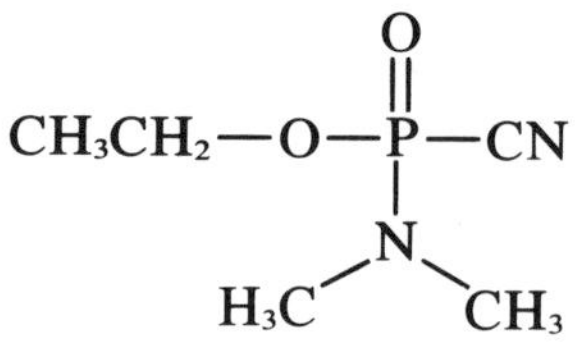 — Tabun

The closely related insecticides Malathion and Parathion are also phosphate esters.

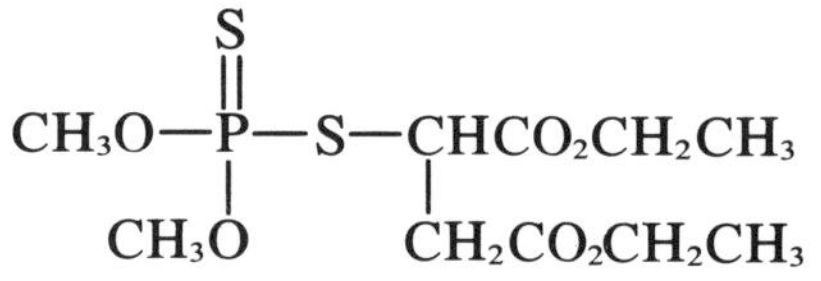

$$CH_3O-\overset{\overset{S}{\|}}{\underset{\underset{CH_3O}{|}}{P}}-S-\underset{\underset{CH_2CO_2CH_2CH_3}{|}}{CH}CO_2CH_2CH_3 \qquad CH_3CH_2O-\overset{\overset{S}{\|}}{\underset{\underset{CH_3CH_2O}{|}}{P}}-O-C_6H_4-NO_2$$

Malathion 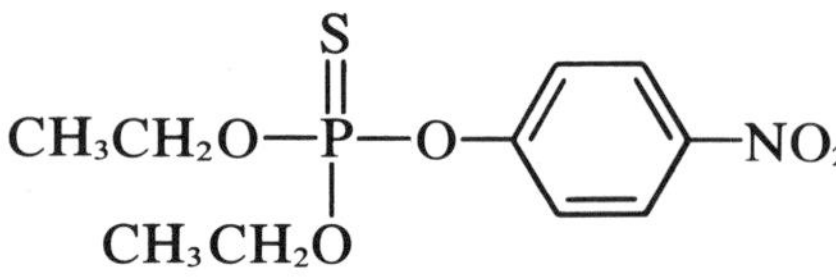 — Parathion

Among the natural acetylcholinesterase inhibitors is a substance contained in a plant extract called curare, a preparation long used by the Amazon Indians on arrow tips to kill or paralyze. In small doses, the extract can relax muscles without causing paralysis or death. A synthetic drug, decamethonium, is considerably less toxic than curare extract and is used medicinally as a muscle relaxant.

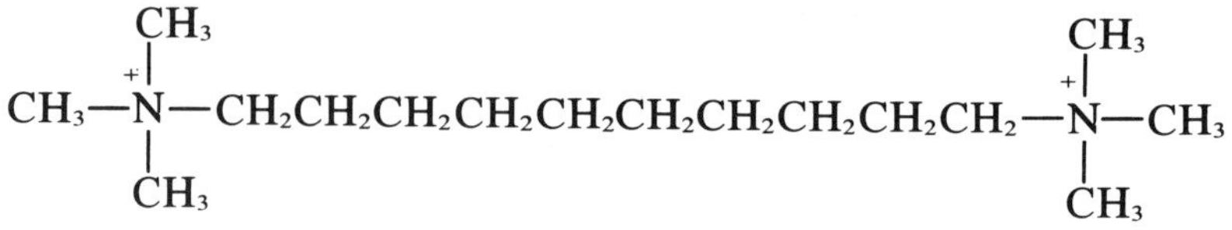

$$CH_3-\overset{\overset{CH_3}{|}}{\underset{\underset{CH_3}{|}}{\overset{+}{N}}}-CH_2CH_2CH_2CH_2CH_2CH_2CH_2CH_2CH_2CH_2-\overset{\overset{CH_3}{|}}{\underset{\underset{CH_3}{|}}{\overset{+}{N}}}-CH_3$$

hexamethyldecamethylenediammonium ion
decamethonium

If the nitrogen atom is incorporated into a ring, the resulting amine is called a heterocyclic amine. In contrast to the carbocyclic rings we have dealt with so far, heterocyclic rings contain one or more atoms other than carbon. Piperidine and pyrrolidine, shown on p. 203, both undergo the reactions of typical secondary amines and consequently we can include them in our discussion of the chemistry and physical properties of aliphatic amines.

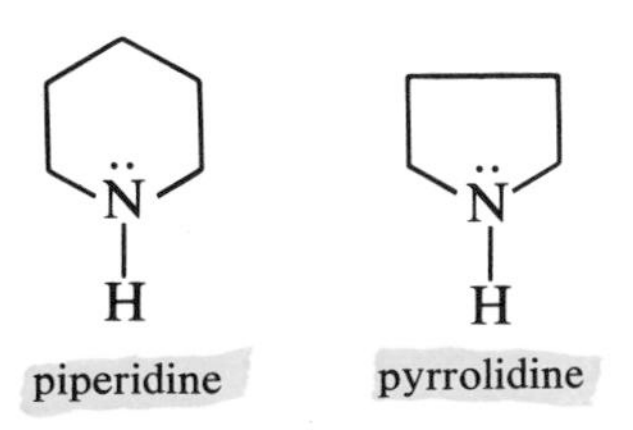

9.4 HETEROCYCLIC AROMATIC AMINES

Of more interest are the heterocyclic amines in which the ring is highly unsaturated. The most important six-membered heterocyclic aromatics are pyridine and pyrimidine. These are shown below along with the numbering system used to specify the location of substituents.

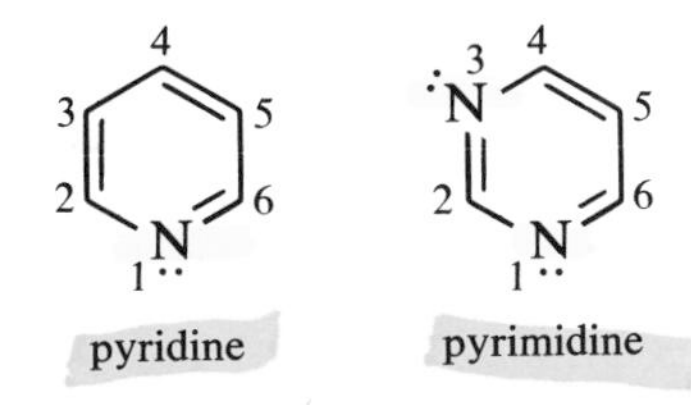

Both pyridine and pyrimidine are heterocyclic analogs of benzene in which one and two carbon atoms of the ring have been replaced by nitrogen atoms. We can represent each of these molecules as a resonance hybrid of two benzene-like contributing structures (Figure 9.1).

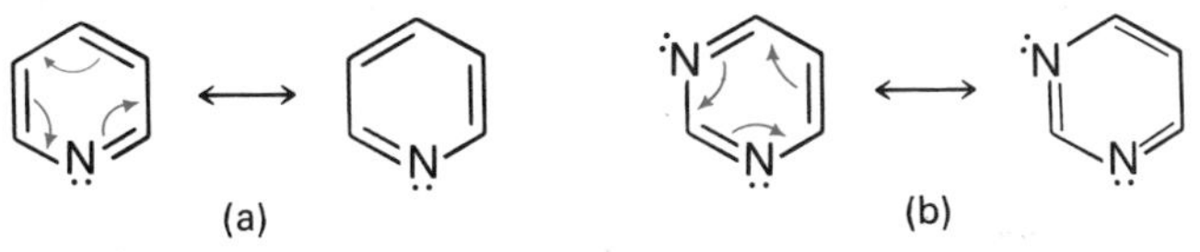

Figure 9.1 Resonance hybrids of two benzene-like contributing structures for (a) pyridine and (b) pyrimidine.

These heterocycles are classified as aromatic compounds because they tend to undergo substitution reactions like benzene rather than addition reactions like alkenes. Further, both pyridine and pyrimidine are resistant to the action of the common laboratory oxidizing and reducing agents.

The pyridine ring and its reduced forms are widespread in nature and are synthesized by plants. Perhaps the most important pyridine derivative is the ubiquitous nicotinamide adenine dinucleotide (NAD^+) and its phosphorylated derivative nicotinamide adenine dinucleotide phosphate ($NADP^+$). These molecules along with the appropriate enzyme systems function in oxidation–reduction reactions in the cells of both plants and animals.

Pyrimidine itself does not occur in nature, but pyrimidine derivatives with substituents at positions 2, 4 and 5 occur widely in nature in nucleic acids (Chapter 14) and in a variety of fused ring systems.

The five-membered heterocyclics pyrrole and imidazole also show some degree of aromatic character.

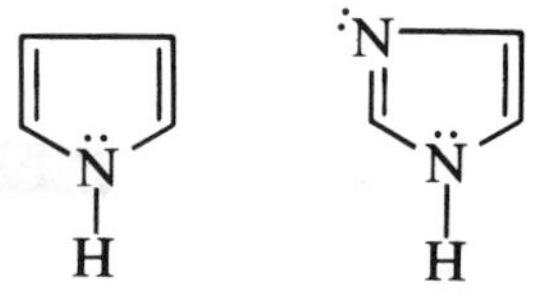

pyrrole imidazole

The imidazole ring is found in nature in the amino acid L-histidine and in histamine. An enzyme, histidine decarboxylase, catalyzes the decarboxylation of L-histidine to form histamine.

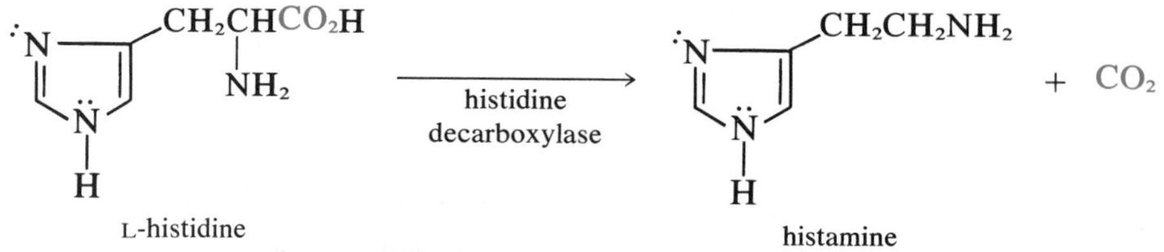

$$\text{L-histidine} \xrightarrow{\text{histidine decarboxylase}} \text{histamine} + CO_2$$

Extensive production of histamine occurs during hypersensitive allergic reactions and the symptoms of this release are unfortunately familiar to most of us, in particular those who suffer from hay fever. The search for antihistamines—drugs that will inhibit the effects of histamine—has led to the synthesis of several drugs whose trade names are well known. Structural formulas for three of the more widely used antihistaminic drugs are shown in Figure 9.2.

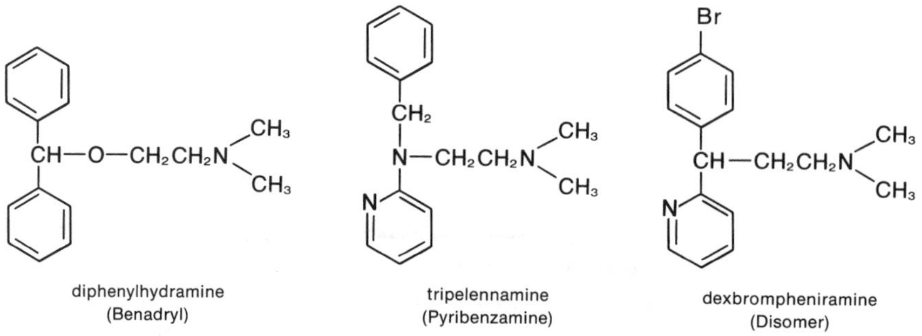

diphenylhydramine (Benadryl) tripelennamine (Pyribenzamine) dexbrompheniramine (Disomer)

Figure 9.2 Three synthetic antihistamines.

Observe the structural similarity in these three drugs: each has (1) two aromatic rings and (2) a dimethylaminoethyl, $—CH_2CH_2N(CH_3)_2$, group. Dexbrompheniramine is one of the most potent of the available antihistaminics. Notice that it has one asymmetric carbon atom. Pharmacological studies have shown clearly that the antihistaminic activity is due almost exclusively to the (+)-isomer. It is nearly twice as potent as the racemic mixture and about thirty times as potent as the (−)-isomer.

Two other heterocyclic ring compounds that are especially important in the biological world are purine and indole.

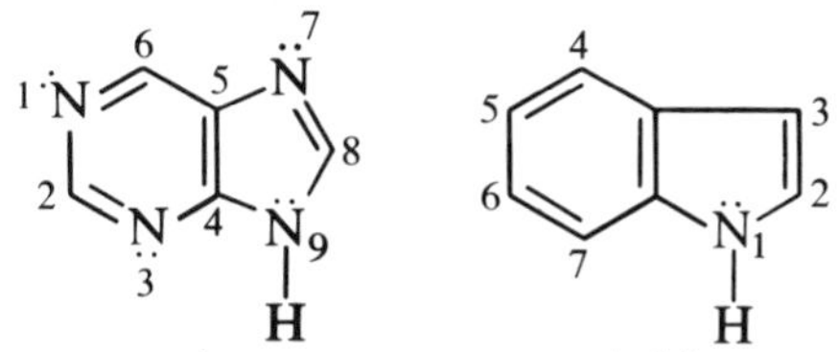

purine indole

Purine contains two fused rings, a six-membered pyrimidine ring and a five-membered imidazole ring. The purine and pyrimidine bases appear in the molecules that comprise the genetic units of heredity, namely deoxyribonucleic acid (DNA) and ribonucleic acid (RNA). DNA is a constituent of the genetic material of the cell nucleus. RNA is found chiefly in the cytoplasm of the cell where it functions in the transfer of information from the nucleus to the sites of protein synthesis. The chemistry and function of DNA and RNA are discussed in Chapter 14.

Indole contains two fused rings, a six-membered benzene ring and a five-membered pyrrole ring. The indole nucleus is found in nature in the amino acid L-tryptophan, in the neurotransmitter substance serotonin, and in many other substances of plant and animal origin.

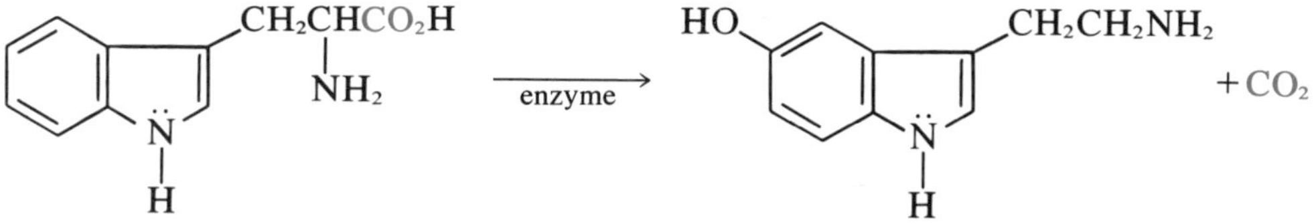

$$\text{L-tryptophan} \xrightarrow{\text{enzyme}} \text{serotonin (5-hydroxytryptamine)} + CO_2$$

L-tryptophan

serotonin
(5-hydroxytryptamine)

Even this brief introduction to amines and amine derivatives leads to the conclusion that these substances deserve attention for their biological activity. We have seen a few examples already. You can read about such familiar amines as atropine, cocaine, Novocaine, morphine, codeine and Demerol in the mini-essay "Alkaloids." There you can also read about several psychoactive amines including mescaline, amphetamine, and LSD. The mini-essay "Biogenic Amines and Emotion" discusses serotonin, norepinephrine and dopamine, all of which are transmitter substances in the central nervous system. In the same mini-essay you can read about the structure and mode of action of some of the major tranquilizers and antidepressants. And in subsequent chapters we shall see many more amines that play vital roles in biological chemistry.

9.5 PHYSICAL PROPERTIES

Amines are polar compounds, and except for tertiary amines, show evidence of hydrogen bonding in the pure liquid. Amines have higher boiling points than nonpolar compounds of comparable molecular weight, but lower boiling points than alcohols and carboxylic acids (see Table 9.1). All classes of amines are capable of forming hydrogen bonds with water.

Table 9.1 Physical properties of amines.

Name	bp (°C)	Solubility (g/100 g H_2O)	pK_b	pK_a
ammonia	−33	90	4.74	9.26
methylamine	−7	very	3.34	10.66
ethylamine	17	very	3.20	10.80
cyclohexylamine	134	slightly	3.34	10.66
benzylamine	185	very	4.67	9.33
dimethylamine	7	very	3.27	10.73
diethylamine	55	very	3.51	10.49
triethylamine	89	1.5	2.99	11.01
aniline	184	3.7	9.37	4.63
pyridine	116	very	8.75	5.25
imidazole	257	very	7.05	6.95

9.6 BASICITY OF AMINES

Amines, though weak bases compared to sodium hydroxide and other metal hydroxides, are strong bases compared to alcohols and water. All amines, whether soluble or insoluble in water, will react quantitatively with mineral acids to form salts. For example, cyclohexylamine, which is only slightly soluble in water, dissolves readily in dilute hydrochloric acid as the cyclohexylammonium salt.

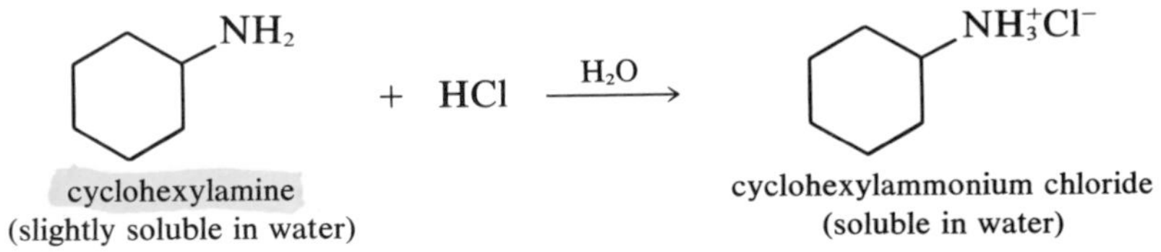

$$C_6H_{11}NH_2 + HCl \xrightarrow{H_2O} C_6H_{11}NH_3^+Cl^-$$

cyclohexylamine (slightly soluble in water) → cyclohexylammonium chloride (soluble in water)

The basicity of amines and the solubility of amine salts can be used to distinguish between amines and non-basic, water-insoluble compounds, and also to separate them. Shown below is a flowchart for the separation of aniline from methyl benzoate.

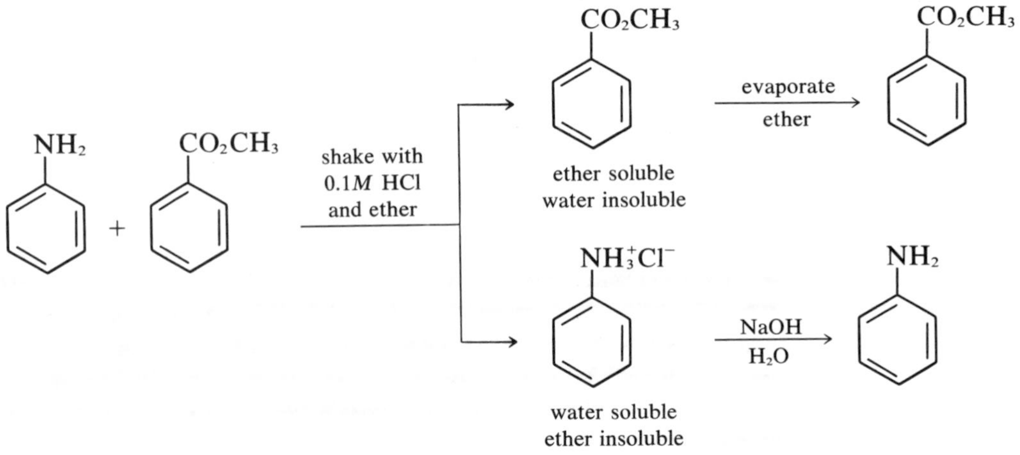

$C_6H_5NH_2$ + $C_6H_5CO_2CH_3$ — shake with 0.1*M* HCl and ether →

- $C_6H_5CO_2CH_3$ (ether soluble, water insoluble) — evaporate ether → $C_6H_5CO_2CH_3$
- $C_6H_5NH_3^+Cl^-$ (water soluble, ether insoluble) — NaOH, H_2O → $C_6H_5NH_2$

Crystalline salts of amines can generally be obtained by reacting one equivalent of amine with an acid and evaporating the solution to dryness.

Amines react with water to give basic solutions. This basicity of amines is due to the presence of the unshared pair of electrons on nitrogen.

$$CH_3NH_2 + H_2O \rightleftharpoons CH_3NH_3^+ + OH^-$$

The equilibrium constant, K_b, for this reaction is given by the expression

$$K_b = \frac{[CH_3NH_3^+][OH^-]}{[CH_3NH_2]}$$

Values of pK_b for some primary, secondary, tertiary, heterocyclic, and aromatic amines are given in Table 9.1.

All aliphatic amines, whether primary, secondary, tertiary, or saturated heterocyclic amines, have about the same base strengths and do not differ appreciably in basicity from ammonia. The K_b for ethylamine is only about 25 times larger than that of ammonia.

Aromatic amines such as aniline are significantly less basic than aliphatic amines. The base dissociation constant of cyclohexylamine is larger than that of aniline by a factor of 10^6.

$$\underset{\text{cyclohexylamine}}{C_6H_{11}NH_2} + H_2O \rightleftharpoons C_6H_{11}NH_3^+ + OH^- \qquad K_b = 5 \times 10^{-4}$$

$$\underset{\text{aniline}}{C_6H_5NH_2} + H_2O \rightleftharpoons C_6H_5NH_3^+ + OH^- \qquad K_b = 4.2 \times 10^{-10}$$

Until recently it was common practice to list only K_b or pK_b values for amines. Now, however, it is more and more common to also list K_a and pK_a values for amines. In fact, it is only K_a and pK_a values that are used in discussing the acid–base properties of the amine group of an amino acid (Chapter 12). For this reason, Table 9.1 lists both pK_b and pK_a values for a variety of common amines.

To illustrate the difference between K_b, pK_b and K_a, pK_a for an amine, consider methylamine.

$$CH_3NH_2 + H_2O \rightleftharpoons CH_3NH_3^+ + OH^- \quad K_b = 4.4 \times 10^{-4} \qquad pK_b = 3.36$$

$$CH_3NH_3^+ \rightleftharpoons CH_3NH_2 + H^+ \quad K_a = 2.7 \times 10^{-11} \qquad pK_a = 10.66$$

Thus, pK_b measures directly the strength of CH_3NH_2 as a base. pK_a measures directly the strength of $CH_3NH_3^+$ as an acid. For perspective, you might compare the pK_a values for acetic acid and for methylamine.

$$CH_3CO_2H \rightleftharpoons CH_3CO_2^- + H^+ \qquad K_a = 1.8 \times 10^{-5} \qquad pK_a = 4.74$$

$$CH_3NH_3^+ \rightleftharpoons CH_3NH_2 + H^+ \qquad K_a = 2.7 \times 10^{-11} \qquad pK_a = 10.66$$

By using K_a and pK_a values for carboxylic acids and amines, we can compare them directly, for in each case we are looking at the dissociation of an acid to form a base and a proton. From a comparison of these K_a and pK_a values, it is obvious that acetic acid is a much stronger acid than methylammonium ion.

9.7 REACTIONS OF AMINES

In our earlier discussions of the chemistry of the functional derivatives of carboxylic acids we have already encountered many of the more important reactions of ammonia. Amines, which resemble ammonia in having an unshared pair of electrons on the nitrogen atom, behave like ammonia in many ways: they form salts with acids, they can catalyze reactions where a basic catalyst is needed, and (except for tertiary amines) they can react with an ester or an acid anhydride to form amides. Below are several examples of reactions typical of amines.

Salt formation with acids:

$$CH_3COOH + (CH_3)_2NH \rightarrow \underset{\text{dimethylammonium acetate}}{(CH_3)_2NH_2^+ + CH_3COO^-}$$

Pyrolysis of ammonium salts of carboxylic acids to give amides:

$$CH_3COO^- + (CH_3)_2NH_2^+ \xrightarrow[200°C]{\text{heat}} \underset{\text{N,N-dimethylacetamide}}{CH_3{-}\overset{O}{\overset{\|}{C}}{-}N(CH_3)_2} + H_2O$$

Ammonolysis of esters to form amides:

$$CH_3{-}\overset{O}{\overset{\|}{C}}{-}OCH_2CH_3 + CH_3NH_2 \longrightarrow \underset{\text{N-methylacetamide}}{CH_3{-}\overset{O}{\overset{\|}{C}}{-}NHCH_3} + CH_3CH_2OH$$

Ammonolysis of anhydrides of carboxylic acids to give amides:

$$CH_3{-}\overset{O}{\overset{\|}{C}}{-}O{-}\overset{O}{\overset{\|}{C}}{-}CH_3 + CH_3NH_2 \longrightarrow \underset{\text{N-methylacetamide}}{CH_3{-}\overset{O}{\overset{\|}{C}}{-}NHCH_3} + CH_3COOH$$

Amines also react with derivatives of sulfuric and phosphoric acids to form amide derivatives. Shown on p. 209 are the structural formulas of benzenesulfonamide and *p*-aminobenzenesulfonamide (more commonly known as sulfanilamide).

The discovery of the medicinal use of sulfanilamide and its derivatives was a milestone in the history of chemotherapy for it represents one of the first rational investigations of synthetic organic molecules as potential drugs to fight infection. Sulfanilamide was first

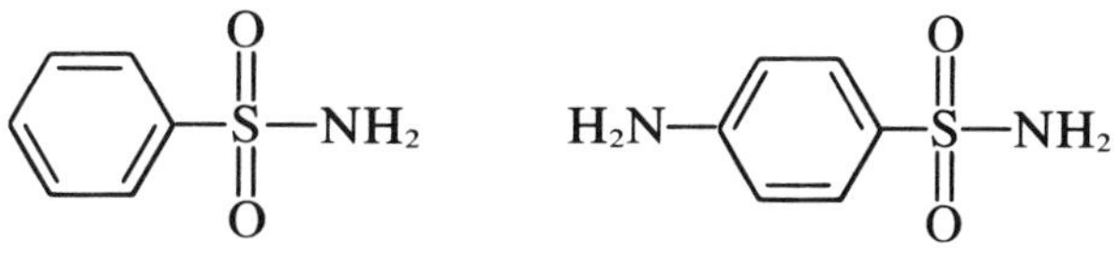

benzenesulfonamide

p-aminobenzenesulfonamide
sulfanilamide

prepared in 1908 in Germany during an investigation of azo dyes (p. 211), but it was not until 1932 that its possible therapeutic value was realized. In that year the azo dye Protonsil was prepared and patented in Germany. In research over the next two years the German scientist G. Domagk observed that mice with streptococcal septicemia (an infection of the blood) could be treated with Protonsil. He also observed the remarkable effectiveness of Protonsil in curing experimental streptococcal and staphylococcal infections in other experimental animals. Domagk further discovered that Protonsil is rapidly reduced in the cell to sulfanilamide, and that it is sulfanilamide and not Protonsil which is the actual antibiotic. His discoveries were honored in 1939 by the Nobel Prize in Medicine.

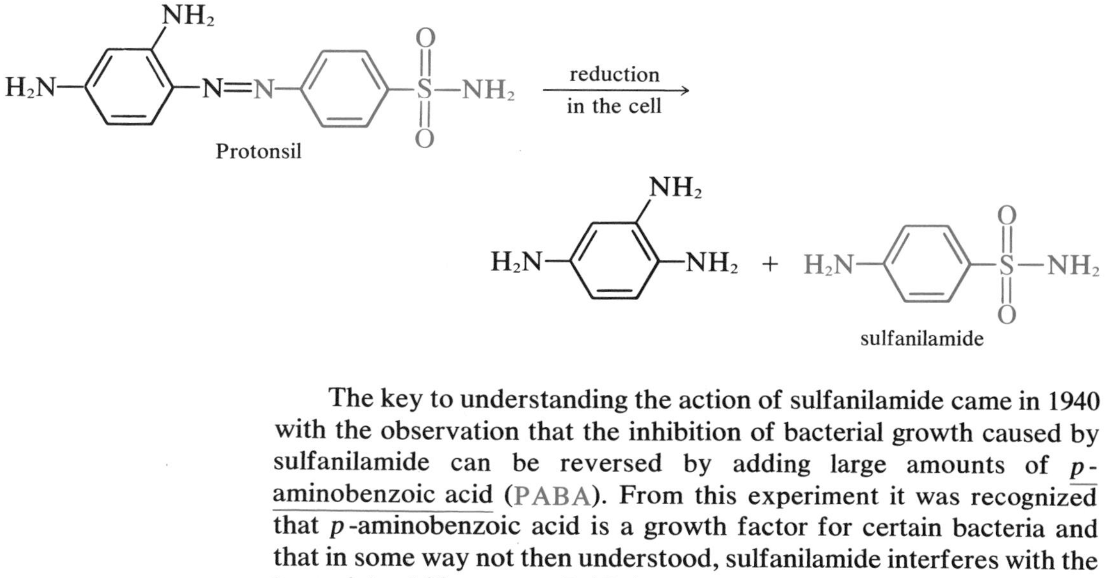

Protonsil $\xrightarrow[\text{in the cell}]{\text{reduction}}$ + sulfanilamide

The key to understanding the action of sulfanilamide came in 1940 with the observation that the inhibition of bacterial growth caused by sulfanilamide can be reversed by adding large amounts of *p*-aminobenzoic acid (PABA). From this experiment it was recognized that *p*-aminobenzoic acid is a growth factor for certain bacteria and that in some way not then understood, sulfanilamide interferes with the bacteria's ability to use PABA.

There are obvious structural similarities between *p*-aminobenzoic acid and sulfanilamide.

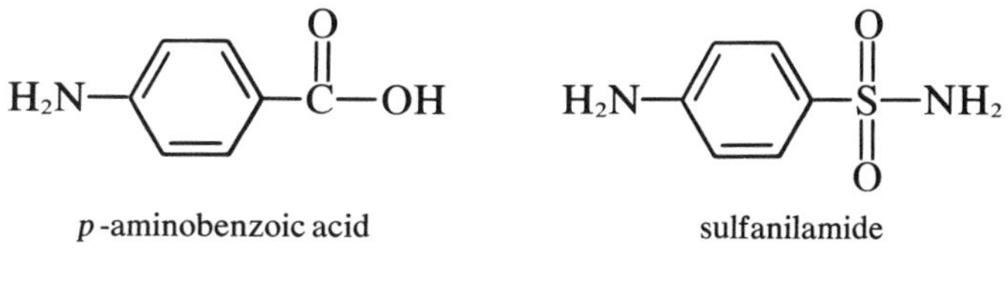

p-aminobenzoic acid

sulfanilamide

It now appears that sulfanilamide drugs inhibit one or more enzyme-catalyzed steps in the synthesis of folic acid from *p*-aminobenzoic acid. The ability of sulfanilamide to combat infections in man probably depends on the fact that man also requires folic acid but does not make it from *p*-aminobenzoic acid. In man folic acid is made by a different biochemical pathway.

In the search for even better sulfa drugs, literally thousands of derivatives of sulfanilamide have been synthesized. Two of the more effective sulfa drugs are shown below.

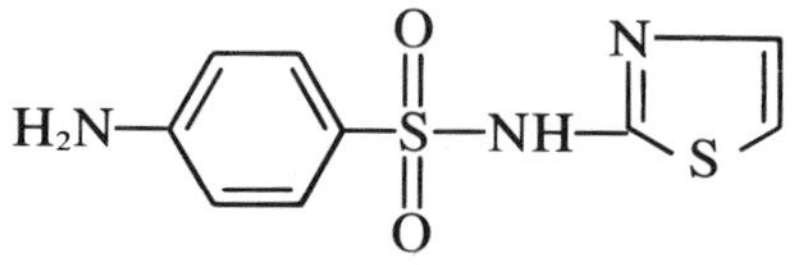

$$H_2N-C_6H_4-SO_2-NH-(\text{thiazol-2-yl})$$

sulfathiazole

$$H_2N-C_6H_4-SO_2-NH-(\text{pyrimidin-2-yl})$$

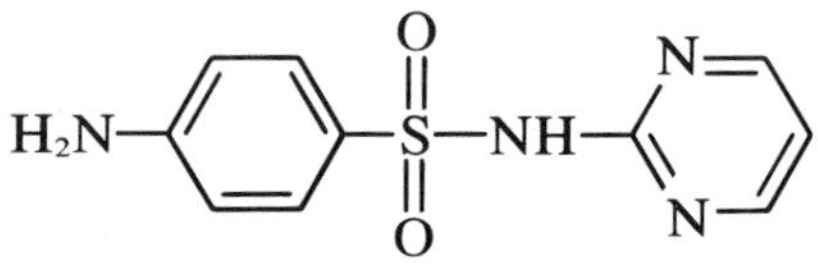

sulfadiazine

Sulfa drugs were found to be effective in the treatment of tuberculosis, pneumonia, and diphtheria and they helped usher in a new era in public health in the United States in the 1930s. During World War II they were routinely sprinkled on wounds to prevent infection. These drugs were among the first of the new "wonder-drugs." As a footnote in history, the use of sulfa drugs to fight bacterial infection has been largely supplanted by an even newer wonder-drug, the penicillins. For the story of the discovery, development, and mode of action of this truly remarkable antibiotic, see the mini-essay "The Penicillins."

9.8 REACTION WITH NITROUS ACID

The reaction of a primary aliphatic amine with nitrous acid at room temperature or even at 0°C gives a quantitative yield of nitrogen gas, N_2, and is the basis for the Van Slyke determination of primary amines. In each molecule of nitrogen produced, one nitrogen atom comes from the primary amine and the other from nitrous acid. The initial reaction produces a diazonium salt which is unstable and decomposes to nitrogen gas and a carbonium ion. The carbonium ion then reacts in typical fashion with water (or other nucleophile) to produce an alcohol (or other derivative depending on the nucleophile), or loses a proton to form an alkene.

$$CH_3CH_2NH_2 + HNO_2 \xrightarrow[\text{water}]{\text{HCl}} \underset{\text{diazonium ion}}{[CH_3CH_2N_2^+]} \rightarrow CH_3CH_2^+ + N_2$$

$$CH_3CH_2^+ \rightarrow CH_3CH_2OH + CH_2{=}CH_2$$

Secondary aliphatic amines react with nitrous acid to form N-nitroso compounds which are insoluble in aqueous acid and therefore separate as an oily layer. Tertiary aliphatic amines do not react with nitrous acid.

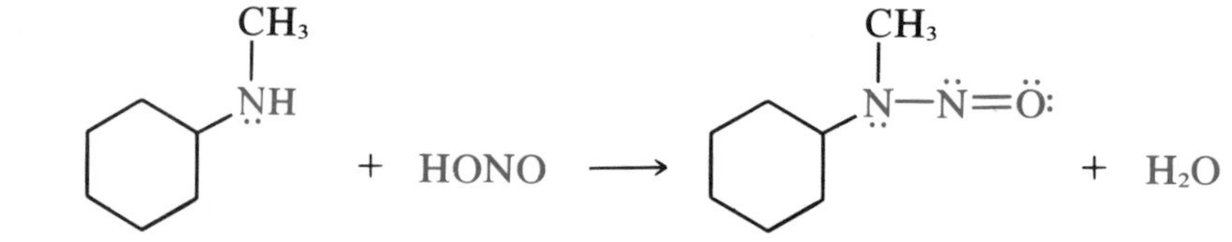

$$C_6H_{11}-\ddot{N}H-CH_3 + HONO \longrightarrow C_6H_{11}-\ddot{N}(CH_3)-\ddot{N}{=}\ddot{O}: + H_2O$$

N-methylcyclohexylamine → N-nitroso-N-methylcyclohexylamine

In the case of aliphatic amines this reaction is of little value except for quantitative or qualitative analysis. Primary, secondary, and tertiary aliphatic amines are soluble in the aqueous media of this test. Evolution of nitrogen is a qualitative test for a primary amine. If the nitrogen gas is collected and measured it can be used as a quantitative test for a primary amine. Secondary amines do not liberate nitrogen but instead are converted into N-nitroso compounds which separate as an oily layer. Tertiary amines neither liberate nitrogen gas nor form N-nitroso compounds; they remain dissolved in the aqueous acid.

Primary aromatic amines react with cold nitrous acid and are converted to diazonium salts in the same way aliphatic amines are. The aromatic diazonium salts are stable at 0°C, but they too decompose with the evolution of nitrogen gas when their aqueous solutions are warmed to room temperature. In water, phenol is the organic product of the decomposition.

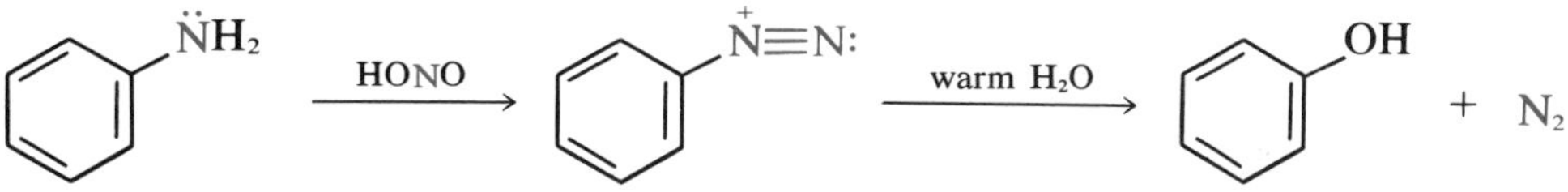

$$\ddot{N}H_2 \xrightarrow{HONO} \overset{+}{N}{\equiv}N: \xrightarrow{\text{warm } H_2O} OH + N_2$$

Aromatic diazonium ions may also attack other aromatic rings and substitute for hydrogen. These reactions are called diazo coupling reactions and the products are called azo compounds. Such azo compounds have found extensive use as dyes and indicators. Below are shown structural formulas for two commercial azo dyes, congo red and para red.

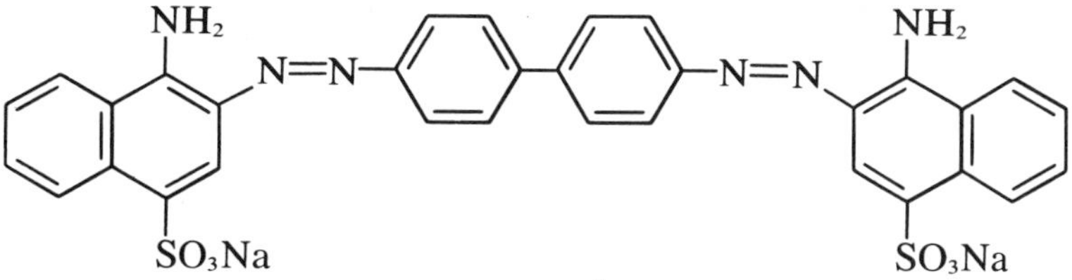

NH_2 — N=N — N=N — NH_2

SO_3Na SO_3Na

congo red

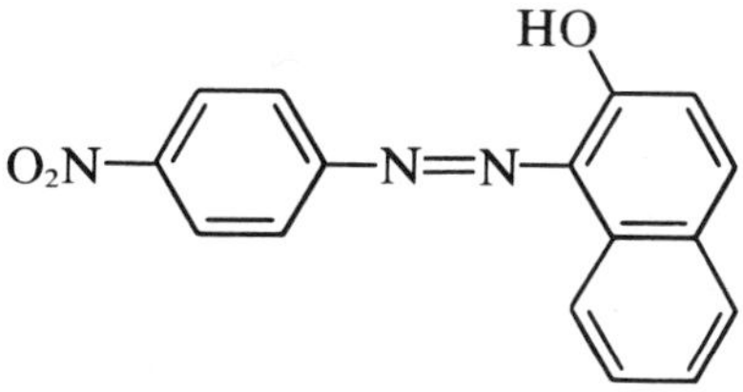

HO

O_2N — N=N —

para red

Azo compounds containing weakly basic amino groups or weakly acidic phenolic hydroxyl groups are useful as acid–base indicators. One such indicator, methyl orange, is formed by the reaction of diazotized sulfanilic acid with N,N-dimethylaniline.

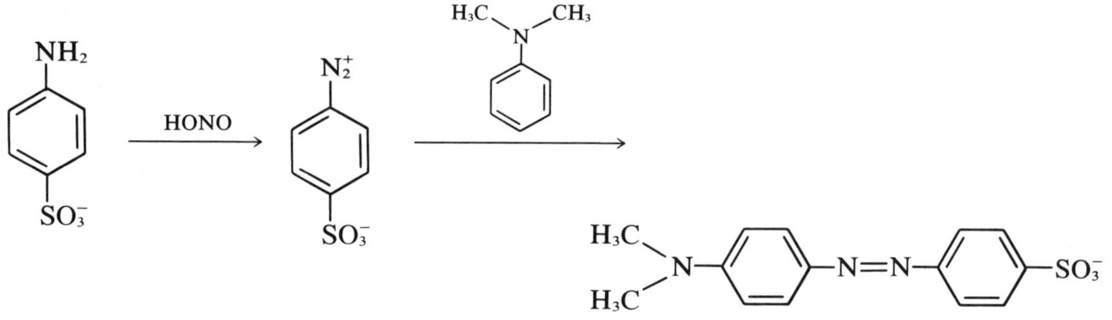

methyl orange

At pH 3.1 methyl orange is red; at pH 4.4 it is yellow-orange. Therefore this azo dye is widely used as an indicator in acid–base titrations where the end point of the titration is in the pH range 3.1 to 4.4.

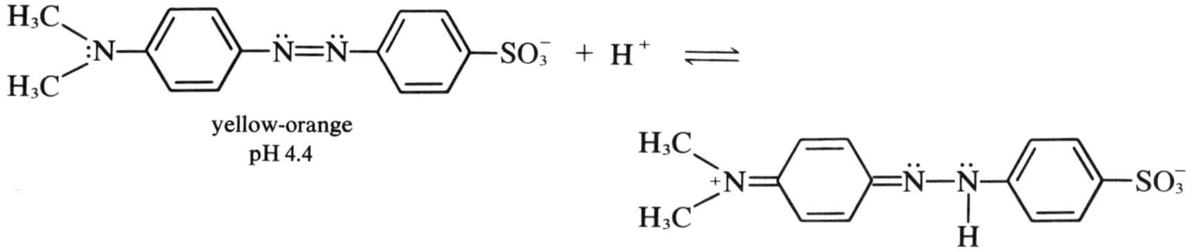

yellow-orange
pH 4.4

red
pH 3.1

9.9 REACTION WITH NINHYDRIN

The detection of very small quantities of amino acids is facilitated by reaction with the reagent ninhydrin. The reaction is very complex. Amino acids in general react to produce a purple-colored anion. The intensity of the color is directly proportional to the concentration of the anion, and hence may be used as a quantitative measure of the concentration of the amino acid in an unknown sample.

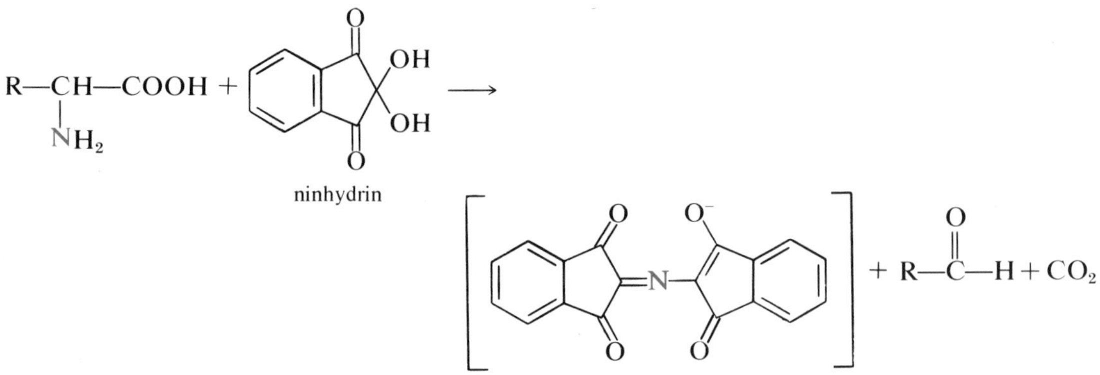

ninhydrin

anion
(purple colored)

Problems

9.1 Write structural formulas for the following compounds.

a diethylamine
b aniline
c cyclohexylamine
d pyrrole
e pyridine
f tetraethylammonium iodide
g 2-aminoethanol (ethanolamine)
h α-aminopropanoic acid (alanine)
i tetramethylammonium hydroxide
j trimethylammonium benzoate
k *p*-methoxyaniline
l N,N-dimethylacetamide
m N-methylaniline

9.2 Give an acceptable name for each of the following compounds.

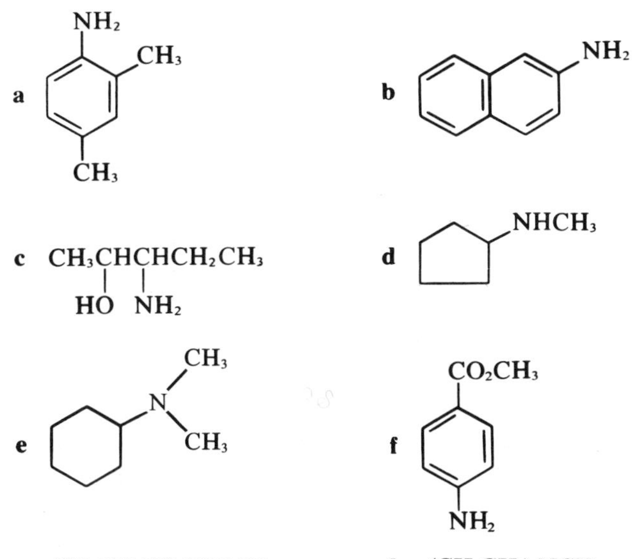

g $CH_3CH_2CH_2CH_2NH_2$
h $(CH_3CH_2)_2NCH_3$
i $(CH_3)_4N^+OH^-$
j $H_2NCH_2CH_2CH_2CH_2CH_2NH_2$

9.3 Draw structural formulas for the eight isomeric amines of molecular formula $C_4H_{11}N$. Name each. Label each as primary, secondary, or tertiary.

9.4 Using the values of pK_b in Table 9.1, determine if methylamine is a stronger or weaker base than triethylamine; than aniline.

9.5 Which of these two compounds would you predict to be the stronger base? Explain your reasoning.

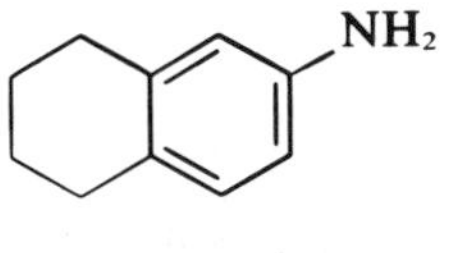

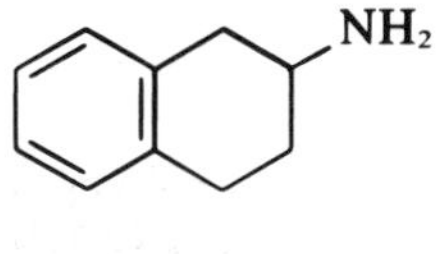

NH₂ NH₂

9.6 Select the stronger acid.

a $CH_3CH_2NH_3^+$ or C_6H_5–NH_3^+
b NH_4^+ or $CH_3NH_3^+$
c (pyridinium, N⁺–H) or (cyclohexenyl–NH_3^+)
d $C_6H_5CH_2NH_3^+$ or $C_6H_5NH_3^+$

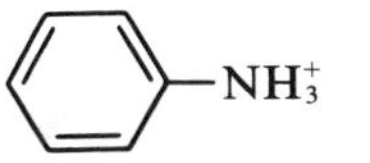

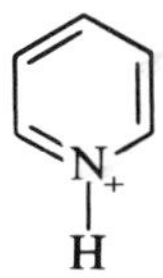

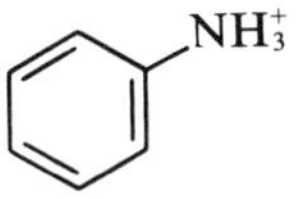

9.7 Select the stronger base.

a $CH_3CH_2NH_2$ or $C_6H_5NH_2$

b NH_3 or CH_3NH_2

c CH_3NH_2 or $CH_3CO_2^-$

d pyridine or aniline (structures: a benzene ring with ring N; a benzene ring bearing NH_2)

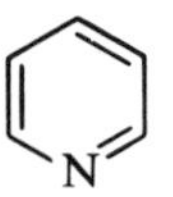

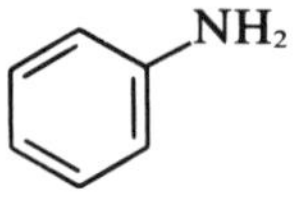

e $C_6H_5CH_2NH_2$ or $C_6H_5NH_2$

9.8 Alanine (α-aminopropanoic acid) is one of the important amino acids found in proteins. Would you expect the structural formula of alanine in aqueous solution to be better represented by I or II below? Explain.

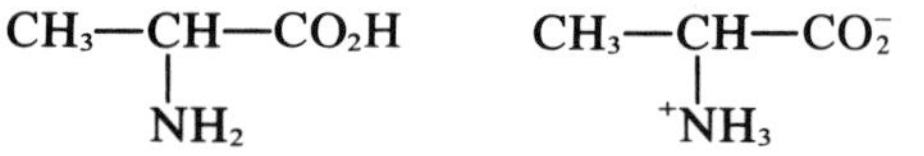

$$\underset{\text{(I)}}{CH_3{-}\underset{\underset{NH_2}{|}}{CH}{-}CO_2H} \qquad \underset{\text{(II)}}{CH_3{-}\underset{\underset{^{+}NH_3}{|}}{CH}{-}CO_2^-}$$

9.9 Calculate the hydroxide ion concentration and the pH of a saturated solution of aniline in water.

9.10 Calculate the hydroxide and hydrogen ion concentrations of a solution containing 10.1 grams of triethylamine per liter of water. Also calculate the concentration of triethylamine and the triethylammonium ion.

9.11 How many milliliters of 0.12M HCl are required to neutralize 0.59 g of trimethylamine?

9.12 Consider the compounds n-butylamine and n-butyl alcohol. One of these compounds has a boiling point of 117°C, the other a boiling point of 78°C. Which of the two would you predict to have the boiling point of 117°C? Explain.

9.13 Would you predict aniline to be most soluble in water, 0.1M NaOH, or 0.1M HCl? Explain.

9.14 Describe a simple chemical test by which you could distinguish between the following pairs of compounds. In each case, state what you would do, what you would expect to observe, and write a balanced equation for all positive results.

a $C_6H_5CH_2OH$ (benzene ring bearing CH_2OH) and $C_6H_5CH_2NH_2$ (benzene ring bearing CH_2NH_2)

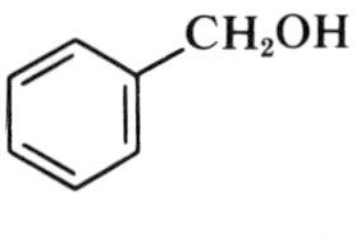

b C_6H_5OH (benzene ring bearing OH) and $C_6H_5NH_2$ (benzene ring bearing NH_2)

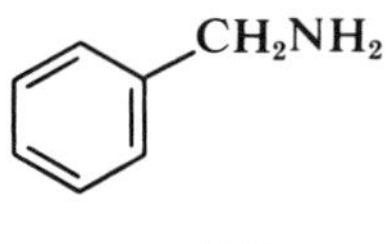

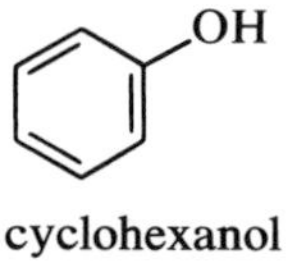

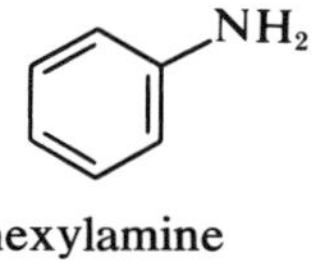

c cyclohexanol and cyclohexylamine

d ammonium butanoate and butanamide

e trimethylacetic acid and 2,2-dimethylpropylamine

9.15 Shown on p. 215 are three compounds involved in the biosynthesis of norepinephrine (noradrenalin), one of the hormones of the adrenal medulla. These steps involve tyrosine, dihydroxyphenylalanine (DOPA), dopamine, and norepinephrine.

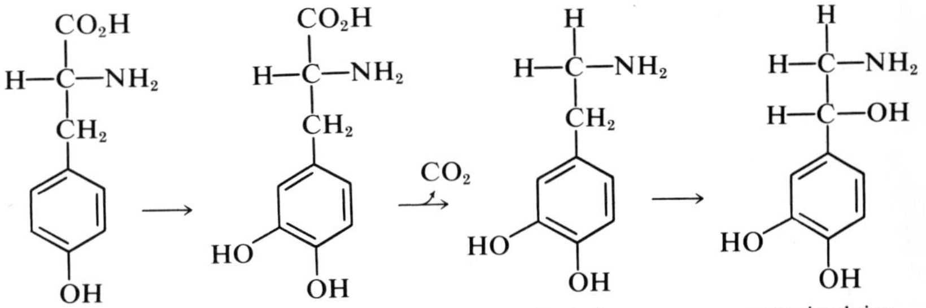

tyrosine $\longrightarrow$ DOPA $\xrightarrow{CO_2}$ dopamine $\longrightarrow$ norepinephrine

Two of the three steps in this sequence involve oxidation. Pick out these two steps and demonstrate by completing a balanced oxidation half-reaction that each is an oxidation.

MINI-ESSAY

ALKALOIDS

The term alkaloid or "alkali-like" is used to include those basic nitrogen-containing compounds of plant origin that manifest significant pharmacological activity. Further, alkaloids usually have rather complex molecular structures with a basic nitrogen atom incorporated into a heterocyclic ring. Today somewhat over 2000 alkaloids are known and it is estimated that alkaloids are present in from 10 to 15% of the vascular plants. Some plant families are not known to produce any alkaloids while others produce large numbers. The opium poppy alone produces some 20 alkaloids. Examples of well-known alkaloids are morphine (from the opium poppy), quinine (from cinchona bark), reserpine (from the rauwolfia shrub), and lysergic acid (from ergot fungus).

The field of alkaloid chemistry is both vast and complex and we shall examine only a few members of this class. In so doing we shall attempt to portray the ranges of structural diversity and physiological activity that characterize the alkaloids, and to indicate the important role of these natural products in chemotherapy and in psychopharmacology.

ATROPINE AND COCAINE

One special subgroup of alkaloids is represented by the tropane alkaloids, so classified because of the unique ring structure common to this group. Although the tropane alkaloids were by no means the first detected in nature, they attracted early serious chemical and pharmaceutical interest. Prior to the elucidation of their structure, the mydriatic action of many of these compounds was noted, e.g., atropine from *Atropa belladonna.* Atropine in dilute solutions (as little as 4.3×10^{-6} gram) causes dilation of the pupil of the eye and is used prior to ophthalmic examination and eye surgery. Other alkaloids, later discovered also to contain the tropane ring, displayed strong anesthetic effects, e.g., cocaine from the leaves of *Erythroxylon coca.*

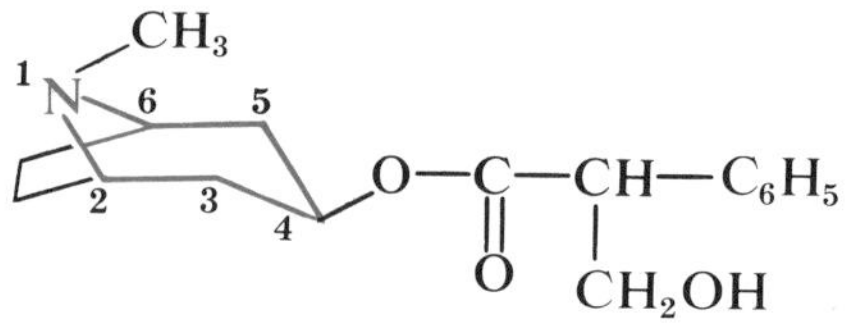

atropine

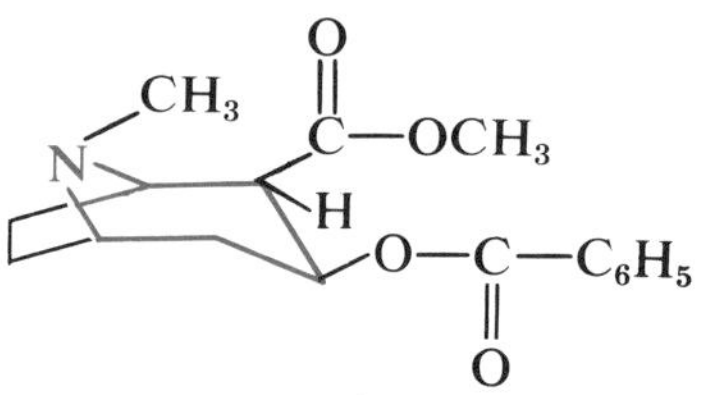

cocaine

Structural determination revealed that both atropine and cocaine have a piperidine ring (p. 203) incorporated into a bicyclic structure known as a tropane ring (shown in color on p. 216). Note the six-membered piperidine ring (atoms 1–6) with a two-carbon bridge fused between carbons 2 and 6. Alternatively we might look upon the tropane ring skeleton as being a pyrrolidine ring (p. 203) incorporated into a more elaborate bicyclic system. The levorotatory isomer of atropine is found in *Atropa belladona* (deadly nightshade).

Cocaine, isolated from the leaves of the South American coca plant, has been used as a local anesthetic and as a stimulant to the central nervous system. Cocaine was at one time a boon to anesthetic surgery, but has since been replaced by other drugs which are less dangerous. Although cocaine in small doses does decrease fatigue and afford a feeling of calm and happiness, its use also produces strong depression which in turn creates a demand for more of the drug. Thus the use of cocaine can create a dependence, and its use in medicine has sometimes lead to unfortunate cases of habituation. With the elucidation of the structural formula of cocaine, the biochemist (or the chemist, or the pharmacologist, etc.) can ask the tantalizing questions: What is the relationship between the structure of cocaine and its physiological function? Is it possible to separate the anesthetic effects from the habituation?

If these questions can be answered, then it should be possible to prepare synthetic alkaloid-like drugs that incorporate only those structural elements essential for anesthetic function, and at the same time eliminate the undesirable effects. In practice the process involves a good bit of (sometimes highly educated) guesswork about what parts of the molecule are essential for each function along with attempts to synthesize a variety of new chemicals, each incorporating one or more of the essential(?) structural features. In the case of cocaine, chemists have duplicated the essential structural features: its benzoate ester, its basic nitrogen, and something of its carbon arrangement. Following the synthetic efforts, pharmacological testing pointed to the presence of a benzoate ester of an aminoalcohol as the one essential feature of the anesthetic action of cocaine. Among the many cocaine substitutes is the widely used local anesthetic procaine.

$$H_2N-C_6H_4-\overset{\overset{\displaystyle O}{\|}}{C}-O-CH_2-CH_2-N\begin{matrix}CH_2CH_3\\CH_2CH_3\end{matrix}$$

procaine

The commonly administered form of the drug, procaine hydrochloride, is marketed under the trade name Novocaine. Seizing on the clues provided by nature, the chemist has been able to synthesize a drug far more suitable for a specific function than anything known to be elaborated by nature itself.

MORPHINE

Of all the alkaloids, probably the most widely known and completely studied are the so-called morphine alkaloids.

Opium, the source of the morphine alkaloids, is obtained from the opium poppy (*Papaver somniferum*) by cutting the unripe seed capsule and collecting the rubbery material that exudes. The milky juice is dried in the air and forms a brownish, gummy mass. This is further dried and powdered to make the opium of commerce, containing well over a score of alkaloids. Of these, only a few, most notably morphine and codeine, have any clinical usefulness. Opium contains about 10% morphine and 0.5% codeine.

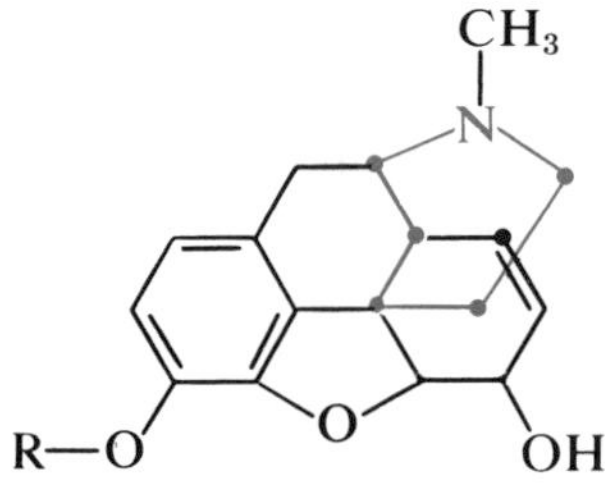

R = H in morphine
R = CH_3 in codeine

Morphine itself has been the object of intensive investigation. From the structural point of view, it is a very complex molecule. In fact, the correct basic structure of the molecule was not suggested until 1925, fully 120 years after its isolation from opium.

The two —OH groups of morphine, one phenolic and the other alcoholic, are of particular importance because many semisynthetic derivatives can be made by relatively simple modifications of the morphine molecule. For example, codeine is methylmorphine, the methyl substitution being on the phenolic —OH group. Diacetylmorphine, or heroin, is made from morphine by acetylation of both the phenolic and hydroxyl groups. Most of the useful semisynthetic morphine alkaloids are derived by chemical modifications of the —OH groups, the isolated double bond, or the replacement of the —CH_3 group on nitrogen by another alkyl group.

For the relief of severe pain of virtually every kind, morphine and its synthetic analogs remain the most potent drugs known. In 1680, Syndenham wrote, "Among the remedies which it has pleased Almighty God to give man to relieve his sufferings, none is so universal and so efficacious as opium." Even today, morphine, the alkaloid which gives opium its analgesic action, remains the standard against which newer analgesics are measured. Many of the newer agents may be considered its equal, but it is doubtful that any is clinically superior.

The stereochemical specificity of morphine is dramatically illustrated by the observation that (+)-morphine, available by laboratory synthesis, has been found to possess none of the pharmacological properties of the naturally occurring (−)-morphine.

It is not yet clear how morphine and related opioids function to produce a state of analgesia, that is, the relief of pain without loss of consciousness. In fact, the relief of pain by morphine and its surrogates is relatively selective in that other sensory modalities (touch, vibration, vision, hearing, etc.) are little changed. With therapeutic doses of morphine the perception of the painful stimulus itself is not always decreased, even in patients who obtain satisfactory pain relief. There is instead an altered reaction to the painful stimulus. Patients frequently report the pain is still present, but that they feel more comfortable about it. It appears then that there are at least two factors in the relief of pain afforded by morphine. The drug elevates the threshold of awareness of pain as a specific sensation. In addition, the drug alters the reaction of the patient to the painful stimulus largely by allaying the reactions (anxiety, fear, panic, and suffering) evoked by the sensation. Thus the patient's ability to tolerate the pain may be markedly increased even while the capacity to perceive the sensation may be relatively unaltered.

We are in a position to ask the same type of question about morphine as we did of cocaine; namely, is it possible, using the structural features of morphine as a guide, to design a drug that possesses the desirable and powerful analgesic effects of morphine and at the same time is free from the undesirable side effects of respiratory depression, hallucinogenic activity, and addiction? During the isolation and purification of morphine for structural studies, a number of closely related compounds became available. From the examination of these and other related substances, it appeared that certain structural features might be required for the analgesic activity of a drug of the morphine type. The most active drugs are characterized by (1) the presence of a tertiary amine with one of the substituent groups being methyl or some other small alkyl group, (2) a central quaternary carbon atom, (3) a phenyl or related aromatic group attached to the quaternary carbon, and (4) separation of the tertiary amino group from the quaternary carbon by two carbons. However, subsequent chemical and pharmacological investigations have shown that all four of these structural features can be omitted (or modified) without losing the analgesic activity.

During these investigations chemists have developed a number of useful synthetic analgesics related to morphine, the oldest and perhaps best known of which is meperidine (also known as Demerol). See Figure 1. Meperidine was at first thought to be free of many of the morphine-like undesirable side effects. However, it is now clearly recognized that meperidine is definitely addictive. In spite of a great deal of determined research, there are as yet no agents as effective as morphine against severe pain that are at the same time entirely free

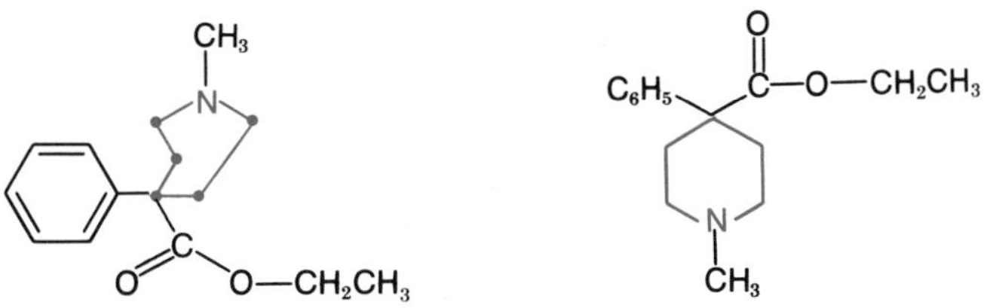

Figure 1 *Meperidine (Demerol) drawn in two different representations, the first suggesting a structural similarity to morphine.*

from the risks of physical and psychological dependence. Still it is a hope that the ideal analgesic will some day be prepared. However, at the present time it seems that treatment and withdrawal from dependency is a more reasonable goal.

INDOLE ALKALOIDS

During the last three decades, a number of naturally occurring indole alkaloids have received a great deal of attention, not only from the chemist but also from the pharmacologist, psychologist, and the physician. This recent interest has evolved from the discovery of the remarkable physiological properties of lysergic acid diethylamide on the one hand and from the sedative properties of reserpine on the other. These alkaloids have been largely responsible for the development of drugs that act on the central nervous system, and have opened new avenues of investigation in neuropharmacology. Current theory holds that each of these compounds in some way alters the action of certain amines in the brain (e.g., serotonin and norepinephrine) and that these amines are in fact the chemical mediators of emotion. (See also the essay "Biogenic Amines and Emotion.") It is this aspect of the physiological activity of certain indole alkaloids that we shall consider in this and the following section. See Figures 2 and 3.

For at least 3000 years the people of India have used the root of the climbing shrub *Rauwolfia serpentina* as a folk remedy to treat a variety of afflictions—as an antidote for snake bites, as a treatment for certain forms of insanity, and as a means of reducing fevers. Rauwolfia is called *sarpagandha* in Sanskrit referring to its use as an antidote for snakebite; it is called *chandra* in Hindi meaning "moon" and referring to its calming effect on certain forms of insanity; and it is called "snakeroot" or "serpentwood" in English referring to the snakelike appearance of its roots.

The modern story of rauwolfia began in the late 1920s when Indian scientists and physicians began to study it and other botanical preparations used by native practitioners. By 1931 chemists had isolated a crystalline powder from dry rauwolfia root and physicians reported that use of this root-powder extract brought relief in cases of acute insomnia accompanied by fits of insanity. According to one report, "Such symptoms as headache, a sense of heat and insomnia disappear quickly and blood pressure can be reduced in a matter of weeks. . . ." Further clinical testing verified the potency of the ancient snakeroot remedy and within a short time rauwolfia became a major drug in India for the treatment of high blood pressure and certain mental cases.

In 1949 Dr. Rustom Jal Vakil, a physician at the King Edward Memorial Hospital in Bombay, reported in the *British Medical Journal* on his research with rauwolfia therapy for patients with high blood pressure. These studies were read with interest by Dr. Robert Wilkins, director of the Hypertension Clinic at Massachusetts Memorial Hospital. He decided to try rauwolfia to see if it would help some of his patients who were not responding to other medications. Wilkins obtained a supply of whole-root extract tablets from India and began experimental tests of the drug. In 1952 he reported that rauwolfia is a slow-acting drug, and confirmed the reports from India of its mildly hypotensive (blood pressure lowering) effect. Further he reported that rauwolfia has a type of sedative action quite different from that of any other drug known at the time. Unlike the barbiturates and other standard sedatives, rauwolfia does not produce grogginess, stupor, or lack of coordination. Patients on rauwolfia therapy appeared to be relaxed, quiet, and tranquil. Reports such as this generated interest among psychiatrists and soon rauwolfia was recognized as an entirely new class of drug for the treatment of mental illness. It was the first of the so-called "tranquilizers." (See also the essay "Biogenic Amines and Emotion.")

Parallel to this clinical research, chemists had undertaken the task of isolating the active substance(s) from rauwolfia. They succeeded in 1952. Because the active component of snakeroot was first isolated from *Rauwolfia serpentina*, the new drug was named reserpine. Reserpine proved to be more than 10,000 times as effective as the same weight of crude snakeroot extract. The structural formula of reserpine (Figure 2) is complex. It contains an indole nucleus as a part of five fused rings and several centers of asymmetry.

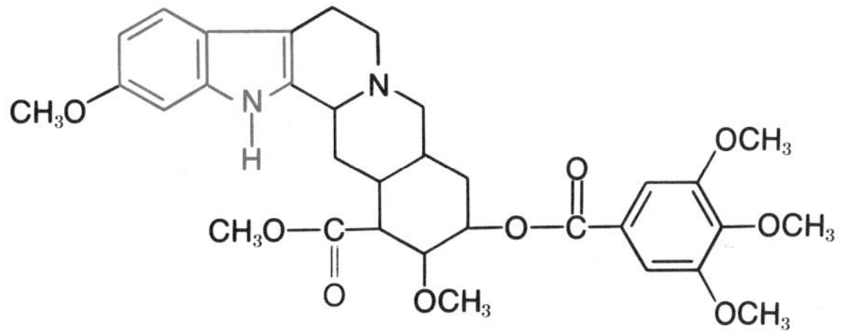

CH_3O N H N OCH_3 O $CH_3O—C$ O O—C OCH_3 OCH_3 OCH_3

Figure 2 *Reserpine, also commonly classed as a rauwolfia alkaloid indicating its natural occurrence. The indole ring is shown in color.*

In 1953 E. R. Squibb & Sons marketed a chemically and biologically standardized preparation of whole-root *Rauwolfia serpentina* extract under the trade name Raudixin. In the same year Ciba Pharmaceutical Company marketed purified reserpine under the trade name Serpasil. In 1954, the first full year of reserpine therapy, two dozen companies in the United States were preparing rauwolfia products and by 1960 prescriptions for these totaled $30 million per year. Within the span of slightly more than two decades, rauwolfia had been advanced from a folk remedy to a widely used and highly effective drug for the treatment of high blood pressure and certain forms of mental illness.

Along with reserpine and reserpine-like alkaloids, another group of organic nitrogenous bases (Figure 3), either containing the indole nucleus or containing structural features suggestive of the indole nucleus, have been the subject of intense research. All of these drugs characteristically produce behavioral aberrations, i.e., hallucinations, delusions, disturbances in thinking and changes in mood. There

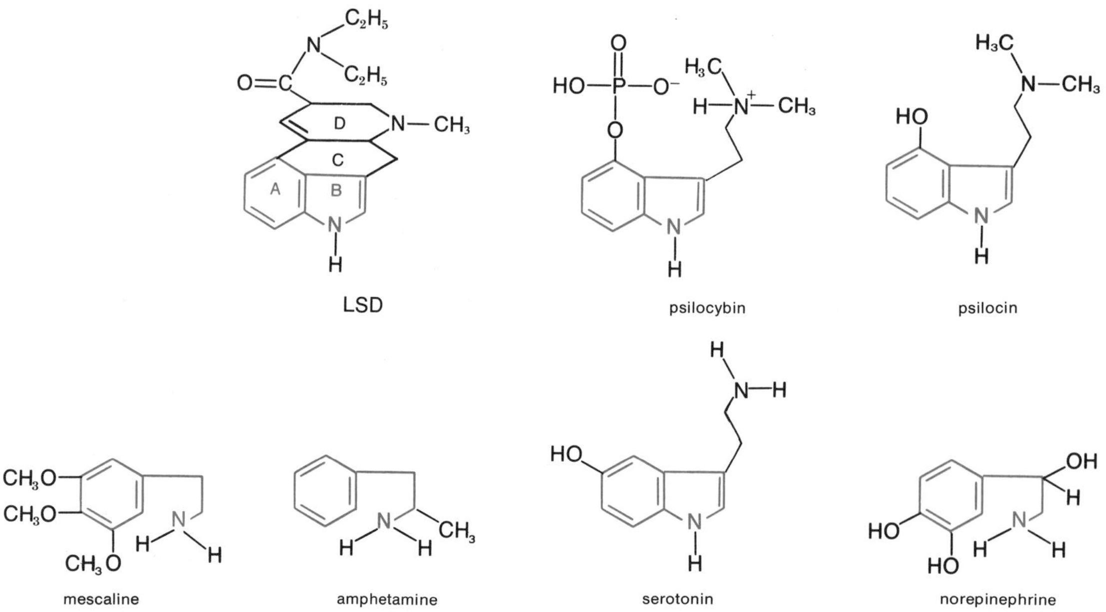

LSD psilocybin psilocin

mescaline amphetamine serotonin norepinephrine

Figure 3 *Structural relationships among several psychotomimetics and neurotransmitters (p. 228). The indole ring appears in LSD, psilocybin, and psilocin. Mescaline and amphetamine are drawn so as to suggest an indole nucleus and norepinephrine is drawn to suggest a relationship to serotonin.*

is some problem in terminology for these drugs and they are known variously as psychotomimetics, hallucinogenics, or psychedelics. Psychotomimetic implies that in some cases the effects of these drugs seem to mimic naturally occurring psychoses. Hallucinogenic suggests that the major action of these drugs is the stimulation of hallucinations, i.e., sensory perceptions having no basis in physical reality. Psychedelic suggests that the states of mind produced by these drugs are characterized by freedom from anxieties, feelings of relaxation, and the realization of unsuspected capacities of imagination. Perhaps the name for the group should be determined by the purpose for which these drugs are used.

Most of the drugs generally included among the psychotomimetics are related either to the indolealkylamines (LSD, psilocybin) or to the phenylethylamines (mescaline and substituted amphetamines).

Mescaline is obtained from the cactus known as peyote or mescal (*Lophophora williamsii*). The drug was named for the Mescalero Apaches of the Great Plains who developed a religious peyote rite. The structure of mescaline was determined in 1918. Mescaline does not properly belong in the class of indole alkaloids, for it does not contain the indole nucleus. Yet there is a suggestive chemical similarity and certainly a strong pharmacological similarity providing us with reason to include it in the same discussion with the indole alkaloid psychotomimetics.

Psilocybin is the hallucinogenic principal of *Psilocybe aztecorum*, the narcotic mushroom of the Aztecs.

By far the most important and clinically useful of the psychotomimetic drugs is lysergic acid diethylamide, LSD. The principal source of lysergic acid is ergot, a fungus that grows parasitically on rye. The rye grains attacked by the fungus develop into dark brown horn-shaped pegs projecting from the ripening ears. Lysergic acid and other alkaloids isolated from ergot are known collectively as ergot alkaloids. LSD, the diethylamide of lysergic acid, was first prepared in 1943 and since its recognition has received a great deal of attention, both popular and scientific.

SEROTONIN AND NOREPINEPHRINE

Parallel to the interest in LSD, reserpine, and other psychotomimetic drugs, another branch of science, namely neurochemistry, was engaged in the study of brain biochemistry. Of the variety of amines found in the brain, serotonin and norepinephrine have attracted special attention, and studies of these amines have played a very important role in currently evolving psychopharmacological concepts. Each of these substances is believed to function in the central nervous system as a transmitter substance (see p. 228), and each is synthesized in neural tissue from a precursor amino acid: norepinephrine from tyrosine (see Problem 9.15) and serotonin from tryptophan, by oxidation followed by decarboxylation.

$$\text{tryptophan (indole-3-}CH_2-CH(CO_2H)-NH_2\text{)} \xrightarrow{\text{(several steps)}} \text{serotonin (5-hydroxytryptamine) (HO-indole-3-}CH_2-CH_2-NH_2\text{)}$$

tryptophan — serotonin (5-hydroxytryptamine)

An examination of the molecular structures of the psychotomimetic drugs reveals that there are certain striking similarities between them and that these similarities parallel certain structural features of serotonin and norepinephrine. Using LSD as a model because it is easily the most powerful of the hallucinogenic drugs studied to date, one theory holds that all known hallucinogenic drugs show the ability to assume a conformation that simulates some combinations of the A, B, and/or C rings of LSD.

As is so often the case in science, one field of research seemed to fertilize another with the demonstration that the excretion of metabolic degradation products of serotonin is greatly increased by the administration of reserpine in large doses. It was further demonstrated that reserpine leads to reduced levels of brain serotonin and norepinephrine. For the first time it was possible to hypothesize a relationship between behavior and the level of concentration of molecules known to be present in the brain. Following these observations many studies have been conducted to associate (or dissociate) the effects of reserpine upon behavior with its ability to deplete serotonin, norepinephrine, or both. Some believe that depletion of brain serotonin is primarily responsible for the behavioral depression. Others believe that it is the depletion of amines like norepinephrine that is responsible for the sedative effect of reserpine.

The possible role of serotonin in brain function is also strikingly suggested in quite another way. Serotonin itself is widely distributed in the body and is known to stimulate the contraction of smooth muscle. A seminal observation was that LSD is a powerful antagonist of serotonin in its effects on smooth muscle. The fact that LSD is a serotonin antagonist on one hand and a powerful affector of mood and behavior on the other has led to speculation on the role of serotonin in the central nervous system, particularly in the intellectual function, perception, and mood. Further, there is a striking degree of overlap between the LSD drug-induced psychoses and the disturbances seen in schizophrenic and manic-depressive psychoses. That this degree of overlap does exist is impressive and certainly worthy of exploration and explanation.

The interest aroused by these exciting discoveries of drugs that act selectively to affect mood and behavior have focused attention on possible chemical bases for disturbances in behavior. Underlying all this research is the realization of the enormous importance of mental illness, and the hope that a tremendous amount of work on the biochemistry and pharmacology of the brain will benefit the understanding and treatment of mental disease. This will require cooperative research from many fields of science and will occupy many of the best minds in the decades ahead.

MINI-
ESSAY

BIOGENIC AMINES AND EMOTION

Over the past several decades, there has been a great deal of study on biochemical correlates of emotion. Initial research in this field concentrated largely on levels of chemical modulators in blood and the excretion of various hormones and metabolites in the urine. The historic studies of Walter B. Cannon shortly after the turn of the century suggested that the biologically active amine adrenalin, also commonly known as epinephrine, was secreted in response to stimuli that produced fear and rage reactions in animals. In the ensuing years, studies in both animals and man clearly indicated an increased secretion of epinephrine as well as norepinephrine in various types of stress including parachute jumping, competitive sports, aggressive behavior, or viewing emotion-laden movies. It appears that secretion of these amines correlates with the intensity rather than the quality of the stress. For example, an increase in urinary excretion of epinephrine and norepinephrine occurs in subjects watching certain motion pictures, the increase being quite similar whether the film is a hilarious comedy or a horror movie.

However, an area of more interest but one far less amenable to direct experimental observation is the pattern of biochemical changes which takes place in the brain itself in relation to affective or emotional states. The reasons that this area is less well studied are obvious. It is far easier to examine blood levels of chemical modulators or the excretion of hormones and metabolites in the urine than it is to examine the biochemistry of the neurons themselves. Yet it has been possible to do just this, in a limited way, in the past three decades, and this research has generated an enormous literature. The catecholamines, norepinephrine and epinephrine (each derived from the parent molecule catechol), and the indole amine serotonin are the monoamines of the brain that have received most attention. In order to sharpen the focus of this essay, we shall concentrate on only one of these amines, norepinephrine, and on its possible role in mediating emotional states. Certainly norepinephrine is not the only monamine and perhaps not even the most important monoamine involved in the central nervous system. Further, much of the evidence for its operation in emotional states by no means excludes the operation of other amines. Therefore we shall discuss norepinephrine not as the biochemical substrate for emotion but simply as a paradigm of an amine in the central nervous system.

First, let us summarize some of the important biochemistry of norepinephrine and then examine a current and admittedly tentative hypothesis on its possible role in modulating emotional states.

Norepinephrine and the other monoamines of the central nervous system are sharply limited to certain cells, where they occur in the cytoplasm, the axon, and in granules called synaptic or storage vesicles. These vesicles are found in highest concentration at the presynaptic nerve endings.

The main source of the norepinephrine of nerve cells is biosynthesis within the neurons themselves. Tyrosine, the precursor, is converted into norepinephrine through the intermediates L-dihydroxyphenylalanine (L-DOPA) and dopamine (Figure 1). The first of these reactions, the hydroxylation of tyrosine, is catalyzed by the enzyme tyrosine hydroxylase. This hydroxylation is a major branch point in the metabolism of tyrosine and it is this step that determines the overall rate at which tyrosine is converted to norepinephrine. Drugs that inhibit this enzyme cause marked decreases in intracellular norepinephrine levels.

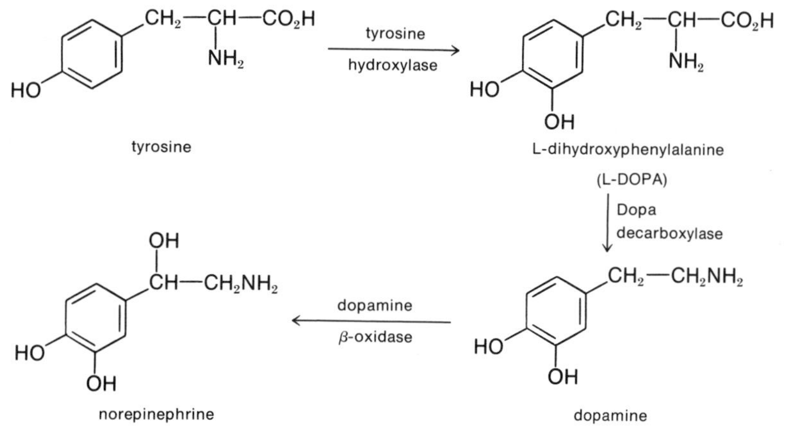

CH_2—CH—CO_2H, NH_2, HO — tyrosine

tyrosine hydroxylase →

CH_2—CH—CO_2H, NH_2, HO, OH — L-dihydroxyphenylalanine (L-DOPA)

↓ Dopa decarboxylase

CH_2—CH_2NH_2, HO, OH — dopamine

← dopamine β-oxidase

OH, CH—CH_2NH_2, HO, OH — norepinephrine

Figure 1 *The biosynthesis of norepinephrine from L-tyrosine.*

Norepinephrine is effectively protected from enzymatic attack while it is bound in the storage vesicles. However, once it is released from the vesicles, it then is susceptible to oxidative degradation by either of two metabolic pathways, one initiated within the neuron itself, the other in the intercellular space between the neurons.

On release from storage vesicles, either spontaneously or by the action of certain drugs, intracellular norepinephrine is metabolized mainly through oxidative deamination catalyzed by intracellular and mitochondrial-bound monoamine oxidase (MAO). The resulting 3,4-dihydroxymandelic acid and subsequent metabolic products leave the cell via the blood and are excreted in the urine. Norepinephrine released from the vesicles and then discharged into the synapse is metabolized principally by methylation catalyzed by the enzyme catechol-O-methyl transferase (COMT). The methylated normetanephrine enters in the blood stream, undergoes oxidative deamination catalyzed by the enzyme monoamine oxidase (MAO) of the liver and kidneys, and finally is excreted in the urine as 3-methoxy-4-hydroxymandelic acid (Figure 2).

The role of norepinephrine in the central nervous system is that of a chemical mediator of communication. It is now well established that

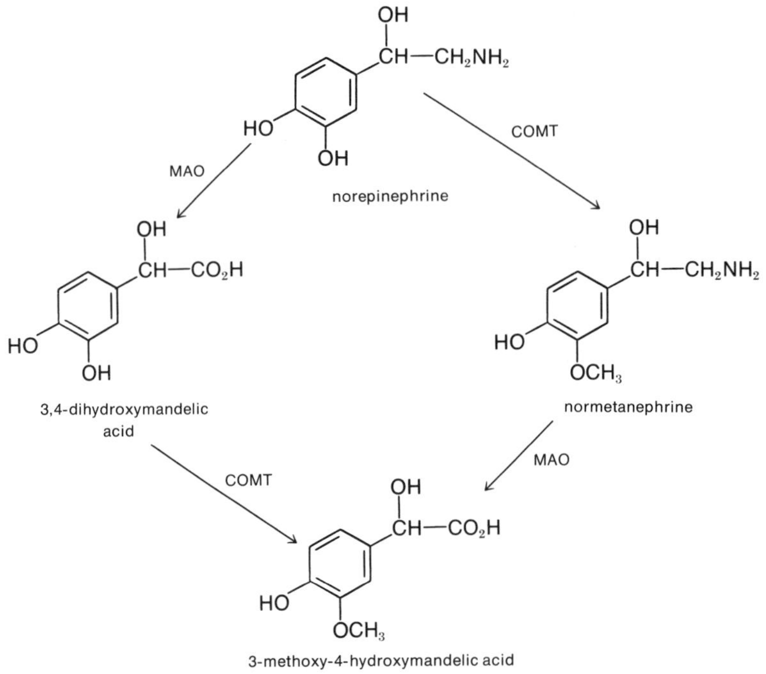

Figure 2 *Two principal biochemical pathways for the metabolism of norepinephrine.*

communication between neurons in the mammalian nervous system is by chemical agents called neurotransmitters, or more simply transmitters. Several different chemicals have been implicated as possible transmitters, among them norepinephrine, epinephrine, dopamine, serotonin, and acetylcholine (p. 201). The evidence for certain of these as transmitters is very convincing; the evidence for others is only tentative. The manner in which these substances function as chemical mediators of interneuronal communication, and for that matter the nature of the neuron itself as a communication device, are topics much too large to develop in even the most rudimentary way in this essay. Therefore we will rely on the background knowledge acquired in a college level course in modern biology. (However, see Section 9.3 for a discussion of acetylcholine as a neurotransmitter.) For our purposes the essentials of transmitter function are that norepinephrine may be released from the storage vesicles and into the synapse or interneuronal space in response to nerve stimulation. The released transmitter then diffuses across the synapse where it interacts with a specific receptor site on an adjacent neuron and thereby influences the excitability of the adjacent neuron. The mechanism of this specific interaction between norepinephrine and the receptor sites on the adjacent neuron (or for any other transmitter and its specific receptors) is only poorly understood.

Norepinephrine discharged into the synapse serves its function of a chemical transmitter and then is inactivated in either of two ways. First, a considerable percentage is taken back into the nerve ending and into the storage vesicles for later reuse. Second, it is metabolized, as previously described, principally by methylation, enters the blood stream, undergoes oxidative deamination in the liver and kidneys, and finally is excreted in the urine as 3-methoxy-4-hydroxymandelic acid.

It is estimated that the half-life of neuronal norepinephrine is from 2–4 hours, i.e., one-half of the norepinephrine present at any time is metabolized within a 2–4 hour period. Obviously, the processes of synthesis, storage, release, metabolism, transmitter–receptor interaction, and inactivation of norepinephrine represent a multiplicity of finely balanced biochemical processes. Any factors, psychological, physiological, or pharmacological, that disturb this dynamic balance could play a vital role in altering brain biochemistry, and as far as we are concerned in this essay, the biochemical substrates for emotional states.

A current hypothesis is that norepinephrine is the neurotransmitter in those parts of the brain that are intimately connected with the affective state and, further, that all of the drugs that have a significant effect on mood have one or another effect on norepinephrine. The major tranquilizer drugs act either by depleting the supply of available norepinephrine or by reducing the effectiveness of its interaction at the specific receptor sites; the major psychic energizers act either by increasing the amount of norepinephrine released into the synapse or by enhancing its accumulation at appropriate receptor sites.

What is the evidence for the hypothesis that norepinephrine has anything to do with emotional states? Several important lines of research evidence are summarized in the references. We will concentrate on a line of research that has emerged from clinical studies of drugs effective in altering emotional states in man.

Although interest in the psychological effects of drugs is almost as old as mankind, the use of chemotherapeutic agents for the treatment of psychiatric disorders has become widespread only since the mid-1950s. Today they are used on a grand scale to change attitudes and emotions of patients. The history of these drugs can probably be traced to the accidental discovery of lysergic acid diethylamine (LSD) in 1943. Among other things, this discovery to a large extent triggered the burgeoning neurochemical and neuropharmacological interest in the biochemical correlates of emotion. Soon after the discovery of LSD, reserpine (p. 221), perhaps the first of the tranquilizing drugs, was discovered. The great wave of enthusiasm for the use of reserpine and other rauwolfia alkaloids for the treatment of mania and excitement came in the mid-1950s. These drugs were described as "tranquilizers," a name since applied to a variety of compounds with similar sedative effects. Reserpine (Rauloydin, Reserpoid, Serpasil and other trade names) produces a marked sedation in animals and tranquilization of overexcited states in man. In some individuals it produces severe depressions of mood that are regarded by many as indistinguishable from naturally occurring depression. The effects of reserpine have been studied in animals and man and suggest that the drug, by some mechanism not as yet understood, acts on the synaptic vesicle storage sites and impares their ability to bind norepinephrine (and also dopamine and serotonin). As a consequence norepinephrine is released, diffuses freely through the cytoplasm and onto mitochondrial-bound MAO, where it is metabolized by oxidative deamination to 3,4-dihydroxymandelic acid. Thus, instead of being stored in vesicles for release on nerve stimulation, the norepinephrine supply may become virtually depleted. Even though there is continuing synthesis, the depletion may last for days or weeks, probably because reserpine attaches itself to the storage sites and impairs uptake of newly synthesized transmitter.

Chloropromazine (Figure 3), which was discovered almost simultaneously, quickly usurped the position of reserpine. It soon became clear that chloropromazine was easier to control and more effective than reserpine. Chloropromazine and its congeners are among the most widely used drugs in medicine today. It is estimated that between 1955 when it was first introduced and 1965 at least 50 million patients received the drug. Chloropromazine (Thorazine, Largactil, Megaphen) and its derivatives (Sparine, Trilafon, Stelazine) produce sedation, emotional quieting, and relaxation without clouding consciousness or intellectual functioning.

The weight of evidence suggests that the principal action of chloropromazine is on the norepinephrine receptor sites in susceptible neurons. The drug combines with these receptor sites and renders them incapable of responding to norepinephrine. The result is similar to a decreased supply of the transmitter substance. Thus although both reserpine and chloropromazine are powerful tranquilizing agents, the mechanism of action of each of these drugs is quite different.

The discovery of the tranquilizing effects of reserpine and chloropromazine was followed within a few years by the discovery of iproniazid, a powerful excitant, euphoriant, and antidepressant. The norepinephrine hypothesis was reinforced with the finding that iproniazid is a powerful inhibitor of monoamine oxidase, the enzyme responsible for the destruction of norepinephrine and other brain monoamines as well. Inhibition of MAO permits an increase in the concentration of norepinephrine at nerve endings and presumably at the synapses as well. It is this accumulation that is thought to produce the antidepressant action of the drug in man. The elucidation of the structure of iproniazid sparked research efforts to synthesize new compounds with even greater clinical effectiveness. Among those synthesized and marketed was isocarboxazid (Marplan).

Imipramine (Tofranil) and other closely related compounds are the drugs most widely used today for the treatment of depressions. It is interesting to note that the ring structure of imipramine (an antidepressant) differs from that of chloropromazine (a tranquilizer) only in the replacement of the sulfur by an ethylene ($—CH_2—CH_2—$) bridge between the two benzene rings (Figures 3 and 4). Imipramine has become very popular for the treatment of depression because it is relatively safer than the MAO inhibitors and more acceptable to patients. The mechanism by which imipramine counteracts depression or produces euphoria and hyperactivity in humans is not known. However, it appears to favor the accumulation of norepinephrine at the receptor sites by inhibiting the reuptake of intercellular norepinephrine. Thus imipramine artificially increases the concentration of norepinephrine at the receptor sites and thereby potentiates its action.

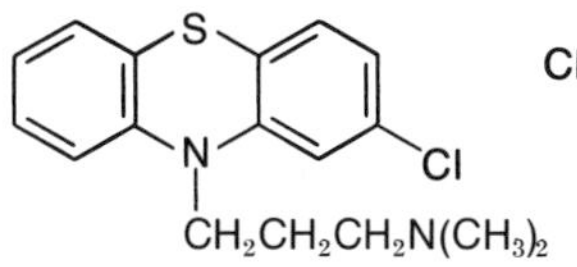

chloropromazine

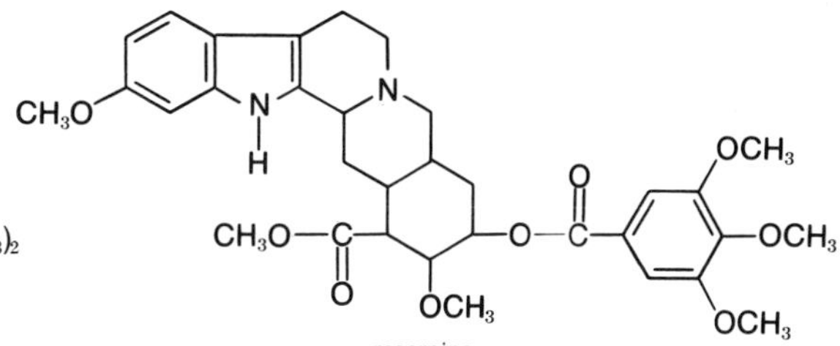

reserpine

Figure 3 *Tranquilizing agents. Chloropromazine impairs norepinephrine-receptor interaction. Reserpine depletes synaptic vesicles of norepinephrine.*

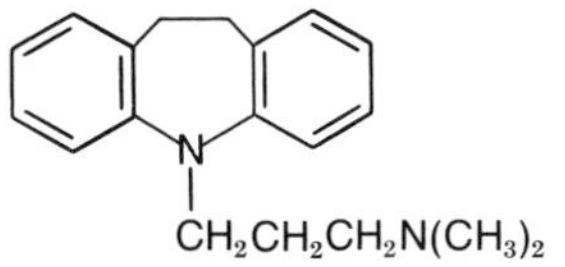

imipramine

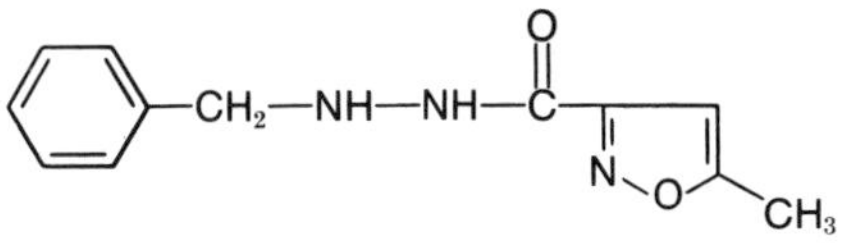

isocarboxazid

Figure 4 *Psychic energizers. Imipramine (Tofranil), an inhibitor of the reuptake of norepinephrine. Isocarboxazid (Marplan), a monoamine oxidase inhibitor.*

In summary, a significant body of pharmacological evidence is at least consistent with the hypothesis that the effects of reserpine, chloropromazine, and similar tranquilizers and of imipramine and other psychic energizers are related to their effects on norepinephrine and that norepinephrine plays an important role in mental and behavioral states. But by no means does this evidence prove the hypothesis. It is important to bear in mind that no drug produces just one effect. Each of the drugs we have presented has the ascribed effect (e.g., reserpine depletes norepinephrine stores) and a host of side effects as well. The problem in the use of drugs as probes in the study of behavior is to separate and identify the various effects of the drug. Finally we must be wary of oversimplification in a subject as complex and multifaceted as emotional behavior. It is unlikely that norepinephrine, or any other single amine for that matter, is entirely responsible for a specific emotional state. Rather it is more likely that the important factors are the interaction of certain amines at particular sites in the central nervous system together with a host of environmental and psychological determinants of emotion. The unravelling of these factors promises man a wealth of information and insight into the relationships between brain and mind.

References

Black, P., Editor, *Physiological Correlates of Emotion* (Academic Press, New York, 1970).

Glass, D. C., Editor, *Neurophysiology and Emotion* (The Rockefeller University Press and Russel Sage Foundation, New York, 1967).

Katz, Bernard, *Nerve, Muscle, and Synapse* (McGraw-Hill Book Company, New York, 1966).

McGeer, Patrick L., "The Chemistry of Mind," *American Scientist*, **59**, 221 (1971).

Quarton, G. C., Melnechuk, T., and Schmitt, T. O., Editors, *The Neurosciences—A Study Program* (The Rockefeller University Press, New York, 1967).

Scheldkraut, J. J., and Kety, S. S., "Biogenic Amines and Emotion," *Science*, **156**, 21 (1967).

10

CARBOHYDRATES

10.1 INTRODUCTION

Carbohydrates are polyhydroxyaldehydes or polyhydroxyketones, or substances which on hydrolysis yield polyhydroxyaldehydes and/or polyhydroxyketones.

These substances are among the most abundant constituents of the plant and animal world and serve many vital functions: as a storehouse of chemical energy (glucose, starch, glycogen); as supportive structural components in plants (cellulose); as essential components in the mechanisms of genetic control of development and growth of living cells (ribose and deoxyribose).

The name carbohydrate is an old one, and was derived from the early observations that many members of this class have the empirical formula $C_n(H_2O)_m$ and were termed "hydrates of carbon." For example, grape sugar (glucose) has the molecular formula $C_6H_{12}O_6$. Cane sugar (sucrose) has the formula $C_{12}H_{22}O_{11}$. Starch and cellulose are very large molecules of variable molecular weight having the empirical formula $(C_6H_{10}O_5)_x$, where x may be as large as several hundred thousand. It soon became clear that not all compounds having the properties of carbohydrates have this general formula. For example, we now know of carbohydrates that contain nitrogen in addition to the elements of carbon, hydrogen and oxygen. Although this term carbohydrate is not fully descriptive, it is firmly rooted in the chemical nomenclature and has persisted as the name of this class of compounds.

Carbohydrates are often referred to as saccharides because of the sweet taste of the simpler members of the family, the sugars (Latin, *saccharum*, sugar).

The chemistry of the carbohydrates is essentially the chemistry of two functional groups, the hydroxyl group and the carbonyl group. Further, we encounter derivatives of these two functional groups in the form of hemiacetals, acetals, esters, and so on. However, the fact that there are only two functional groups in these substances belies the complexity and challenge of carbohydrate chemistry. All but the most simple carbohydrates contain multiple centers of asymmetry and it has been a challenge to the chemist to develop techniques to study the stereochemistry of these substances. Further, all carbohydrates contain multiple hydroxyl groups and accordingly multiple reactive sites. It has also been a challenge to develop techniques for reaction at selected functional groups within the molecule.

10.2 MONOSACCHARIDES

Monosaccharides, or the simple sugars, are carbohydrates that have the general formula $C_nH_{2n}O_n$, where n varies from three to eight. They are grouped together according to the number of carbons they contain.

$C_3H_6O_3$	triose	$C_6H_{12}O_6$	hexose
$C_4H_8O_4$	tetrose	$C_7H_{14}O_7$	heptose
$C_5H_{10}O_5$	pentose	$C_8H_{16}O_8$	octose

While examples of heptoses and octoses do occur in nature, they are far less abundant than the smaller monosaccharides.

Glyceraldehyde (I) and dihydroxyacetone (II) are the smallest molecules termed carbohydrates and are the only possible trioses.

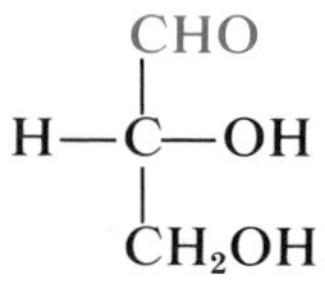

$$\begin{array}{c} CHO \\ | \\ H-C-OH \\ | \\ CH_2OH \end{array} \qquad \begin{array}{c} CH_2OH \\ | \\ C{=}O \\ | \\ CH_2OH \end{array}$$

glyceraldehyde I — dihydroxyacetone II

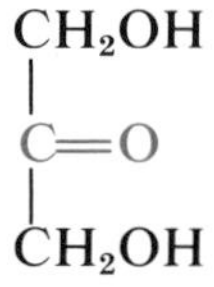

Glyceraldehyde has one asymmetric carbon atom and therefore can exist as a pair of enantiomers, Ia and Ib:

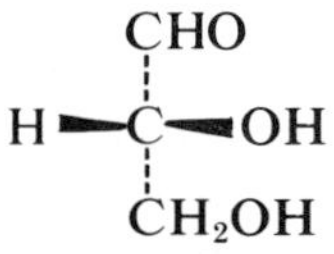

$$\begin{array}{c} CHO \\ \vdots \\ H \blacktriangleright C \blacktriangleleft OH \\ \vdots \\ CH_2OH \end{array} \qquad \begin{array}{c} CHO \\ \vdots \\ HO \blacktriangleright C \blacktriangleleft H \\ \vdots \\ CH_2OH \end{array}$$

Ia 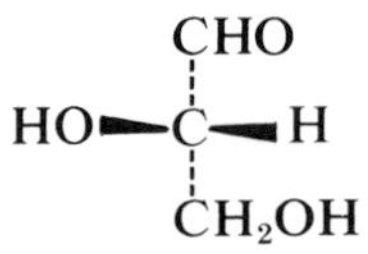 — Ib

Because the representations of three-dimensional structures are often cumbersome, a convention for picturing the molecules in the plane of the paper has been adopted. It is the so-called Fischer projection method. According to this convention, the carbon chain is written vertically on the page with carbon 1 toward the top of the page. Horizontal groups (those written to the left and right of any asymmetric carbon atom) are understood to project above the plane of the paper; vertical groups (those written above and below any asymmetric carbon atom) are understood to project below the plane of the paper. Applying the Fischer projection rules to the two three-dimensional representations of glyceraldehyde gives structures Ia′ and Ib′.

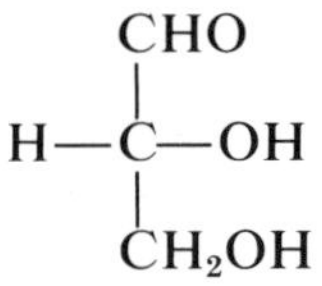

$$\begin{array}{c} CHO \\ | \\ H-C-OH \\ | \\ CH_2OH \end{array} \qquad \begin{array}{c} CHO \\ | \\ HO-C-H \\ | \\ CH_2OH \end{array}$$

Ia′ 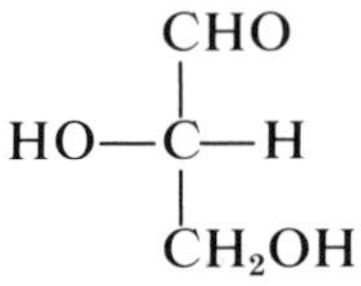 — Ib′

Structures Ia and Ib are mirror images and show identical reactivities except in the presence of reagents which themselves are asymmetric. One of the enantiomers has a specific rotation of +20.9°

while the other has a specific rotation of $-20.9°$. The question is, which of the two structures corresponds to the enantiomer having the specific rotation of $+20.9°$? Is Ia dextrorotatory and Ib levorotatory, or is the situation reversed? The answer is that Ia is the structure of the enantiomer that is dextrorotatory. This assignment was made quite arbitrarily by the German chemist Emil Fischer in 1891. (Of course, he had a 50:50 chance of being correct in this assignment.) In 1952 his choice was proved to be the correct one by a Dutch scientist using a special technique of X-ray analysis. By convention, structure Ia is called D-glyceraldehyde where the D refers to the configuration at the asymmetric carbon. Structure Ib is designated as L-glyceraldehyde. The configuration of D-glyceraldehyde serves as a reference point for the assignment of configuration to other carbohydrates.

D-Glyceraldehyde is the parent member of the aldose (aldehyde + -ose) family of monosaccharides. Aldoses contain one or more hydroxymethylene (—CHOH) groups between the aldehyde carbonyl and the terminal —CH_2OH group.

Dihydroxyacetone is the parent member of the ketose (ketone + -ose) family of monosaccharides. General formulas for D-aldoses and D-ketoses are shown below.

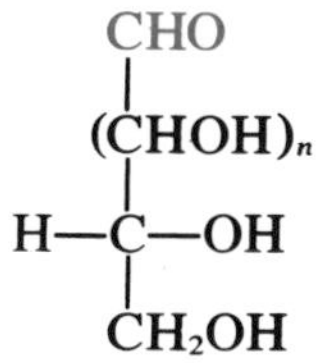

```
    CHO                 CH2OH
     |                    |
  (CHOH)n                C=O
     |                    |
 H—C—OH               (CHOH)n
     |                    |
   CH2OH              H—C—OH
                          |
                        CH2OH
  D-aldose            D-ketose
```

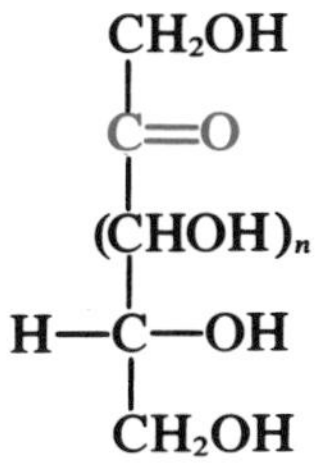

Tables 10.1 and 10.2 (see pp. 236–237) show the names and structural formulas for all trioses, tetroses, pentoses, and hexoses of the D-series.

D-Ribose and 2-deoxy-D-ribose are the most important pentoses. These pentoses are present as intermediates in metabolic pathways and are important building blocks in nucleic acids, D-ribose in ribonucleic acid (RNA) and 2-deoxy-D-ribose in deoxyribonucleic acid (DNA).

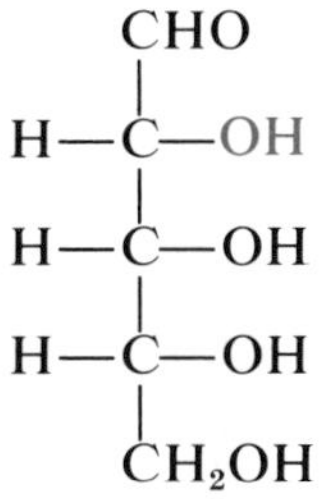

```
     CHO                  CHO
      |                    |
  H—C—OH               H—C—H
      |                    |
  H—C—OH               H—C—OH
      |                    |
  H—C—OH               H—C—OH
      |                    |
    CH2OH                CH2OH
  D-ribose          2-deoxy-D-ribose
```

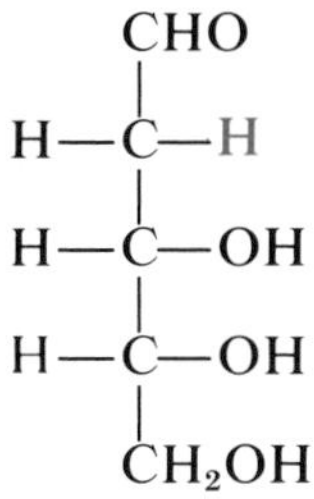

Hexoses are very common in nature and play major physiological roles. Glucose is by far the most common hexose monosaccharide. It is known as dextrose because it is optically active and dextrorotatory. It is also referred to as grape sugar, blood sugar, and corn sugar, names that clearly indicate its source in an uncombined state. Glucose is the carbohydrate used by cells; mannose, galactose and fructose must first be converted into glucose for utilization as an energy source. Glucose is

Table 10.1 *The isomeric D-aldotetroses, D-aldopentoses and D-aldohexoses derived from D-glyceraldehyde.*

D-glyceraldehyde
CHO
H—C—OH
CH_2OH

D-erythrose	D-threose
CHO	CHO
H—C—OH	HO—C—H
H—C—OH	H—C—OH
CH_2OH	CH_2OH

D-ribose	D-arabinose	D-xylose	D-lyxose
CHO	CHO	CHO	CHO
H—C—OH	HO—C—H	H—C—OH	HO—C—H
H—C—OH	H—C—OH	HO—C—H	HO—C—H
H—C—OH	H—C—OH	H—C—OH	H—C—OH
CH_2OH	CH_2OH	CH_2OH	CH_2OH

D-allose	D-altrose	D-glucose	D-mannose
CHO	CHO	CHO	CHO
H—C—OH	HO—C—H	H—C—OH	HO—C—H
H—C—OH	H—C—OH	HO—C—H	HO—C—H
H—C—OH	H—C—OH	H—C—OH	H—C—OH
H—C—OH	H—C—OH	H—C—OH	H—C—OH
CH_2OH	CH_2OH	CH_2OH	CH_2OH

D-gulose	D-idose	D-galactose	D-talose
CHO	CHO	CHO	CHO
H—C—OH	HO—C—H	H—C—OH	HO—C—H
H—C—OH	H—C—OH	HO—C—H	HO—C—H
HO—C—H	HO—C—H	HO—C—H	HO—C—H
H—C—OH	H—C—OH	H—C—OH	H—C—OH
CH_2OH	CH_2OH	CH_2OH	CH_2OH

stored in the liver as the polysaccharide glycogen (Section 10.9). The blood normally contains 65–110 mg of glucose per 100 ml (0.06–0.1%) but in diabetics the level may be much higher.

The structural formula of D-glucose (see p. 238) reveals that the molecule contains four asymmetric carbons. D-Glucose and its mirror image L-glucose represent two of the sixteen ($2^4 = 16$) possible optical isomers of this structure. Of these sixteen, only D-glucose, D-galactose, and D-mannose are widely distributed in nature. The other thirteen isomers have been prepared in the laboratory and characterized.

The structural difference between D-fructose and the other hexoses is readily apparent. D-Fructose is a 2-ketohexose rather than an aldohexose.

Table 10.2 *The isomeric D-ketopentoses and D-ketohexoses derived from dihydroxyacetone and D-erythrulose.*

dihydroxyacetone
CH_2OH
CO
CH_2OH

D-erythrulose
CH_2OH
CO
H—C—OH
CH_2OH

D-ribulose	D-xylulose
CH_2OH	CH_2OH
CO	CO
H—C—OH	HO—C—H
H—C—OH	H—C—OH
CH_2OH	CH_2OH

D-allose (D-psicose)	D-fructose	D-sorbose	D-tagatose
CH_2OH	CH_2OH	CH_2OH	CH_2OH
CO	CO	CO	CO
H—C—OH	HO—C—H	H—C—OH	HO—C—H
H—C—OH	H—C—OH	HO—C—H	HO—C—H
H—C—OH	H—C—OH	H—C—OH	H—C—OH
CH_2OH	CH_2OH	CH_2OH	CH_2OH

$$\begin{array}{c} CHO \\ | \\ H-C-OH \\ | \\ HO-C-H \\ | \\ H-C-OH \\ | \\ H-C-OH \\ | \\ CH_2OH \\ \text{D-glucose} \end{array} \qquad \begin{array}{c} CHO \\ | \\ H-C-OH \\ | \\ HO-C-H \\ | \\ HO-C-H \\ | \\ H-C-OH \\ | \\ CH_2OH \\ \text{D-galactose} \end{array} \qquad \begin{array}{c} CHO \\ | \\ HO-C-H \\ | \\ HO-C-H \\ | \\ H-C-OH \\ | \\ H-C-OH \\ | \\ CH_2OH \\ \text{D-mannose} \end{array} \qquad \begin{array}{c} CH_2OH \\ | \\ C=O \\ | \\ HO-C-H \\ | \\ H-C-OH \\ | \\ H-C-OH \\ | \\ CH_2OH \\ \text{D-fructose} \end{array}$$

D-Galactose is found combined with glucose in the disaccharide lactose (Section 10.8). It is interesting that lactose appears nowhere else except in milk, where it is present to the extent of about 5%. The enzyme system in glycolysis and the tricarboxylic acid cycle cannot metabolize galactose directly and it must first be converted by isomerization at carbon 4 into glucose. There is an interesting inherited disease, galactosemia, among human infants which manifests itself by the inability to metabolize galactose, which therefore accumulates in various tissues including the central nervous system and causes damage to cells. Without treatment, the infant either suffers permanent damage to its organs including the brain, or at worst, succumbs. This disease occurs because one of the enzymes involved in the isomerization of galactose to glucose is not synthesized in the liver of these infants, due to lack of the necessary gene. Since we are unable to alter the genetic mechanism to restore a nonfunctioning gene, it is not possible to correct this enzyme deficiency. Yet it is possible to treat the disease in a surprisingly simple manner, providing diagnosis is made early enough; since milk is the only source of the galactose, it is excluded from the infant's diet. Our ability to remedy this disease marked an important milestone, for it is the first hereditary disease to be controlled through a knowledge of genetics and biochemistry.

10.3 ASCORBIC ACID

The structural formula of ascorbic acid (vitamin C) resembles that of a monosaccharide, and in fact this vitamin is synthesized biochemically from D-glucose. Enzyme-catalyzed oxidation of D-glucose by NAD^+ gives D-glucuronic acid. This four-electron oxidation requires two moles of NAD^+ per mole of D-glucose. A selective enzyme-catalyzed reduction of the —CHO group of D-glucuronic acid gives L-gulonic acid. The carbon atom 1 of what was D-glucose is now $—CH_2OH$, and carbon atom 6 of D-glucose is now $—CO_2H$. According to the Fischer convention, the molecule should be renumbered so that the $—CO_2H$ group is carbon atom 1. Therefore the Fischer projection formula is turned 180° in the plane of the paper so that the $—CO_2H$ group is

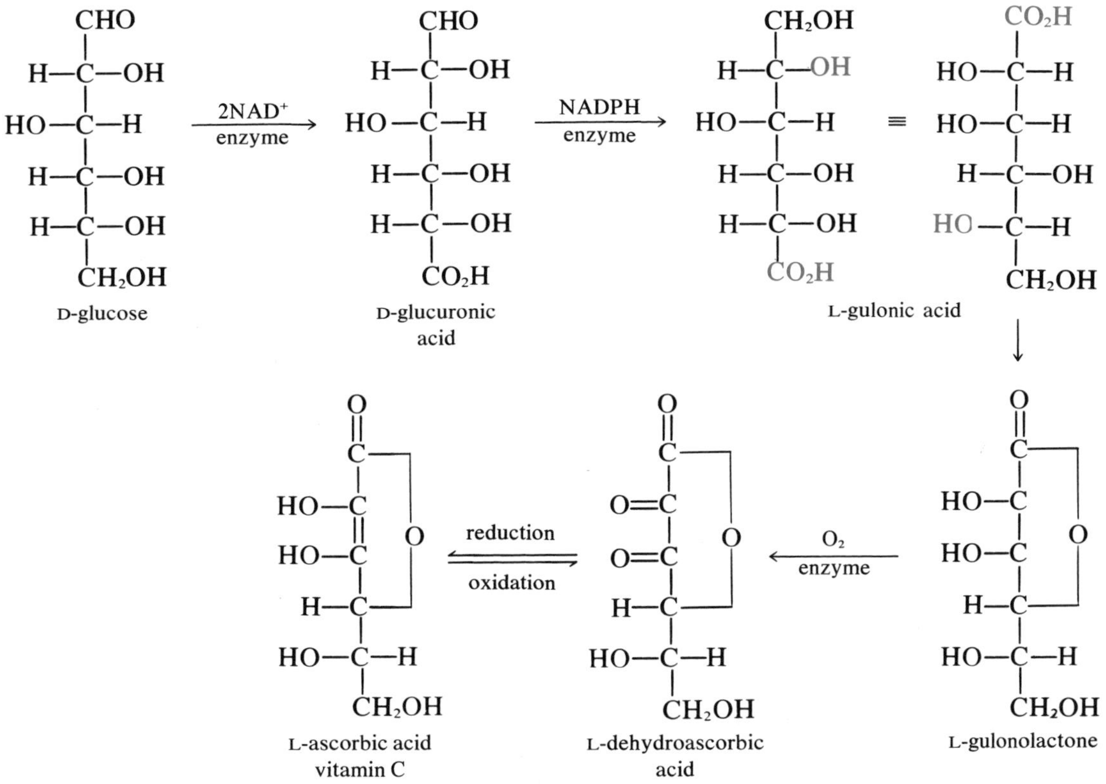

D-glucose $\xrightarrow[\text{enzyme}]{2NAD^+}$ D-glucuronic acid $\xrightarrow[\text{enzyme}]{NADPH}$ L-gulonic acid

L-gulonic acid $\downarrow$ L-gulonolactone $\xrightarrow[\text{enzyme}]{O_2}$ L-dehydroascorbic acid $\underset{\text{oxidation}}{\overset{\text{reduction}}{\rightleftharpoons}}$ L-ascorbic acid vitamin C

uppermost and appears as carbon 1. This molecule, L-gulonic acid, in turn forms a cyclic ester (a lactone), and is oxidized by molecular oxygen in an enzyme-catalyzed reaction to L-ascorbic acid. Man and other primates, and guinea pigs, lack the enzyme system necessary to convert L-gulonic acid to L-ascorbic acid and therefore are unable to synthesize ascorbic acid. For humans, ascorbic acid is therefore a necessary dietary supplement.

Vitamin C is readily oxidized to dehydroascorbic acid. Both forms are physiologically active and are found in body fluids.

The biochemical function of ascorbic acid is still not completely understood. It probably functions as a biological oxidation–reduction reagent. Whatever the biological roles of ascorbic acid may be, its most clearly established function is in the maintenance of collagen (Section 12.12), an intercellular material of bone, dentine, cartilage and connective tissue.

Severe deficiency of ascorbic acid produces scurvy. A most vivid description of this vitamin deficiency disease was given by Jacques Cartier in 1536 when it afflicted his men during the exploration of the Saint Lawrence River.

> Some did lose all their strength, and could not stand on their feet.... Others also had all their skin spotted with spots of blood of a purple

> color; then did it ascend up to their ankles, knees, thighs, shoulders, arms, and necks. Their mouths became stinking, their gums so rotten that all of the flesh did fall off, even to the roots of the teeth, which did also almost fall out.

This is scurvy! Green or fresh vegetables and ripe fruits are the best remedies and the best means of preventing the disease. (These foods are of course sources of vitamin C.) This was recognized by the English physician James Lind, who in 1753 urged the inclusion of lemon or lime juice in the diet of British sailors, hence the nickname limeys.

There is good evidence that severe ascorbic acid deficiency as seen in scurvy is the result of impaired synthesis of collagen, the protein material of bone and connective tissue. Collagen synthesized in the absence of ascorbic acid cannot properly form fibers, thereby resulting in such typical symptoms as fragility of blood vessels and hemorrhaging, loosening of the teeth, poor wound healing, etc. We will discuss the structure and function of collagen more fully in Chapter 12.

10.4 AMINO SUGARS

Amino sugars are formed through the substitution of an hydroxyl group by an amino group on a monosaccharide. There are three naturally occurring amino sugars, D-glucosamine, D-mannosamine, and D-galactosamine. Of these, D-glucosamine is the most common.

$$\begin{array}{c} CHO \\ | \\ H-C-NH_2 \\ | \\ HO-C-H \\ | \\ H-C-OH \\ | \\ H-C-OH \\ | \\ CH_2OH \end{array} \qquad \begin{array}{c} CHO \\ | \\ H_2N-C-H \\ | \\ HO-C-H \\ | \\ H-C-OH \\ | \\ H-C-OH \\ | \\ CH_2OH \end{array} \qquad \begin{array}{c} CHO \\ | \\ H-C-NH_2 \\ | \\ HO-C-H \\ | \\ HO-C-H \\ | \\ H-C-OH \\ | \\ CH_2OH \end{array}$$

D-glucosamine D-mannosamine D-galactosamine

In D-glucosamine, and in D-mannosamine and D-galactosamine as well, the —OH group on carbon 2 is replaced by an $—NH_2$ group. In most cases where these monosaccharides occur in nature, the $—NH_2$ groups are acetylated. D-Amino sugars are essential parts of many polysaccharides including chitin, the hard, shell-like covering of lobsters, crab, shrimp and other crustaceans. Further they are an integral part of the cell wall structure of certain bacteria and of heparin, a naturally occurring blood anticoagulant.

Since D-glucose occupies a central position in carbohydrate metabolism and since it is the most common and most thoroughly investigated carbohydrate, we shall examine its structure and properties in some detail. Much of this discussion applies to other sugars as well.

10.5 MUTAROTATION AND THE CYCLIC STRUCTURE OF GLUCOSE

Although glucose shows many of the typical reactions of aldehydes, it fails to give a positive Schiff's test (a qualitative color reaction test characteristic of aldehydes). This failure led Tollens in 1883 to propose a cyclic hemiacetal structure for glucose. This formula was not generally accepted due to the (then) well-known instability of hemiacetals.

In 1893, Fischer attempted to convert glucose into a dimethyl acetal (p. 131) by treatment with methyl alcohol in accordance with the known reaction:

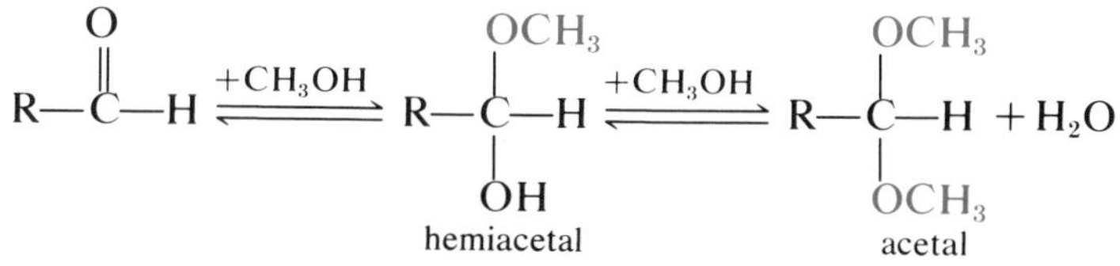

$$R-\overset{O}{\overset{\|}{C}}-H \underset{}{\overset{+CH_3OH}{\rightleftharpoons}} \underset{\text{hemiacetal}}{R-\overset{OCH_3}{\overset{|}{\underset{\underset{OH}{|}}{C}}}-H} \overset{+CH_3OH}{\rightleftharpoons} \underset{\text{acetal}}{R-\overset{OCH_3}{\overset{|}{\underset{\underset{OCH_3}{|}}{C}}}-H} + H_2O$$

Surprisingly, the product obtained had the formula $C_7H_{14}O_6$ rather than $C_8H_{18}O_7$ expected for the dimethyl acetal and contained only one methyl group. It was called a methyl glucoside. This substance showed no aldehydic properties and Fischer correctly regarded it as a true acetal, a cyclic acetal structure involving one of the hydroxyl groups from the sugar itself and one molecule of methyl alcohol. The proposed partial structures shown below indicate the presence of one new asymmetric carbon called the anomeric carbon and suggest that glucose should yield two isomeric methyl glucosides. The second methyl glucoside was isolated the following year by van Eckenstein. The designation α-methyl D-glucoside was applied to the isomer having the higher (more positive) specific rotation. The other was designated β-methyl D-glucoside. For the moment we shall represent these two methyl glucosides in partial formulas. In each the stereochemistry at carbon 1, the new center of asymmetry, is represented according to the Fischer convention.

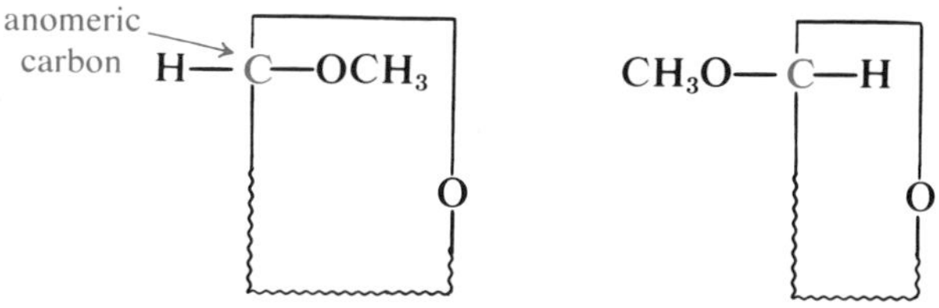

anomeric carbon → H—C—OCH_3 (ring closed through O)

CH_3O—C—H (ring closed through O)

α-methyl glucoside
mp 169°C
$[\alpha] + 159°$

β-methyl glucoside
mp 107°C
$[\alpha] - 34°$

Fischer himself was careful to point out that the cyclic structure of the glucosides did not constitute evidence for similar structures of glucose itself. Yet the failure of glucose to show certain reactions characteristic of aldehydes certainly suggested a cyclic hemiacetal structure for glucose. Other monosaccharides also react with alcohols in the presence of a trace of mineral acid to form anomeric α- and β-cyclic acetal structures. The term glycoside is a general name for this type of substance. Particular glycosides are named by dropping the suffix -e from the name of the parent saccharide and adding -ide, e.g., glucoside, riboside, galactoside.

The next indication of a cyclic structure of the free sugar came from observations of another phenomenon, namely mutarotation, and from the isolation of two isomeric forms of D-glucose. One form (α), best isolated by crystallization of glucose from alcohol–water below 30°C, has a specific rotation of +112°. The other (β), obtained when an aqueous solution of glucose is evaporated at temperatures above 98°C, has a specific rotation of +19°.

A freshly prepared solution of α-glucose shows an initial rotation of +112°, which gradually decreases to a constant value of +52°. This change, first observed in 1846, is known as mutarotation. A solution of β-glucose also undergoes mutarotation, the specific rotation changing from an initial value of +19° to an equilibrium value of +52°. This equilibrium solution of rotation +52° can be evaporated above 98°C and pure β-glucose isolated. Alternatively it can be concentrated below 30°C and pure α-glucose can be isolated. These results demonstrate that α- and β-glucose differ only at carbon 1, the aldehyde carbon. This can be accounted for by the cyclic hemiacetal structure (p. 131). The molecular event responsible for the observed change in the optical rotation is attributed to the interconversion of the two forms in solution with the free aldehyde as an intermediate.

It was demonstrated in 1901 that the configuration at carbon 1 of α-methyl glucoside is the same as that of α-glucose, and that of β-methyl glucoside corresponds to β-glucose. Fischer had observed that the hydrolysis of α-methyl glucoside is catalyzed by the enzyme maltase, but not emulsin, and that the hydrolysis of β-methyl glucoside is catalyzed by emulsin but not maltase. It was subsequently shown that α-methyl glucoside is hydrolyzed to α-glucose.

The next question is to decide the size of the ring, in other words, which oxygen of the carbon chain is bonded to the carbon 1 in the cyclic hemiacetal and in the cyclic acetal. Conclusive evidence that the ring is actually six-membered was presented in 1926. Thus we may complete the formulas of α-D-glucose, β-D-glucose and the corresponding methyl glucosides as shown below. Also shown is the open-chain or free aldehyde form of D-glucose. This form is intermediate in

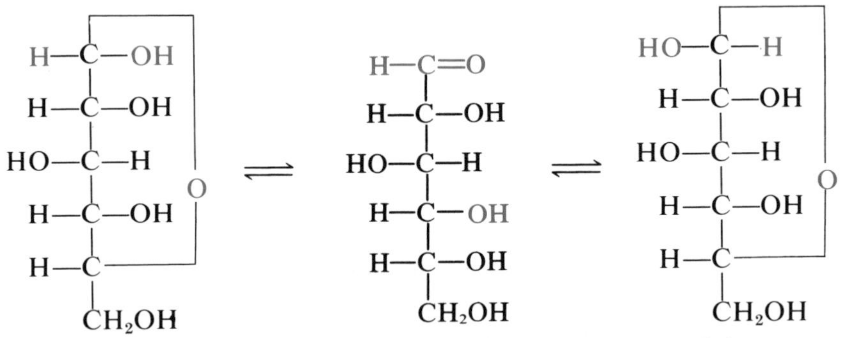

H—C—OH
H—C—OH
HO—C—H
H—C—OH
H—C (—O— ring to C-1)
CH_2OH

α-glucose
α-D-glucopyranose
mp 146°C
[α] +112°

$\rightleftharpoons$

H—C=O
H—C—OH
HO—C—H
H—C—OH
H—C—OH
CH_2OH

open-chain or
free aldehyde form
of D-glucose

$\rightleftharpoons$

HO—C—H
H—C—OH
HO—C—H
H—C—OH
H—C (—O— ring to C-1)
CH_2OH

β-glucose
β-D-glucopyranose
mp 150°C
[α] +19°

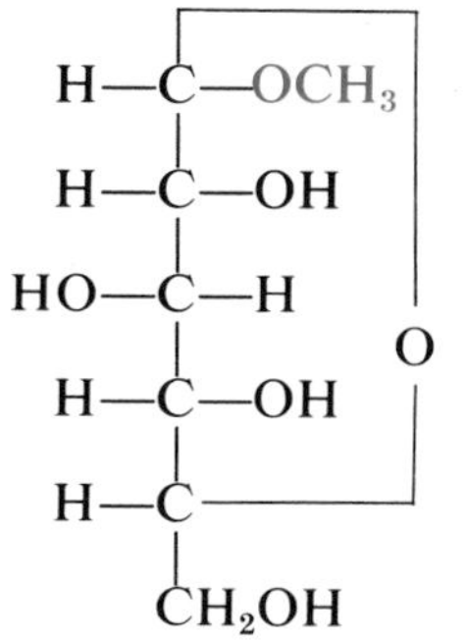

α-methyl glucoside
α-methyl D-glucopyranoside
[α] + 159°

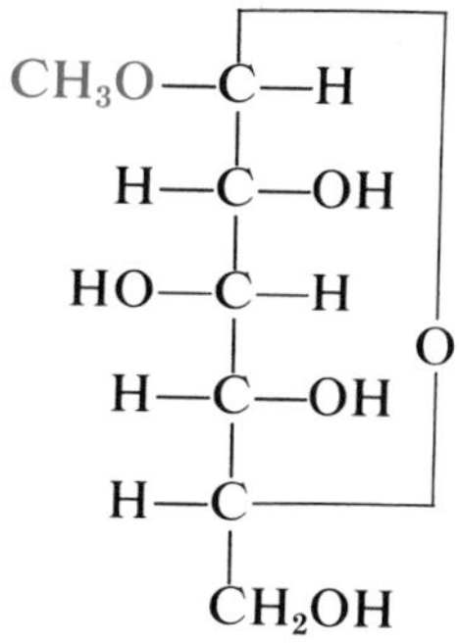

β-methyl glucoside
β-methyl D-glucopyranoside
[α] − 34°

the interconversion between α-glucose and β-glucose. This interconversion is the molecular change underlying mutarotation.

Both α- and β-glucose have the six-membered ring structure of pyran, and in the systematic nomenclature are designated as glucopyranoses, e.g., α-D-glucopyranose. The glucosides are properly designated as pyranosides, e.g., α-methyl D-glucopyranoside. A sugar having a five-membered cyclic structure is designated as a furanose (in the case of a hemiacetal) or a furanoside (in the case of an acetal). For example, ribose as it is found combined in RNA is a five-membered ring and is properly called a β-D-ribofuranoside (Figure 14.2).

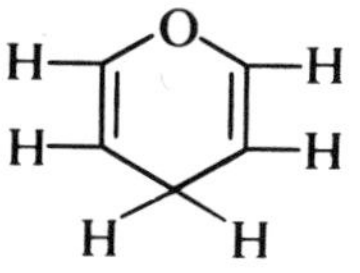

pyran

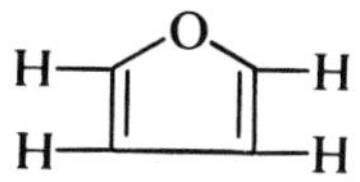

furan

10.6 STEREO-REPRESENTATIONS OF SUGARS

There are three common ways for representing the stereochemistry of glucose and other carbohydrates. We have already encountered the Fischer projection representation. While this type of representation does show clearly the configuration at each carbon, it is grossly inaccurate in its representation of bond angles and the geometry of the molecule. In the Haworth projection, the cyclic forms of carbohydrates are represented as regular hexagons or pentagons, as shown below for glucose, ribose and fructose.

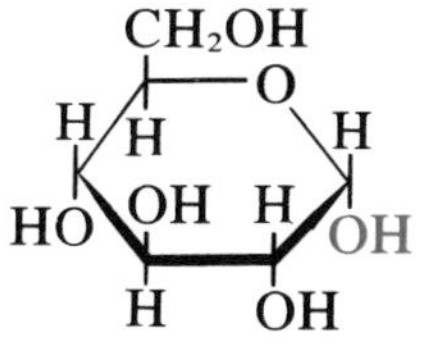

α-D-glucose
α-D-glucopyranose

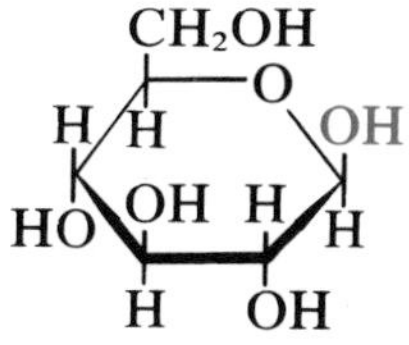

β-D-glucose
β-D-glucopyranose

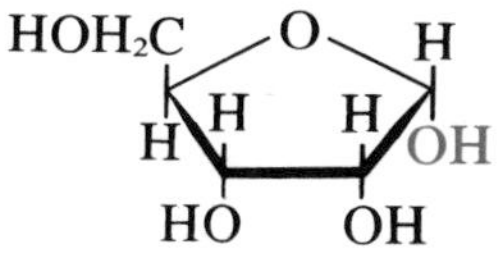

α-D-ribose
α-D-ribofuranose

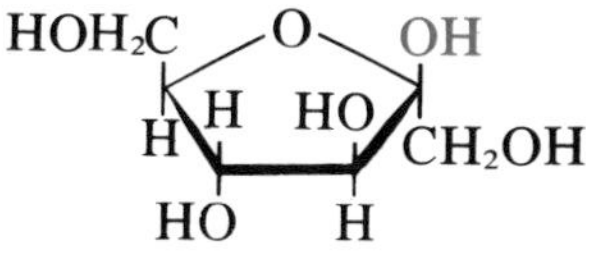

β-D-fructose
β-D-fructofuranose

Although useful, the Haworth projection formulas are not accurate in that they show the six-membered ring as being planar. The ring is more accurately represented as a strain-free chair conformation in which the substituents on carbons 2, 3, 4, and 5 of the ring all lie in equatorial positions. The structural formulas of α-glucose and β-glucose and the phenomenon of mutarotation are shown in this type of representation.

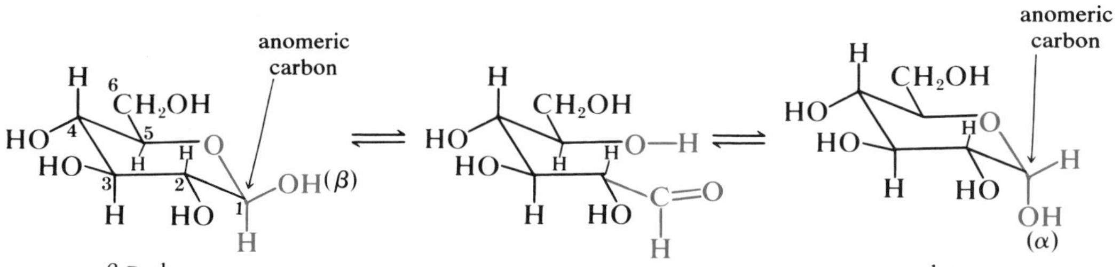

anomeric carbon

OH(β)

β-D-glucose
β-D-glucopyranose

anomeric carbon

(α)

α-D-glucose
α-D-glucopyranose

10.7 PROPERTIES OF MONOSACCHARIDES

The monosaccharides are colorless crystalline solids and sweet to the taste. Because of the possibilities for hydrogen bonding they are very soluble in water but only slightly soluble in alcohol and quite insoluble in non-hydroxylic solvents such as ether, chloroform, and benzene.

Monosaccharides (and carbohydrates in general) are conveniently classified as reducing or as nonreducing sugars according to their behavior toward Cu^{2+} (Benedict's solution, Fehling's solution) or toward Ag^{+} in ammonium hydroxide (Tollens' solution). Reducing sugars contain a free or potentially free aldehyde or α-hydroxyketone. As we saw in Chapter 6, an aldehyde is very susceptible to oxidation to the corresponding acid, even by such weak oxidizing agents as silver ion. In Tollens' test, the deposition of a silver mirror indicates the presence of an aldehyde group. A second reagent, Fehling's solution, is prepared by mixing a solution of copper sulfate with an alkaline solution of a salt of tartaric acid. The resulting deep blue solution contains copper ion complexed with tartrate. When this solution reacts with a sugar containing an aldehyde or an α-hydroxyketone, the cupric ion is reduced to cuprous ion, which precipitates as brick red cuprous oxide.

$$RCHO + Cu^{2+} \longrightarrow RCO_2^- + \underset{\text{brick red precipitate}}{Cu_2O}$$

Benedict's solution contains cupric ion complexed with citrate and serves the same function as Fehling's solution.

Even though monosaccharides are predominantly in the cyclic hemiacetal form, the cyclic forms are in equilibrium with the free aldehyde and are therefore susceptible to oxidation by the above-mentioned solutions. All monosaccharides are reducing sugars.

hemiacetal form ⇌ open-chain form $\xrightarrow{\text{oxidation}}$ carboxylic acid

Glycosides of monosaccharides are nonreducing sugars, since the acetal linkage is stable in the alkaline solution used to test reducing power.

Monosaccharides also undergo reaction with a variety of other oxidizing and reducing agents. The classical reducing agent and the one so commonly used in the early investigations of carbohydrate structure was sodium amalgam in aqueous acid. Such reductions can now be accomplished by a variety of laboratory methods.

$$\underset{\text{D-glucose}}{\begin{matrix} CHO \\ | \\ (CHOH)_4 \\ | \\ CH_2OH \end{matrix}} \xrightarrow[\text{reduction}]{\text{Na/Hg at pH4}} \underset{\substack{\text{D-glucitol} \\ \text{sorbitol}}}{\begin{matrix} CH_2OH \\ | \\ (CHOH)_4 \\ | \\ CH_2OH \end{matrix}}$$

These reduction products are known as alditols; e.g., D-glucose forms D-glucitol (also known as sorbitol), D-threose forms D-threitol.

Oxidation of aldoses by Fehling's, Benedict's, or Tollens' solutions forms the corresponding carboxylic acids. This selective oxidation of the aldehyde group can also be accomplished by using a halogen (most commonly bromine or iodine) in base. The monocarboxylic acid products are known as aldonic acids; e.g., D-glucose forms D-gluconic acid, D-threose yields D-threonic acid.

$$\underset{\text{D-glucose}}{\begin{matrix} CHO \\ | \\ (CHOH)_4 \\ | \\ CH_2OH \end{matrix}} \xrightarrow[\text{or } Br_2 \text{ in base}]{Ag^+ \text{ or } Cu^{2+}} \underset{\text{D-gluconic acid}}{\begin{matrix} CO_2H \\ | \\ (CHOH)_4 \\ | \\ CH_2OH \end{matrix}}$$

More vigorous oxidation with warm nitric acid leads to oxidation of the —CHO group and also the terminal $—CH_2OH$ group. These dicarboxylic acids are known as aldaric acids, e.g., nitric acid oxidation of either D-glucose or D-gluconic acid yields D-glucaric acid.

$$\underset{\text{D-glucose}}{\begin{matrix} CHO \\ | \\ (CHOH)_4 \\ | \\ CH_2OH \end{matrix}} \xrightarrow[\text{heat}]{HNO_3} \underset{\text{D-glucaric acid}}{\begin{matrix} CO_2H \\ | \\ (CHOH)_4 \\ | \\ CO_2H \end{matrix}}$$

Devising accurate methods for both qualitative and quantitative determinations of carbohydrates in blood, urine, and other biological fluids has long challenged clinical chemists. This interest has been stimulated by the high incidence of diabetes mellitus and its serious clinical consequences. What is needed is a method that will measure accurately the concentration of glucose but will not give false positive results with any other substances that may also be present in these fluids. Over the last 50 years or more, a great many glucose tests have been developed. We shall discuss three of these in order to illustrate the problems involved in developing such procedures for use in the clinical chemistry laboratory and how these problems can be solved.

The first method is based on the fact that glucose is a reducing sugar, and specifically that it will reduce ferricyanide ion (yellow in aqueous solution) to ferrocyanide ion (colorless in aqueous solution).

$$\underset{}{\text{glucose}} + \underset{\substack{\text{ferricyanide ion}\\\text{(yellow)}}}{2Fe(CN)_6^{3-}} \longrightarrow \underset{}{\text{gluconic acid}} + \underset{\substack{\text{ferrocyanide ion}\\\text{(colorless)}}}{2Fe(CN)_6^{4-}}$$

The decrease in intensity of the yellow color is proportional to the concentration of glucose in the sample. A modification of this method, introduced in 1928 and for many years the standard procedure for the determination of blood glucose, involves carrying out the reaction in the presence of excess ferric ion, Fe^{3+}. Under these conditions, the ferrocyanide ion formed on oxidation of glucose reacts further to form ferric ferrocyanide, or as it is more commonly known, Prussian blue.

$$\underset{\substack{\text{ferrocyanide ion}\\\text{(colorless)}}}{3Fe(CN)_6^{4-}} + 4Fe^{3+} \longrightarrow \underset{\substack{\text{ferric ferrocyanide}\\\text{(Prussian blue)}}}{Fe_4[Fe(CN)_6]_3}$$

The intensity of the blue color is directly proportional to the concentration of glucose. While this and most other oxidation methods do measure glucose concentrations accurately, they suffer from the disadvantage that they give false positive results in the presence of other reducing substances (ascorbic acid, uric acid, certain amino acids, phenols, other aldoses, etc.) also present in normal blood.

Part of this problem of interference can be avoided by using methods that take advantage of a type of chemical reactivity of glucose other than its property as a reducing sugar. One of the most useful of these nonoxidative methods involves reaction of glucose with *o*-toluidine to form a Schiff base whose blue-green color is directly proportional to the concentration of glucose in the sample.

$$\underset{\text{glucose}}{R{-}CH{=}O} + \underset{o\text{-toluidine}}{H_2N{-}C_6H_4{-}CH_3} \longrightarrow \underset{\substack{\text{a Schiff base}\\\text{(blue-green)}}}{R{-}CH{=}N{-}C_6H_4{-}CH_3} + H_2O$$

The procedure calls for mixing the test sample with *o*-toluidine in glacial acetic acid, heating it in a boiling water bath for

8–10 minutes, and then measuring the intensity of the blue-green color. This method can be applied directly to serum, plasma, cerebrospinal fluid, and urine. It does not give false positive tests with other reducing substances because the procedure itself does not involve oxidation. However, galactose and mannose also react to give colored Schiff bases. This is generally not a problem though, because these monosaccharides are normally present in serum and plasma in very low concentrations.

The search for even greater specificity in glucose determinations led to the introduction of enzyme-based procedures. What is needed is an enzyme that will catalyze some reaction of glucose but will not catalyze a comparable reaction for any other substance that may be in biological fluids. The enzyme glucose oxidase does just this. It catalyzes the oxidation of glucose by molecular oxygen.

$$\text{D-glucose} + O_2 \xrightarrow{\text{glucose oxidase}} \text{D-gluconic acid} + H_2O_2$$

The rate of oxidation of glucose catalyzed by glucose oxidase is about 1,000 times more rapid than that of any other mono- or disaccharide that might also be in plasma, serum, or urine. Direct determination can be made quantitative by coupling this oxidation with another reaction in which the H_2O_2 generated is used to oxidize *o*-toluidine to a colored product. In practice, the enzyme peroxidase is used to catalyze this second oxidation.

A number of commercially available filter paper strip tests use this procedure for qualitative determination of glucose. One of these, Clinistix (Ames Laboratories), consists of a filter paper strip impregnated with glucose oxidase, peroxidase, and *o*-toluidine. The test strip is dipped in urine and the color of the strip compared to a color chart.

10.8 DISACCHARIDES

Most carbohydrates in nature contain more than one hexose unit. Sucrose (cane sugar), lactose (milk sugar), maltose, and cellobiose each consist of two hexose units and are called disaccharides. In a disaccharide the two monosaccharides are linked together by acetal formation (Section 6.8). Condensation of the hydroxyl function of the hemiacetal group of one monosaccharide with an hydroxyl group of another monosaccharide forms the bond (called a glycosidic bond) joining the two saccharide units. Since glycosides are acetals they may be hydrolyzed to the constituent monosaccharides by dilute mineral acid or by the appropriate enzymes such as maltase and emulsin. We shall look at the structures of four disaccharides: maltose, cellobiose, lactose, and sucrose.

Maltose derives its name from the fact that it occurs mainly in malt liquors, the juice expressed from sprouted barley and other cereal grains. The enzyme amylase in yeast or malt hydrolyzes the starch in the seeds to maltose in about 18% yield. Further hydrolysis of maltose by the enzyme maltase yields only glucose. Maltose is formed from two molecules of glucose joined together by an α-glycosidic bond between carbon 1 of one glucose and the carbon 4 of the second glucose unit. In maltose, glucose is joined to glucose by an α-1,4-glycosidic bond.

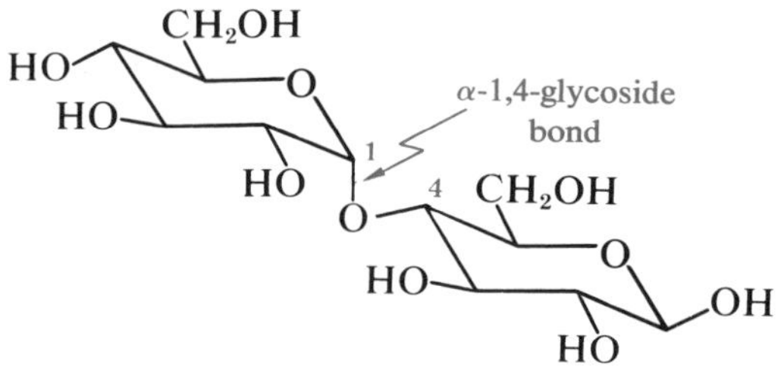

α-1,4-glycoside bond

α-D-glucopyranose unit — β-D-glucopyranose unit

maltose (from degradation of starch)

Cellobiose, a reducing sugar, is one of the major fragments isolated after extensive hydrolysis of cellulose (cotton fiber for example). Further hydrolysis affords two molecules of glucose. Cellobiose differs from maltose only in that the two D-glucose units are joined by a β-1,4-glycosidic linkage.

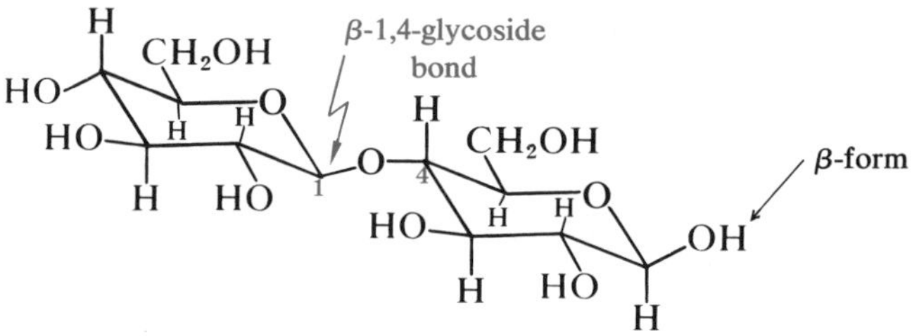

β-1,4-glycoside bond

β-form

β-D-glucopyranose unit — β-D-glucopyranose unit

cellobiose (from controlled hydrolysis of cellulose)

Another disaccharide of special interest is lactose. It is the major sugar of milk and is present to the extent of 5–8% in human milk and 4–6% in cow's milk. Hydrolysis of lactose affords equal amounts of D-glucose and D-galactose. In lactose, galactose is joined to glucose by a β-1,4-glycosidic bond.

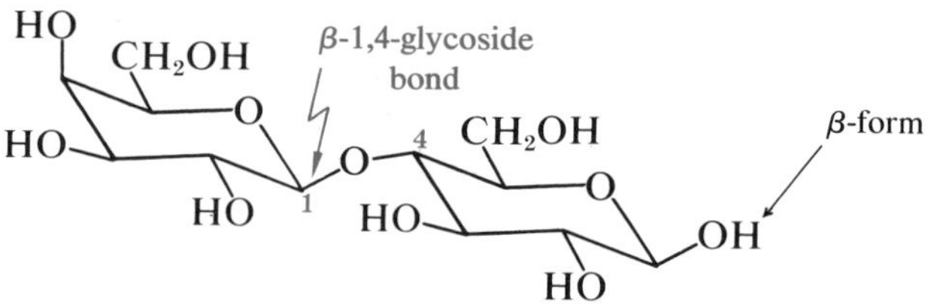

β-1,4-glycoside bond

β-form

β-D-galactopyranose unit — β-D-glucopyranose unit

β-lactose (from milk of mammals)

Sucrose is the most abundant disaccharide. It is common table sugar and is readily obtained from the juice of cane sugar or sugar beet. Hydrolysis of sucrose affords one molecule of D-glucose and one of D-fructose. Sucrose is not a reducing sugar. Since both glucose and fructose are reducing sugars, both monosaccharides must be present as the glycosides, i.e., the anomeric hemiacetal carbons of both glucose and fructose must be involved in the formation of the glycosidic bond in sucrose.

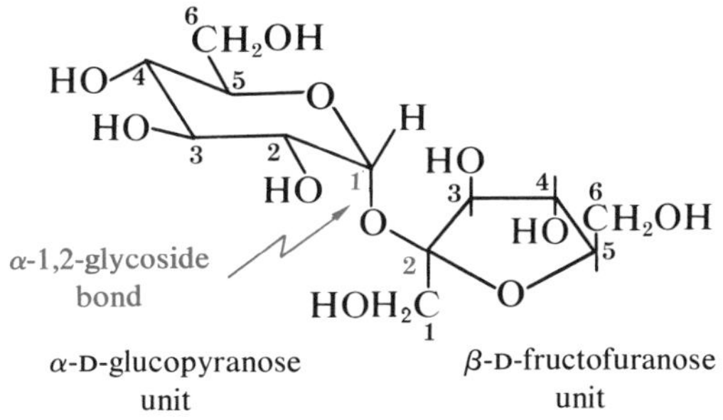

α-1,2-glycoside bond

α-D-glucopyranose unit

β-D-fructofuranose unit

sucrose (cane sugar)

Sucrose is a glycoside in which the carbon 1 of glucose is bonded by an α-glycosidic bond to the carbon 2 of fructose, i.e., in sucrose, glucose is joined to fructose by an α-1,2-glycosidic bond. Glucose is in the pyranoside form while fructose is in the furanoside form.

10.9 POLYSACCHARIDES

Polysaccharides are grouped together into two general classes: those that are insoluble and form the skeletal structure of plants, and those that constitute the reserve sources of simple sugars and are liberated as required by certain enzymes in the organism. Both types are high-molecular-weight polymers.

Starch is the reserve carbohydrate for plants. It is found in all plant seeds and tubers, and is the form in which glucose is stored for use by the plant in its metabolic processes. Starch can be separated into two fractions by making a paste with water and warming it to 60–80°C. One fraction, amylose, comprising about 20% of the starch, is soluble in the hot water. The water-insoluble fraction, amylopectin, can be separated by centrifugation. Although amylose and amylopectin are both polymers of glucose, and yield maltose on partial hydrolysis and only glucose on complete hydrolysis, they differ in several respects. Amylose has a molecular weight range of 10,000–50,000 (60 to 300 glucose units) and amylopectin has a molecular weight range of 50,000–1,000,000 (300 to 6000 glucose units). X-ray diffraction studies of amylose show a continuous, unbranched chain of glucose units with the carbon 1 of one glucose linked by an α-glycosidic bond to the carbon 4 of the adjacent glucose.

Amylopectin has a highly branched structure. It contains the same type of repetitive sequence of α-1,4-glycosidic bonds as does amylose but the chain lengths vary only from about 24 to 30 units. In addition, there is considerable branching from this linear network. At branch points a new chain is started by an α-1,6-glycosidic linkage between the carbon 1 of one glucose unit and a carbon 6 hydroxyl of another glucose unit.

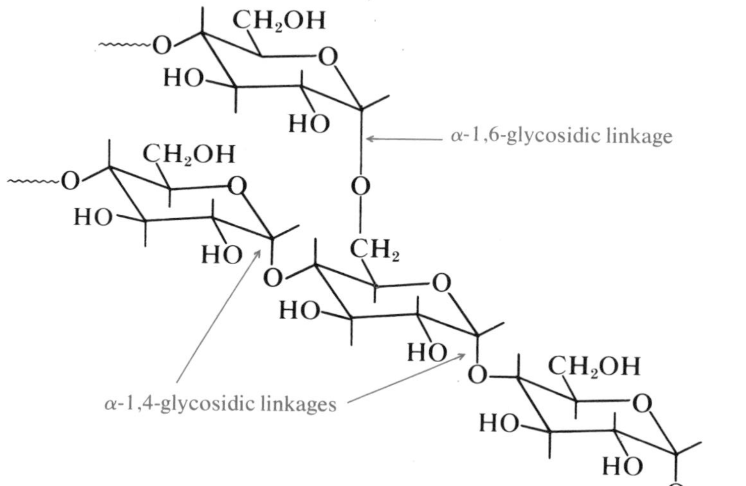

α-1,6-glycosidic linkage

α-1,4-glycosidic linkages

amylopectin

Plants and animals synthesize these large molecules at the expense of considerable energy. Why? If a plant seed requires a large amount of food as stored carbohydrate, why is it not sufficient to pack into it a certain amount of glucose? The reason is rooted in a very elementary principle of physical chemistry. Osmotic pressure is proportional to the molar concentration, not the molecular weight, of a solute. If we assume that 1000 glucose monomers are assembled into a starch macromolecule, then we can predict that a solution of starch containing 1 gram of starch per 10 ml of solution will have only 1/1000 the osmotic pressure of 1 gram of glucose in the same volume of solution. This feat of packaging is of tremendous advantage because it reduces the strain on various membranes enclosing such macromolecules.

Glycogen serves as the reserve carbohydrate for animals. Like amylopectin, glycogen is a nonlinear polymer of glucose units joined by α-1,4- and α-1,6-glycoside bonds, but it has a lower molecular weight and a more highly branched structure. The chief source of glycogen is ingested starch or glycogen, which is then hydrolyzed by amylases in the intestinal tract to glucose. Since the amount of glucose absorbed into the blood stream during digestion is greater than that required for immediate use by the body, most of the excess glucose is converted into glycogen (a process called glycogenesis) and stored in the liver and muscles. The total amount of glycogen in the body of a well-nourished adult is about 350 grams divided nearly equally between the liver and muscle.

Cellulose is the most widely distributed skeletal polysaccharide. It constitutes almost half of the cell wall material of wood. Cotton is almost pure cellulose. Controlled hydrolysis of cellulose affords cellobiose as the only disaccharide. Vigorous hydrolysis affords only D-glucose. This and other evidence demonstrates that cellulose is a linear polymer of glucose units that are joined together by β-1,4-glycosidic linkages. The molecular weight of cellulose is approximately 400,000, corresponding to about 2800 glucose units. X-ray analysis indicates that cellulose fiber consists of bundles of parallel-oriented chains in which the pyranose rings lie alternately in different orientations as shown in the following diagram. The chains are held together by hydrogen bonding between the hydroxyls of the adjacent chains. This type of arrangement of parallel chains into bundles and the resultant hydrogen bonding gives cellulose fibers a high mechanical strength and a chemical stability.

cellobiose unit

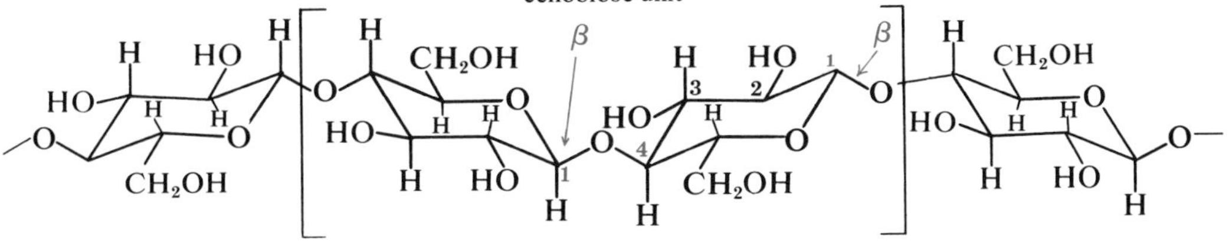

cellulose chain

Man and other carnivorous animals are unable to utilize cellulose as a foodstuff even though, like starch, it is a polymer of glucose. In the evolutionary scheme of things man either lost or never carried the genetic information required to make the enzyme system capable of catalyzing the hydrolysis of the β-glycosidic linkage. Our digestive systems contain only the α-amylases and hence we must use starch and glycogen as our sources of glucose. On the other hand, many bacteria and microorganisms do possess the enzyme systems capable of digesting cellulose. Termites are fortunate to have such bacteria in their intestine and can use wood as their principal food. Ruminants (cud-chewing animals) can also digest grasses and wood because of the presence of microorganisms within the specially constructed alimentary system.

Cellulose, whether from cotton, wood pulp, or other natural source, is such a readily available and relatively inexpensive raw material that besides processing it into cotton and linen fibers for textiles and into paper goods, it has been possible to modify it chemically to produce certain widely used semisynthetic materials. The most important of these modified celluloses are cellulose nitrate and cellulose acetate.

Nitration of cellulose by a mixture of nitric and sulfuric acids yields a polynitrate ester called cellulose nitrate or nitrocellulose. The extent of nitration varies with the conditions of the process used. Below is shown a partial structural formula of fully nitrated cellulose. Note that there are three nitrate esters per glucose unit.

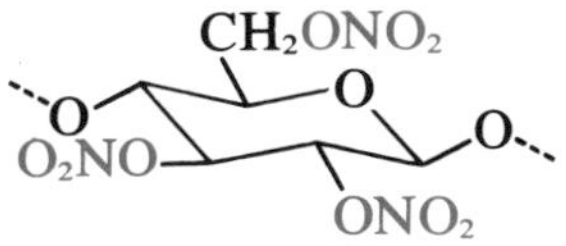

cellulose nitrate
(partial formula)

Guncotton approaches this degree of nitration. Less completely nitrated celluloses are used in the manufacture of the common moldable plastic celluloid. In this process, partially nitrated cellulose is mixed with alcohol and camphor and then molded or rolled into sheets. The molded or rolled article is heated at a slightly elevated temperature to evaporate the alcohol. The finished material is a tough, hard plastic. One of the first uses for this plastic was a substitute for ivory billiard balls. Celluloid deserves special mention in the chemistry and technology of plastics for it was the first and until about 1920 the only plastic to be manufactured on an industrial scale. Its major disadvantage is that it is highly flammable.

Fibers made from regenerated and chemically modified cellulose were the first of the man-made fibers to become and remain commercially important. Because of its insolubility in any of the convenient and inexpensive solvents, cellulose must be chemically modified and made soluble before it can be processed into a fiber. If the modifying groups are subsequently removed in the processing, the resulting fiber is called regenerated cellulose. If the modifying groups appear in the resulting fiber, this is indicated in the name of the fiber, as for example in the name acetate rayon.

In one industrial process, cellulose is reacted with carbon disulfide to form an alkali-soluble xanthate ester. A solution of cellulose xanthate is then extruded into dilute sulfuric acid which brings about hydrolysis of the xanthate ester and precipitation of the free or regenerated cellulose. Extruding the regenerated cellulose as a filament produces "viscose rayon" threads; extruding it as sheets produces cellophane.

In another industrial process, cellulose is acetylated with acetic anhydride. The extent of acetylation varies with the conditions. Acetylated cellulose is dissolved in a suitable solvent, precipitated, and then drawn into fibers. This material is known as acetate rayon. Cellulose acetate, acetylated to the extent of about 80%, became commercial in Europe about 1920 and in the United States a few years later. Cellulose triacetate, which has about 97% of the hydroxyls acetylated, became commercial in the United States in 1954. Acetate fibers rank fourth in production in the United States, surpassed only by polyester, nylon, and rayon fibers.

Problems

10.1 Define carbohydrate in terms of the functional groups present. Literally the term carbohydrate is derived from "hydrates of carbon." Show by reference to molecular formulas the origin of this term.

10.2 Draw perspective formulas for D-glyceraldehyde and L-glyceraldehyde. Explain the meaning of the designation D- and L- as used to specify the stereochemistry of glyceraldehyde.

10.3 List the rules for drawing the so-called Fischer projection formulas. Draw Fischer projection formulas for D- and L-glyceraldehyde.

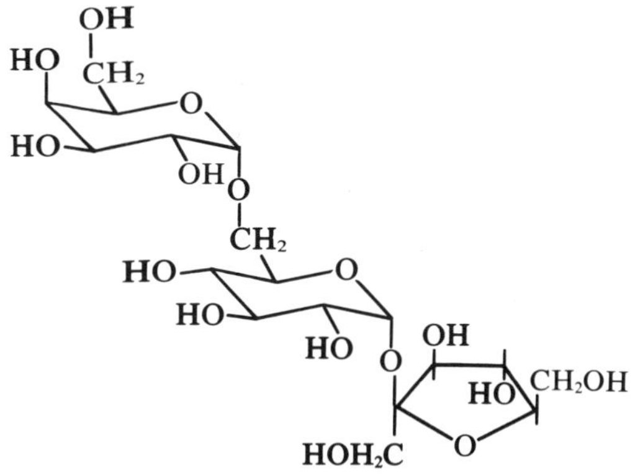

raffinose

10.25 Draw Haworth and chair representations for the α- and β-pyranose forms of D-glucosamine, D-mannosamine, and D-galactosamine.

10.26 Shown below is a Fischer projection formula for N-acetyl-D-glucosamine.

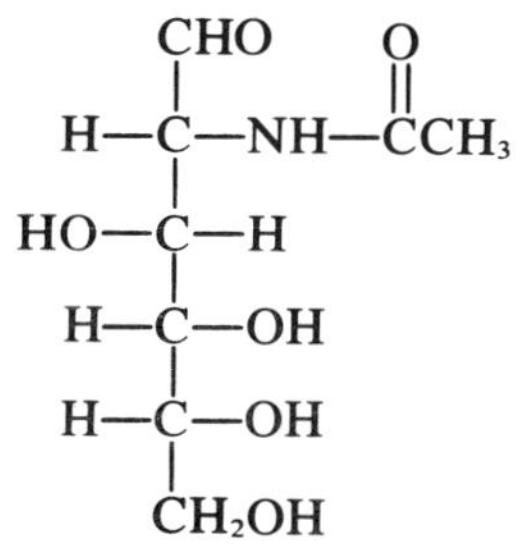

```
        CHO      O
        |        ||
     H—C—NH—CCH3
        |
   HO—C—H
        |
     H—C—OH
        |
     H—C—OH
        |
       CH2OH
```

N-acetyl-D-glucosamine

a Draw Haworth and chair representations for the β-pyranose form of this sugar.

b Draw Haworth and chair representations for the disaccharide formed by joining two β-pyranose units of N-acetyl-D-glucosamine together by a β-1,4-glycoside bond. (If you have done this correctly, you have drawn a structural formula for the repeating dimer of chitin, the polysaccharide component of chitin.)

10.4 Draw the four stereoisomers of 2,3,4-trihydroxybutanal. Label them I, II, III, and IV. Which are enantiomers? diastereomers? (To check your answer, refer to Section 4.7.)

10.5 What is the difference in structure between D-ribose and 2-deoxy-D-ribose? Draw structural formulas for the open-chain form of each.

10.6 A careful study of D-ribose in aqueous solution shows that it exists as an equilibrium mixture of both furanoses and pyranoses. Draw suitable stereorepresentations of these four possible cyclic hemiacetals:

a α-D-ribofuranose **b** β-D-ribofuranose
c α-D-ribopyranose **d** β-D-ribopyranose

10.7 One of the important techniques for establishing relative configurations among the isomeric straight-chain aldoses has been to convert both terminal carbon atoms into the same functional group. This can be done either by selective oxidation or reduction. As a specific example, nitric acid oxidation of D-erythrose leads to the formation of optically inactive meso-tartaric acid. Oxidation of D-threose under similar conditions leads to the formation of optically active D-tartaric acid.

$$\text{D-erythrose} \xrightarrow[\text{oxidation}]{HNO_3} \text{meso-tartaric acid}$$

$$\text{D-threose} \xrightarrow[\text{oxidation}]{HNO_3} \text{D-tartaric acid}$$

Using this information, show which of the structural formulas (a) or (b) is D-erythrose and which is D-threose. Check your answer by referring to Table 10.1.

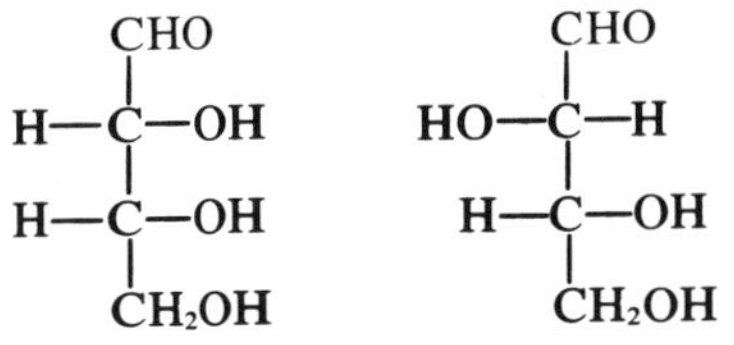

```
      CHO             CHO
      |               |
   H—C—OH        HO—C—H
      |               |
   H—C—OH         H—C—OH
      |               |
     CH2OH           CH2OH
      (a)             (b)
```

10.8 As you can see from Table 10.1, there are four isomeric D-aldopentoses. Suppose that each is reduced with H_2 in the presence of a platinum catalyst. Which of the four will yield optically inactive D-alditols; which will yield optically active D-alditols?

10.9 Draw a Fischer projection formula for D-glucose. State the total number of possible optical isomers of this structural formula. Of these only D-glucose, D-galactose, and D-mannose occur in nature. Draw Fischer projection formulas for D-galactose and D-mannose.

10.10 These are three common conventions for representing the stereochemistry of carbohydrates. They are (1) the Fischer projection, (2) the Haworth projection, and (3) the chair/boat projection formulas. Draw α-D-glucose according to the rules of each of these conventions. Do the same for β-D-glucose.

10.11 Using a set of molecular models, build the following:

a D-glucose in the open-chain form.
b Using this molecular model, show the reaction of the —OH on carbon 5 with the aldehyde of carbon 1 to form a cyclic hemiacetal.

c Show that either α-D-glucose or β-D-glucose can be formed depending on the direction from which the —OH group interacts with the aldehyde group.
d Below is drawn the chair representation of β-D-glucose.

Draw the mirror image of this molecule. Show, using your molecular model from part **b** above, that the structural formula drawn here corresponds to β-D-glucose, and that the mirror image you have drawn corresponds to β-L-glucose.

10.12 Explain the conventions α- and β- as used to designate the stereochemistry of the cyclic forms of carbohydrates.

10.13 Draw structural formulas for the open-chain and cyclic forms of D-fructose.

10.14 Explain the phenomenon of mutarotation with reference to carbohydrates. By what means is it detected?

10.15 A freshly prepared solution of α-D-glucose has a specific rotation of $+112°$; a similarly prepared solution of β-D-glucose has a specific rotation of $+19°$. On mutarotation, the specific rotation of either solution changes to an equilibrium value of $+52°$. Calculate the percentage of β-D-glucose in the equilibrium mixture.

10.16 Treatment of D-glucose in dilute aqueous alkali at room temperature yields an equilibrium mixture of D-glucose, D-mannose, and D-fructose. Account for this transformation.

10.17 Ketones cannot be oxidized by mild oxidizing agents. However, both dihydroxyacetone and fructose give a positive Benedict's test and are therefore classed as reducing sugars. Using structural formulas show how these monosaccharides react to give a positive test.

10.18 Fischer attempted to convert D-glucose into its dimethyl acetal according to the following reaction:

$$\text{(i)} \quad \underset{\text{D-glucose}}{C_6H_{12}O_6} + 2CH_3OH \xrightarrow{H^+} \underset{\text{D-glucose dimethyl acetal}}{C_8H_{18}O_7} + H_2O$$

However, the reaction that takes place is instead

$$\text{(ii)} \quad \underset{\text{D-glucose}}{C_6H_{12}O_6} + CH_3OH \xrightarrow{H^+} \underset{\text{methyl D-glucoside}}{C_7H_{14}O_6} + H_2O$$

Draw a structural formula for the expected product, $C_8H_{18}O_7$, and the observed product, $C_7H_{14}O_6$. Reaction (ii) above gives two isomeric methyl glucosides designated as α-methyl D-glucoside and β-methyl D-glucoside. Draw Fischer, Haworth, and chair projection formulas for each of these and label them accordingly.

10.19 Draw Haworth and chair projection formulas for the disaccharides sucrose, lactose, cellobiose and maltose. Name the constituent monosaccharides and label as α or β the stereochemistry of the glycosidic linkages.

State which of these dis
mutarotation. State one

10.20 Draw Haworth a
linked in the polysaccha
each of these polysacch

10.21 What is the diffe
glycoside?

10.22 Propose a likely s

a Alginic acid, isolated
ice cream and other fo
D-mannuronic acid units ir
joined together by β-1,4-g

b Pectic acid is the main
of jellies from fruits and
D-galacturonic acid units in
joined together by α-1,4-gl

D-mannuronic

10.23 Trehalose is found in y
in the blood of certain insects.
glucose units joined by an α-1,

On the basis of its structural forn

a be a reducing sugar?
b undergo mutarotation?

10.24 Raffinose (shown on p. 256) is

a Name each of the three monosa
b There are two glycoside bonds
already done for other disaccharides
β-1,4-glycoside bond in cellobiose).

11

LIPIDS

11.1 INTRODUCTION

The term lipid designates a rather heterogeneous class of naturally occurring organic substances. Unlike the various classes of organic compounds we have studied so far (e.g., alcohols, carboxylic acids, ketones, amines, carbohydrates), lipids as a class are grouped together not by the presence of a distinguishing functional group or structural feature, but rather on the basis of common solubility properties. Lipids are all insoluble in water and highly soluble in one or more organic solvents including ether, chloroform, benzene, and acetone. In fact, these four solvents are often referred to as "lipid-solvents" or "fat-solvents."

Lipids are widely distributed in the biological world and play a wide variety of roles in both plant and animal tissue. In the human body, lipids function as storage forms of energy, metabolic fuels, structural components of cell membranes, emulsifying agents, vitamins, and regulators of metabolism. This heterogeneous class of compounds is generally divided into four major groups, based primarily on certain common structural features.

1. Simple lipids: esters of fatty acids with various alcohols. These include the neutral fats, oils, and waxes.
2. Compound lipids: esters of fatty acids with alcohols plus other groups. These include the phospholipids, glycolipids, and lipoproteins.
3. Derived lipids: substances derived from hydrolysis of compounds in groups 1 and 2 that still retain the general physical characteristics of lipids. These include the saturated and unsaturated fatty acids, as well as alcohols and steroids.
4. Miscellaneous lipids: These include the fat-soluble vitamins E and K as well as certain terpenes such as the carotenes.

In this chapter we shall discuss the structure and biological function of representative members of each group.

11.2 FATS AND OILS

We are all familiar with fats and oils, for we encounter them every day in such things as milk, cream, butter, lard, oleomargarine, the liquid vegetable oils such as corn, cottonseed, and soybean oils, and many other foods. The common neutral fats and oils are triesters of glycerol and are the most abundant naturally occurring lipids. The only distinction between a fat and an oil is one of melting point; fats are by definition solids at room temperature while oils are liquid at room temperature.

Since neutral fats and oils are triesters of glycerol, they are more correctly called triglycerides or triacylglycerols. There are many different types of triglycerides depending on the identity and the position of the three fatty acid components. Those containing only one kind of fatty acid are called simple triglycerides. Examples are glyceryl tripalmitate and glyceryl tristearate, commonly known as tripalmitin and tristearin. Simple triglycerides rarely occur in nature.

$$
\begin{array}{ll}
CH_2-O-\overset{O}{\overset{\|}{C}}-(CH_2)_{14}CH_3 & CH_2-O-\overset{O}{\overset{\|}{C}}-(CH_2)_{16}CH_3 \\
| & | \\
CH-O-\overset{O}{\overset{\|}{C}}-(CH_2)_{14}CH_3 & CH-O-\overset{O}{\overset{\|}{C}}-(CH_2)_{16}CH_3 \\
| & | \\
CH_2-O-\overset{O}{\overset{\|}{C}}-(CH_2)_{14}CH_3 & CH_2-O-\overset{O}{\overset{\|}{C}}-(CH_2)_{16}CH_3
\end{array}
$$

glyceryl tripalmitate
tripalmitin

glyceryl tristearate
tristearin

Those containing two or more fatty acid components are called mixed triglycerides. In general, a fat or oil does not consist of one pure glyceride, but rather is a mixture (often complex) of glycerides. The composition of a fat is usually expressed in terms of the acids obtained from it by hydrolysis. Some fats and oils contain mainly one or two fatty acids: olive oil is 83% oleic acid, 6% palmitic acid, 4% stearic acid, and 7% linoleic acid; palm oil is 43% palmitic acid, 43% oleic acid, and 10% linoleic acid. Butter fat consists of a complex mixture of esters of at least 14 different acids; it differs from most other fats in having appreciable amounts of lower-molecular-weight fatty acids (Table 11.1).

Table 11.1 *Fatty acid composition of butter.*

Carbon Atoms	Fatty Acid	%	Carbon Atoms	Fatty Acid	%
4	butanoic	3.0	18	stearic	9.2
6	hexanoic	1.4	20	arachidic	1.3
8	octanoic	1.5	12	lauroleic	0.4
10	decanoic	2.7	14	myristoleic	1.6
12	lauric	0.7	16	palmitoleic	4.0
14	myristic	12.1	18	oleic	29.6
16	palmitic	25.3	18	linoleic	3.6

The melting points of fats are determined by their fatty acid components. In general, the melting point increases with the number of carbons in the hydrocarbon chain and with the degree of saturation of the acid components. Triglycerides rich in low-melting fatty acids such as oleic, linoleic, or linolenic acid (Table 7.3) generally are liquid at room temperature and accordingly are called oils. Those rich in saturated fatty acids such as palmitic or stearic acid generally are semisolid or solid at room temperature and accordingly are called fats.

For a variety of reasons, in part convenience and in part dietary preference, a major industry has developed for the conversion of oils to fats. The process involves catalytic hydrogenation of some or all of the double bonds of the fatty acid constituents of a triglyceride and is called "hardening." If all of the double bonds are saturated, the product is hard and brittle. In practice, the degree of hardening is carefully controlled to produce a fat of the desired consistency. The resulting fats are sold for kitchen use (Crisco, Spry, etc.). Oleomargarine and other butter substitutes are prepared by hydrogenation of cottonseed, soybean, corn, or peanut oils. The resulting product is often churned with milk and artificially colored to give it a flavor and consistency resembling that of butter.

On exposure to air, most triglycerides develop an unpleasant odor and flavor and are said to become rancid. In part this is the result of slight hydrolysis of the fat and the production of lower-molecular-weight fatty acids. For example, the odor of rancid butter is due to the presence of butanoic acid formed by the hydrolysis of butterfat. These same low-molecular-weight, volatile substances can be formed by oxidation and cleavage of double bonds in the unsaturated fatty acid side chains. The rate of oxidation and the production of rancidity vary with the individual fat, largely because of the presence of certain naturally occurring lipid-soluble substances which inhibit this oxidation. These substances are known as antioxidants. One of the most common lipid antioxidants is vitamin E.

11.3 PHOSPHOLIPIDS

These lipids, known alternatively as phospholipids, phosphoglycerides, or phosphatides are the second most abundant kind of naturally occurring lipid. They are found almost exclusively in plant and animal cell membranes, which typically consist of about 40–50% phospholipid and 50–60% protein.

The most abundant phospholipids contain glycerol and fatty acids, as do the simple fats. In addition they also contain phosphoric acid and a low-molecular-weight alcohol. The most common of these low-molecular-weight alcohols are choline, ethanolamine, serine, and inositol.

$HOCH_2CH_2\overset{+}{N}(CH_3)_3$ — choline

$HOCH_2CH_2NH_3^+$ — ethanolamine

$HOCH_2CH(NH_3^+)CO_2^-$ — serine

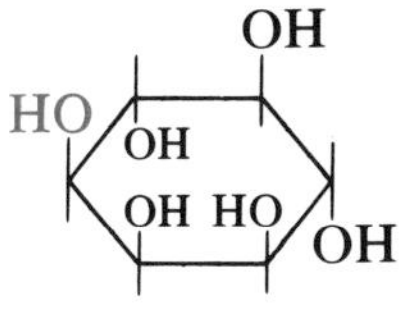

inositol (cyclohexane ring bearing six OH groups)

The most abundant phosphoglycerides in higher plants and animals are phosphatidyl choline, more commonly known as lecithin, and phosphatidyl ethanolamine or cephalin.

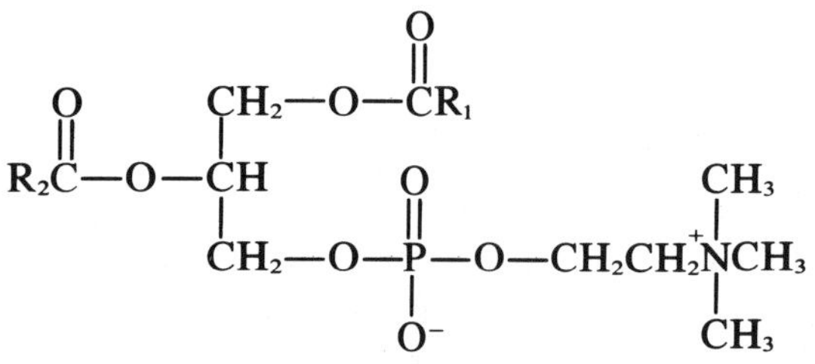

$$\begin{array}{l} CH_2{-}O{-}\overset{O}{\overset{\|}{C}}R_1 \\ R_2\overset{O}{\overset{\|}{C}}{-}O{-}CH \\ CH_2{-}O{-}\underset{O^-}{\underset{|}{\overset{O}{\overset{\|}{P}}}}{-}O{-}CH_2CH_2\overset{+}{N}(CH_3)_3 \end{array}$$

a phosphatidyl choline
a lecithin

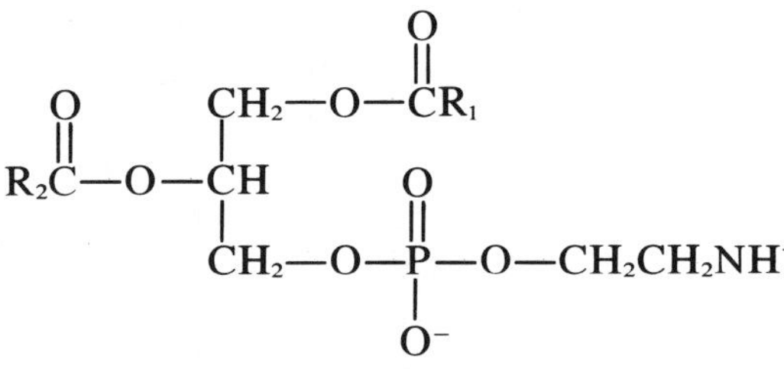

$$\begin{array}{l} CH_2{-}O{-}\overset{O}{\overset{\|}{C}}R_1 \\ R_2\overset{O}{\overset{\|}{C}}{-}O{-}CH \\ CH_2{-}O{-}\underset{O^-}{\underset{|}{\overset{O}{\overset{\|}{P}}}}{-}O{-}CH_2CH_2NH_3^+ \end{array}$$

a phosphatidyl ethanolamine
a cephalin

The lecithins and cephalins are a family of compounds that differ only in the nature of the fatty acid components. The fatty acids most common in these membrane phosphoglycerides are palmitic and stearic acids (both fully saturated) and oleic acid (one double bond in the hydrocarbon chain).

11.4 CELL MEMBRANES

The cell membrane is an important feature of cell structure and serves a number of functions essential to the life of the cell. First, it is a mechanical support which separates the contents of the cell from the external environment. Second, it controls the passage of substances into and out of the cell. For example, essential nutrients are transported into the cell and metabolic wastes out of the cell through the membrane. The membrane also helps to regulate the concentrations of molecules and ions within the cell. Third, the cell membrane provides structural support for certain proteins. Some of these proteins are "receptors" for hormone-carried messages; others are specific enzyme complexes. Obviously the cell membrane is more than an impervious, mechanical barrier separating the cell contents from the external environment. It is a highly specialized structure that performs a multitude of tasks with great precision and accuracy.

We know that the membranes of plant and animal cells are typically composed of 40–50% phospholipid and 50–60% protein. Yet there are wide variations in phospholipid–protein content even between different types of cells within the same organism. For example, myelin, the membrane that surrounds specific types of nerve cells and serves as an insulator, contains only about 18% protein. Membranes that are active in transporting specific molecules into and out of cells contain about 50% protein. The membranes actively involved in the transformation of energy, such as those of mitochondria and chloroplasts, contain up to 75% protein.

The question of the detailed molecular structure of cell membranes is one of the most challenging problems in biochemistry today and despite intensive research, many aspects of cell membrane structure and activity still are not understood. Let us begin the discussion of a possible model for cell membrane structure by first considering the shapes of phospholipid molecules and the possible arrangement of phospholipids in aqueous solution.

Phospholipids are elongated, almost rod-like molecules with the two nonpolar fatty acid hydrocarbon chains lying essentially parallel to one another and with the polar choline, serine, ethanolamine or inositol and phosphate groups pointing in the other direction (Figure 11.1).

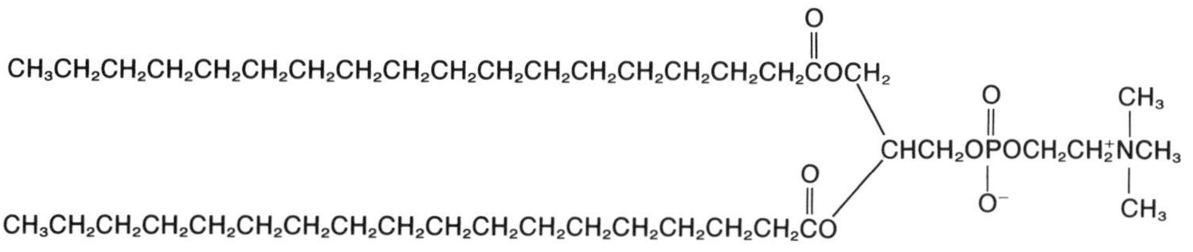

$$CH_3CH_2CH_2CH_2CH_2CH_2CH_2CH_2CH_2CH_2CH_2CH_2CH_2CH_2CH_2CH_2CH_2\overset{O}{\overset{\|}{C}}OCH_2$$

$$\qquad\qquad\qquad\qquad\qquad\qquad\qquad\qquad\qquad\qquad \backslash \qquad \overset{O}{\overset{\|}{}} \qquad\qquad \overset{CH_3}{|}$$

$$\qquad\qquad\qquad\qquad\qquad\qquad\qquad\qquad\qquad\qquad CHCH_2O\underset{\underset{O^-}{|}}{P}OCH_2CH_2^+\underset{\underset{CH_3}{|}}{N}CH_3$$

$$CH_3CH_2CH_2CH_2CH_2CH_2CH_2CH_2CH_2CH_2CH_2CH_2CH_2CH_2CH_2CH_2CH_2\overset{O}{\overset{\|}{C}}O /$$

Figure 11.1 *A phospholipid molecule showing the polar head and nonpolar hydrocarbon tails.*

Next let us consider what happens when phospholipid molecules are placed in an aqueous medium. Recall that when soap molecules are placed in water, they form micelles (Section 7.7) in which the polar head groups interact with water molecules and the nonpolar hydrocarbon tails cluster within the micelle and are removed from contact with water. One possible arrangement for phospholipids also is micelle formation (Figure 11.2).

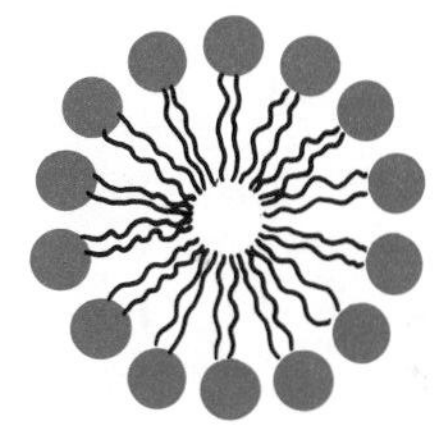

Figure 11.2 *Micelle formation of phospholipids in aqueous medium.*

Another arrangement which also satisfies the requirement that polar groups interact with water and nonpolar groups avoid water is the so-called lipid bilayer or bimolecular sheet. A schematic diagram of a lipid bilayer is shown in Figure 11.3.

The favored structure for most phospholipids in aqueous solution is the lipid bilayer rather than the micelle because micelles can only grow to a limited size before holes begin to appear in the outer polar

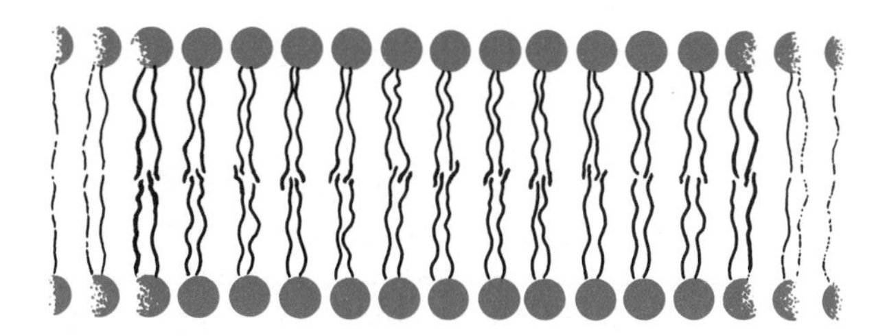

Figure 11.3 *Lipid bilayer or bimolecular sheet formed by phospholipid molecules in aqueous medium.*

surface. Lipid bilayers can grow to almost infinite size and provide a boundary surface for the cell, whatever its size.

It is important to realize that the self-assembly of phospholipid molecules into the lipid bilayer is a spontaneous process and the information necessary to assemble individual molecules into bilayers is inherent in the very structures of the phospholipid molecules themselves. Specifically, this information is contained in the particular combination of a polar head and nonpolar tails in each molecule. This spontaneous self-assembly is driven by two types of noncovalent interactions: (1) hydrophobic interactions when nonpolar chains unite to exclude water molecules, and (2) electrostatic interactions and hydrogen bonding between the polar head groups and water molecules.

Hydrophobic interactions are not bonds in the usual sense of the word because the hydrocarbon chains do not interact or bond among themselves. Rather, hydrophobic interactions result from the particular properties of water molecules. Placing a nonpolar molecule in water will cause the surrounding water molecules to become more ordered and restricted in their motions, i.e., the entropy of the solution is decreased. Removing the nonpolar molecule from water will leave the water molecules more free to move about, i.e., the entropy of the solution is increased. Thus, clustering together of nonpolar molecules or groups occurs not because the nonpolar groups interact with themselves but rather because the surrounding water molecules can return to a more random and disordered state.

As we might expect from these structural characteristics, lipid bilayers are highly impermeable to ions and most polar molecules, for it would take a great deal of energy (work) to transport an ion or a polar molecule through the nonpolar interior of the bilayer. One exception is water, which readily passes in and out of the lipid bilayer. Glucose passes through lipid bilayers 10,000 times more slowly than water, and sodium ion 1,000,000,000 times more slowly than water.

It is the protein components that are responsible for transporting ions and polar molecules across cell membranes and for the multitude of processes by which a cell communicates with its external environment. Proteins act as the "pumps" and "gates" to transport certain molecules across the membranes but exclude others. Proteins also act as receptor sites by which certain molecules on the outside of the cell communicate messages to the inside of the cell. For example, the polypeptide hormone insulin (Section 16.3) regulates the cellular uptake of glucose in certain target cells yet does not cross the cell membrane. Instead insulin interacts with specific receptor proteins on the outer surface of the membrane, and the receptor proteins in turn communicate the insulin-borne message to the inside of the cell. The lipid bilayer surface also supports enzymes for specific intracellular processes. For example, the process of photosynthesis is catalyzed by a series of enzymes fixed to the inner membrane of chloroplasts. Oxidative phosphorylation is catalyzed by a series of enzymes located on the inner surface of the mitochondrion.

The most satisfactory current model for the arrangement of proteins and phospholipids in plant and animal cell membranes is the so-called fluid-mosaic model. According to this model, the membrane phospholipids form a lipid bilayer which is highly impermeable to virtually all polar molecules and ions. The membrane proteins are imbedded in this bilayer. Some proteins are exposed to the aqueous environment on the outer surface of the membrane, others provide channels that penetrate from the outer to the inner surface of the membrane, while still others are imbedded within the lipid bilayer. Four possible protein arrangements are shown schematically in Figure 11.4.

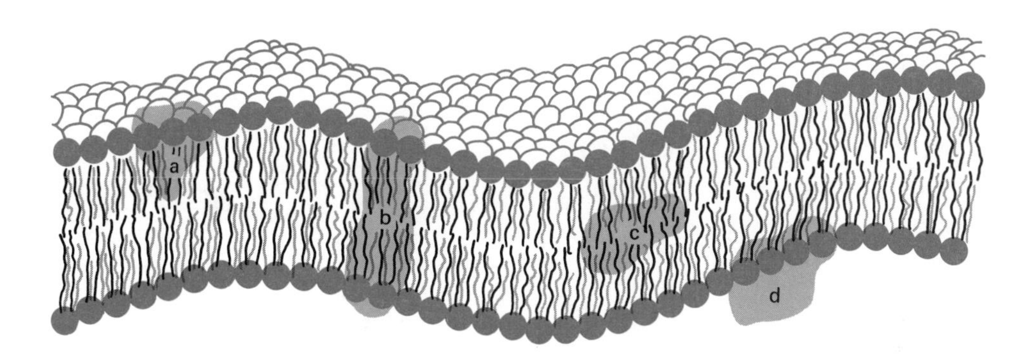

Figure 11.4 *Fluid-mosaic model of a cell membrane showing the lipid bilayer and membrane proteins oriented (a) on the outer surface, (b) penetrating the entire thickness of the membrane, (c) embedded within the membrane, and (d) on the inner surface of the membrane.*

The fluid-mosaic model is consistent with the evidence provided by chemical analysis and electron microscope pictures of cell membranes. However, this model does not explain just how membrane proteins act as pumps and gates for the transport of ions and molecules across the membrane, or how they act as receptors for hormone-borne messages and communications between one cell and another. Nor does it explain how enzymes bound on membrane surfaces catalyze the reactions they do. All of these questions are very active areas of research today.

11.5 FAT-SOLUBLE VITAMINS

Vitamins are divided into two broad classes on the basis of solubility: those that are water-soluble, and those that are fat-soluble. The water-soluble vitamins include ascorbic acid, nicotinamide, pantothenic acid, thiamine, riboflavin, pyridoxine, lipoic acid, biotin, the folic acid group, and vitamin B_{12}. In most cases, the water-soluble vitamins serve as components of coenzymes and the role of these coenzymes is quite well understood. We shall discuss the structure and function of several of these vitamin-derived coenzymes in Chapters 15–18. The fat-soluble vitamins include vitamins A, D, E, and K. At the present time, the molecular basis of their function is only poorly understood.

Vitamin A, or retinol, is a primary alcohol of molecular formula

$C_{20}H_{30}O$. Notice that its structural formula can be divided into four isoprene units and it therefore can be classed as a terpene. Vitamin A alcohol occurs only in the animal world, where the best sources are cod-liver oil and other fish-liver oils, animal liver, and dairy products. Vitamin A in the form of a precursor or provitamin is found in the plant world in a group of plant pigments called carotenes. The carotenes are part of the pigments of many green and yellow vegetables. One such carotene, β-carotene, is shown in Section 3.6 in the discussion of the terpene hydrocarbons. The carotenes themselves have no vitamin-A activity. However, after ingestion in the diet, they are cleaved in the intestinal mucosa at the central double bond into vitamin A.

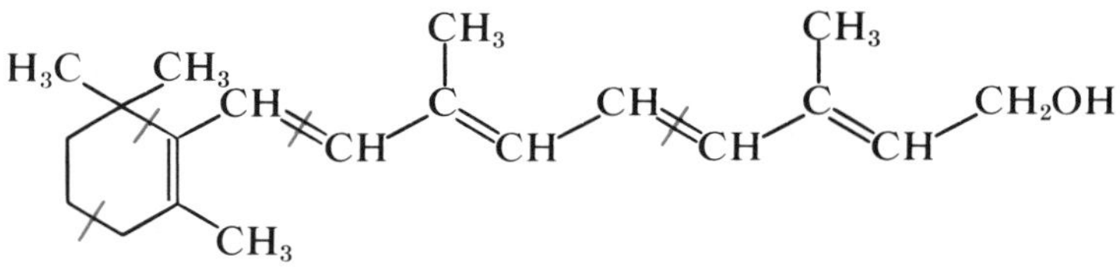

vitamin A
retinol

A deficiency of vitamin A or vitamin-A precursors leads to a slowing or stopping of growth. Probably the major action of this vitamin is on epithelial cells, particularly those of the mucous membranes of the eye, respiratory tract, and genitourinary tract. Without adequate supplies of vitamin A, these mucous membranes become hard and dry, a process known as keratinization. One of the first and most obvious effects of vitamin-A deficiency is on the eye. The cells of the tear glands become keratinized and stop secreting tears, and the external surface of the eye becomes dry, dull, and often scaly. Without tears to remove bacteria, the eye is much more susceptible to serious infection. If this condition is not treated in time, blindness results. The mucous membranes of the respiratory, digestive, and urinary tracts also become keratinized in vitamin-A deficiency and susceptible to infection.

A less serious condition but one frequently seen in humans whose diets contain insufficient vitamin A is night blindness. This is the inability to see in dim light or to adapt to a decrease in light intensity. The role of vitamin A in the visual process is discussed in the mini-essay "The Chemistry of Vision."

Vitamin D is a secondary alcohol of molecular formula $C_{28}H_{44}O$. Its primary effect is on calcium metabolism. Vitamin D increases the absorption of calcium ions (Ca^{2+}) from the intestinal tract and is necessary for normal calcification of bone tissue. Rickets is a vitamin-D deficiency disease in which the growing parts of bones are affected, producing bowlegs, knock-knees, and enlarged joints. Because of its relationship to rickets, vitamin D is often called the antirachitic vitamin.

There are at least ten different compounds with antirachitic activity. These are designated D_1, D_2, etc. The two of greatest importance are D_2 or ergocalciferol of vegetable origin, and D_3 or cholecalciferol of animal origin.

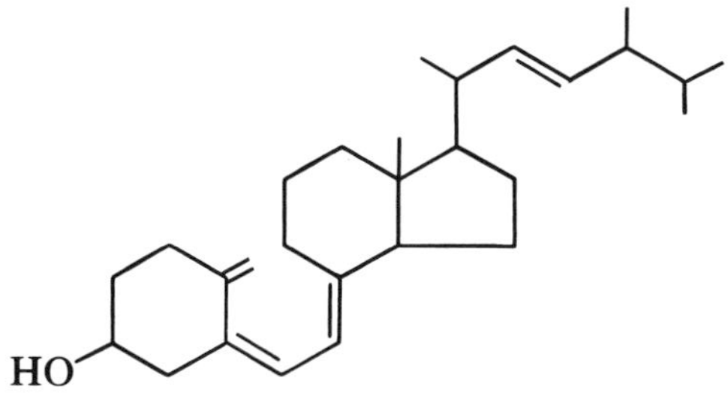

ergocalciferol
D_2

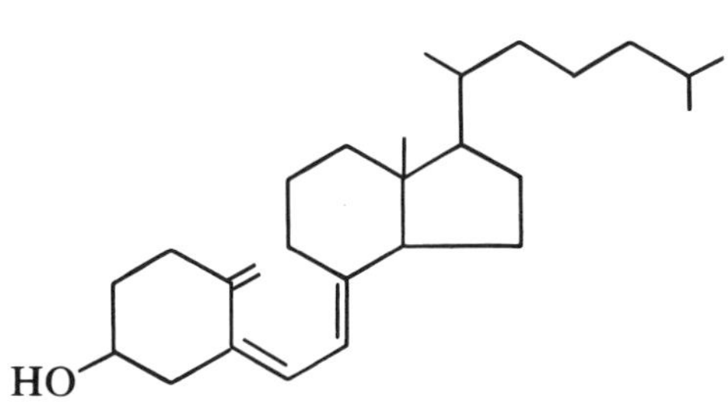

cholecalciferol
D_3

D_3 is the form of the vitamin found in fish-liver oils, the richest natural source. Vitamin D_3 is also produced in the skin by irradiation of 7-dehydrocholesterol. It is for this reason that exposure to sunlight greatly increases the vitamin-D content of the body.

Vitamin E is actually a group of about seven compounds of similar structure. Of these, α-tocopherol has the greatest potency. Notice that a part of the structural formula of α-tocopherol can be divided into four isoprene units.

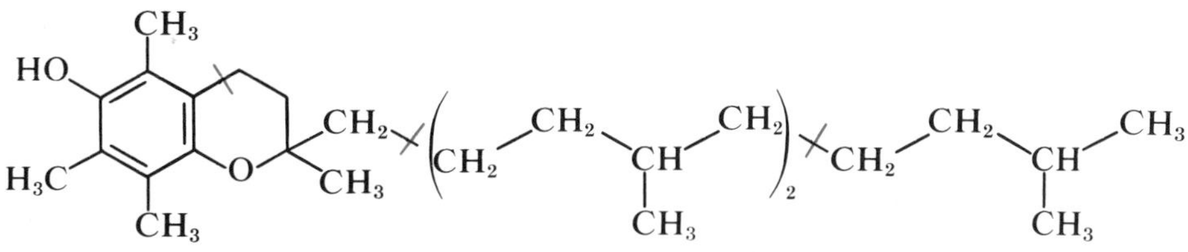

vitamin E (α-tocopherol)

Vitamin E was first recognized in 1922 as a dietary substance essential for normal reproduction in rats. From this observation came its name (Greek, *tocopherol*, promoter of childbirth). Vitamin E occurs in fish oil, in other oils such as cottonseed and peanut oil, and in green leafy vegetables. The richest source of vitamin E is wheat germ oil. In the body, vitamin E functions as an antioxidant in that it inhibits the oxidation of unsaturated lipids by molecular oxygen. In addition it is necessary for the proper development and functioning of membranes in red blood cells, muscle cells, etc.

Vitamin K was discovered in 1935 as a result of a study of newly hatched chicks that had a fatal hemorrhagic disease. This condition could be prevented and cured by the administration of a substance found in hog liver and in alfalfa. It was later discovered that the delayed clotting time of the blood was caused by a deficiency of

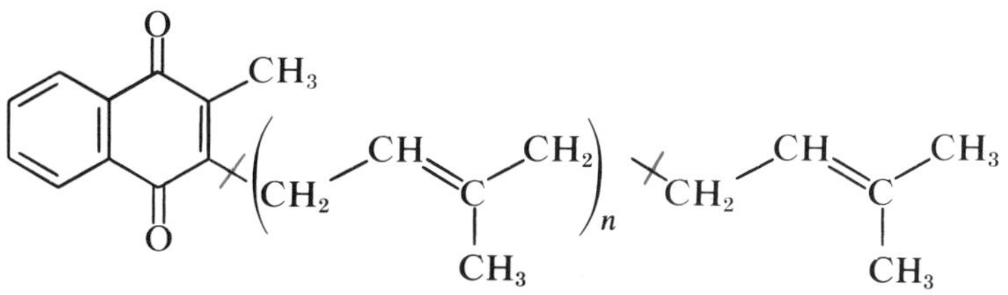

vitamin K_2 (n may be 5, 6, or 8)

prothrombin. It is now known that vitamin K is essential for the synthesis of prothrombin in the liver. The natural vitamin has a long, branched alkyl chain of usually 20 or 30 carbon atoms.

The natural vitamins of the K family have for the most part been replaced by synthetic preparations. Menadione, one such synthetic material with vitamin-K activity, has only hydrogen in the place of the alkyl chain.

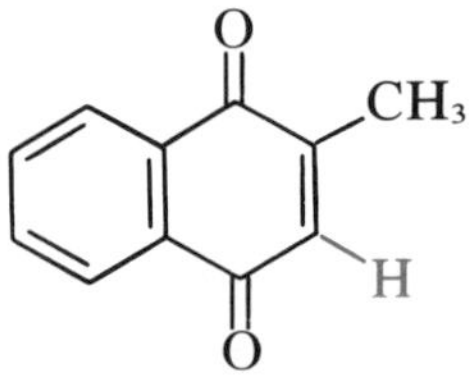

menadione

11.6 STEROIDS

All steroids contain four fused carbon rings: three rings of six carbon atoms and one ring of five carbon atoms. These 17 carbon atoms make up the structural unit known as the steroid nucleus. Figure 11.5 shows both the numbering system and the letter designation for the steroid nucleus.

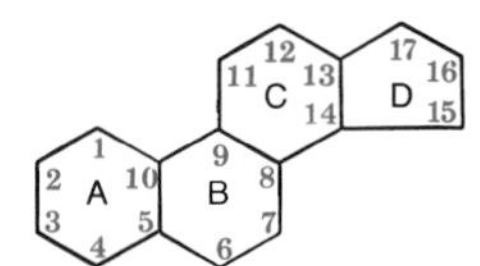

Figure 11.5 *The steroid nucleus.*

The steroid nucleus is found in a number of extremely important biomolecules. For our discussion, we will divide these into four groups: cholesterol, the adrenocorticoid hormones, the sex hormones, and bile acids.

11.7 CHOLESTEROL

Cholesterol is a white, optically active solid found in varying amounts in practically all living organisms except bacteria. In animal cells it serves (1) as an essential component of membrane structures, and (2) as the precursor of bile acids, steroid hormones, and vitamin D. In

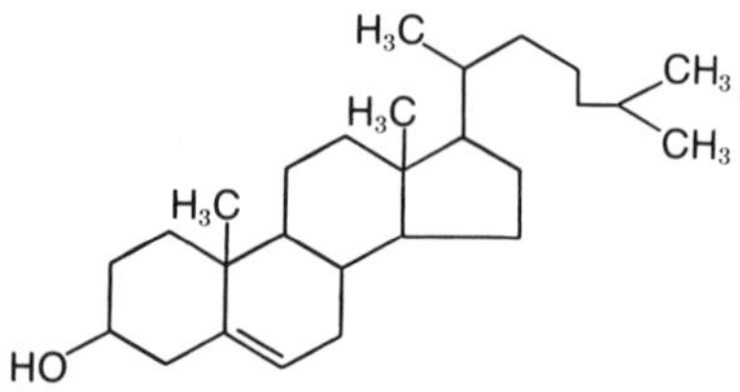

Figure 11.6 *Cholesterol.*

man, the central and peripheral nervous systems have a very high cholesterol content (about 10% of dry brain weight). Human plasma contains an average of 50 mg of free cholesterol per 100 ml and about 170 mg of cholesterol esterified with fatty acids. Gallstones are almost pure cholesterol.

Most organisms, including man, have the ability to synthesize cholesterol in a complex series of reactions beginning with acetate molecules in the form of acetyl coenzyme A.

$$\text{acetate units in the form of acetyl CoA} \xrightarrow[\text{enzyme-catalyzed reactions}]{\text{complex series of}} \text{cholesterol}$$

In fact, biosynthesis of cholesterol provides the major portion of man's and other animals' need for this steroid.

Since it is relatively easy to measure the concentration of cholesterol in serum, a great deal of information has been collected in attempts to correlate serum levels with various diseases. One of these diseases, arteriosclerosis or "hardening of the arteries," is among the most common diseases of aging. With increasing age, humans normally develop decreased capacity to metabolize fat, and therefore cholesterol concentration in the tissues increases. When arteriosclerosis is accompanied by the build-up of cholesterol and other lipids on the inner surfaces of arteries, the condition is known as atherosclerosis and results in a decrease in the diameter of the channels through which blood must flow. This decreased diameter together with increased turbulence leads to a greater probability of clot formation within the channel. If the channel is blocked by a clot, cells may be deprived of oxygen and die. The death of tissue in this way is called infarction.

Infarction can occur in many different tissues and the clinical symptoms depend upon which vessels and tissues are involved. Myocardial infarction, which is the most common, involves the arteries of the heart. When tissues die, the cells rupture and enzymes from them leak into the surrounding fluids. Measurement of the blood plasma levels of these enzymes provides a valuable and simple method to tell if and when infarction of heart, muscle, or other tissue has occurred. We shall discuss the use of enzymes in clinical chemistry in Section 13.9.

It is well known that those suffering from atherosclerosis usually have high blood cholesterol levels. However, this is not always the case. Not all patients with atherosclerosis have high blood cholesterol levels, and not all persons with high cholesterol levels have atherosclerosis. Consequently there are different theories on the origin and treatment of this disease. Blood cholesterol levels do respond directly to the amount and type of fat in the diet. People on low-fat diets have lower cholesterol levels than those on higher fat diets and the incidence of atherosclerosis in these people is lower. In addition, the amount of carbohydrate and the relative proportion of unsaturated to saturated fatty acids in the diet significantly affect blood cholesterol levels in man.

11.8 GENERAL CHARACTERISTICS OF HORMONES

Communication is one of the most important problems in any multicellular organism. For example, in the human body it is absolutely essential that the various tissues communicate with each other so that each can perform its particular role with maximum efficiency. For this communication there are two prime channels. One is the central nervous system. The other is by chemical means and involves the synthesis and release of substances called hormones. Actually these two systems, the neurologic and the hormonal, are closely interrelated; working together they regulate our metabolism, growth, and development.

Hormones are produced by special glands or cells such as the adrenals, ovaries, testes, pituitary, and the thyroid. They are generally classified into three groups:

1. Steroids. These all have the four-ring steroid nucleus and are derived from cholesterol.
2. Derivatives of amino acids. These include thyroxine and epinephrine which are derived from the aromatic amino acid tyrosine.
3. Polypeptides and proteins (Section 12.7).

Before we discuss the structure and function of steroid hormones in this chapter and polypeptide and protein hormones in Chapter 12, let us briefly describe the general characteristics of hormones. First, hormones act by regulating a critical or rate-limiting step in a metabolic pathway. For example, release of adrenocorticotropic hormone (ACTH) stimulates increased production of enzymes that catalyze the biosynthesis of steroid hormones. Another hormone, glucagon, increases blood glucose levels by activating enzymes that catalyze the hydrolysis of glycogen.

Second, hormones are not secreted at a uniform rate. Usually they are secreted only in response to some biological or environmental change. For example, both insulin and glucagon are secreted in response to changes in the plasma concentration of glucose—insulin when blood glucose is high, glucagon when blood glucose is low. Aldosterone is secreted in response to changes in the plasma levels of sodium ion.

Third, hormones exert their effects at very low or catalytic concentrations, as little as 10^{-6} molar. Further, there is a tremendous biological magnification of the hormone-borne message. A single liver cell may respond to just a few molecules of glucagon, yet the cell will produce hundreds or even thousands of glucose molecules in response to this signal.

Fourth, the turnover or half-life of a hormone, once it is secreted into plasma, is very short. This must be, for otherwise accumulation of hormone would lead to a continuing and even increasing response. It is estimated that the half-life of insulin, once it is secreted from the beta cells of the pancreas, is 7–10 minutes.

Fifth, only certain cells or tissues will respond to a given hormone. In the presence of the same hormone, other cells simply do not respond. Those cells responding are called target cells.

Sixth, some hormones cause responses from target cells without ever penetrating the cell membrane. Instead, they activate receptor proteins in the cell membrane (Section 11.4) which transmit the message to the interior of the cell and trigger a response in the cytoplasm. Insulin is one such hormone. Other hormones penetrate not only the cell membrane but also the nuclear membrane and act at the gene level. For example, the glucocorticoid hormones penetrate the cell and nuclear membranes and act at the chromosomal level to stimulate the production of messenger RNA. This mRNA in turn directs the synthesis of specific enzyme proteins. Finally, some hormones may have both cytoplasmic and nuclear (chromosomal) effects.

With this background let us now look at two classes of steroid hormones, the adrenocorticoid hormones and the sex hormones.

11.9 ADRENOCORTICOID HORMONES

The cortex of the adrenal gland is stimulated by the pituitary hormone ACTH to synthesize several hormones that affect (1) water and electrolyte balance and (2) carbohydrate and protein metabolism. Those that control mineral balance are called mineralocorticoids; those that control glucose and carbohydrate balance are called glucocorticoid hormones. Both groups of hormones are derived from cholesterol and have the four-ring nucleus common to all steroids.

Cortisol or hydrocortisone is the major glucocorticoid hormone in man, and aldosterone is the major mineralocorticoid hormone (Figure 11.7). Cortisol and other glucocorticoid hormones act to increase the

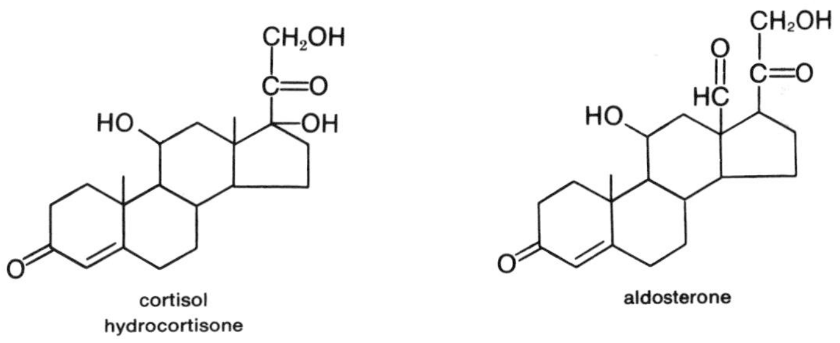

cortisol
hydrocortisone

aldosterone

Figure 11.7 Adrenocorticoid hormones: cortisol and aldosterone.

supply of glucose and glycogen at the expense of body protein. In the presence of cortisol, the synthesis of protein in muscle tissue is depressed, protein degradation is increased, there is an increase in free amino acids in muscle cells as well as in blood plasma, and increased synthesis of glucose. The synthesis of glucose from noncarbohydrate precursors, including amino acids, is called gluconeogenesis and is discussed further in Section 16.3.

Aldosterone is the most effective of the mineralocorticoid hormones in regulating electrolyte balance. This hormone acts on the kidneys to stimulate sodium reabsorption, and the overall effect is Na^+ retention and K^+ excretion.

11.10 SEX HORMONES

The testes in the male and ovaries in the female, besides producing spermatozoa or ova, also produce steroid hormones which control secondary sex characteristics, the reproductive cycle, and the growth and development of accessory reproductive organs.

Of the male sex hormones or androgens, testosterone is the most important. It is produced in the testes from cholesterol. The chief function of testosterone is to promote normal growth of the male reproductive organs and development of the characteristic deep voice, pattern of facial and body hair, and male type of musculature.

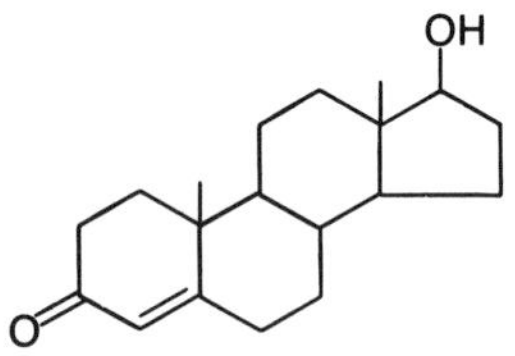

Figure 11.8 Testosterone.

In the female there are two types of sex hormones of particular importance, progesterone and a group of hormones known as estrogens. Changing rates of secretion of these hormones cause the periodic change in the ovaries and uterus known as the menstrual cycle. Immediately following menstrual flow, increased estrogen secretion causes growth of the lining of the uterus and ripening of the ovum. Estradiol is one of the most important estrogens, which are also responsible for development of the female secondary sex characteristics.

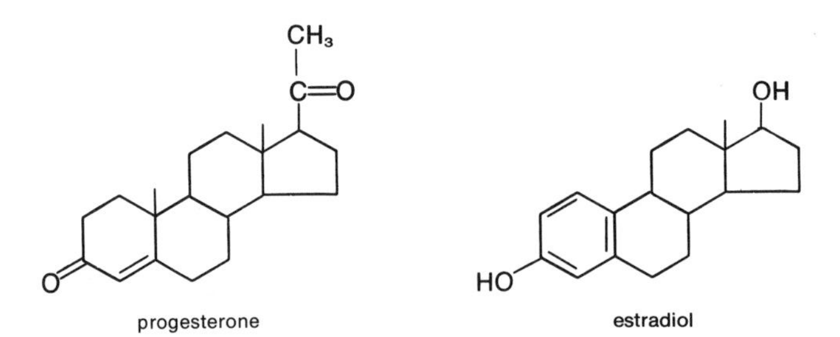

Figure 11.9 Progesterone and estradiol, two female sex hormones.

Progesterone is synthesized by the oxidation of cholesterol. Its secretion just prior to ovulation prevents other ova from ripening and also prepares the uterus for implantation and maintenance of a fertilized egg. If conception does not occur, progesterone production decreases and menstruation occurs. If fertilization and implantation do occur, production of progesterone continues and helps to maintain the pregnancy. One of the consequences of continued progesterone production is prevention of ovulation during pregnancy.

Once the role of progesterone in inhibiting ovulation was understood, its potential as a possible contraceptive drug was realized. Unfortunately, progesterone itself is relatively ineffective when taken orally and injection often produces local irritation. As a result of massive research programs, a large number of synthetic steroids that could be administered orally became available in the early 1960s. When taken regularly, these drugs prevent ovulation yet allow most women a normal menstrual cycle. Some of the most effective contain a progesterone-like analog such as ethynodiol diacetate (Figure 11.10) combined with a smaller amount of an estrogen-like material. The small amount of estrogen prevents irregular menstrual flow ("break-through bleeding") during prolonged use of contraceptive pills.

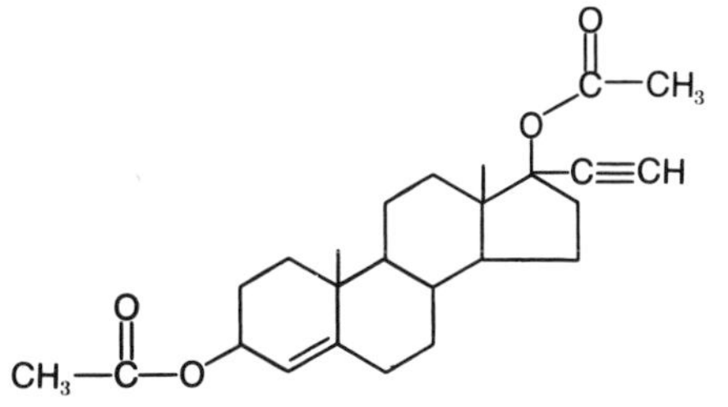

Figure 11.10 Ethynodiol diacetate, a progesterone analog widely used in oral contraceptive preparations.

11.11 BILE ACIDS

These important compounds are synthesized in the liver from cholesterol and then stored in the gallbladder. During digestion, the gallbladder contracts and supplies bile to the small intestine by way of the bile duct. The primary bile acid in humans is cholic acid (Figure 11.11).

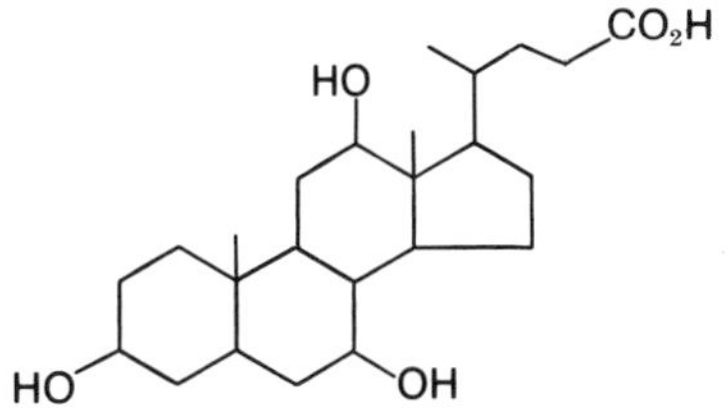

Figure 11.11 Cholic acid, an important constituent of human bile.

Bile acids have several important functions. First, they are products of the breakdown of cholesterol and thus are a major pathway for the elimination of cholesterol from the body via the feces. Second, because they are able to emulsify fats in the intestine, bile acids aid in the digestion and absorption of dietary fats. Third, they can dissolve cholesterol by the formation of cholesterol–bile salt micelles or cholesterol–lecithin–bile salt micelles. In this way cholesterol, whether it is from the diet, synthesized in the liver, or removed from circulation by the liver, can be solubilized.

Problems

11.1 List six major functions of lipids in the human body. Name and draw a structural formula for a lipid representing each function.

11.2 How many isomers (including stereoisomers) are possible for a triglyceride containing one molecule each of palmitic, stearic, and oleic acid?

11.3 What is meant by the term "hardening" as applied to fats and oils?

11.4 Saponification (Section 7.7) is the alkaline hydrolysis of naturally occurring fats and oils. A saponification number is the number of mg of potassium hydroxide required to saponify 1 g of a fat or oil. Calculate the saponification number of tristearin (p. 258), molecular weight 890.

11.5 The saponification number of butter is about 230; that of oleomargarine is about 195.

a Calculate the average molecular weight of butter; of oleomargarine.
b Show that these molecular weight values are in agreement with the fact that butter differs from oleomargarine in having appreciable amounts of lower-molecular-weight fatty acids (Table 11.1).

11.6 Calculate the percentage of unsaturated fatty acids in butter fat. Compare this with the percentage in olive oil.

11.7 Draw a structural formula for a phosphatidyl serine; a phosphatidyl inositol.

11.8 Draw structural formulas for the products of complete hydrolysis of a lecithin; a cephalin.

11.9 Two of the major noncovalent forces directing the organization of biomolecules in aqueous solution are the tendencies to (1) arrange polar groups so that they can interact with water by hydrogen bonding, and (2) arrange nonpolar molecules or groups so that they are shielded from water. Show how these forces direct micelle formation by soap molecules and lipid bilayer formation by phospholipids.

11.10 Describe the major features of the fluid-mosaic model of cell membrane structure.

11.11 Describe three major functions of the protein components of cell membranes.

11.12 Examine the structural formula of vitamin A and state the number of geometric isomers possible for this molecule.

11.13 Describe the symptoms of severe vitamin-A deficiency.

11.14 Examine the structural formulas of vitamins A, D_2 and D_3, E, and K_2. Based on their structural formulas, would you expect them to be more soluble in water or in olive oil? Would you expect them to be soluble in blood plasma?

11.15 Draw the structural formula of cholesterol, and number the carbon atoms in accord with the IUPAC convention. Label all centers of asymmetry and state the total number of optical isomers that could exist.

11.16 Are humans able to synthesize cholesterol? Explain.

11.17 Esters of cholesterol and fatty acids are normal constituents of blood plasma. The fatty acids esterified with cholesterol are generally unsaturated. Draw the structural formula for cholesteryl oleate.

11.18 Cholesterol is an important component of the lipid fraction of cell membranes. However, its precise function in membranes is unknown. How do you think a cholesterol molecule might be oriented in a cell membrane according to the fluid-mosaic model?

11.19 Describe six general characteristics of hormones.

11.20 Name the six functional groups in cortisol; in aldosterone.

11.21 Examine the structural formulas of testosterone, a male sex hormone, and progesterone, a female sex hormone. What are the similarities in structure between the two? the differences?

11.22 Why are progesterone and the estrogens called "sex hormones"?

11.23 Describe how a combination of progesterone and estrogen analogs functions as an oral contraceptive.

11.24 Examine the structural formula of cholic acid and account for the fact that this and other bile acids are able to emulsify fats and oils.

12

AMINO ACIDS AND PROTEINS

12.1 INTRODUCTION

Proteins, as much as any other class of compounds, are inseparable from life itself. These remarkable molecules function in a variety of separate and distinct ways. Certain fibrous proteins serve as major elements of the structural support system for living organisms. For example, the keratins are the structural proteins of hair, skin, nails, and claws; collagen is the structural protein of the connective tissue of flesh, tendons, and muscle; myosin and actin are major components of the myofilaments of skeletal muscle. Other proteins, the enzymes, catalyze the myriad cellular reactions vital to maintenance and growth of an organism. Certain large globular proteins known as antibodies function as one of the major defense mechanisms of the body. Polypeptide hormones like insulin, glucagon, and oxytocin play essential roles in the regulation of physiological processes.

In this chapter we examine a few of the more important properties of proteins and explore the question of what it is that gives these molecules such a wide range of properties. To do this we shall have to look at what proteins are and how they are put together.

12.2 AMINO ACIDS

Amino acids are substances that contain both a carboxyl group and an amino group. While many types of amino acids are known in nature, it is the α-amino acids that are most significant in the biological world for they are the fundamental units from which proteins are constructed. This can be demonstrated by refluxing any common protein in strong aqueous acid for a number of hours. Under these conditions the protein is hydrolyzed to a mixture of as many as 20 different α-amino acids. The general formula of an α-amino acid is shown in Figure 12.1.

Although Figure 12.1(a) is a common way of writing structural formulas for amino acids, it is not an accurate representation (review your answer to Problem 9.8) for it shows an un-ionized acidic

(a) $R-CH(NH_2)-CO_2H$

(b) $R-CH(NH_3^+)-CO_2^-$

Figure 12.1 General formula for α-amino acids.

group (—CO_2H) and an unprotonated basic group (—NH_2) within the same molecule. These acidic and basic groups will of course react to form an internal salt or "zwitterion," Figure 12.1(b). Note that the zwitterion form of an amino acid has no net charge for it contains one positive and one negative charge. In the remainder of this text we will use the zwitterion formulas for amino acids. Furthermore, when discussing amino acids and proteins important in biological systems, we will show all amino and carboxyl groups as they are ionized at pH 7.4, the physiological pH.

It is readily apparent from Figure 12.1(b) that if the R— group is something other than hydrogen, the amino acids of protein origin contain an asymmetric carbon adjacent to the carboxylic acid. For example, the amino acid serine, where R is —CH_2OH, may exist in two forms, one the mirror image of the other. It has been established that all the amino acids that occur naturally in proteins have the same configuration about the asymmetric α-carbon. With D-glyceraldehyde as a standard, the naturally occurring amino acids have the opposite, or L-configuration. This relationship is illustrated for L-serine.

$$\begin{array}{c} CHO \\ | \\ H—C—OH \\ | \\ CH_2OH \end{array} \qquad\qquad \begin{array}{c} CO_2^- \\ | \\ H_3\overset{+}{N}—C—H \\ | \\ CH_2OH \end{array}$$

D-glyceraldehyde L-serine

While D-amino acids are not found in proteins and are not a part of the metabolism of higher organisms, several are important in the structure and metabolism of lower forms of life. As an example, both D-alanine and D-glutamic acid are structural components of the cell walls of certain bacteria. For a discussion of the cell wall structure of these bacteria, see the mini-essay "The Penicillins."

Table 12.1 shows the names, structural formulas, and standard three-letter abbreviations for each of the 20 common amino acids found in proteins. In this table the amino acids are grouped into three categories according to the nature of their side chains. The nonpolar category includes glycine, alanine, valine, leucine, isoleucine, and proline with aliphatic hydrocarbon chains; phenylalanine with an aromatic side chain; and methionine. The polar but neutral category includes serine and threonine with hydroxyl groups; asparagine and glutamine with amide groups; tyrosine with a phenolic side chain; tryptophan with a heterocyclic aromatic amine side chain; and cysteine with a sulfhydryl group. In the charged polar category are aspartic acid and glutamic acid with acidic side chains; and histidine, lysine, and arginine with basic side chains. Note that at 7.4, the pH of cellular fluids and blood plasma, aspartic acid and glutamic acid bear net negative charges because of the ionization of the side-chain carboxyl groups; lysine and arginine bear net positive charges because of the protonation of the side-chain amino groups.

A few proteins contain special amino acids. For example, both L-hydroxyproline and L-5-hydroxylysine are important components of collagen but are found in very few other proteins. These and all other

Table 12.1 *The 20 common amino acids of protein origin, grouped by categories.*

Nonpolar Side Chains

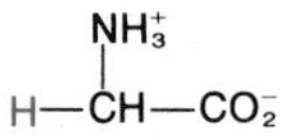

$H-CH(NH_3^+)-CO_2^-$

glycine (gly)

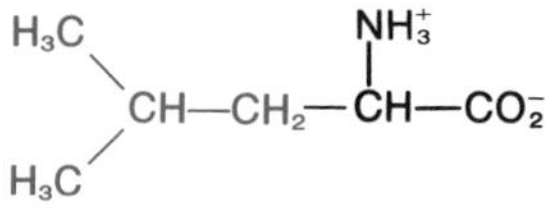

$(H_3C)_2CH-CH_2-CH(NH_3^+)-CO_2^-$

L-leucine (leu)

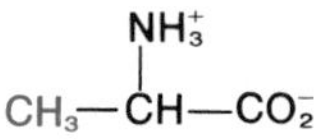

$CH_3-CH(NH_3^+)-CO_2^-$

L-alanine (ala)

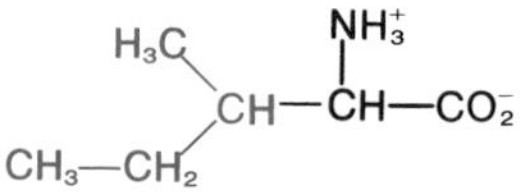

$H_3C(CH_3-CH_2)CH-CH(NH_3^+)-CO_2^-$

L-isoleucine (ile)

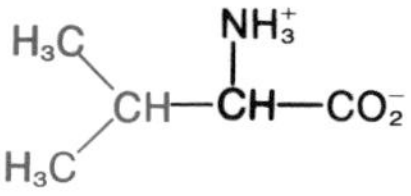

$(H_3C)_2CH-CH(NH_3^+)-CO_2^-$

L-valine (val)

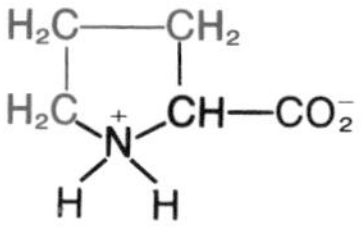

H_2C-CH_2, H_2C, $CH-CO_2^-$, $\overset{+}{N}$, H H (ring)

L-proline (pro)

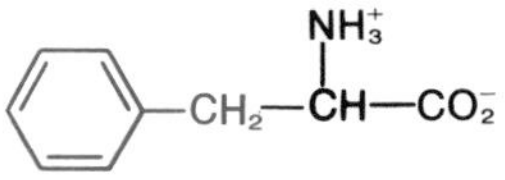

$C_6H_5-CH_2-CH(NH_3^+)-CO_2^-$

L-phenylalanine (phe)

$CH_3-S-CH_2-CH_2-CH(NH_3^+)-CO_2^-$

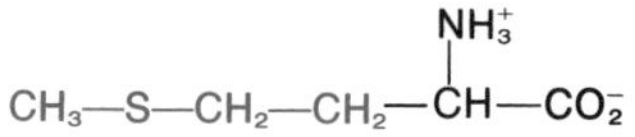

L-methionine (met)

Polar but Uncharged Side Chains

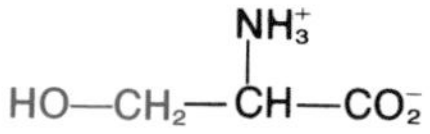

$HO-CH_2-CH(NH_3^+)-CO_2^-$

L-serine (ser)

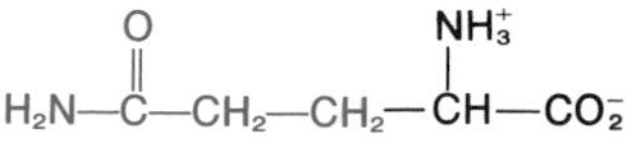

$H_2N-C(=O)-CH_2-CH_2-CH(NH_3^+)-CO_2^-$

L-glutamine (gln)

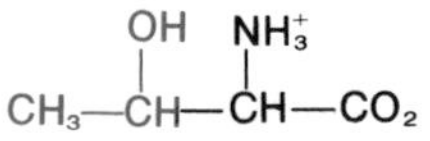

$CH_3-CH(OH)-CH(NH_3^+)-CO_2$

L-threonine (thr)

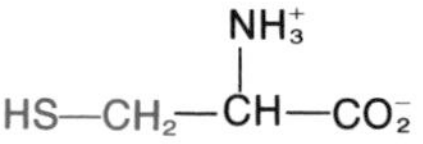

$HS-CH_2-CH(NH_3^+)-CO_2^-$

L-cysteine (cys)

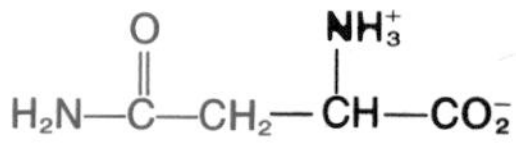

$H_2N-C(=O)-CH_2-CH(NH_3^+)-CO_2^-$

L-asparagine (asn)

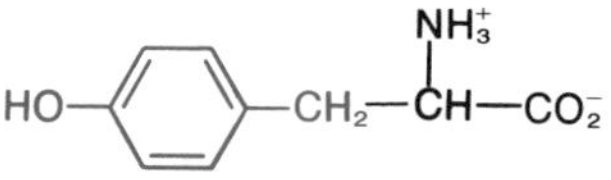

$HO-C_6H_4-CH_2-CH(NH_3^+)-CO_2^-$

L-tyrosine (tyr)

(indole, N–H)$-CH_2-CH(NH_3^+)-CO_2^-$

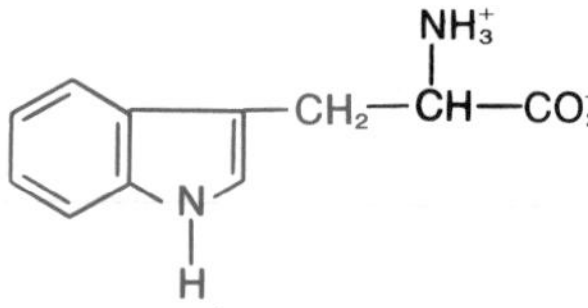

L-tryptophan (trp)

Polar Charged Side Chains

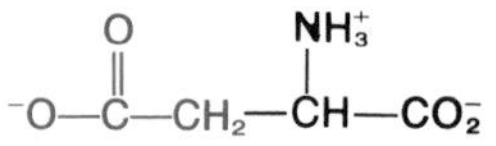

$^-O-C(=O)-CH_2-CH(NH_3^+)-CO_2^-$

L-aspartic acid (asp)

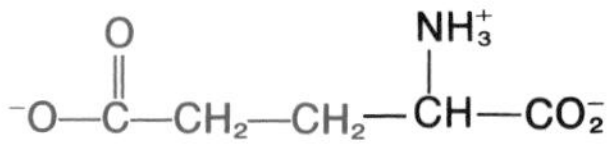

$^-O-C(=O)-CH_2-CH_2-CH(NH_3^+)-CO_2^-$

L-glutamic acid (glu)

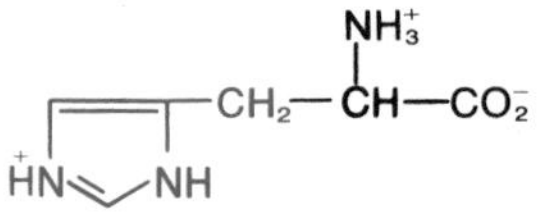

(imidazole, $H\overset{+}{N}$, NH)$-CH_2-CH(NH_3^+)-CO_2^-$

L-histidine (his)

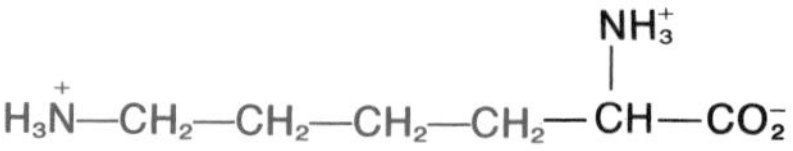

$H_3\overset{+}{N}-CH_2-CH_2-CH_2-CH_2-CH(NH_3^+)-CO_2^-$

L-lysine (lys)

$H_2N-C(=NH_2^+)-NH-CH_2-CH_2-CH_2-CH(NH_3^+)-CO_2^-$

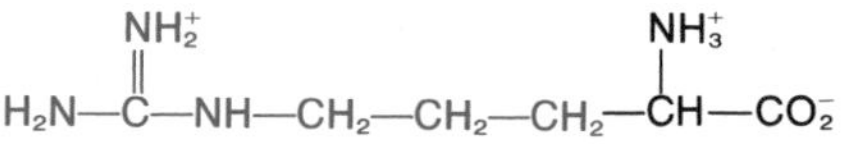

L-arginine (arg)

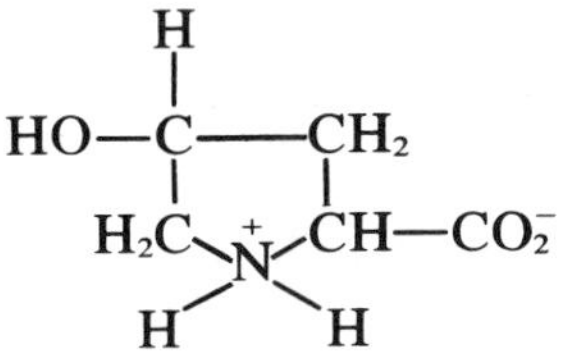

L-hydroxyproline

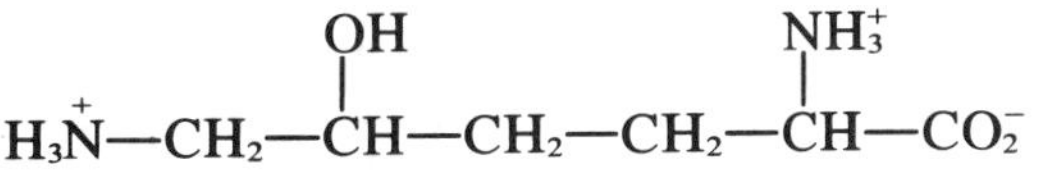

$$H_3\overset{+}{N}-CH_2-\underset{}{\overset{OH}{CH}}-CH_2-CH_2-\overset{NH_3^+}{CH}-CO_2^-$$

L-5-hydroxylysine

special amino acids are formed after the protein is constructed by modification of one of the 20 common amino acids already incorporated into the protein.

In addition to those amino acids listed in Table 12.1, there are a number of important nonprotein-derived amino acids, many of which are either metabolic intermediates or parts of nonprotein biomolecules. Several of these are shown in Table 12.2. Ornithine and citrulline each are part of the metabolic pathway that converts excess ammonia to urea. Gamma-aminobutyric acid (GABA) is present in the free state in brain tissue. Its function in the brain is as yet largely unknown.

Table 12.2 Several nonprotein-derived amino acids.

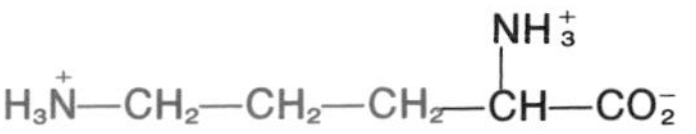

$$H_3\overset{+}{N}-CH_2-CH_2-CH_2-\overset{NH_3^+}{CH}-CO_2^-$$

ornithine (orn)

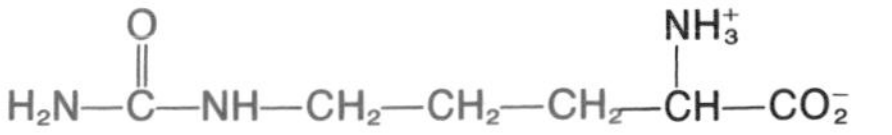

$$H_2N-\overset{O}{\overset{\|}{C}}-NH-CH_2-CH_2-CH_2-\overset{NH_3^+}{CH}-CO_2^-$$

citrulline

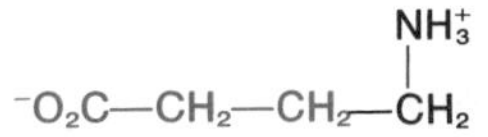

$$^-O_2C-CH_2-CH_2-\overset{NH_3^+}{CH_2}$$

γ-aminobutyric acid (GABA)

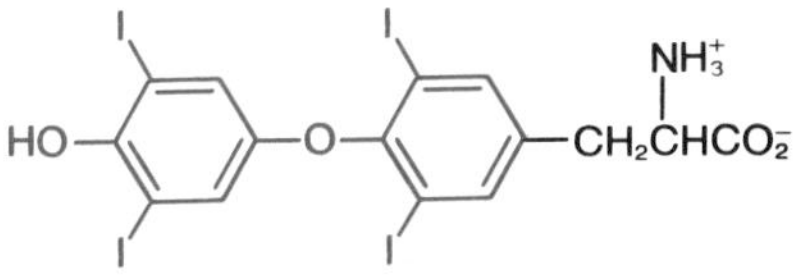

thyroxine
tetraiodothyronine or T_4

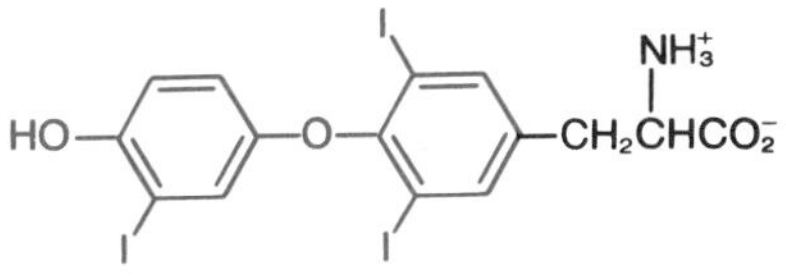

triiodothyronine or T_3

Thyroxine, one of several hormones derived from the amino acid tyrosine, was first isolated from thyroid tissue in 1914. In 1952, triiodothyronine, a compound identical to thyroxine except that it contains only three atoms of iodine, was also discovered in the thyroid. Triiodothyronine is even more potent than thyroxine. Although the exact mechanism of action of these thyroid hormones is not known, they are essential for the proper regulation of cellular metabolism. The levorotatory isomer of each is significantly more active than the dextrorotatory isomer.

12.3 ESSENTIAL AMINO ACIDS

Of the 20 amino acids required by the body for the production of proteins, adequate amounts of about 10 can be synthesized by enzyme-catalyzed reactions starting from carbohydrate or lipid fragments and a source of nitrogen. For the remaining amino acids, either there are no biochemical pathways available for their synthesis, or the available pathways do not provide adequate amounts for proper nutrition. Accordingly, these amino acids must be supplied in the diet, and are therefore called "essential" amino acids. In reality all amino acids are essential for normal tissue growth and development. However, the term "essential" is reserved for those that must be supplied in the diet.

During the late 1950s, C. W. Rose and his coworkers at the University of Illinois determined which amino acids are essential for fully grown young men by first feeding them a well-balanced mixture of the known amino acids in pure form (instead of in proteins) in an otherwise adequate diet. Next, different amino acids were left out of the diet, one at a time, and the effect on nitrogen balance was observed. Rose determined that for fully grown young men, eight amino acids are essential. The estimated minimum daily requirements of these are given in Table 12.3. Note that since phenylalanine is the precursor of tyrosine, the figures given for phenylalanine assume the presence of adequate tyrosine. Similarly, the figures for methionine assume the presence of adequate cysteine. Anyone who consumes about 40–55 grams of protein daily in the form of meat, fish, cheese, or eggs certainly satisfies his need for these essential amino acids.

Table 12.3 *Estimated daily requirements of the essential amino acids.*

Essential Amino Acids	Minimum Daily Requirement infants (mg/kg)	women (grams)	men (grams)
isoleucine	126	0.45	0.70
leucine	150	0.62	1.10
lysine	103	0.50	0.80
methionine	45	0.55	1.01
phenylalanine	90	1.12	1.40
threonine	87	0.30	0.50
tryptophan	22	0.16	0.25
valine	105	0.65	0.80

In later studies, histidine was shown to be essential for growth in infants and there is some evidence that it may be needed by adults as well. Arginine can be synthesized by adults but apparently the rate of internal synthesis is not fast enough to meet the needs of the body during periods of rapid growth and protein synthesis. Therefore, depending on the age and state of health, either eight, nine or ten amino acids may be essential for humans.

The relative usefulness of a dietary protein depends on how well its amino acid pattern matches that required for the formation of tissue protein in humans. For proper tissue maintenance and growth, all amino acids, both essential and nonessential, must be present at the same time. In this sense, tissue growth is an all-or-nothing process; if even one amino acid is missing, no protein at all is made.

The biological value of a dietary protein is a measure of the percentage that is absorbed and can be used to build body tissue. Some of the first information on the biological value of dietary proteins came from studies on rats. In one series of experiments, young rats were fed diets containing 18% protein in the form of either casein (a milk protein), gliadin (a wheat protein), or zein (a corn protein). With casein as the sole source of protein, the rats remained healthy and grew normally. Those fed gliadin maintained their weight but did not grow much. Those fed zein not only failed to grow but began to lose weight, and they eventually died if kept on this diet. Since casein evidently supplies all the required amino acids in the correct proportions needed for growth, it is called a complete protein. Analysis of the other two proteins revealed that gliadin contains too little lysine and that zein is low in both lysine and tryptophan. When the gliadin diet was supplemented with lysine, or the zein diet with lysine and tryptophan, the test animals grew normally and thrived.

Table 12.4 shows the biological value for rats of some common dietary sources of protein. Note that the mixture of proteins in egg is the best quality natural protein for the maintenance of tissue. In comparison, the mixed proteins of milk rank 84, those of meats and soybeans about 74. The legumes, vegetables, and cereal grains are in the range 50–70.

Table 12.4 *The biological value for rats of some common sources of dietary protein. Biological value is the portion of absorbed protein that is retained as body tissue.*

	Food Protein			
Food	% as purchased	% of dry solid	Kcal, % of total Kcal	Biological Value, %
hen's egg, whole	13	48	33	94
cow's milk, whole	4	27	23	84
fish	19	72	61	83
beef	18	45	29	74
soybeans	38	41	39	73
dry beans, common	22	25	22	58
wheat, whole grain	12	14	13	65
corn, whole grain	10	11	7	59
rice, brown	8	9	7	73
potato, white	2	9	7	67

Plant proteins generally vary more from the amino acid pattern we require than do animal proteins. Fortunately, however, not all plant proteins are deficient in the same amino acids. For example, beans are low in the sulfur-containing amino acids cysteine and methionine yet are high in lysine. Wheat has just the opposite pattern. By eating wheat and beans together, it is possible to increase by 33% the usable protein you would get by eating each of these foods separately.

The provision of a diet adequate in protein and the essential amino acids is a grave problem in the world today, especially in areas of Asia, Africa, and Latin America. The overriding dimension of this problem is poverty and the inability to select foods of adequate protein and caloric content. The best overall source of calories is the cereal grains, which not only provide calories but proteins as well. When these are supplemented by animal protein or the right selection of plant protein, the diet is adequate for even the most vulnerable. However, as income decreases, there is less animal protein in the diet and even the cereal grains are often replaced by cheaper sources of calories such as sugar or tubers, which have either very little or no protein. The poorest 25% of the people in the world consume diets with caloric and protein content that falls below, often dangerously below, the calculated minimum daily requirements.

Those most apt to show the symptoms of too little food, too little protein, or both are young children in the years immediately following weaning. There is failure to grow properly and a wasting of tissue. This sickness is called marasmus, a name derived from a Greek word meaning "to waste away." The muscles become atrophied and the face develops a wizened "old man" look. Another disease, kwashiorkor, leads to tragically high death rates among children. As long as a child is breast fed it is healthy. At weaning (often forced when a second child is born), the first child's diet suddenly is switched to starch and inadequate sources of protein. Such children develop bloated bellies and patchy, discolored skin, and are often doomed to short lives.

One attack on the problem of quantity and quality of protein is the breeding of new varieties of cereal grains with higher protein content, better protein quality, or both. Alternatively, it is possible to supplement present cereals or their derived products with the deficient amino acids—principally lysine for wheat, lysine or lysine plus threonine for rice, and lysine plus tryptophan for corn. New methods of chemical synthesis and fermentation now provide a cheap source of these amino acids, thus making the economics of food fortification entirely practical. In another attack on the problem of protein malnutrition, several special high-protein, low-cost infant foods have been developed by teams of nutritionists. These take advantage of locally available foods that have supplementary amino acid compositions and conform as closely as possible to cultural food preferences. Clearly, advances in chemistry and food technology have provided the means to eradicate most hunger and malnutrition. What remains is for political and social systems to put this knowledge into practice.

12.4 TITRATION OF AMINO ACIDS

Glycine and all other monoamino monocarboxylic amino acids contain —COOH and $—NH_3^+$ groups that can ionize in aqueous solution, and each group will have a specific ionization or dissociation constant. The value of these dissociation constants can be determined by titration. In order to illustrate how this might be done, imagine that a solution contains 1 mole of glycine and that enough strong acid is added so that each ionizable group on glycine is fully protonated. Next this solution is titrated with aqueous sodium hydroxide; the volume of base added and the pH of the resulting solution are recorded and then plotted as shown in Figure 12.2.

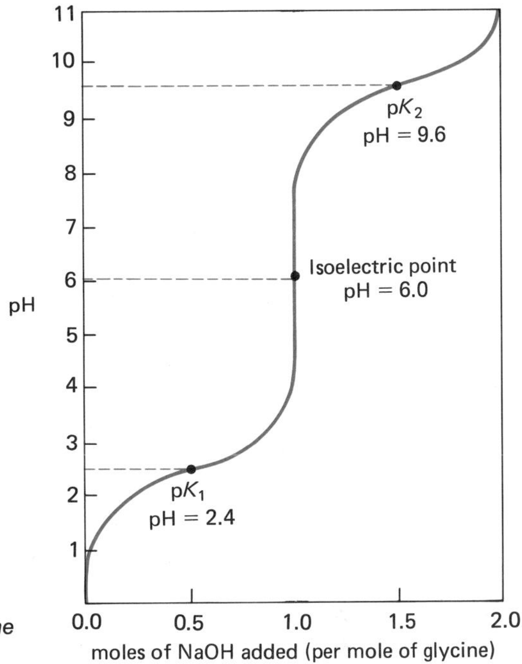

Figure 12.2 *Titration of glycine with sodium hydroxide.*

The most acidic group and the one to react first with added sodium hydroxide is —COOH. At pH 2.4, enough sodium hydroxide has been added to half-neutralize the —COOH group. Recall from Section 7.13 that the pH at which this half-neutralization occurs is equal to the pK_a of the acid; accordingly, the pK_1 of glycine (the pK_a of the —COOH group) is 2.4. When 1 full mole of sodium hydroxide has been added, glycine is in the zwitterion form and has no net charge. The pH at which a molecule has no net charge is called its isoelectric point, and is denoted by pI. Accordingly, the isoelectric point of glycine is 6.0. Addition of more sodium hydroxide converts the —NH_3^+ group to —NH_2. Half-neutralization of —NH_3^+ occurs at pH 9.6 and accordingly the pK_2 of glycine (the pK_a of the —NH_3^+ group) is 9.6. This acid–base behavior of the glycine molecule is shown in Figure 12.3.

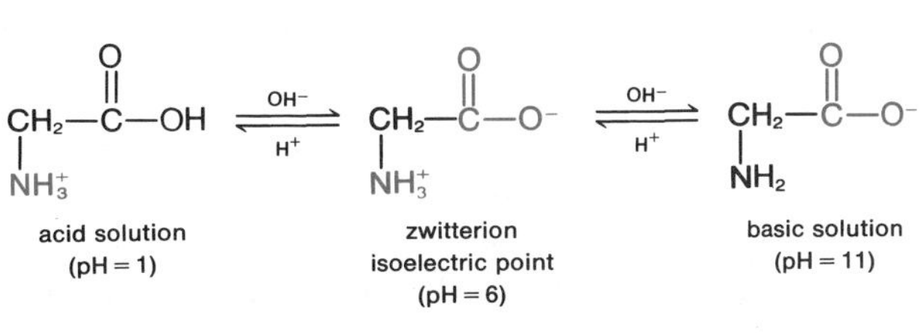

$$\underset{NH_3^+}{\underset{|}{CH_2}}\text{—}\overset{O}{\overset{\|}{C}}\text{—OH} \;\underset{H^+}{\overset{OH^-}{\rightleftharpoons}}\; \underset{NH_3^+}{\underset{|}{CH_2}}\text{—}\overset{O}{\overset{\|}{C}}\text{—}O^- \;\underset{H^+}{\overset{OH^-}{\rightleftharpoons}}\; \underset{NH_2}{\underset{|}{CH_2}}\text{—}\overset{O}{\overset{\|}{C}}\text{—}O^-$$

acid solution (pH = 1) | zwitterion isoelectric point (pH = 6) | basic solution (pH = 11)

Figure 12.3 *Acid-base behavior of glycine.*

The monoamino monocarboxylic acids have isoelectric points in the pH range 5.5–6.5 (6.0 for glycine). The basic amino acids have isoelectric points at higher pH values (e.g., 9.74 for lysine), and the acidic amino acids have isoelectric points at lower pH values (e.g., 3.22 for glutamic acid).

An important feature of amino acids is that they can act as buffers not at just one pH but at two, and in the case of those with ionizable groups on their side chains, at three different pH ranges. As we have just illustrated, glycine can act as a buffer around pH 2.4 and also around pH 9.6. The imidazole group on the side chain of histidine has a pK_a of 6.0 and is therefore a particularly effective buffer around pH 7.1–7.4, the pH of cellular fluids and blood plasma.

An understanding of this acid–base behavior of amino acids and proteins is important for two reasons. First, it helps us to understand the solubility of these molecules as a function of pH. While amino acids generally are quite soluble in water, their solubility reaches a minimum at the isoelectric point. Put another way, the zwitterion is the least soluble form of an amino acid. In order to crystallize an amino acid or protein, the pH of an aqueous solution is adjusted to the pI and the substance is precipitated, filtered, and collected. This process is known as isoelectric precipitation.

Second, knowing the isoelectric point can also be useful for separating mixtures of amino acids or proteins since it permits us to predict the way each component of the mixture will migrate in an electrical field. This process of separating substances on the basis of their electrical charges is called electrophoresis and is illustrated in Figure 12.4.

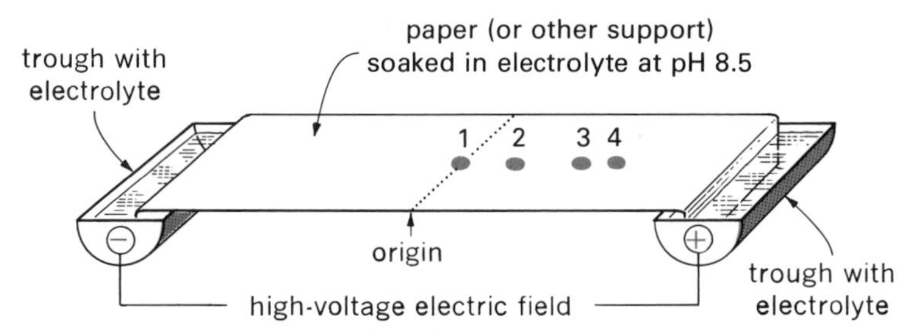

Figure 12.4 Paper electrophoresis of a protein mixture carried out at pH 8.5. Component 1 does not migrate from the origin and therefore must have an isoelectric point of 8.5. The other three components of the mixture move toward the positive electrode and therefore have net negative charges at pH 8.5. Component 4 has the greatest negative charge density. From David S. Page, Principles of Biological Chemistry (Willard Grant Press, Boston, 1976).

In paper electrophoresis, a paper strip saturated with an aqueous buffer of predetermined pH serves as a bridge between two electrode vessels. A sample of protein or amino acid is applied as a spot. When an electrical potential is applied, the protein or amino acid molecules migrate toward the electrode carrying the opposite charge. Molecules having a greater density of either positive or negative charge will move more rapidly to the opposite electrode. Any molecule already at its isoelectric point will remain stationary at the origin. After separation is complete, the strip is dried; to fix the separated components on the paper and to render them visible, the paper is generally sprayed with a dye. For example, ninhydrin can be used to make amino acids visible as purple spots.

Electrophoretic separations also can be carried out using starch, agar, certain plastics, and cellulose acetate as solid supports. This technique is extremely important in biochemical research and also is an invaluable tool in the clinical chemistry laboratory. For a discussion of the electrophoretic screening of blood samples for sickle-cell anemia, see the mini-essay "Abnormal Human Hemoglobins."

12.5 THE PEPTIDE BOND

We owe our earliest formulation of the manner in which amino acids are bonded in proteins to the great German chemist Emil Fischer. He postulated in 1902 that proteins are long sequences of amino acids joined in linear fashion by amide linkages between the carboxyl group of one amino acid and the α-amine of another. Such linkages between amino acids are called peptide bonds. Figure 12.5 illustrates the peptide bond formed between glycine and alanine in the peptide glycylalanine.

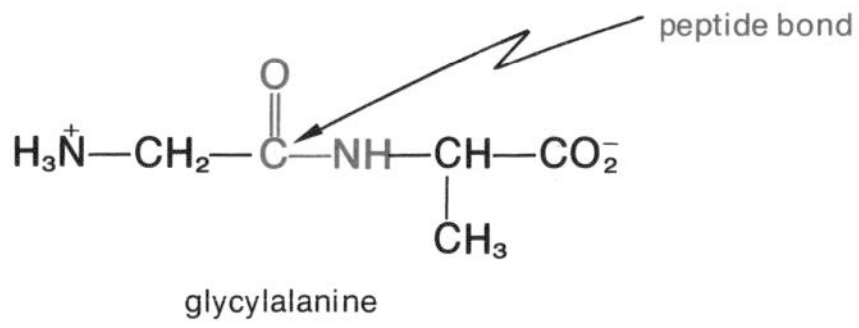

$$H_3\overset{+}{N}-CH_2-\overset{O}{\overset{\|}{C}}-NH-\underset{CH_3}{\underset{|}{CH}}-CO_2^-$$

peptide bond

glycylalanine

Figure 12.5 The peptide bond in glycylalanine.

A peptide such as glycylalanine contains two amino acids and is called a dipeptide. Those containing larger numbers of amino acids are called tripeptides, tetrapeptides, pentapeptides, etc. Peptides containing 10 or more amino acids generally are called polypeptides.

Proteins are biological macromolecules, generally of molecular weight 5,000 or greater, that consist of one or more polypeptide chains. In addition, many proteins also contain nonamino acid groups as integral parts of their structure. For example, myoglobin (Figure 12.17) has within its structure a porphyrin ring which is essential for its biological activity.

By convention, polypeptides generally are written beginning with the free $-NH_3^+$ group on the left and proceeding to the right toward the free terminal $-CO_2^-$ group. Alternatively, the standard abbreviations of the amino acids may be used, each connected by an arrow. The tail of the arrow indicates the amino acid contributing the carboxyl group to the peptide bond and the head of the arrow indicates the amino acid contributing the amino group.

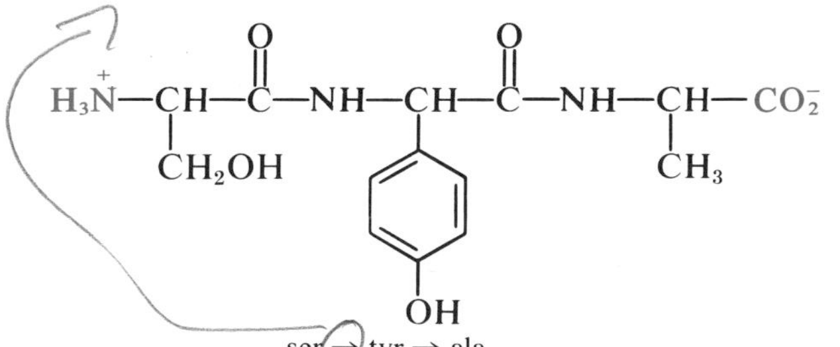

$$H_3\overset{+}{N}-\underset{CH_2OH}{\underset{|}{CH}}-\overset{O}{\overset{\|}{C}}-NH-\underset{C_6H_4OH}{\underset{|}{CH}}-\overset{O}{\overset{\|}{C}}-NH-\underset{CH_3}{\underset{|}{CH}}-CO_2^-$$

ser → tyr → ala

Given the fact that proteins are linear sequences of amino acids joined by peptide bonds, the next questions we might ask are: What is the sequence of amino acids along the polypeptide chain? And what is the detailed three-dimensional arrangement of the polypeptide chain within the protein molecule?

12.6 AMINO ACID SEQUENCE

To appreciate the problem of deciphering the sequence of amino acids along a polypeptide chain, we need only imagine the incredibly large number of different chemical words (proteins) that can be constructed with a 20-letter alphabet, where words can range from under 10 letters to well into hundreds of letters. With only three amino acids there are six entirely different tripeptides possible. If we choose glycine, alanine, and serine, the six possible tripeptides are:

gly → ala → ser	ala → ser → gly
gly → ser → ala	ser → gly → ala
ala → gly → ser	ser → ala → gly

For a polypeptide containing one each of the 20 different amino acids, the number of possible molecules runs to $20 \times 19 \times 18 \cdots \times 2 \times 1$ or about 2×10^{18}. With larger polypeptides and proteins, the number of possible arrangements becomes truly astronomical! Until recently the possibility of decoding the exact amino acid sequence for any protein seemed staggeringly complicated.

The first major attack on the problem of sequence was undertaken by Frederick Sanger in the early 1940s for the hormone insulin. It was known at the time that insulin contains 16 of the 20 known amino acids and has a molecular weight of 6000. It was also known that the insulin molecule contains a total of 51 amino acid units or residues.

Sanger's first approach was to determine the amino acid residue at the N-terminal end of the chain, and for this purpose he introduced the reagent 1-fluoro-2,4-dinitrobenzene. This reagent, now known as Sanger's reagent, reacts readily in mildly alkaline solution with free amino groups to form a linkage that is stable to hydrolysis.

$$O_2N\text{-}C_6H_3(NO_2)\text{-}F + H_2N\text{-}CHR\text{-protein} \rightarrow O_2N\text{-}C_6H_3(NO_2)\text{-}NH\text{-}CHR\text{-protein} + HF$$

1-fluoro-2,4-dinitrobenzene (Sanger's reagent)

hydrolysis in strong acid ↓

$$O_2N\text{-}C_6H_3(NO_2)\text{-}NH\text{-}CHR\text{-}CO_2H + \text{amino acids}$$

dinitrophenylamino acid

Insulin and Sanger's reagent were reacted and the resulting product was hydrolyzed in strong acid. The labeled dinitrophenylamino acid is yellow, ether soluble, and easily identified. Sanger was able to show that insulin contained not one but two different labeled amino acids. This could only mean that insulin contains not one but two polypeptide chains, one (Chain B) with phenylalanine and the other (Chain A) with glycine as the N-terminal amino acids. Sanger further reasoned that since insulin contains six cysteine residues, the two chains must be cross-linked by at least one disulfide bridge. The two chains were separated by breaking the disulfide bonds and each chain was isolated. The glycine chain was found to contain 21 amino acid residues and the phenylalanine chain 30 residues.

Since there were no methods for taking the chains apart one amino acid at a time, Sanger undertook to hydrolyze each chain into a number of smaller polypeptides and then determine the sequence of amino acids in these smaller subunits. By knowing the structure of these smaller polypeptides, Sanger was able to determine the amino acid sequence in both chains. An example of the technique of matching overlapping units is illustrated with an octapeptide from the A chain. By matching the peptide fragments 1–7, he was able to deduce the structure of the octapeptide:

1. ile → val → glu → gln
2. gln → cys → cys → ala
3. gly → ile → val → glu
4. cys → cys → ala
5. glu → gln
6. val → glu
7. ile → val

---

8. gly → ile → val → glu → gln → cys → cys → ala

After a decade of work, enough peptides were identified to permit Sanger to draw the complete structure of the bovine insulin molecule (Figure 12.6). The sequence of amino acids in insulin reveals no particular pattern; it appears to be just one of an almost infinite variety of possible sequences. Yet somewhere in the sequence lies the key to its physiological potency in the regulation of blood glucose levels. Where in the sequence this potency lies we do not know.

The structures of human, beef, pig, sheep, horse, and sperm whale insulin have been determined, and they reveal some interesting species characteristics. Chain B is the same in each of these insulins. The only differences are in the amino acids at positions 8, 9, and 10 in the A chain (Table 12.5). Apparently these amino acids do not have unique roles and substitution is possible without radically altering the physiological behavior of insulin. This is fortunate since diabetics who become allergic to one type of insulin may begin using insulin from another source and thereby avoid allergic reactions.

In the decades since Sanger's pioneering work on the structure of the insulin molecule, several new techniques have been devised for

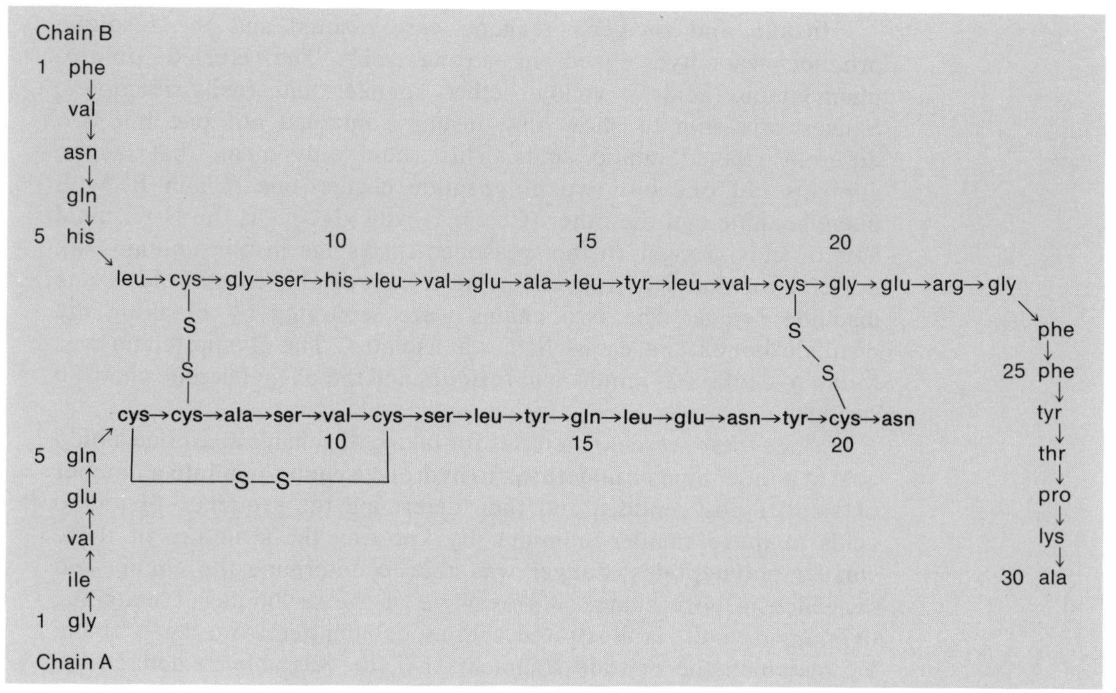

Figure 12.6 *The amino acid sequence of bovine insulin and the position of the —S—S— cross-linkages. Chain A is composed of 21 amino acids, has glycine (gly) at the* $—NH_3^+$ *terminus and asparagine (asn) at the* $—CO_2^-$ *terminus. Chain B is composed of 30 amino acids, has phenylalanine (phe) at the* $—NH_3^+$ *terminus and alanine (ala) at the* $—CO_2^-$ *terminus.*

Table 12.5 *Amino acid variations in insulin Chain A.*

Animal	Amino acid residues of Chain A 8	9	10
man	threonine	serine	valine
beef	alanine	serine	valine
pig	threonine	serine	isoleucine
sheep	alanine	glycine	valine
horse	threonine	glycine	isoleucine
sperm whale	threonine	serine	isoleucine

sequence determinations. It is now possible to cleave amino acid residues one at a time from the protein chain. The enzyme carboxypeptidase selectively hydrolyzes the peptide bond adjacent to the free $—CO_2^-$ on the C-terminal end of the chain. The enzyme aminopeptidase selectively cleaves the peptide bond adjacent to the free $—NH_3^+$ group on the N-terminal end of the chain. The enzymes chymotrypsin, trypsin, and pepsin catalyze the hydrolysis of polypeptide chains at specific points, thus leading to polypeptide fragments that are easier to isolate in pure form. These and other useful methods have enabled biochemists to determine the sequence of amino acids in an increasingly large number of proteins.

12.7 POLYPEPTIDE AND PROTEIN HORMONES

In this section we shall look at the structure and function of five polypeptide and protein hormones: vasopressin, oxytocin, glucagon, ACTH, and human growth hormone. These are by no means all of the important human polypeptide and protein hormones. However, they will illustrate something of the structure and function of this class.

Both oxytocin and vasopressin (Figure 12.7) are nonapeptide hormones secreted by the pituitary gland. The primary function of

cys→tyr→ile→gln→asn→cys→pro→leu→glyNH$_2$
S—S (linking the two cys residues)
bovine oxytocin

cys→tyr→phe→gln→asn→cys→pro→arg→glyNH$_2$
S—S (linking the two cys residues)
bovine vasopressin

Figure 12.7 *Two peptide hormones. In each the carboxyl terminus is glycinamide, indicated above by glyNH$_2$.*

vasopressin is to increase blood pressure by regulating the excretion of water through the kidneys. The hormone itself is synthesized in the hypothalamus and then migrates to the pituitary gland where it is stored in granules. In response to a rise in osmotic pressure or a decrease in total blood volume, vasopressin is released into the blood stream and transported to the kidneys where it interacts with certain cells to decrease loss of fluid in the urine. Conversely, a decrease in secretion of the hormone permits more water to be lost in the urine. Damage to the hypothalamus (the site of vasopressin synthesis) or to the pituitary (the site of vasopressin storage) can result in the excretion of up to several liters of urine per day. This condition is known as diabetes insipidus.

Oxytocin acts on the smooth muscles of the uterus to enhance contraction and is often administered at childbirth to induce delivery. Oxytocin also acts on the smooth muscles of lactating mammary glands to stimulate milk ejection.

Notice that the structures of vasopressin and oxytocin are similar. In fact, they differ only in the amino acids at positions 3 and 8. Because of this similarity, their physiological actions overlap. For example, vasopressin stimulates contraction of the pregnant uterus, but not nearly so strongly as oxytocin. Vasopressin also stimulates milk ejection, but again not nearly so strongly as oxytocin. By comparing the physiological functions of these two polypeptide hormones, we can see that only slight changes in structure can lead to major changes in function.

Another polypeptide hormone, glucagon (Figure 12.8) has a profound effect on glycogen metabolism. Glucagon is secreted by the alpha cells of the pancreas during the fasting state when blood glucose levels are decreasing. This hormone stimulates the enzymes that catalyze the hydrolysis of glycogen to glucose and thus helps to maintain blood glucose levels within a normal concentration range. Glucagon contains 29 amino acids and has a molecular weight of about 3,500.

1 5 10 15
his—ser—glu—gly—thr—phe—thr—ser—asp—tyr—ser—lys—tyr—leu—asp—

16 20 25 29
—ser—arg—arg—ala—gln—asp—phe—val—gln—trp—leu—met—asn—thr

glucagon

Figure 12.8 The amino acid sequence of glucagon.

Adrenocorticotropic hormone (ACTH) is one of the most important hormones synthesized by the anterior lobe of the pituitary gland. It consists of a single polypeptide chain of 39 amino acids with no disulfide cross-links and has a molecular weight of 4,500 (Figure 12.9).

1 10 20
ser—tyr—ser—met—glu—his—phe—arg—trp—gly—lys—pro—val—gly—lys—lys—arg—arg—pro—val—

21 24 25 30 39
—lys—val—tyr—pro—asp—ala—gly—glu—asp—gln—ser—ala—glu—ala—phe—pro—leu—glu—phe

bovine ACTH

Figure 12.9 The amino acid sequence of bovine ACTH.

The major function of ACTH is to stimulate the biosynthesis of steroid hormones in the cortex of the adrenal gland. This hormone has been prepared in pure form from the pituitary glands of sheep, pigs, and cattle. It has been possible to cleave ACTH at specific sites and to study the biological activity of various fragments. These studies have revealed that the full biological activity of the hormone is present in the amino acid sequence 1–24. In other words, the amino acid sequence 25–39 is not necessary for the hormonal action of ACTH. The pituitary gland stores about 300×10^{-6} grams of ACTH and secretes it at the rate of 10×10^{-6} grams per day. The level of ACTH in circulating blood plasma is 0.03×10^{-9} grams per ml.

Human growth hormone (HGH, or more simply GH) is another hormone of the anterior lobe of the pituitary. It consists of a single polypeptide chain of 190 amino acids cross-linked in two places by disulfide bonds. Unlike ACTH, which is quite specific, growth hormone acts on many tissues including muscle, adipose tissue, cartilage, and connective tissue. It is known to stimulate the synthesis of collagen (Section 12.12) as well as the metabolism of proteins, carbohydrates, and lipids. The normal concentration of human growth hormone in circulating plasma is about 2.5×10^{-9} grams per ml.

12.8 CONFORMATIONS OF PROTEIN CHAINS

In the late 1930s, Linus Pauling and his collaborators set out to determine by means of X-ray crystallographic studies the geometry of the peptide bond and to gain a clearer understanding of the patterns of protein chain folding. Pauling began with the study of crystalline amino acids and simple di- and tripeptides. One of his first discoveries was that the peptide bond itself is planar. As shown in Figure 12.10, the four

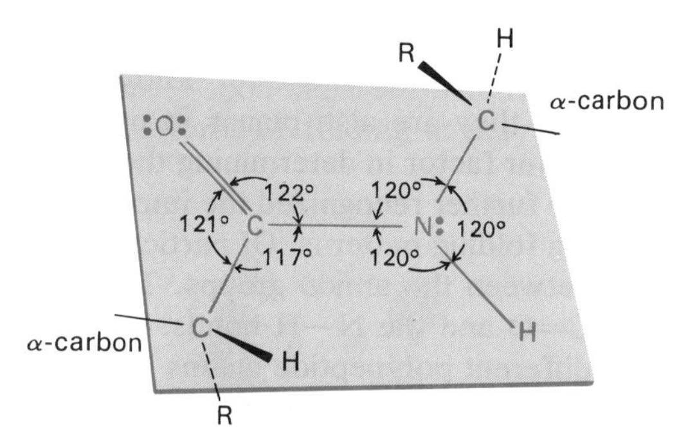

Figure 12.10 *Planarity of the peptide bond. Bond angles about the carbonyl carbon and the amide nitrogen are all approximately 120°.*

atoms of the peptide bond and the two alpha carbons joined to the peptide bond all lie in the same plane.

Had we been asked in Chapter 1 to describe the geometry of the peptide bond, we would have predicted bond angles of approximately 120° about the carbonyl carbon and 109.5° about the amide nitrogen. This prediction is in good agreement with the observed bond angles of 120° for the carbonyl carbon; however, the geometry of the amide nitrogen is unexpected. To account for this observed geometry, Pauling proposed that the peptide bond is more accurately represented as a resonance hybrid of two important contributing structures.

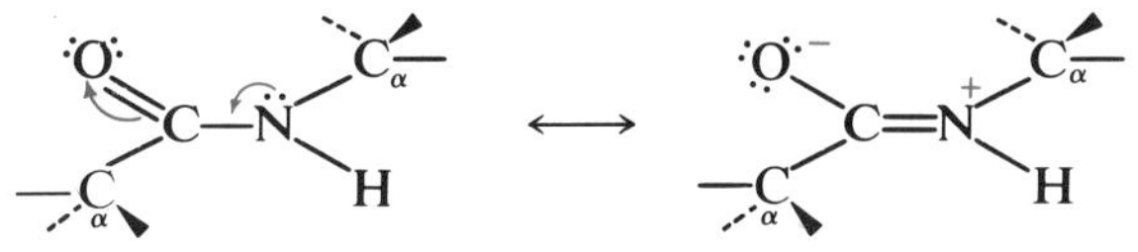

I ⟷ II

The hybrid, of course, is neither of these, but in it the carbon–nitrogen bond has at least partial double-bond character. Accordingly, in the hybrid the six-atom group is planar.

There are two possible planar arrangements of the atoms of the peptide bond. In one configuration, the two α-carbons are at diagonally opposite corners of the amide plane. This is called a *trans* configuration.

cis *trans*

In the other, a *cis* configuration, the two α-carbons are *cis* to each other in the amide plane. The *cis* configuration appears slightly less favorable, probably because the bulkier α-carbons are in closer proximity. In the *trans* configuration, the bulky α-carbons are maximally separated from each other. The *cis* configuration is found in only a few polypeptides.

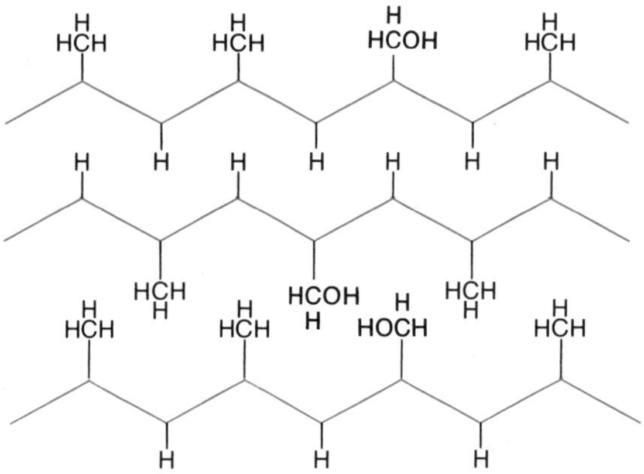

Figure 12.14 Schematic representation of three polypeptide chains in the β-pleated sheet conformation of silk. The conformation of each β-sheet is as represented in Figure 12.13. In this perspective, each is viewed from the side looking through the plane formed by the atoms of the peptide bonds.

side of the sheet are methyl or hydroxymethyl groups. This is represented schematically in Figure 12.14.

In silk fiber, the sheets are stacked with glycine facing glycine (hydrogen against hydrogen) and with the alanines and serines also facing each other. The bulk properties of silk are a consequence of this molecular structure and arrangement. The silk fiber is very strong because tension is borne by the covalent bonds of the polypeptide chains themselves. Silk is resistant to stretching because the chain is already extended as far as it can go. Since there are no strong forces of attraction between the side chains of the amino acid units, the sheets themselves are quite flexible and slip past each other easily.

12.11 WOOL—THE α-HELIX STRUCTURE

The second great class of fibrous proteins is illustrated by hair and wool. This type of fiber is very flexible and in contrast to silk is quite extensible under tension and elastic, so that when tension is released the fiber snaps back to its original condition. At the molecular level, the basic structural unit of hair is the polypeptide chain wound in an α-helix conformation. Furthermore, there are several levels of structural organization built from the simple α-helix. First, it appears that three α-helices are arranged together to form a larger interwoven coil called a protofibril. In the protofibril, each α-helix is itself coiled slightly in much the manner that three strands of fiber are twisted together in rope or cable (Figure 12.15). These protofibrils are then arranged in bundles to form an 11-stranded cable called a microfibril. These in turn are imbedded in a larger matrix that ultimately forms the hair fiber. The α-helices themselves are cross-linked by disulfide bonds between cysteine side chains.

Regarding the macroscopic properties of the fiber itself, we have already noted that wool can be stretched and will spring back on release of tension. What happens in the stretching process is an

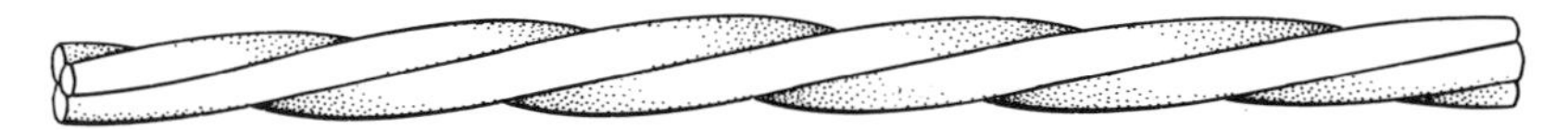

Figure 12.15 The supracoiling of three α-helices in hair and wool to form a protofibril.

elongation of the hydrogen bonds along turns of the α-helix. The major force causing the stretched wool fiber to return to its original length is the reformation of the hydrogen bonds in the α-helices. The α-keratins of horns and claws have essentially the same structure but with a much higher content of cysteine and a greater degree of disulfide bridge cross-linking between the helices. These additional disulfide bonds greatly increase the resistance to stretch and produce the hard keratins of horn and claw.

12.12 COLLAGEN—THE TRIPLE HELIX

The third major class of fibrous proteins contains the collagens and elastins. Collagens are constituents of skin, bone, teeth, blood vessels, tendons, cartilage, and connective tissue. In fact, they are the most abundant of all proteins in higher vertebrates, making up almost 30% of the total body protein mass in humans. The distinctive physical property of collagen is that it forms long, insoluble fibers of very high tensile strength. Table 12.6 lists the collagen content of several tissues. Note that bone, Achilles' tendon, skin, and the cornea of the eye are largely collagen.

Table 12.6 Collagen and elastin content of some tissues.

Tissue	Collagen (% dry weight)	Elastin (% dry weight)
bone, mineral free	88	
Achilles' tendon	86	4.4
skin	72	0.6
cornea	68	
cartilage	46–63	
ligament	17	75
aorta	12–24	28–32

Elastin is also an important structural element of skin, arteries, ligaments, and connective tissue. The distinctive physical property of elastin is that it also forms long, insoluble fibers, but unlike collagen, elastin fibers can be stretched several times their length and will snap back almost like a piece of rubber. The elastin content of several tissues is also listed in Table 12.6. Note that elastic tissues like ligaments and the walls of the aorta are especially rich in elastin.

Because of its abundance and wide distribution in vertebrates and because it is associated with a variety of diseases and the problems of aging, more is known about collagen than probably any other fibrous protein. The collagen molecule is very large and it has a distinctive amino acid composition. One-third of the amino acids in collagen are glycine and another 20% are proline and hydroxyproline. Tyrosine is present in very low amounts and the essential amino acid tryptophan is absent entirely. Cysteine is also absent, so there are no disulfide cross-links in collagen. When collagen fibers are boiled in water, they are converted into soluble gelatins. Gelatin itself has no biological food value because it lacks the essential amino acid tryptophan.

Collagen chains contain an unusually high percentage of proline and hydroxyproline, amino acids in which the α-amino groups are incorporated into a five-membered ring. This ring formation places a constraint on the rotation about the α-carbon and as a consequence, proline and hydroxyproline molecules cannot be rotated properly in a polypeptide chain to fit into an α-helix. Furthermore, because of the relatively high concentrations of amino acids with bulky side chains, these proteins of collagen cannot pack together in the form of β-pleated sheets. However, they can fold into another type of conformation that is particularly stable and unique to collagen. In this conformation, three protein strands wrap around each other to form a left-handed superhelix which looks much like a three-stranded rope.

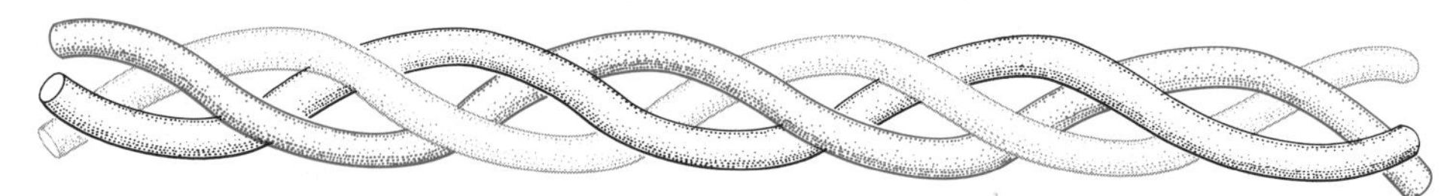

Figure 12.16 The collagen triple helix.

This three-stranded conformation is called the collagen triple helix and the unit itself is called tropocollagen. Collagen fibers are formed when many tropocollagen molecules line up side-by-side in a regular pattern and are then cross-linked by the formation of new covalent bonds. Recall from Section 10.3 that one of the effects of severe ascorbic acid deficiency is impaired synthesis of collagen. Without adequate supplies of vitamin C, cross-linking of tropocollagen strands is inhibited with the result that they do not unite to form stable, physically tough fibers.

The extent and type of cross-linking vary with age and physiological conditions. For example, the collagen of rat Achilles' tendon is highly cross-linked while that of the more flexible tendon of rat tail is much less highly cross-linked. Further, it is not clear when, if ever, the process of cross-linking is completed. Some believe it is a process that continues throughout life, producing increasingly stiffer skin, blood vessels, and other tissues which then contribute to the medical problems of aging and the aged.

12.13 MYOGLOBIN—A GLOBULAR PROTEIN

Myoglobin is the first of the globular proteins for which a three-dimensional structure was determined. This feat represented a milestone in the study of the molecular architecture of proteins, and for this pioneering research, J. C. Kendrew shared in the Nobel Prize in

Chemistry in 1963. Myoglobin is a relatively small globular protein (molecular weight 16,700) which consists of a single polypeptide chain of 153 amino acids. The complete amino acid sequence (the primary structure) of the chain is known. Myoglobin also contains an iron–porphyrin unit (Figure 16.13). It is found in cells of skeletal muscle and is particularly abundant in diving mammals such as seals, whales, and porpoises. Myoglobin and its structural relative hemoglobin (p. 304) are the oxygen transport and storage molecules of vertebrates. Hemoglobin binds molecular oxygen in the lungs and transports it to myoglobin in the muscle cells. Myoglobin stores the molecular oxygen until it is required for metabolic oxidation.

The three-dimensional structure of myoglobin was deduced from X-ray analysis. The first analysis, completed in 1957, revealed that the single peptide chain of myoglobin is bent and folded into a very compact shape with little empty space in the interior of the molecule. A more detailed analysis revealed the exact location of the atoms of the peptide backbone and also of the R— groups. The important structural features of myoglobin are:

1. The backbone consists of eight relatively straight sections of α-helix, each separated by a bend in the polypeptide chain. The longest α-helix section has 23 amino acid residues, the shortest has 7. Some 75% of the amino acid residues are found in these eight regions of α-helix.
2. The peptide bonds are planar with each peptide N—H and C=O lying *trans* to each other.
3. The inside of the molecule consists almost entirely of nonpolar side chains such as those of leucine, valine, methionine, and phenylalanine. Thus hydrophobic interactions appear to be a major factor in determining the three-dimensional shape of myoglobin, i.e., its tertiary structure.
4. The polar side chains of glutamate, aspartate, glutamine, asparagine, lysine, and arginine are on the outside of the molecule and in contact with the aqueous environment. The only two polar side chains that point toward the interior of the molecule are those of two histidines. These side chains can be seen in Figure 12.17 as five-membered rings that point inward toward the heme group.

The three-dimensional structures of several other globular proteins have also been determined and their secondary and tertiary structures analyzed. It is clear that globular proteins contain α-helix and β-pleated sheet structure but that there is wide variation in the relative amounts of each. Lysozyme with 129 amino acids in a single polypeptide chain has only about 25% of its amino acids in α-helix regions. Cytochrome with 104 amino acids in a single polypeptide chain has no regions of α-helix structure but does contain several regions of β-pleated sheet conformation. Yet whatever the proportions of α-helix and β-pleated sheet structure, virtually all nonpolar side chains of globular proteins are directed toward the interior of the molecule while polar side chains are on the outer surface of the molecule and in contact with water. Note that this arrangement of polar and nonpolar

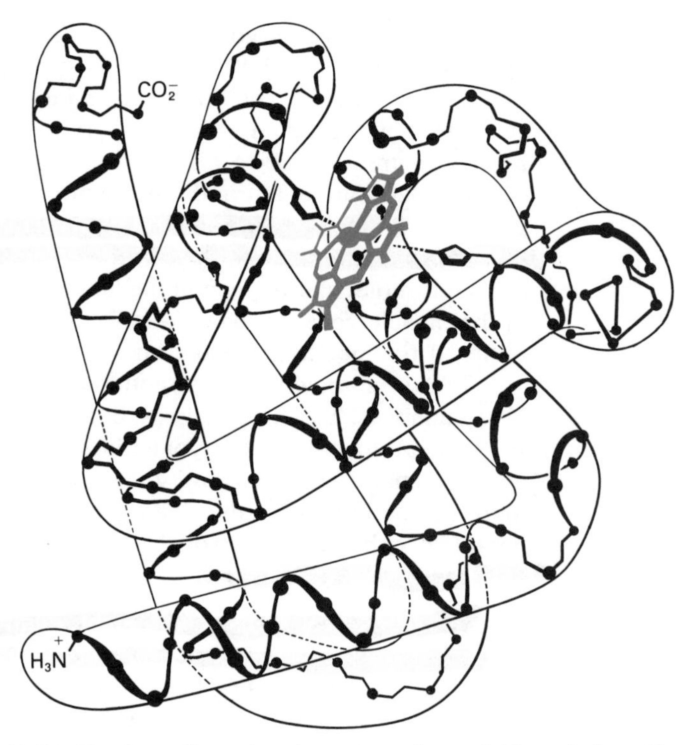

Figure 12.17 *The three-dimensional structure of myoglobin as deduced from X-ray analysis. The heme group is shown in color. The —NH_3^+ terminus is at the lower left, the —CO_2^- terminus at the upper left. Reproduced from R. E. Dickerson, in H. Neurath, Ed., The Proteins, Vol. II (Academic Press, New York, 1964).*

groups in globular proteins very much resembles the arrangements of polar and nonpolar groups of soap molecules in micelles (Figure 7.4) and of phospholipid molecules in lipid bilayers (Figure 11.3).

12.14 QUATERNARY STRUCTURE

Many protein molecules exist as oligomers, or large molecules formed by the assembly of either identical or closely related subunits. This arrangement of protein monomers into an aggregation is known as quaternary structure (Section 12.9). Most proteins of molecular weight greater than 50,000 consist of two or more noncovalently linked chains. Hemoglobin is a good example of such a protein. It exists as a tetramer of four separate protein monomers, two so-called alpha chains of 141 amino acids each and two so-called beta chains of 146 amino acids each. The molecular weights, number of monomeric subunits, and biological functions of several proteins with quaternary structure are shown in Table 12.7.

While there is some stabilization of these quaternary structures by ionic interactions and hydrogen bonding, the major factor stabilizing the aggregation of protein subunits is hydrophobic interaction. Even when the separate monomers fold into a compact three-dimensional shape so as to expose polar side chains to the aqueous environment

Table 12.7 *Quaternary structure of selected proteins.*

Protein	Mol Wt	Number of Subunits	Subunit Mol Wt	Biological Function
insulin	11,466	2	5,733	a hormone regulating glucose metabolism
hemoglobin	64,500	4	16,400	oxygen transport in blood plasma
alcohol dehydrogenase	80,000	4	20,000	an enzyme of alcoholic fermentation
lactic dehydrogenase	134,000	4	33,500	an enzyme of anaerobic glycolysis
aldolase	150,000	4	37,500	an enzyme of anaerobic glycolysis
fumarase	194,000	4	48,500	an enzyme of the tricarboxylic acid cycle
tobacco mosaic virus	40,000,000	2,130	17,500	plant virus coat

and at the same time shield the nonpolar side chains from water, there are still one or more hydrophobic "patches" on the surface and in contact with water. These patches can be shielded from water if two or more monomers assemble so that their hydrophobic patches are in contact. Thus hydrophobic interactions, which are so important in directing the self-assembly of lipid monomers into lipid bilayers, also direct the self-assembly of protein monomers. In addition, this information that directs self-assembly is inherent in the primary structure of the protein monomers themselves.

12.15 DENATURATION

Globular proteins as found in living organisms usually are soluble substances with definite three-dimensional shapes and definite biological activity. Globular proteins are also remarkably sensitive to changes in their environment. Even relatively mild changes in temperature or solvent composition for only short periods of time will cause them to lose some or all of their biological activity, that is, to denature.

Denaturation is a physical change, the most observable result of which is loss of biological activity. At the molecular level, it is the result of unfolding or disruption of secondary, tertiary, and quaternary structure of the biologically active conformation of the native protein. With the exception of cleavage of disulfide bonds, denaturation stems from changes in conformation through disruption of noncovalent interactions, e.g., hydrogen bonds, salt bridges, and hydrophobic interactions. Common denaturing agents include:

1. Heat. Most globular proteins denature when heated above 50–60°C. For example, boiling or frying an egg causes the egg white protein to denature and form an insoluble mass.
2. Large changes in pH. Adding concentrated acid or alkali to a

protein in aqueous solution causes changes in the charged character of the ionizable side chains and interferes with ionic or salt interactions. For example, in certain clinical chemistry tests it is necessary to first remove any protein material. This is done by adding trichloroacetic acid (a strong organic acid) to denature and precipitate any protein present.

3. Detergents. Treatment of protein in aqueous solution with sodium dodecylsulfate (SDS), a common detergent, causes the native conformation to unfold and exposes the nonpolar protein side chains. These side chains are then stabilized by hydrophobic interaction with the long hydrocarbon chain of the detergent.
4. Organic solvents such as alcohols, acetone, or ether. These solvents can participate in hydrogen bonding and can disrupt the hydrogen bonding in the native protein.
5. Mechanical treatment. Most globular proteins denature in aqueous solution if they are stirred or shaken vigorously. An example is the whipping of egg whites to make a meringue.
6. Urea and guanidine hydrochloride.

$$H_2N-\overset{\overset{\displaystyle O}{\|}}{C}-NH_2 \qquad\qquad H_2N-\overset{\overset{\displaystyle NH_2^+\ Cl^-}{\|}}{C}-NH_2$$

urea — guanidine hydrochloride

These reagents cause disruption of protein hydrogen bonding and hydrophobic interactions.

Denaturation may be partial or it may be complete. It may also be reversible or irreversible. The hormone insulin can be denatured with $8M$ urea and the three disulfide bonds cleaved. If the urea is then removed and the disulfide bonds reformed, the resulting molecule has less than 1% of its former biological activity. In this case, the denaturation has been both complete and irreversible. As another example, treating the enzyme aldolase, a tetramer (Table 12.7), with $4M$ urea separates and fully denatures the four subunits. If the urea is then removed from the solution, the protein recovers about 70% of its biological activity. In this example, the subunits not only refold with the proper secondary and tertiary conformations but also assemble with the proper quaternary structure; the denaturation has been complete but reversible.

Problems

12.1 Define and give an example of the following.

a α-amino acid **b** tripeptide **c** peptide bond
d fibrous protein **e** polypeptide **f** globular protein
g nonprotein-derived amino acid **h** zwitterion **i** isoelectric point

12.2 Amino acids of protein origin are designated as L-amino acids. Explain the meaning of the designation L- as it is used to indicate the stereochemistry of amino acids. What is the structural relationship between L-serine and D-glyceraldehyde? L-phenylalanine? D-glucose? D-serine?

12.3 Draw structural formulas for two amino acids that contain more than one center of asymmetry.

12.4 Name the essential amino acids for man. Why are they termed "essential"? Compare meat, fish, and the cereal grains in terms of their ability to supply the essential amino acids.

12.5 Consider the amino acids alanine, lysine, and aspartic acid. Draw the structural formulas for the form of each that you would expect to predominate at 7.4, the physiological pH; at pH 2.0; at pH 10.0.

12.6 Glutamic acid is an acidic amino acid with three ionizable groups. The pK_a values for each are shown below.

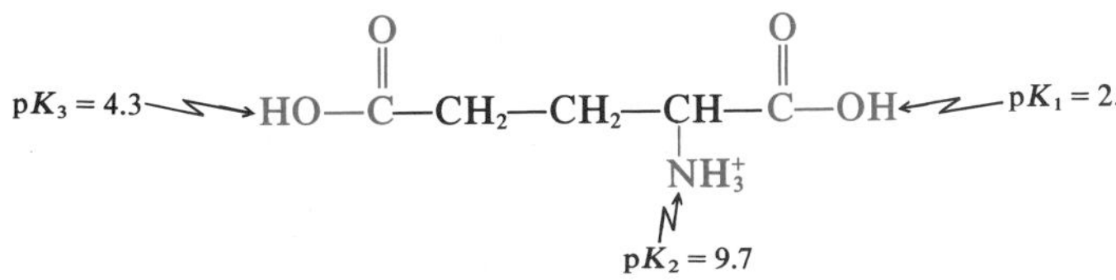

$$pK_3 = 4.3 \rightarrow HO{-}\overset{O}{\overset{\|}{C}}{-}CH_2{-}CH_2{-}\underset{NH_3^+}{\underset{|}{CH}}{-}\overset{O}{\overset{\|}{C}}{-}OH \leftarrow pK_1 = 2.2$$

$$pK_2 = 9.7 \text{ (}NH_3^+\text{)}$$

a Which of these ionizable groups is the strongest acid; the weakest acid?
b Estimate the net charge on glutamic acid at pH 1.0; at pH 3.25; at pH 7.4; at pH 11.0.
c In what pH range(s) will glutamic acid function as a buffer?

12.7 Would you expect an aqueous solution of lysine to be acidic, basic, or neutral? Explain your reasoning. (In thinking about this problem, first consider the effect on pH of the carboxyl and the α-amino groups together in the zwitterion form and then the effect of the terminal amino group.)

12.8 Draw structural formulas for glutamic acid and glutamine. One of these amino acids has an isoelectric point of 5.7, the other has an isoelectric point of 3.2. Which amino acid has which isoelectric point? Explain your reasoning.

12.9 Write the structural formula for the tripeptide, glycylserylaspartic acid. Write the structural formula for an isomeric tripeptide.

12.10 How many tetrapeptides can be constructed from the 20 amino acids (a) if each of the amino acids is used only once in the tetrapeptide; (b) if each amino acid can be used up to four times in the tetrapeptide?

12.11 Examine the amino acid composition of bovine insulin (Figure 12.6) and list the total number and kind of acidic and basic groups in the molecule. Would you predict the isoelectric point of insulin to be nearer that of the acidic amino acids (pH 2.0–3.0), the neutral amino acids (5.5–6.5), or the basic amino acids (9.5–11.0)? Do the same for glucagon and bovine ACTH.

12.12 Would you predict bovine oxytocin or bovine vasopressin to have the higher isoelectric point?

12.13 The isoelectric point of pepsin (one of the digestive enzymes of the stomach) is somewhat less than 1.0. The isoelectric point of hemoglobin is 6.8. What information does this give you about the amino acid composition of each of these proteins?

12.14 Using structural formulas, show how the theory of resonance accounts for the fact that the peptide bond is planar.

12.15 In constructing models of arrangements of polypeptide chains that would be particularly stable, Pauling assumed that for maximum stability (1) all amide bonds would be *trans* and coplanar, and (2) there would be a maximum of hydrogen bonding between amide groups. Examine the β-pleated sheet (Figure 12.13) and the α-helix (Figure 12.12) and convince yourself that in each of these, and amide bonds are planar and that each carbonyl is hydrogen bonded to an amide hydrogen.

12.16 Examine the structure of the α-helix. Are the amino acid side chains arranged all inside the helix, all outside the helix, or randomly oriented?

12.17 Characterize the amino acid composition of silk. Silk fiber is very strong and quite resistant to stretching. How are these properties accounted for in terms of the molecular structure?

12.18 Compare and contrast silk and nylon in terms of molecular structure and bulk physical properties.

12.19 If a significant number of the glycines in silk were replaced by amino acids having much larger side chains, what effect would you expect this substitution to have on the stacking of the pleated sheets?

12.20 To what aspects of protein structure do each of the following terms refer: primary structure; secondary structure; tertiary structure; quaternary structure?

12.21 What is the function of collagen? Describe (a) the macroscopic physical properties and (b) the molecular structure of collagen.

12.22 Examine the structure of myoglobin (Figure 12.17) and identify (a) the amino terminal end of the polypeptide chain, (b) the carboxyl terminal end of the polypeptide chain, and (c) the eight relatively straight sections of α-helix.

12.23 Five types of interactions stabilize protein conformations. Draw partial structural formulas to illustrate each of these.

a Hydrogen bonding between the peptide bonds of the sequences glycylalanine and serylvaline.
b Hydrogen bonding between the side chains of glutamine and serine; serine and tyrosine.
c Ionic interactions between the side chains of lysine and glutamic acid; aspartic acid and arginine.
d Disulfide bonding between the side chains of two cysteines.
e Hydrophobic interactions between two phenylalanine side chains.

12.24 Consider a typical globular protein in aqueous medium at pH 7.4. Which of the following amino acids would you expect to find on the outside and in contact with water; which on the inside?

a glutamic acid **b** glutamine **c** arginine
d serine **e** valine **f** phenylalanine
g lysine **h** isoleucine **i** threonine

12.25 Assume that the same globular protein of Problem 12.24 is imbedded in a lipid bilayer, as in Figure 11.4. Now which amino acids would you expect to find on the outside in contact with the interior of the lipid bilayer; which on the inside?

12.26 What is meant by the term denaturation?

12.27 Account for the fact that the solubility of globular proteins is a function of pH, and that solubility is a minimum when the pH of the solution equals the isoelectric point (pI) of the protein.

12.28 Insulin, glucagon, and ACTH are all globular proteins, soluble in aqueous media. Calculate the percentage of polar amino acids (both neutral and charged) in each of these proteins. Also calculate the percentage of polar amino acids in the repeating hexapeptide of silk (Section 12.10). What generalization might you make about the relative percentages of polar groups in globular proteins compared to fibrous proteins?

12.29 Would you expect to be able to separate beef and human insulin by paper electrophoresis; sperm whale and human insulin?

12.30 Myoglobin and hemoglobin are globular proteins. Myoglobin consists of a single polypeptide chain of 153 amino acids. Hemoglobin is composed of four polypeptide chains, two of molecular weight 141 and two of molecular weight 146. The three-dimensional structure of myoglobin and hemoglobin polypeptide chains is very similar. Yet, myoglobin exists as a monomer in aqueous solution while the four polypeptide chains of hemoglobin self-assemble to form a tetramer. Which polypeptide chains, those of myoglobin or hemoglobin, would you predict to have a higher percentage of nonpolar amino acids?

MINI-ESSAY

ABNORMAL HUMAN HEMOGLOBINS

Hemoglobin is one of the most plentiful proteins in the body. In the blood stream there are 5 billion red cells or erythrocytes per milliliter and each is packed with 280 million molecules of hemoglobin. The role of hemoglobin is to pick up molecular oxygen in the lungs and deliver it to all parts of the body for use in metabolic oxidation. The main component of normal adult hemoglobin (Hemoglobin A or Hb A) is the protein globin, molecular weight 64,500. Globin consists of four polypeptide chains, two identical alpha chains (141 amino acid residues) and two identical beta chains (146 residues); the sequence of amino acids in each type of chain is known. One heme group (Figure 16.13) is bound to each polypeptide chain. The complete three-dimensional structure of hemoglobin was elucidated by J. C. Kendrew and for this pioneering work he shared in the Nobel Prize in Chemistry in 1963.

The abnormal hemoglobins have attracted medical attention because of the anemias they produce. When combined with oxygen, the red blood cells of certain individuals have the flat, disk-like conformation of normal erythrocytes, but when oxygen pressure is reduced, these cells tend to assume a crescent-like shape and become quite rigid. This phenomenon is known as sickling. An increase in oxygen pressure reverses the sickling. In most individuals whose cells are capable of sickling, less than 1% of the red cells in venous circulation are normally sickled and there are no harmful effects ascribable to the condition. These individuals are said to have sickle-cell trait. However, about 2% of those individuals whose cells are capable of sickling suffer from severe, chronic anemia resulting from excessive destruction of erythrocytes, a condition known as sickle-cell anemia. From 30–60% of the red blood cells in circulation in these individuals are sickled. The sickling phenomenon is due to the presence of a single gene which, when inherited from one parent, produces the benign sickle-cell trait, but when inherited from both parents produces the crippling sickle-cell anemia.

The molecular basis for this abnormality was discovered by Linus Pauling. He isolated hemoglobin (Hemoglobin S or Hb S) from the blood of sickle-cell anemic patients, separated the colored heme from the globin, and compared each with corresponding fractions from the hemoglobin of normal individuals. The heme from the two sources was identical. Next he examined the protein from each hemoglobin by paper electrophoresis (see Section 12.4).

Paper electrophoresis of normal and sickle-cell anemia hemoglobin at pH 6.9 indicated a significant difference between the two; normal hemoglobin moves as a negative ion while sickle-cell hemoglobin moves as a positive ion. The two types are quite distinct. The hemoglobin from individuals with sickle-cell trait is a mixture: 60% normal hemoglobin (Hb A) and 40% sickle-cell hemoglobin (Hb S).

Pauling's discovery revealed the cause of the sickling phenomenon. The abnormal hemoglobin functions perfectly satisfactorily in transporting molecular oxygen to the cells, but after it has discharged oxygen it changes shape; the hemoglobin molecules cluster together in stacks and fail to fill out the cell membrane, which then collapses. The erythrocyte appears to sickle, and becomes more rigid and more susceptible to rupture and enzymatic destruction. Further, these studies confirmed a direct relationship between the gene present and the characteristics of the hemoglobin found. Persons with sickle-cell trait, heterozygotes for the sickle-cell gene, synthesize both normal hemoglobin and sickle-cell hemoglobin. Those with sickle-cell anemia, homozygotes for the gene, produce only abnormal hemoglobin. Thus sickle-cell anemia is a molecular disease of genetic origin.

Vernon Ingram of Cambridge, England, discovered that sickle-cell hemoglobin differs from normal hemoglobin in only a single amino acid residue; the alpha chains of the two are identical but glutamic acid at position 6 of the normal beta chain is replaced by valine in Hb S. This result is remarkable for its simplicity. A single amino acid substitution accounts for the differences in charge between the two hemoglobins. At pH 6.9, where both glutamic acid carboxyl groups of the beta chain are ionized, normal hemoglobin has a more negative charge than the abnormal hemoglobin. Also, this single amino acid substitution makes the difference between healthy and sickling red cells. The clinical symptoms of sickle-cell anemia result from two stresses. First, the body eliminates sickled red cells in large numbers and its synthesizing capacity to produce new red cells is sorely taxed. There is overactivity in bone marrow and general development is poor. Second, the sickled cells tend to clump and interfere with blood circulation, causing a host of syndromes ranging from kidney to brain damage. That such a host of clinical symptoms should result from this single amino acid substitution is truly amazing.

The fact that there is so much natural selection pressure against this abnormal gene raises the question of why the sickle-cell trait is so common in populations in certain parts of the world. Assuming that the gene resulted from a single mutation, one must conclude, from the current wide distribution of the disease, that the deleterious gene mutation occurred long ago. Why then was it not erased from the pool of human genes long ago? The curious geographic distribution of the disease provides an important clue. The incidence of the disease is high (15% and over) among the populations of central Africa and southern India, areas where falciparum malaria is endemic. There is now considerable evidence that heterozygotes for the hemoglobin S gene have a greater resistance to malarial infection. This is a rare case of a deleterious mutation providing advantage in a hostile environment.

The dramatic success in discovering the genetic and molecular basis for sickle-cell anemia spurred interest in searching for other abnormal hemoglobins. To date about 150 have been isolated and, in over 100 of these, the changes in primary structure have been determined. In the vast majority, there is but a single amino acid residue change in either the alpha or the beta chain and each substitution is consistent with the change of a single nucleotide in one DNA codon (Section 14.11). Several abnormal hemoglobins are listed in Table 1.

Table 1 *Abnormal human hemoglobins. Many of these names are derived from the location of their discovery.*

Hemoglobin Variant	Amino Acid Substitution Position	From	To
	alpha chain		
J-Paris	12	ala	asp
G-Philadelphia	68	asn	lys
M-Boston	58	his	tyr
Dakar	112	his	gln
	beta chain		
S	6	glu	val
J-Trinidad	16	gly	asp
E	26	glu	lys
M-Hamburg	63	his	tyr

These abnormal hemoglobins are the clearest examples in man of the expression of mutant forms of a single gene and there can be no doubt that many other kinds of physiological and morphological variations will be found to have similar structural bases. In particular, it is probable that some and perhaps most of the specific enzyme deficiencies underlying the inborn errors of metabolism (p. 410) will turn out to be due to specific amino acid substitutions in critical enzymes and consequent loss of their activity. Conditions such as sickle-cell anemia, alcaptonuria, and phenylketonuria may be simply extreme examples of biochemical variation present everywhere in minor degree.

References

Beale, D., and Lehmann, H., "Abnormal Hemoglobins and the Genetic Code," *Nature*, **207**, 259 (1965).

Dayhoff, M. O., "Atlas of Protein Sequence and Structure," Vol. 4, National Biomedical Research Foundation, 1969.

Ingram, V., "Gene Mutations in Human Hemoglobin. The Chemical Difference Between Normal and Sickle-Cell Hemoglobin," *Nature*, **180**, 326 (1957).

Pauling, L., Itano, H. A., Singer, S. J., and Wells, I. C., "Sickle-Cell Anemia, a Molecular Disease," *Science*, **110**, 543 (1949).

Perutz, M. F., "The Hemoglobin Molecule," *Scientific American*, November 1964.

Perutz, M. F., and Lehmann, H., "Molecular Pathology of Human Hemoglobin," *Nature*, **219**, 902 (1968).

13

ENZYMES

13.1 INTRODUCTION

One of the unique characteristics of the living cell is its ability to carry out complex reactions rapidly and with remarkable specificity. The principal agents responsible for these transformations are a group of protein biocatalysts called enzymes.

It is astonishing how recent our knowledge is of these biological catalysts and how they function. Enzyme research in the late 1800s and early 1900s was devoted largely to the isolation of subcellular enzyme systems and to the study of the chemical reactions which these systems catalyze. As late as 1911, the British biochemist Bayliss stated confidently that enzymes were not proteins. In 1926, James Sumner made a major advance in the field of enzymology with the demonstration that at least one enzyme could be isolated in pure crystalline form. The enzyme was urease, which catalyzes the hydrolysis of urea to ammonia and carbon dioxide.

$$H_2N-\overset{\overset{\displaystyle O}{\|}}{C}-NH_2 + H_2O \xrightarrow{\text{urease}} 2NH_3 + CO_2$$

Further, Sumner proved that urease is a protein. Since that time scores of enzymes have been isolated in pure crystalline form and every one of them has turned out to be a protein. Some of these enzymes function as catalysts by themselves, while others require the presence of a metal ion (e.g. Mg^{2+}, Zn^{2+}) or some other nonprotein molecule (e.g. NAD^+).

It is now clear that all enzymes are proteins, that the particular properties of any given enzyme are determined by the sequence of its amino acid residues, and that the catalytic action is carried out within a discrete region of the enzyme known as the active site. The most challenging questions in enzyme research today are (1) what is the primary structure and detailed three-dimensional conformation of the enzyme under study, (2) where in this three-dimensional conformation is the active site, (3) how does the conformation of the active site lead to the enormously rapid and stereospecific enhancement of chemical reactions, and (4) how does the cell regulate the activities of its many enzymes so as to maintain the proper balance between the multitude of vital reactions it must perform? In addition, there is a rapidly expanding field of research that deals with diagnostic enzymology. This research has shown that the enzyme content of human sera may change significantly in certain pathological conditions. The study of these changes in concentration has proved to be an especially important diagnostic tool for the physician.

13.2 ENZYME CATALYSIS

Basically enzymes function as catalysts in much the same way as the common inorganic or organic laboratory catalysts. A catalyst, whatever the kind, combines with a reactant to "activate" it. In the case of enzyme catalysis, the reactant or reactants are generally referred to as substrates. This activated complex then undergoes a chemical change to form product(s) and regenerate the enzyme.

$$\underset{\text{enzyme}}{E} + \underset{\text{substrate}}{S} \longrightarrow \underset{\text{enzyme–substrate complex}}{E\text{—}S} \longrightarrow \underset{\text{enzyme}}{E} + \underset{\text{product}}{P}$$

As catalysts, enzymes are far superior to their nonbiological laboratory counterparts in three major ways: (1) enzymes have enormous catalytic power, (2) they are able to discriminate between very closely related molecules, and (3) the activity of many enzymes is regulated. Let us look at each of these unique characteristics in more detail.

First, enzymes have enormous power to increase the rate of chemical reactions. In fact, most of the reactions that occur readily in living cells would occur too slowly to support life in the absence of these biocatalysts. Consider, for example, the enzyme carbonic anhydrase, which catalyzes the reaction of carbon dioxide and water to produce carbonic acid. At physiological pH (7.1–7.4), carbonic acid ionizes to form H^+ and bicarbonate ion.

$$CO_2 + H_2O \xrightleftharpoons{\text{carbonic anhydrase}} H_2CO_3$$

Carbonic anhydrase increases the rate of hydration of carbon dioxide almost 10^7 times compared to the uncatalyzed reaction. Red blood cells are especially rich in this enzyme and for this reason they are able to promote the rapid interconversion of carbon dioxide and bicarbonate ion.

Consider also hexokinase, an important enzyme in the metabolism of glucose (Chapter 15), which catalyzes the reaction of α-glucose and adenosine triphosphate (ATP) to form glucose 6-phosphate and ADP.

$$\alpha\text{-glucose} + ATP \xrightleftharpoons{\text{hexokinase}} \alpha\text{-glucose-6-phosphate} + ADP$$

(α-glucose: ring with CH_2OH, O, OH, HO, OH, OH; α-glucose-6-phosphate: $CH_2—O—P(=O)(O^-)—O^-$ on the same ring)

It has been estimated that hexokinase accelerates the phosphorylation of glucose by a factor of at least 10^{10}.

Second among their unique properties is that enzymes are highly specific in the reactions they catalyze. A given enzyme will generally catalyze only one reaction or a single type of reaction. In other words, competing reactions and by-products such as we find under laboratory conditions are not observed in enzyme-catalyzed reactions. We have already discussed in Chapter 4 how an enzyme might catalyze a reaction of (+)-glyceraldehyde but not of its enantiomer (−)-glyceraldehyde, and how an enzyme might catalyze the reduction of pyruvate to form (+)-lactate but not (−)-lactate. These are but two examples of specificity. In more general terms, there are ranges of enzyme specificity.

Some enzymes act on one substrate and no others. In such cases we say the enzyme has absolute specificity. For example, succinate dehydrogenase, a key enzyme of the tricarboxylic acid cycle, catalyzes the oxidation of succinate to fumarate. In this reaction, FAD (flavine adenine dinucleotide) is the oxidizing agent.

$$\underset{\text{succinate}}{^{-}O_2CCH_2CH_2CO_2^-} + FAD \xrightleftharpoons{\text{succinate dehydrogenase}} \underset{\text{fumarate}}{(^{-}O_2C)(H)C{=}C(H)(CO_2^-)} + FADH_2$$

Succinate dehydrogenase will not catalyze the oxidation of other dicarboxylic acids even though they are very closely related in structure to succinate. Note that there is another element of specificity in this reaction. The product is exclusively fumarate (the *trans* geometric isomer) rather than maleate (the *cis* isomer).

Other enzymes show optical specificity. For example, the enzyme lactate dehydrogenase (LDH) catalyzes the oxidation of (+)-lactate but not that of (−)-lactate.

$$\underset{\text{(+)-lactate}}{CH_3{-}CH(OH){-}CO_2^-} + NAD^+ \xrightleftharpoons{\text{LDH}} \underset{\text{pyruvate}}{CH_3{-}C(=O){-}CO_2^-} + NADH + H^+$$

Still other enzymes show group or linkage specificity. The enzymes chymotrypsin and trypsin catalyze the hydrolysis of amide bonds of proteins and polypeptides and for this reason are called proteases.

$$-NH{-}CH(R_1){-}C(=O){-}NH{-}CH(R_2){-}C(=O){-} + H_2O \xrightleftharpoons{\text{protease}} -NH{-}CH(R_1){-}C(=O){-}O^- + H_3\overset{+}{N}{-}CH(R_2){-}C(=O){-}$$

Glycosidases like maltase and emulsin catalyze the hydrolysis of glycoside bonds. Lipases catalyze the hydrolysis of ester bonds in fats and oils. Further, there are varying degrees of group or linkage specificity. Maltase will catalyze the hydrolysis of alpha (as in starch) but not beta (as in cellulose) glycoside bonds. Emulsin has just the opposite specificity; it will catalyze the hydrolysis of beta but not alpha glycoside bonds. While trypsin will catalyze the hydrolysis of most amide bonds of polypeptides, this hydrolysis is rapid only when the side chain of the carboxyl-containing amino acid is that of either arginine or lysine.

Third among their unique properties is the fact that the activities of many enzymes can be regulated. We shall discuss this characteristic in more detail in Section 13.8. For the moment, we will simply note that mechanisms exist in living cells to both increase and decrease the activities of many enzymes. Most often, these mechanisms are in the form of signals from certain small molecules.

The potential of enzymes for enormous catalytic power, for great specificity, and for regulation have important consequences for the living cell. Any living cell contains literally thousands of different molecules and there is an almost infinite number of chemical reactions that are thermodynamically possible in this mix. Yet the cell, by virtue of its enzymes, can not only select which chemical reactions will take place but, by regulating the activities of key enzymes, it can also control the rates of these reactions and how much of any given product is formed. In this regard, enzymes are truly remarkable catalysts!

13.3 COENZYMES AND VITAMINS

Among the enzymes that act as biocatalysts, there is considerable diversity of structure. Many enzymes are what we call simple proteins. By this we mean that the protein itself is the true catalyst. Still other enzymes catalyze reactions of their substrates only in the presence of specific nonprotein organic molecules called coenzymes. When coenzymes are required, the complete or holoenzyme consists of the protein part, or apoenzyme, and a nonprotein part, or coenzyme.

$$\underset{\text{(apoenzyme)}}{\text{protein part}} + \underset{\text{(coenzyme)}}{\text{nonprotein part}} \rightleftharpoons \text{holoenzyme}$$

Both the apoenzyme and coenzyme parts by themselves are inactive; it is the holoenzyme that is the active biocatalyst.

In one sense we can regard a coenzyme as a second substrate. For example the enzyme lactate dehydrogenase (LDH) requires nicotinamide adenine dinucleotide (NAD^+) for activity.

$$\underset{\text{(+)-lactate}}{CH_3{-}\overset{\overset{\large OH}{|}}{C}H{-}CO_2^-} + NAD^+ \underset{}{\overset{LDH}{\rightleftharpoons}} \underset{\text{pyruvate}}{CH_3{-}\overset{\overset{\large O}{\|}}{C}{-}CO_2^-} + NADH + H^+$$

It should be obvious why NAD^+ is required; NAD^+ is the molecule that oxidizes lactate to pyruvate and is itself reduced to NADH.

Most water-soluble vitamins either are coenzymes themselves or they are small molecules from which coenzymes can be synthesized.

Man and a great many other organisms do not have the enzymes necessary to synthesize certain essential coenzymes from the simpler substances already in the cell and therefore must obtain either the coenzyme or a key component of the coenzyme from an organism that is able to make it. Eleven essential coenzymes are listed in Table 13.1 along with their vitamin precursors and biological functions. We will present structural formulas for these coenzymes in later chapters when we discuss their biological function.

Table 13.1 *Eleven essential coenzymes, their vitamin precursors and biological functions.*

Coenzyme	Vitamin Precursor	Biological Function
NAD^+, $NADP^+$	nicotinic acid (niacin) or nicotinamide (niacinamide)	oxidation-reduction
FAD, FMN	riboflavin or vitamin B_2	oxidation-reduction
lipoic acid	none	oxidation-reduction
thiamine pyrophosphate	thiamine or vitamin B_1	decarboxylation
pyridoxal phosphate	pyridoxine or vitamin B_6	group transfer reactions, transamination
coenzyme A	pantothenic acid	transfer of $CH_3CO—$ groups
tetrahydrofolic acid	folic acid	transfer of $—CH_3$, $—CH_2OH$, and —CHO groups
biotin	none	carboxylation reactions
cobamide	cobalamin or vitamin B_{12}	transfer of $—CH_3$ groups

In addition to these coenzymes, many enzymes also require specific metal ions for full catalytic activity. For example, the enzyme carbonic anhydrase requires one atom of Zn^{2+} per molecule for activity. In some cases, the metal ions appear to be rather loosely associated with the active enzyme and can be removed from them rather easily. In other instances, the enzymes retain the metal ions throughout normal isolation and purification procedures, and the metal ions can be regarded as integral parts of the enzyme structure. Most often, however, if the metal ion is removed, enzymatic activity is lost. The most common metal ion cofactors are Mg^{2+}, Ca^{2+}, Mn^{2+}, Co^{2+}, and Zn^{2+}. The metal molybdenum is required by the enzymes that "fix" atmospheric nitrogen, i.e., convert N_2 to nitrates and ammonia.

13.4 CATALYSIS AND ENERGY OF ACTIVATION

We have already examined the function of catalysts in several organic reactions, e.g., hydration of alkenes, Fischer esterification, acid-catalyzed hydrolysis of esters, and the aldol condensation. In each example we suggested that in the absence of either acid or base catalysis, as the case may be, these reactions proceed only slowly toward equilibrium. With a catalyst, equilibrium is attained much more rapidly. The catalyst in no way affects the position of the equilibrium but rather enters into the reaction in a very specific way and is regenerated unchanged on completion of the reaction. Further, for each of the above reactions we were able to specify how the catalyst functioned at the molecular level. The same features are true of enzyme catalysis. An enzyme is a catalyst and consequently it cannot change the position of equilibrium. It simply accelerates the rate at which equilibrium is reached.

Recall from your study of general chemistry that the tendency of a reaction to take place is measured by the change in free energy, ΔG This quantity is the difference (Δ) between the free energy (G) of the products and the reactants. When a chemical reaction proceeds toward an equilibrium state, the free energy of the system decreases and work can be done by the system. Thus, free energy is a measure of the maximum useful work that can be obtained from a chemical reaction. The greater the decrease in free energy, the greater the useful work that can be obtained during the course of the reaction. Work, in a biological system, includes such things as muscular contraction, nervous transmission, cell division, biosynthesis of macromolecules, etc.

Before we look at the energetics of enzyme catalysis, we must adopt certain conventions for handling free energy data. For most purposes, chemists have agreed to describe the energetics of chemical reactions in terms of the change in standard free energy, ΔG^0. ΔG^0 is the change in free energy when a reaction takes place at 25°C, at 1 atm pressure, and when both reactants and products are maintained at 1 molar concentration. Because most biochemical reactions occur at or near pH 7, it is more convenient to use $\Delta G^{0\prime}$ to indicate the standard free energy change at pH 7.0.

Reactions that occur with a negative change in free energy are said to be exergonic. Exergonic reactions include the breakdown and oxidation of carbohydrates, fats, and proteins. For example, the standard free energy change at pH 7.0 for the oxidation of glucose to carbon dioxide and water is $-686{,}000$ cal/mole.

$$\text{glucose} + 6O_2 \longrightarrow 6CO_2 + 6H_2O \qquad \Delta G^{0\prime} = -686{,}000 \text{ cal/mole}$$

Exergonic reactions such as the oxidation of glucose are energetically possible because they occur with a decrease in free energy. However, it is important to recognize that although the sign of the free energy change tells us a reaction can occur, it gives us no indication of the rate at which the reaction will occur.

Reactions that occur with a positive change in free energy are said to be endergonic. Endergonic reactions include photosynthesis, muscle contraction, and the biosynthesis of macromolecules. For example, the standard free energy change at pH 7.0 for the synthesis of glucose from carbon dioxide and water is $+686{,}000$ cal/mole.

$$6CO_2 + 6H_2O \longrightarrow \text{glucose} + 6O_2 \qquad \Delta G^{0\prime} = +686{,}000 \text{ cal/mole}$$

Endergonic reactions are not spontaneous in the thermodynamic sense and will not go unless energy to drive them is supplied by some external source. The energy to drive photosynthesis is of course supplied in the form of radiant energy from the sun.

With this concept of free energy, we can now look at the energetics of enzyme catalysis in more quantitative terms. Surely the oxidation of glucose by oxygen to carbon dioxide and water is a spontaneous reaction, for it occurs with a large decrease in free energy, −686,000 cal/mole. Yet we know that glucose in a bottle or in an aqueous solution in equilibrium with oxygen is stable almost indefinitely. This reaction is spontaneous in the thermodynamic sense yet it does not occur at any measurable rate under the conditions just described. Similarly, the hydrolysis of a peptide bond such as that in glycylglycine is an exergonic reaction and therefore spontaneous in the thermodynamic sense.

$$H_3\overset{+}{N}CH_2\overset{\overset{O}{\|}}{C}\text{—}NHCH_2CO_2^- + H_2O \longrightarrow 2H_3\overset{+}{N}CH_2CO_2^-$$

$$\Delta G^{0\prime} = -4{,}600 \text{ cal/mole}$$

Yet if this reaction and the hydrolysis of other peptide bonds of cellular protein material were to occur at any appreciable rate in the absence of specific catalysts, life would be all too brief!

By the end of the 19th century, chemists understood that although thermodynamics tells us whether or not a reaction will occur with a decrease in free energy, it does not tell us anything about how fast it will occur. Although two molecules may collide, they may not react because they do not have sufficient energy to react. They must acquire an extra amount of energy, the energy of activation, before they can react. The activation energy, $\Delta G^{\ddagger}$, is the difference in free energy between the substrate (reactant) and the transition state (Figure 13.1).

The function of a catalyst is to provide an alternative pathway for the reaction, and more importantly, a pathway with a lower energy of

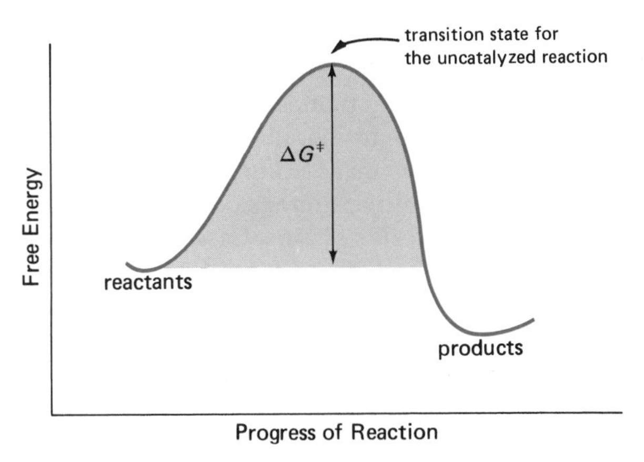

Figure 13.1 *Energetics of an uncatalyzed reaction showing the free energy of activation,* $\Delta G^{\ddagger}$.

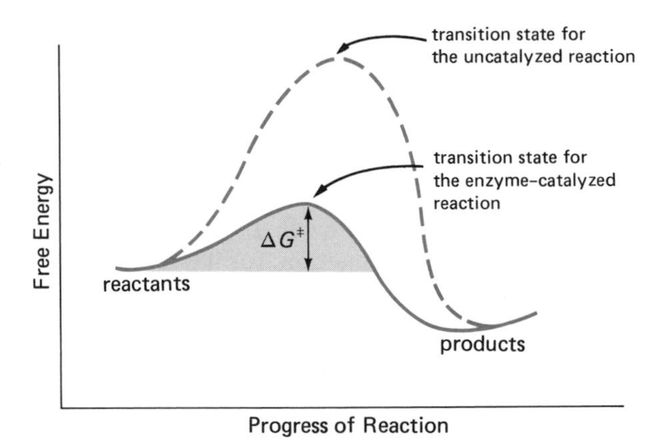

Figure 13.2 *Energetics of an enzyme-catalyzed reaction showing the free energy of activation, $\Delta G^{\ddagger}$.*

activation (Figure 13.2). For example, glycylglycine dissolved in water is stable because the energy of activation for hydrolysis is large. In the presence of a suitable enzyme catalyst, however, the reaction proceeds quite readily because the enzyme provides an alternative pathway for the reaction, one with a substantially lower energy of activation.

13.5 MECHANISM OF ENZYME CATALYSIS

The formation of an enzyme–substrate complex (Section 13.2) is the first and crucial step in enzyme catalysis. Virtually all enzymes are globular proteins and even the simplest have molecular weights ranging from 12,000 to 40,000 (i.e., they consist of from 100 to 400 amino acids). Because enzymes are so large compared to the molecules whose reactions they catalyze, it was proposed that substrate and enzyme interact at only a small portion or region of the enzyme surface. We call this region of interaction the active site.

To date, a large number of enzymes have been prepared in pure crystalline form and studied by X-ray crystallography. For several of these enzymes we have the same type of detailed information on three-dimensional structure as we have for myoglobin (Section 12.13). In all cases where three-dimensional conformations have been determined and the interactions between enzyme and substrate studied, the active site has been found, just as predicted, to be a relatively small portion of the enzyme surface. Furthermore, the active site has been found to be a specific three-dimensional region having a unique arrangement of amino acid side chains. Often these side chains so close together in the active site are contributed by amino acids quite far apart on the linear sequence of the polypeptide chain.

One enzyme that has been studied in detail is hen egg-white lysozyme, which consists of a single polypeptide chain of 129 amino acids and catalyzes the hydrolysis of specific polysaccharide components of bacterial cell walls. The active site of lysozyme is constructed of the side chains of glutamic acid-35, aspartic acid-52, tryptophan-62, tryptophan-63, and aspartic acid-101. (The number after each amino acid refers to its location along the polypeptide chain.) It is precisely this unique combination of side chains at the active site of lysozyme that accounts for its specificity.

Emil Fischer in 1890 likened the binding of an enzyme and its substrate to the interaction of lock and key. According to this model, shown schematically in Figure 13.3, the substrate and enzyme have complementary shapes and "fit" together. Recall from our discussion of the significance of asymmetry in the biological world (Section 4.10) that we were able to account for the remarkable ability of enzymes to distinguish between enantiomers by proposing that an enzyme and its substrate must interact through at least three specific binding sites on the surface of the enzyme. In Figure 13.3, these three binding sites are labeled *a*, *b*, and *c*. The complementary regions of the substrate are labeled *a′*, *b′*, and *c′*.

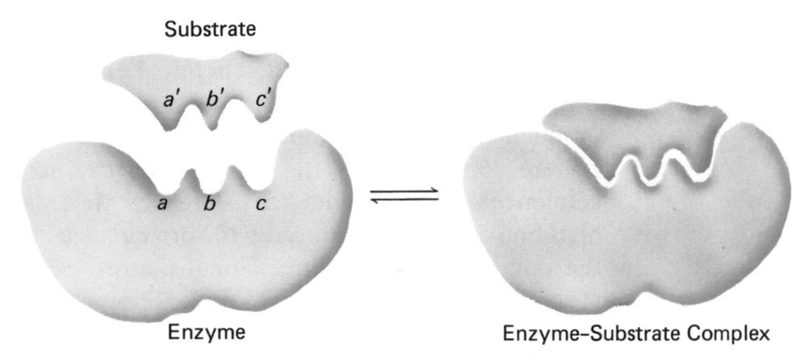

Figure 13.3 *Lock-and-key model of the interaction of enzyme and substrate.*

Once a substrate molecule has been recognized and bound to the active site of the enzyme, certain groups of the active site participate directly in the making and breaking of chemical bonds. These are called catalytic groups. For lysozyme, the catalytic groups include glutamic acid-35 and aspartic acid-52. In a sense then, the active site of an enzyme is a unique combination of both binding groups and catalytic groups. While this lock-and-key model of enzyme catalysis has been refined and modified in the years since it was first proposed by Fischer, it is still regarded as essentially correct.

We have learned a great deal about the nature of the active site and the reactions involved in the formation of the enzyme–substrate complex by studying the interaction of enzymes with synthetic substrates and inhibitors. Enzyme inhibitors are broadly classed as irreversible or reversible. We shall discuss irreversible inhibition here and reversible inhibition in Section 13.8.

Irreversible inhibitors usually produce a permanent alteration of the catalytic activity of an enzyme by the destruction or modification of one or more groups (either binding or catalytic) at the active site. A classic example is the irreversible inhibition of the enzyme acetylcholine esterase by diisopropylfluorophosphate (DFP), which we saw in Section 9.3.

DFP inhibits not only acetylcholine esterase but also a number of other enzymes, including chymotrypsin. After reaction of DFP with chymotrypsin, followed by isolation and degradation of the inhibited enzyme, it was found that the phosphorus appears on serine in the form of an organophosphorus ester of serine.

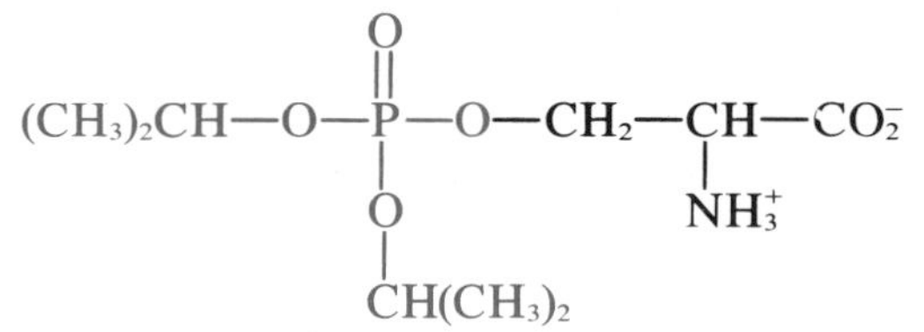

$$(CH_3)_2CH{-}O{-}\overset{\overset{\displaystyle O}{\|}}{\underset{\underset{\displaystyle CH(CH_3)_2}{|}}{\underset{|}{\underset{\displaystyle O}{P}}}}{-}O{-}CH_2{-}\underset{\underset{\displaystyle NH_3^+}{|}}{CH}{-}CO_2^-$$

This observation led to the conclusion that serine is a part of the active site. Yet the problem is not as simple as it would seem, for there are 26 residues of serine on one molecule of chymotrypsin. One and only one of the serines reacts with DFP. Thus, one of the serine molecules must in some way be unique. The problem then is to specify which of the serines reacts with DFP, the chemical environment of that serine, and to explain what it is in this environment that confers a chemical uniqueness to the serine. The serine has been identified as ser-195 (where 195 refers to the position of the serine in the amino acid sequence). Other evidence suggests that in addition to ser-195, histidine-57 and aspartic acid-102 are necessary for catalytic activity of the enzyme. A three-dimensional picture of the enzyme from X-ray studies shows that these three groups are indeed close to each other.

At least a partial picture of the mechanism of chymotrypsin action on a peptide bond is suggested in Figure 13.4. Interaction of histidine-57 with serine-195 removes a proton from the serine and creates a negatively charged oxygen, $—O^-$. This very strong nucleophile then attacks the amide carbonyl and forms a tetrahedral carbonyl addition intermediate. The intermediate breaks apart to form the acyl enzyme and the $H_2N—$ polypeptide fragment. Hydrolysis of the acyl enzyme regenerates the free enzyme.

Research over the last few decades has given us a great deal of information on many enzymes. As a result, for at least some of these enzymes we can answer two of the questions posed in the introduction to this chapter; namely what is the structure and three-dimensional conformation of the enzyme under study, and where in this three-dimensional conformation is the active site? However, we still have only partial answers to the third question, namely how does the conformation of the active site lead to the enormously rapid and

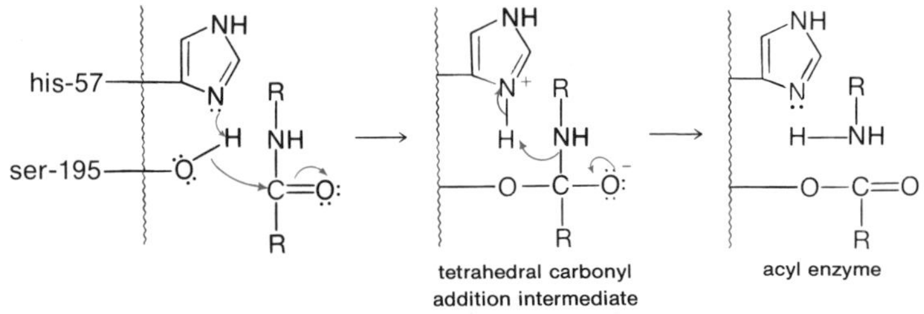

Figure 13.4 *A partial mechanism for chymotrypsin action.*

stereospecific enhancement of reaction rate? Although we understand the factors involved in the stereospecificity of binding, we still do not have an adequate explanation of the enormous catalytic power of enzymes.

13.6 KINETICS OF ENZYME CATALYSIS

Let us look in more detail at several factors that affect the rate of enzyme catalysis. The first of these is the concentration of the enzyme itself. For the reaction

$$\text{substrate} \xrightarrow{\text{enzyme}} \text{product}$$

increasing the concentration of enzyme increases the rate of conversion of substrate to product (Figure 13.5).

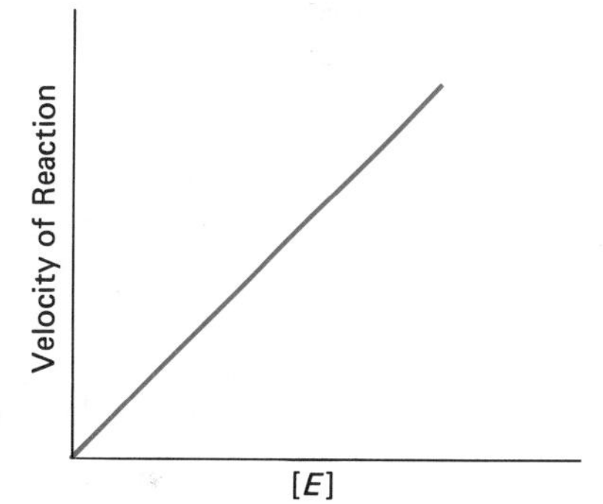

Figure 13.5 Dependence of the initial velocity of an enzyme-catalyzed reaction on the concentration of enzyme.

The second factor is the quantity of substrate. Suppose we were to mix a given concentration of enzyme and substrate and then measure the initial velocity of the reaction. In a second experiment, we now mix the same concentration of enzyme but increase the concentration of substrate and again measure the initial velocity. This can be repeated with a single enzyme concentration and a range of substrate concentrations. Figure 13.6 shows the relationship between initial reaction velocity and concentration of substrate.

From this graph you can see that in the region labeled (a), any increase in substrate concentration results in a direct increase in the reaction velocity. Beyond (a), further increases in substrate concentration also result in increased reaction velocity, but the effect is increasingly small. Finally, in region (b), increasing the substrate concentration has no effect at all on the reaction velocity. This is because in region (b) all enzyme active sites are bound with substrate and all enzyme molecules are in continuous operation catalyzing the conversion of substrate to product. At this point we say that the enzyme is saturated. This phenomenon of saturation of enzyme with substrate occurs with all enzymes and provides strong support for the notion of an enzyme–substrate complex.

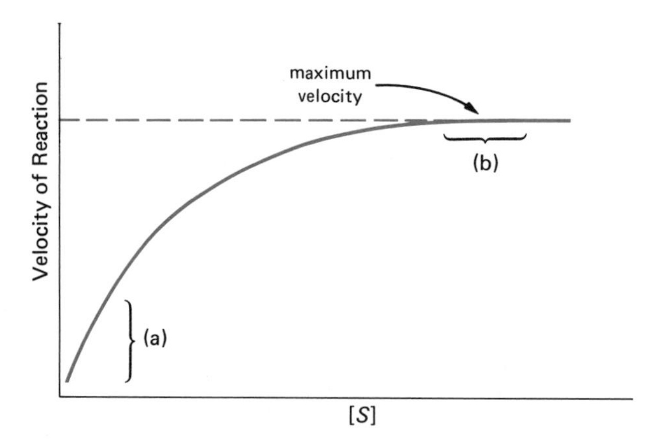

Figure 13.6 *Dependence of initial velocity of an enzyme-catalyzed reaction on the concentration of substrate. For an explanation of regions (a) and (b) see text.*

13.7 THE EFFECT OF pH ON ENZYME CATALYSIS

Most enzymes have a maximum catalytic activity in the pH range 4 to 8. However, some enzymes have maximum activity outside this range. One such example, pepsin, a digestive enzyme of the stomach, has a maximum catalytic activity around pH 1.5, the pH of gastric juices.

The catalytic activity of most enzymes falls off rapidly at pH values on either side of their particular pH maximum. The fact that the catalytic activity of enzymes is so very sensitive to pH is one reason why regulation of pH is so critical to an organism. Figure 13.7 shows a typical plot of enzyme activity as a function of pH.

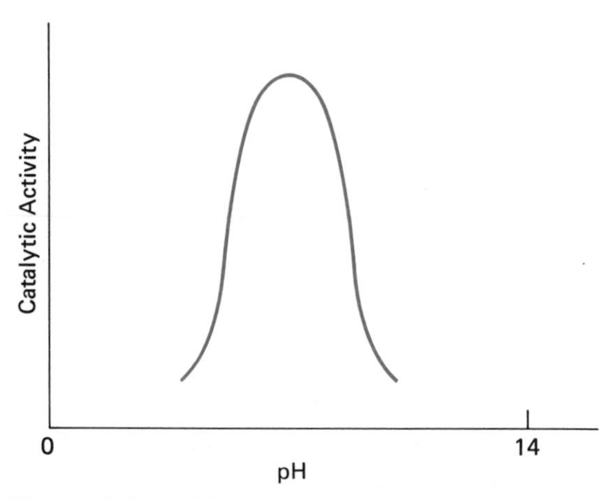

Figure 13.7 *The catalytic activity of a typical enzyme as a function of pH.*

Variations in the catalytic activity of enzymes with small changes in pH are the result of changes in the charged character of ionizable amino acid side chains on the surface of the enzyme. For example, suppose the active site of a particular enzyme contains a side chain of lysine and another of glutamic acid, and that for maximum catalytic

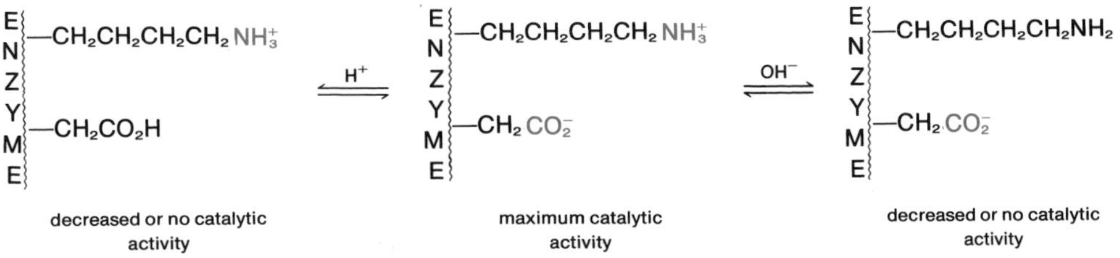

$$\text{ENZYME}\begin{cases}-CH_2CH_2CH_2CH_2NH_3^+ \\ -CH_2CO_2H\end{cases} \underset{H^+}{\rightleftharpoons} \text{ENZYME}\begin{cases}-CH_2CH_2CH_2CH_2NH_3^+ \\ -CH_2CO_2^-\end{cases} \overset{OH^-}{\rightleftharpoons} \text{ENZYME}\begin{cases}-CH_2CH_2CH_2CH_2NH_2 \\ -CH_2CO_2^-\end{cases}$$

decreased or no catalytic activity — maximum catalytic activity — decreased or no catalytic activity

Figure 13.8 *Catalytic activity as a function of pH for an enzyme requiring lys—NH_3^+ and glu—CO_2^- for maximum activity.*

activity, each of these side chain groups must be fully ionized (Figure 13.8). A decrease in pH changes the charged character of the glutamic acid side chain so that it is no longer ionized. An increase in pH changes the charged character of the lysine side chain so that it is no longer charged. In either case, the charged character of the enzyme is altered and its catalytic activity is decreased or even lost entirely. This decrease may be because (1) the binding groups or sites are no longer able to accommodate and bind substrate, (2) the catalytic groups are no longer able to interact properly with substrate and participate in the making and breaking of chemical bonds, or (3) the conformation of the enzyme is altered.

13.8 ENZYME INHIBITORS

As we saw in Section 13.5, the fact that enzymes can be inhibited by specific molecules can give us valuable information about the nature of the active site of the enzyme itself. However, the study of enzyme inhibition has even broader significance for the following reason: certain molecules present in the cell during the normal course of metabolism inhibit specific reactions and thereby provide a means for the internal regulation of cellular metabolism. Furthermore, many drugs and poisons act by inhibiting key enzymes. Let us look at two types of reversible inhibition and see how each can provide the cell with an invaluable regulatory mechanism.

One type of reversible inhibition occurs when a substance is sufficiently like the true substrate in structure that it combines with the enzyme at the active site and forms an enzyme–inhibitor complex (Figure 13.9). Since the inhibitor "competes" with the substrate for the enzyme, this is called competitive inhibition.

The difference between E–S and E–I is that the enzyme–inhibitor complex cannot give product. A classic example of this type of inhibition is that of succinate dehydrogenase by malonate, $^-O_2CCH_2CO_2^-$.

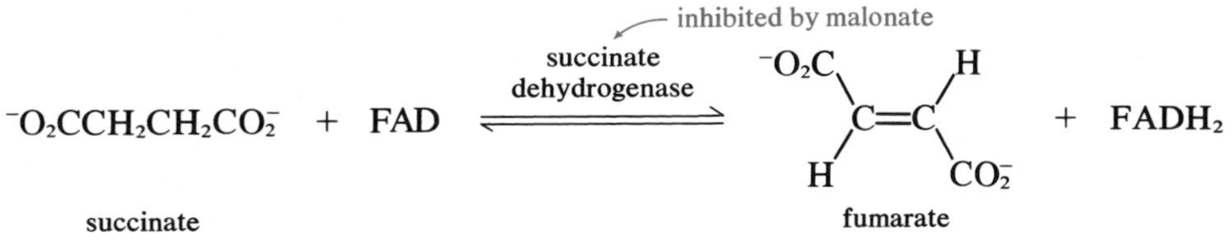

$$\underset{\text{succinate}}{^-O_2CCH_2CH_2CO_2^-} + FAD \xrightleftharpoons[]{\text{succinate dehydrogenase (inhibited by malonate)}} \underset{\text{fumarate}}{{}^-O_2C(H)C{=}C(H)CO_2^-} + FADH_2$$

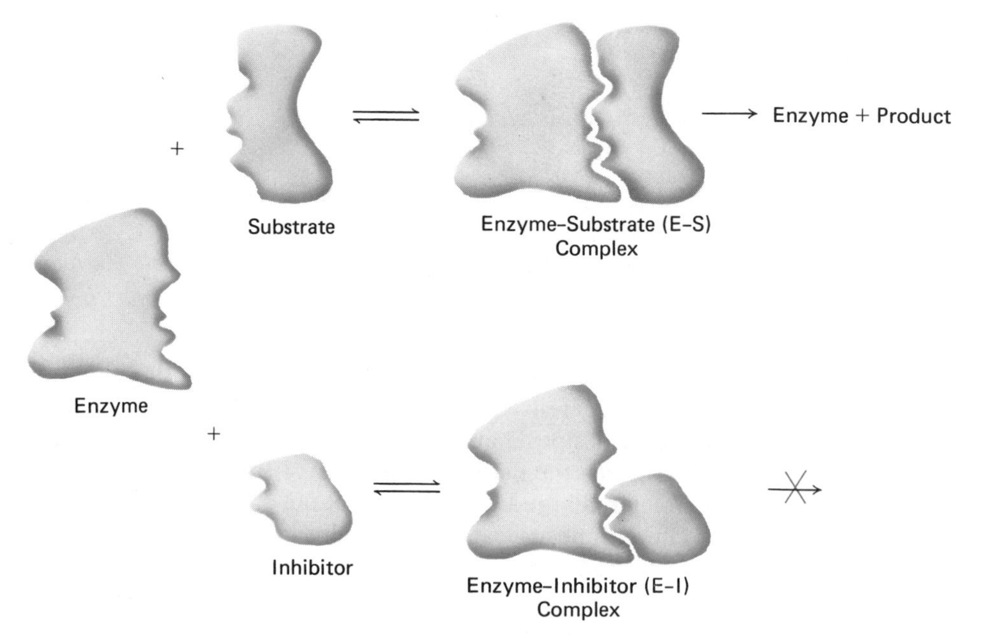

Figure 13.9 *Competitive inhibition. Substrate and inhibitor compete for the active site of the enzyme.*

If malonate is added to the reaction mixture, this dicarboxylate anion competes with succinate for the active site of the enzyme. A characteristic feature of competitive inhibition is that it can be reversed by adding excess substrate. Thus, if more succinate is added to the solution containing succinate dehydrogenase and malonate, the binding of enzyme and inhibitor can be reversed and the rate of formation of fumarate can be increased to what it would have been in the absence of inhibitor.

Competitive inhibitors that block enzyme-catalyzed reactions in a parasite or microorganism are potent chemotherapeutic agents. One such example is sulfanilamide (Section 9.7). A structural analog of *p*-aminobenzoic acid, sulfanilamide will inhibit one or more key steps in the biosynthesis of folic acid; the resulting deficiency of this essential vitamin is fatal to microorganisms. Man lacks the enzymes to make folic acid from *p*-aminobenzoic acid and therefore requires folic acid itself in the diet. Another example is penicillin, perhaps the most remarkable of all antibiotics (see the mini-essay "The Penicillins"). This drug is an inhibitor of the enzyme system that catalyzes final steps in the construction of certain bacterial cell walls.

A second type of reversible inhibition occurs when an enzyme binds an inhibitor at a site quite different from the binding site of its substrate (Fig. 13.10). Because enzyme and inhibitor do not compete for the same binding site, this is called noncompetitive inhibition.

In noncompetitive inhibition, the catalytic activity of the enzyme is reduced because the nature of the active site is changed and either (1) the enzyme is no longer able to bind substrate at the active site, or (2) the enzyme binds substrate but the catalytic groups are no longer able to participate in the making and breaking of chemical bonds.

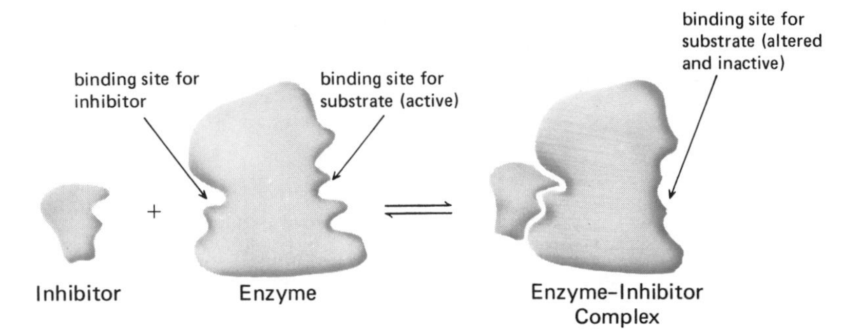

Figure 13.10 *Noncompetitive inhibition. Substrate and inhibitor do not compete for the same binding site.*

Unlike competitive inhibition, this type of inhibition cannot be reversed by the addition of more substrate.

The fact that an enzyme can be inhibited by the binding of a small molecule at the active site or at a site quite distinct from the active site represents one of the key mechanisms by which the internal environment of the cell can be controlled. Most constituents of the cell are maintained within a fairly narrow concentration range. This range represents a balance between the processes that supply the particular biomolecule and those that degrade or remove it. For economy and efficiency in the cell, there must be adequate amounts of these essential biomolecules but not an oversupply. Many factors—physical, chemical, and physiological—interact to regulate the concentration of various biomolecules and the reactions they undergo.

The first clearly recognized example of regulation by enzyme inhibition was discovered in 1957 and involves the synthesis of isoleucine from threonine in the bacterium *E. coli*. This process, which begins with threonine, involves five separate steps and five different enzymes (Figure 13.11).

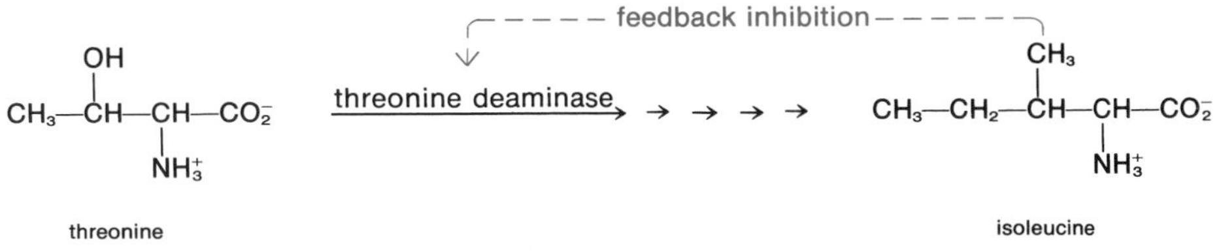

feedback inhibition

$$CH_3-\underset{\underset{}{}}{\overset{OH}{\overset{|}{C}H}}-\underset{\underset{NH_3^+}{|}}{CH}-CO_2^- \xrightarrow{\text{threonine deaminase}} \rightarrow \rightarrow \rightarrow \rightarrow CH_3-CH_2-\overset{CH_3}{\overset{|}{C}H}-\underset{\underset{NH_3^+}{|}}{CH}-CO_2^-$$

threonine　　　　　　　　　　　　isoleucine

Figure 13.11 *Feedback inhibition in the biosynthesis of isoleucine from threonine. In this pathway, the enzyme threonine deaminase is inhibited by isoleucine.*

The first step in the pathway is catalyzed by the enzyme threonine deaminase. This enzyme is inhibited by isoleucine, the final product of the pathway. Thus, when the concentration of isoleucine reaches a level adequate to meet the needs of the cell, threonine deaminase is inhibited and further synthesis of isoleucine is therefore inhibited. This type of regulation of enzyme activity is known as feedback inhibition. By having the end product of a given biosynthetic pathway inhibit the very first step in the pathway for its formation, the cell avoids the buildup of intermediates for which it has no immediate use. We shall see more examples of this type of metabolic regulation in Chapters 15–18.

13.9 ENZYMES IN CLINICAL DIAGNOSIS

Certain enzymes, such as those involved in blood coagulation, are normal constituents of plasma and the concentration of these enzymes in plasma is high compared to their concentration in cells. Other enzymes are normally present almost exclusively in cells and are released into the blood and other biological fluids only as the result of routine destruction of cells. Plasma levels of these enzymes are low, often up to a million times lower than cell levels. However, the plasma concentrations of these cellular enzymes may be elevated significantly in cases of cell injury and destruction or excess cell proliferation, as in cancer. By choosing appropriate enzymes for examination, we can use serum enzyme levels not only to detect cell damage but also to suggest the site of the damage. Further, the degree of elevation of the plasma concentration often can be used to determine the extent of cellular damage. For these reasons, the measurement of enzyme concentrations in blood plasma and other biological fluids has become a major diagnostic tool, particularly for diseases of the heart, liver, pancreas, skeletal muscle, and bone, and for malignant diseases. In fact, certain enzyme determinations are performed so often that they have become almost routine in the clinical chemistry laboratory. Several of these enzymes and the principal clinical conditions in which they are used are shown in Table 13.2.

Table 13.2 *Some enzymes used in diagnostic enzymology.*

Enzyme	Principal clinical condition in which the enzyme determination is used
lactate dehydrogenase (LDH)	heart or skeletal muscle damage, myocardial infarction
alkaline phosphatase	liver and bone disease
acid phosphatase	cancer of the prostate
serum glutamate oxaloacetate transaminase (SGOT)	liver and heart disease
creatine phosphokinase (CPK)	myocardial infarction and muscle disease
α-amylase	pancreatitis

Problems

13.1 Define or give a brief explanation of the following terms.

a active site
b enzyme–substrate complex
c substrate
d absolute specificity

e optical specificity
f linkage specificity
g apoenzyme
h coenzyme
i haloenzyme
j free energy of activation
k exergonic reaction
l endergonic reaction
m reversible inhibition
n irreversible inhibition
o competitive inhibition
p noncompetitive inhibition
q feedback inhibition

13.2 Malonate is an inhibitor of succinate dehydrogenase. Explain how you might determine experimentally whether malonate is a competitive or a noncompetitive inhibitor of succinate dehydrogenase.

13.3 Refer to Figure 13.6 and account for the fact that in region (a), an increase in substrate concentration results in a direct increase in reaction velocity. Also account for the fact that in region (b), an increase in substrate concentration has no effect on the reaction velocity.

13.4 Most enzymes have a maximum catalytic activity in a very narrow pH range. Explain why catalytic activity falls off rapidly on either side of this pH maximum.

13.5 The catalytic activity of the enzyme lysozyme is a maximum at pH 5.0 and falls off on either side of this pH. The side chains of aspartic acid-52 and glutamic acid-35 are at the active site and involved in lysozyme catalysis. Of the three ionization states for these groups shown below, which do you think is the catalytically active form of lysozyme? Explain your reasoning.

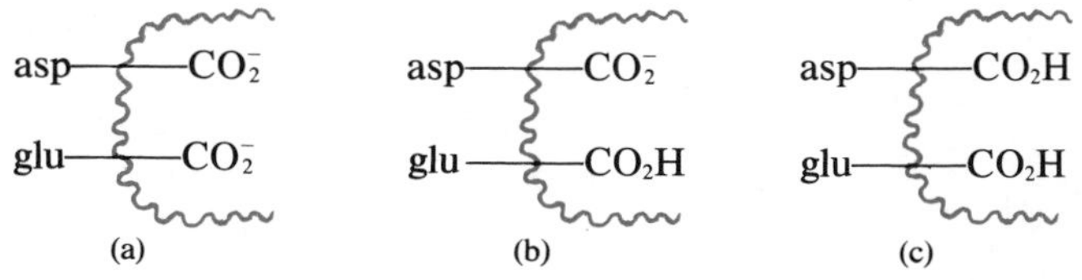

asp—CO_2^- glu—CO_2^- (a)

asp—CO_2^- glu—CO_2H (b)

asp—CO_2H glu—CO_2H (c)

13.6 Explain why it is that most enzymes lose catalytic activity when heated above 50–60°C.

13.7 Is diisopropylfluorophosphate (DFP) a competitive or noncompetitive inhibitor of chymotrypsin? Explain.

13.8 Figure 13.10 shows an example of noncompetitive inhibition of enzyme activity. Other small molecules may function as noncompetitive activators. Using appropriate diagrams, show an enzyme, an enzyme activator, and an enzyme–activator complex. Explain how your diagram accounts for enzyme activation.

MINI-
ESSAY

THE PENICILLINS

The phenomenon of the inhibition of growth of one sort of microorganism by the metabolic products of another was observed very early in the history of microbiology. Several specific substances were isolated and shown to stop the growth of a number of species of bacteria, and in some cases even to kill them. These substances were properly regarded as antibiotics, at least in the literal sense of the word, for they did inhibit bacterial growth. However, these substances usually did not distinguish between microorganisms and mammalian tissue, and as a consequence were often highly toxic to animals. The term antibiotic, as it is understood today, implies that the substance not only shows toxicity to microorganisms but also that the substance is more toxic to microorganisms than to mammalian tissues. Of course, this distinction is sharper in some classes of antibiotics than in others.

It is common knowledge that the most successful of the antibiotics are the penicillins. These truly remarkable drugs are almost completely innocuous to all living materials except for certain groups of bacteria. As a result of the tremendous research that has been done on the structure, the chemistry, the large-scale commercial manufacture, and the mechanism of action of the penicillins, it is safe to say that we have a clearer understanding of the penicillins than of any other antibiotic.

The discovery of penicillin was purely fortuitous but its development and therapeutic application represent the results of a well-planned and carefully executed program that brought about one of the major advances in medical science. In 1929 Sir Alexander Fleming published his now famous observations that colonies of staphylococci lysed on a plate which had become contaminated with the mold *Penicillium notatum.* Some unknown substance from the mold was bactericidal not only for staphylococci but for many other common disease-causing organisms as well. Because the mold belonged to the genus *Penicillium*, Fleming named the antibacterial substance penicillin. His efforts to extract the penicillin failed and very little was made of this discovery for nearly a decade. In 1939, the outbreak of hostilities stimulated an intensive search for new chemotherapeutic agents, and in Great Britain the potential of Fleming's penicillin was reinvestigated. Within a few months, crude preparations were available and many of the chemical, physical, and antibacterial properties of penicillin were determined. Preliminary results were so impressive that production was undertaken. Because large-scale manufacture in Great Britain was impossible due to the exigencies of the war, a vast cooperative research program was undertaken in the United States. By 1943, penicillin was authorized for use by the medical service of the U.S. Armed Forces, and by 1949 the antibiotic was available for widespread use in almost unlimited quantity. Thus, penicillin progressed in the span of two decades from a chance observation in a research laboratory to a therapeutic agent of enormous importance.

Preliminary investigations of the structure and chemistry of penicillin presented a confusing picture because of discrepancies in the analytical and degradative results obtained in different laboratories. These discrepancies were resolved once it was discovered that *P. notatum* produces different kinds of penicillin depending on the nature of the medium in which the mold is grown. Initially six different penicillins were recognized, and all proved ultimately to be acyl derivatives of 6-aminopenicillanic acid. Today the term penicillin is used as a generic name to include all acyl derivatives of 6-aminopenicillanic acid.

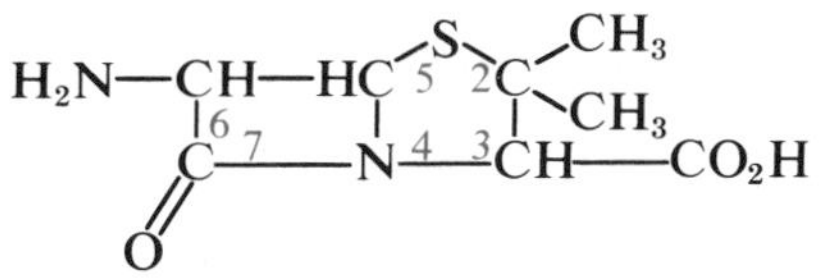

H_2N—CH—HC—S—C(CH_3)$_2$; C(=O)—N—CH—CO_2H (ring positions 2–7)

6-aminopenicillanic acid

The basic structure of penicillin consists of a five-membered ring containing a nitrogen and a sulfur (a thiazolidine ring) fused to a four-membered ring containing a cyclic amide (a β-lactam). The structural integrity of these two rings is essential for the biological activity of penicillin and cleavage of either ring leads to products devoid of antibacterial activity.

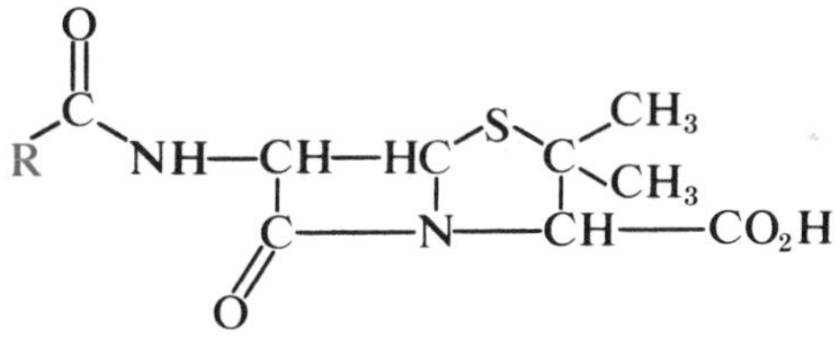

R—C(=O)—NH—CH—HC—S—C(CH_3)$_2$; C(=O)—N—CH—CO_2H

penicillin

R = C_6H_5—CH_2—; benzylpenicillin

R = CH_3CH_2—CH=CH—CH_2—; 2-pentenylpenicillin

Penicillin F (2-pentenylpenicillin) was the predominant component of British penicillin where *P. notatum* was grown in a synthetic medium. In the United States, where a medium of corn-steep liquor was used, penicillin G (benzylpenicillin) was the principal product. Penicillin G is the most widely used of the penicillins and is the standard against which others are compared.

Penicillin undergoes a variety of chemical reactions, some of which are very complex. We shall look at only two of these reactions, each of which has important consequences for therapeutic use of this antibiotic. Treatment of penicillin with strong mineral acid brings about hydrolysis to penicillamine and a penaldic acid (Figure 1). The rupture of the β-lactam ring involves hydrolysis of an amide bond, and the rupture of the thiazolidine ring can be formulated as the hydrolysis of

penicillin $\xrightarrow{H^+}$ penaldic acid + penicillamine

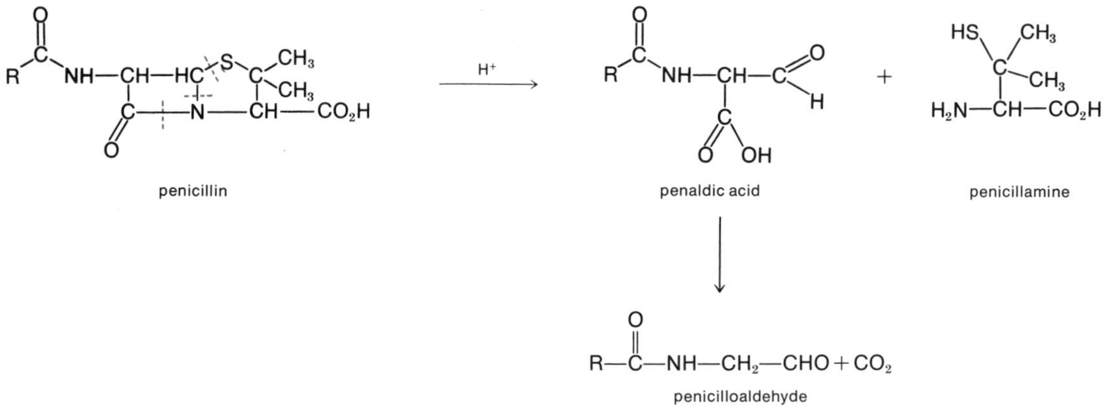

$$R{-}\overset{O}{\overset{\|}{C}}{-}NH{-}CH_2{-}CHO + CO_2$$

penicilloaldehyde

Figure 1 *Hydrolysis of penicillin by strong acid.*

the nitrogen–sulfur analog of an acetal. Penaldic acid, like other compounds with a carboxylic acid beta to a carbonyl group, undergoes ready decarboxylation, and gives CO_2 and a penicilloaldehyde. Penicillamine, penaldic acid, and penicilloaldehyde are all devoid of antibacterial activity. Because penicillin G is rapidly inactivated by this type of hydrolysis in the acid conditions of the stomach, it cannot be administered orally and it is used most effectively by injection.

A second important reaction is the selective hydrolysis of the β-lactam ring catalyzed by the enzyme penicillinase (Figure 2).

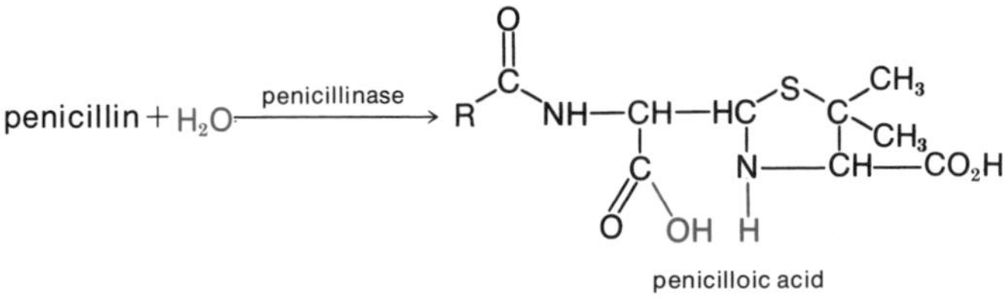

penicillin + H_2O $\xrightarrow{\text{penicillinase}}$ penicilloic acid

Figure 2 *Hydrolysis of penicillin catalyzed by penicillinase.*

Penicilloic acid, like other ring cleavage products, has no antibacterial activity. The main basis for natural bacterial resistance to penicillin is the ability of resistant strains to synthesize penicillinase.

All penicillins owe their antibacterial activity to a common mechanism which inhibits the biosynthesis of a vital part of the bacterial cell wall. Essentially all bacteria are surrounded by a layer of strengthening material known as mucopeptide or glycopeptide. This mucopeptide is the principal mechanical support of the entire cell and the viability of the bacterial cell depends on the integrity of the cell wall. It is well known that the concentration of smaller-molecular-weight substances within the cell is often considerably greater than in the surrounding medium, giving rise to a high osmotic pressure within the cell. In certain bacterial cells, this osmotic pressure has been estimated to be as high as 10–20 atm. The cell wall, in providing a mechanical support, prevents the disruption of the cell.

The simplest type of mucopeptide is made up of a backbone of long polysaccharide chains cross-linked by short peptide chains (Figure 3). The polysaccharide chain itself is made up of alternating units of N-acetyl-D-glucosamine and N-acetyl-D-muramic acid. Each of the N-acetyl-D-muramic acids has a carboxyl group and in the mucopeptide, each carboxyl group is linked by an amide bond to a pentapeptide beginning with L-alanine.

polysaccharide backbone

$C{=}O$	
NH	
$CH{-}CH_3$	L-alanine
$C{=}O$	
NH	
$CH{-}CH_2{-}CH_2{-}CO_2H$	D-glutamic acid
$C{=}O$	
NH	
$CH{-}CH_2{-}CH_2{-}CH_2{-}CH_2{-}NH_2$	L-lysine
$C{=}O$	
NH	
$CH{-}CH_3$	D-alanine
$C{=}O$	
NH	
$CH{-}CH_3$	D-alanine
CO_2H	

Figure 3 *The polysaccharide backbone from mucopeptide with pentapeptides attached. The pentapeptide is shown on the left by structural formula and on the right is indicated by the names of the constituent amino acids. Note that three of the amino acids are of the rare D series.*

The final reaction sequence in construction of the cell wall matrix is formation of amide bonds between carboxyl and amino groups from adjacent mucopeptide chains. In staphylococci this is thought to occur first by hydrolysis of the terminal D-alanine, and then peptide bond formation between the remaining D-alanine of one chain and the amino group of lysine from another chain. Thus each long polysaccharide chain is cross-linked to other polysaccharide chains by short lengths of polypeptide chain and this matrix of each cell wall probably forms one enormous molecule—a "bag-shaped macromolecule."

It is the formation of the final cross-linked mucopeptide macromolecule that is inhibited by the penicillins, and a number of hypotheses about the selective action of the antibiotic have been suggested. One theory is that penicillin is structurally similar to D-alanyl-D-alanine, the terminal amino acid of the peptide units that must be cross-linked (Figure 4). Penicillin is thought to bind selectively to the active site of the enzyme complex that catalyzes peptide bond hydrolysis of the terminal D-alanine and peptide bond formation in cross-linking, thus making the enzyme complex unavailable for cell wall synthesis. Hence penicillin's antibacterial activity at the molecular level appears to derive from selective enzyme inhibition.

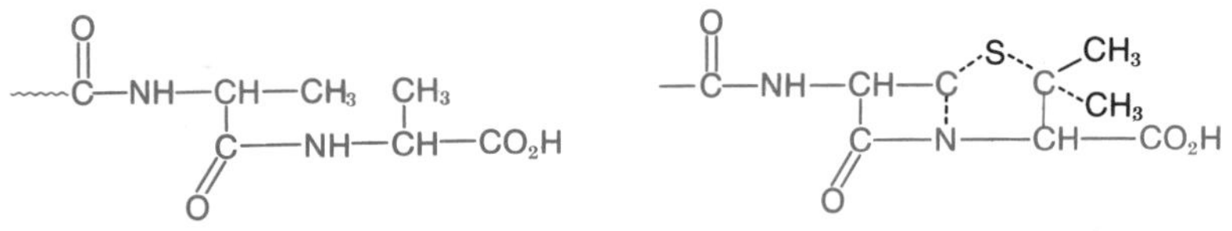

D-alanyl-D-alanine

penicillin

Figure 4 *Penicillin and D-alanyl-D-alanine, drawn to suggest a structural similarity between the two.*

The fact that this type of rigid cell wall construction is unique to bacteria and is not present in mammalian cells no doubt accounts for the lack of toxicity of penicillin to patients even when it is administered in massive doses. Penicillin is a remarkable drug that does its work with great efficiency and selectivity. However, severe allergic reaction in some patients is a serious problem. A significant proportion of the population (perhaps as high as 5–10% in some countries) has become hypersensitive to penicillin and this has tended to limit use of the drug. The factor responsible for the allergic reaction is not penicillin itself but rather certain degradation products of the drug, particularly 6-aminopenicillanic acid. This allergic sensitivity is not regarded as toxicity in the usual pharmacological sense and is not related to the drug's effects on bacterial cell walls.

The susceptibility of penicillin G to hydrolysis by acid and the emergence of penicillinase-producing strains have provided incentive for the discovery and clinical development of various other natural and semisynthetic penicillins. The nature of the R—group has a major effect on the pharmacological properties, and a large number of new synthetic penicillins have been prepared either by the addition of substituted acetic acids (R—CH_2—COOH) to the fermentation medium, or by the direct acylation of 6-aminopenicillanic acid. By these means, new penicillins have been prepared which (1) possess greater acid stability than penicillin G, (2) are more resistant to the action of penicillinase, and (3) possess a broader spectrum of antibacterial activity than penicillin G. Of those more stable to hydrolysis by acid, penicillin V (α-phenoxymethylpenicillin) is the most widely used. Its acid stability is sufficient to permit oral administration. Methicillin (2,6-dimethoxyphenylpenicillin) is inactivated by penicillinase 100 times more slowly than penicillin G and hence may be used in the treatment of infections caused by resistant staphylococci.

Despite the development of many new antibiotics, penicillin remains one of the most important of the anti-infective agents. Since the possibilities for adding new acyl groups to the essential unit, 6-aminopenicillanic acid, are manifold, new penicillins which are even more important chemotherapeutic agents may yet be developed.

References

Weinstein, L., "The Penicillins," in *The Pharmacological Basis of Therapeutics*, L. S. Goodman and A. Gilman, Editors (The Macmillan Company, New York, 1970).

Fleming, A., "History and Development of Penicillin," in *Penicillin: Its Practical Applications* (The Blakiston Company, Philadelphia, 1946).

14

NUCLEIC ACIDS

14.1 INTRODUCTION

Nucleic acids are a third great class of macromolecules or biopolymers which, like proteins and polysaccharides, are vital components of living materials. Of all the biopolymers, probably none are so important as the nucleic acids, for they are the genetic material itself. In this chapter we shall examine the structure of the mononucleotides, and the manner in which these small building blocks are bonded together to form giant nucleic acid molecules. Then we look at the three-dimensional structure of nucleic acids, and finally the manner in which genetic information is coded, used in the synthesis of proteins, and transmitted from cell to cell.

Controlled hydrolysis, catalyzed by aqueous acid, base, or specific enzymes, breaks nucleic acids into successively smaller fragments and ultimately into three characteristic components: (1) phosphoric acid, (2) a pentose, and (3) heterocyclic bases. Nucleic acids which on hydrolysis yield the pentose ribose (p. 235) are termed ribonucleic acids (RNAs); those yielding deoxyribose (p. 235) are termed deoxyribonucleic acids (DNAs).

nucleic acids (DNA,RNA)

$\downarrow$ H_2O catalyst

mononucleotides

$\downarrow$ H_2O catalyst

nucleosides + phosphoric acid

$\downarrow$ H_2O catalyst

nitrogenous bases + D-ribose or 2-deoxy-D-ribose

Figure 14.1 Hydrolysis products of nucleic acids.

Two classes of heterocyclic bases are found on hydrolysis of nucleic acids: those related to pyrimidine and those related to purine. The heterocyclic bases related to pyrimidine are cytosine (C), uracil (U), and thymine (T).

Cytosine has been found in all nucleic acids examined thus far. Uracil is present in all samples of RNA but is usually absent from DNA. Conversely, thymine is found in samples of DNA but is only rarely found in RNA. It often is possible to distinguish DNA from RNA by the presence or absence of thymine. In addition to these three major pyrimidines, several others (e.g. 5-methylcytosine) occur in lesser amounts.

pyrimidine   cytosine (C)   uracil (U)   thymine (T)

The heterocyclic bases related to purine are adenine (A) and guanine (G).

purine   adenine (A)   guanine (G)

Adenine and guanine are universally distributed in all nucleic acids. Other purines, e.g. N^6-methyladenine, occur in small amounts in certain specific nucleic acids. Note that the designation N^6 specifies that the methyl substituent is attached to the nitrogen atom on carbon-6 of the adenine ring.

14.2 NUCLEOSIDES

Nucleosides are a combination of a purine or pyrimidine base and a pentose, either D-ribose or 2-deoxy-D-ribose. Two nucleosides, adenosine and 2′-deoxycytidine, are shown in Figure 14.2. Natural nucleosides are glycosides in which a nitrogen (N-9 of the purines or N-1 of the pyrimidines) is bonded by a glycoside linkage to the carbon 1′ of ribose or deoxyribose. The pentose is always in the furanose form

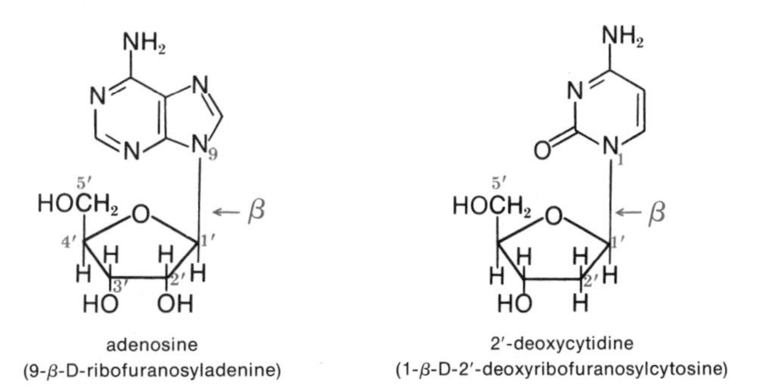

Figure 14.2 Nucleosides: adenosine and 2′-deoxycytidine. Unprimed numbers are used for atoms of the purine and pyrimidine bases; primed numbers are used for atoms of the pentoses.

and the glycoside linkage is always beta (see Section 10.6). Like all glycosides, the nucleosides are relatively stable to aqueous alkali but undergo hydrolysis in acid to form a base and a pentose.

14.3 MONONUCLEOTIDES

The mononucleotides are phosphoric acid esters of nucleosides in which the phosphoric acid is esterified with one of the hydroxyls of the pentose. There are three hydroxyl groups in ribose and two free hydroxyl groups in deoxyribose and, in principle, the phosphoric ester can be formed with any one of these hydroxyls. Although all variations do exist in cells, the predominant ester is the 5′-phosphate. These compounds, illustrated below by the 5′-esters of adenosine, are named either as esters (adenosine monophosphate, abbreviated AMP) or as acids (adenylic acid). Note that the two protons on each phosphate ester are fully ionized.

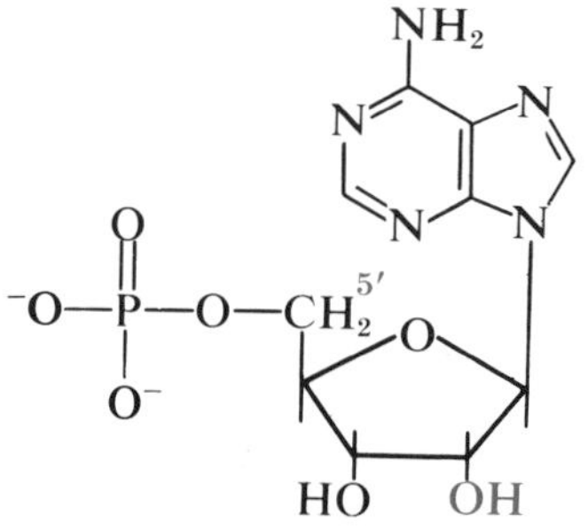

adenosine 5′-monophosphate

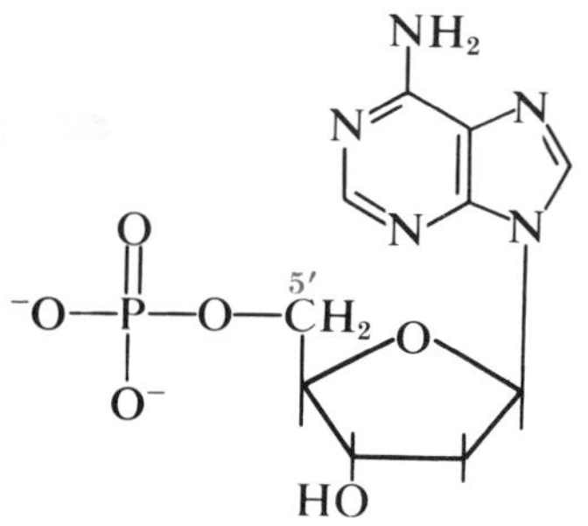

deoxyadenosine 5′-monophosphate

Table 14.1 lists the names of the major mononucleotides derived from ribonucleic and deoxyribonucleic acids.

All nucleoside monophosphates may be further phosphorylated to form nucleoside diphosphates and nucleoside triphosphates. In the case of the di- and triphosphates, the second and third phosphate groups are

Table 14.1 *The major mononucleotides derived from DNA and RNA. Each is named as a monophosphate, as an acid, and by a three- or four-letter abbreviation.*

Mononucleotides Derived from Ribonucleic acids	Mononucleotides Derived from Deoxyribonucleic acids
adenosine monophosphate adenylic acid, AMP	deoxyadenosine monophosphate deoxyadenylic acid, dAMP
guanosine monophosphate guanylic acid, GMP	deoxyguanosine monophosphate deoxyguanylic acid, dGMP
cytidine monophosphate cytidylic acid, CMP	deoxycytidine monophosphate deoxycytidylic acid, dCMP
uridine monophosphate uridylic acid, UMP	(uridine monophosphate not a major component of DNA)
(thymidine monophosphate not a major component of RNA)	deoxythymidine monophosphate deoxythymidylic acid, dTMP

joined by anhydride bonds. Attachment of one molecule of phosphate to adenosine by an anhydride bond forms adenosine diphosphate (ADP); attachment of two molecules of phosphate forms adenosine triphosphate (ATP), shown below.

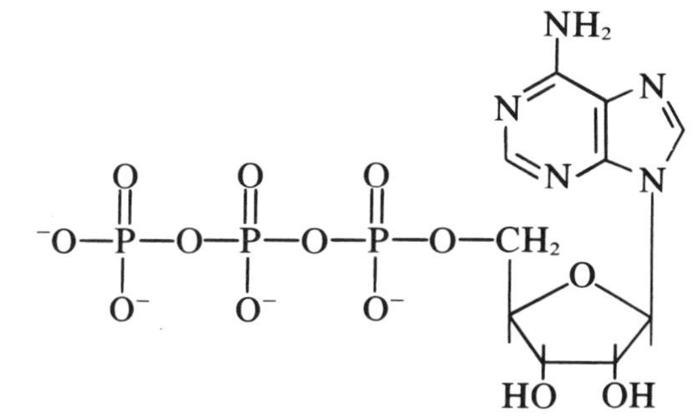

adenosine triphosphate (ATP)

At the physiological pH, all four phosphate protons are fully ionized and the molecule has a net charge of -4.

14.4 DNA—COVALENT STRUCTURE

Deoxyribonucleic acid (DNA) consists of a backbone of alternating units of deoxyribose and phosphate in which the 3′-hydroxyl of one deoxyribose is joined to the 5′-hydroxyl of the next deoxyribose by a phosphodiester bond (Figure 14.3). This backbone is constant throughout the entire DNA molecule. A heterocyclic base, either adenine, guanine, thymine, or cytosine, is then attached to each deoxyribose by a β-glycoside bond.

Here we might compare the primary structures of DNA and a protein molecule and note that each consists of a backbone or constant portion and a variable portion. In DNA, the backbone consists of deoxyriboses linked by phosphodiester bonds; in a protein, the backbone consists of repeating peptide bonds. In DNA, the variable portion is the sequence of four heterocyclic bases along the backbone; in proteins, the variable portion is the sequence of amino acid side chains along the backbone.

Another characteristic of DNA molecules is that one end of the chain has the 5′-hydroxyl of deoxyribose free while the other end has the 3′-hydroxyl free. By convention, nucleic acid chains are always written beginning with the free 5′-hydroxyl group and proceeding toward the free 3′-hydroxyl group. Given this way of writing polynucleotide chains, it should be obvious that ATC and CTA are different trinucleotides, just as gly—ala—ser and ser—ala—gly are different tripeptides.

Naturally occurring DNA molecules are long and thread-like, and have very high molecular weights. For example, the chromosome of the bacteria *E. coli* is a single DNA molecule of molecular weight 2,300,000,000 (2.3 trillion). Its diameter is only 20 angstroms (20×10^{-8} cm) but its total length is 0.12 cm. By comparison, a molecule of collagen, one of the longest proteins, has a length of only 0.00003 cm.

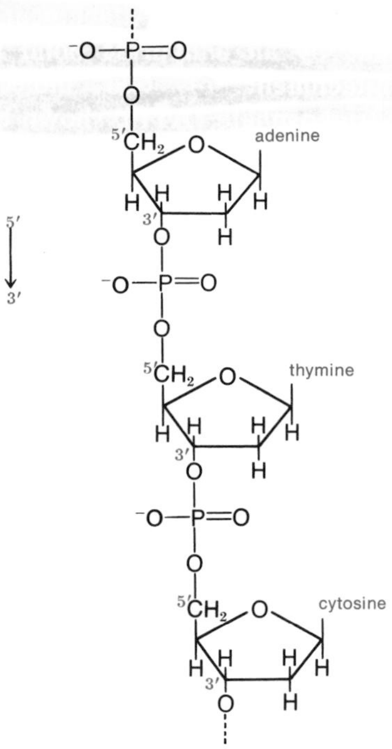

adenine

thymine

cytosine

5′ → 3′

----ApTpCp-------
abbreviated formula
for this trinucleotide
sequence

Figure 14.3 *Deoxyribonucleic acid (DNA). Partial structural formula showing a trinucleotide sequence. In the abbreviated sequence shown (right), the bases of the trinucleotide are read from the 5′ end of the chain to the 3′ end as indicated by the arrow (far left).*

14.5 DNA—BIOLOGICAL FUNCTION

Nucleic acids were discovered in 1869 by the Swiss chemist Miescher in the nuclei of pus cells. However, it took many decades before the major function of DNA was determined. One of the first clues to its biological function grew out of an observation by the British physician Fred Griffith. Griffith had isolated and cultured the *Pneumococcus* bacterium which causes pneumonia in humans and other susceptible animals. These bacteria are normally surrounded by a polysaccharide wall or capsule made up of alternating units of glucose and glucuronic acid (Section 10.3). This polysaccharide capsule is essential for pathogenicity, and mutants without it are nonpathogenic. Normal, infectious pneumococci with the polysaccharide capsule form smooth colonies and for this reason are designated S. Those mutants without the polysaccharide capsule form much smaller, rough-looking colonies and are designated R.

In 1928, Griffith discovered that a particular type of noninfectious R mutant could be transformed into a strain of infectious S bacteria in the following manner. First he injected live mice with the R mutant and showed that it was noninfectious. Next he injected live mice with heat-killed S bacteria, which also proved noninfectious. Finally, Griffith injected mice with a mixture of live R bacteria and heat-killed S bacteria, and made the startling discovery that this mixture was infectious and lethal to the mice. Neither the live R bacteria nor the heat-killed S bacteria was lethal by itself, but the mixture of the two was! When Griffith isolated and cultured the pathogenic bacteria from the dead mice, he found that they were live S pneumococci. In the animal, the R bacteria had been transformed into S bacteria.

Other biologists pursued the investigation of this transformation. They fractionated the components of S cells and tested the ability of each to transform R cells into S cells. In 1944, Oswald Avery, Colin MacLeod, and Maclyn McCarty reported that "a nucleic acid of the deoxyribose type is the fundamental unit of the transforming principle of the *Pneumococcus* Type III." Moreover, they found that DNA-induced transformation of R cells into S cells is a permanent, heritable characteristic. Thus, for the first time it was realized that genetic information is stored in DNA. This discovery has been confirmed over and over again by a wide range of biochemical investigations and today there is no doubt about the fact that genes are made of DNA.

14.6 THE THREE-DIMENSIONAL STRUCTURE OF DNA

By 1950, there was general agreement that DNA molecules consist of chains of alternating units of deoxyribose and phosphate groups linked by phosphodiester bonds and with a heterocyclic base attached to each deoxyribose. Although the general structural formula of DNA was known, the precise sequence of bases along the chain was completely unknown for any particular DNA molecule. At one time it was thought that the heterocyclic bases occurred in equal ratios and that perhaps the four major bases might repeat in a regular pattern along the pentose–phosphate backbone of the molecule. However, more precise determinations of base composition (Table 14.2) revealed that the bases do not occur in equal ratios.

Table 14.2 *Comparison of base composition (in mole percent) of DNAs from several organisms.*

Organism	A	G	C	T	A/T	G/C	purines/pyrimidines
man	30.9	19.9	19.8	29.4	1.05	1.00	1.04
sheep	29.3	21.4	21.0	28.3	1.03	1.02	1.03
sea urchin	32.8	17.7	17.3	32.1	1.02	1.02	1.02
marine crab	47.3	2.7	2.7	47.3	1.00	1.00	1.00
yeast	31.3	18.7	18.7	32.9	0.95	1.09	1.00
E. coli	24.7	26.0	26.0	23.6	1.04	1.01	1.03

From consideration of data such as these, the following conclusions emerged:

1. The mole percent base composition of DNA in any organism is the same in all cells and is characteristic of the organism.
2. Base compositions vary from one organism to another. The DNAs of closely related species have similar base compositions whereas the DNAs of widely differing species are likely to have widely different base compositions.

3. In nearly all DNAs, the mole percent of adenine equals that of thymine, and the mole percent of guanine equals that of cytosine.
4. The total number of purine residues (A + G) equals the total number of pyrimidine residues (C + T).

Additional information on the structure of DNA emerged from analysis of X-ray diffraction photographs of DNA fibers taken by Rosalind Franklin and Maurice Wilkins. It became clear that DNA molecules are long, fairly straight, and not more than a dozen atoms thick. Further, despite the fact that the base composition of DNAs isolated from different organisms varied over a rather wide range, the DNA molecules themselves appeared to be remarkably uniform in thickness. Herein lay one of the major problems to be solved. How could the structure appear so regular in molecular dimensions when the ratios of bases differed so widely? There was also another problem to be solved. In what form is the genetic information stored in DNA molecules and how is it transmitted or replicated from one generation or cell to the next?

With this accumulated information the stage was set for the development of a hypothesis about DNA conformation. In 1953 F. H. C. Crick, a British physicist, and James D. Watson, an American biologist, postulated a precise model of the three-dimensional structure of the DNA molecule, a model in accord with the known dimensions of the structural units, the equivalences of certain base ratios, and the measured dimensions of intact molecules. This model not only accounted for many of the observed physical and chemical properties of DNA but also suggested a mechanism by which genetic information could be repeatedly and accurately replicated. Watson, Crick, and Wilkins shared the 1962 Nobel Prize in Physiology and Medicine for "their discoveries concerning the molecular structure of nucleic acids, and its significance for information transfer in living material."

The heart of their model is the postulate that a molecule of DNA consists of two polynucleotide chains both coiled in a right-handed manner about the same axis to form a double helix. To account for the observed base ratios and the constant thickness of the molecular strand, Watson and Crick postulated that the purine and pyrimidine bases project inward toward the axis of the helix and that the bases always pair in a very specific manner. Thymine is always hydrogen-bonded to adenine and cytosine is always hydrogen-bonded to guanine. The hydrogen bonding between base pairs is a major factor stabilizing the three-dimensional conformation of DNA (see Figure 14.4).

According to scale models, the dimensions of the thymine–adenine base pair are identical with those of the cytosine–guanine base pair, and the length of each couple is consistent with the thickness of the DNA strand. A significant fact arising from this model building is that no other base pairing is consistent with the observed regularities of DNA structure. A pair of pyrimidines would be too small to account for the observed thickness of the DNA molecule; a pair of purine bases would be too big to fit into the observed thickness.

The formation of such hydrogen-bonded base pairs with identical overall dimensions resolved the paradox presented by the X-ray data:

Figure 14.4 *Hydrogen-bonded interaction between thymine and adenine; cytosine and guanine. The first couple is abbreviated as T═A (showing two hydrogen bonds) and the second couple as C≡G (showing three hydrogen bonds).*

the emergence of repetitive structure out of the disorder of random base composition. According to the model, the repeating units are not single bases of differing dimensions, but specific base pairs of identical dimensions.

To account for the periodicity observed from X-ray data, Watson and Crick postulated that the bases are closely stacked perpendicular to the axis of the helix at a distance of 3.4 angstroms between base pairs, and exactly ten base pairs are coiled in one complete turn of the helix. The distance (34 Å) between the repeat pattern in the crystal corresponds to one complete turn of the helix.

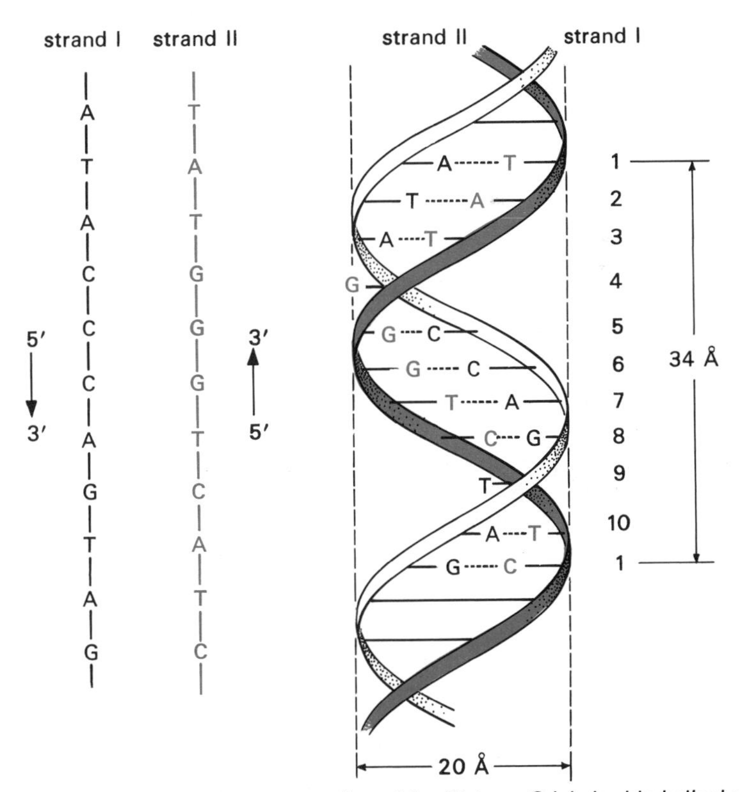

Figure 14.5 *Abbreviated representation of the Watson–Crick double-helical model of DNA. On the left are shown two uncoiled complementary antiparallel polynucleotide strands. On the right the strands are twisted in a double helix of thickness 20 angstroms and a repeat distance of 34 angstroms along the axis of the double helix. There are 10 base pairs per complete turn of the helix.*

14.7 DNA REPLICATION

At the time Watson and Crick postulated the model for the conformation of DNA, biologists were amassing evidence that DNA was in fact the hereditary material. Detailed studies of living cells revealed that during cell division there is an exact copying or duplication of chromosomes, and that chromosomes contain DNA and some protein. The challenge posed to molecular biologists was this: How does the genetic material duplicate itself with such unerring fidelity?

One of the exciting things about the double helix model is that it immediately suggested how DNA might produce an exact copy of itself. The double helix consists of two parts, each the precise complement of the other. If the two strands separate and each serves as a template for the construction of its own new complement, then each new double strand is an exact replica of the original DNA. Because each new double-stranded DNA molecule contains one strand from the parent molecule and one newly synthesized strand, the process is said to be a semiconservative replication. This process is illustrated schematically in Figure 14.6.

This hypothesis of DNA replication was simple and at the same time revolutionary. Yet it was at this stage only a hypothesis. Could it be confirmed experimentally? Within a few years, two separate and ingenious lines of research led to the experimental confirmation of this bold hypothesis. Meselson and Stahl in 1957–1958 demonstrated conclusively that in living intact bacteria, *Escherichia coli (E. coli)*, DNA

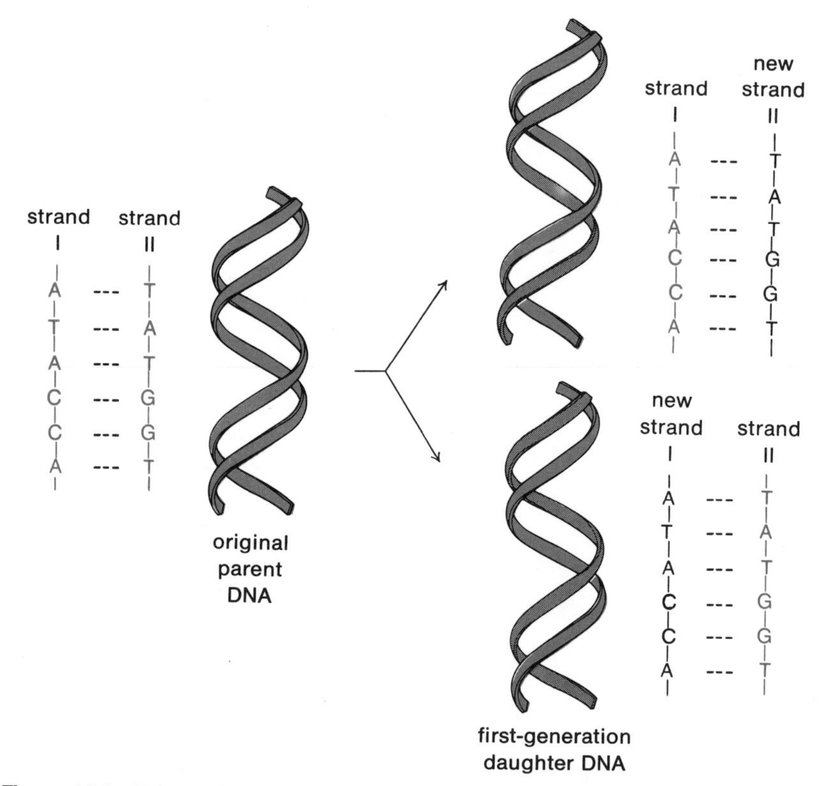

Figure 14.6 *Schematic diagram of the replication mechanism by which DNA might duplicate itself. The double helix uncoils and each chain of the parent serves as a template for the synthesis of its complement. Each daughter DNA contains one strand from the original DNA and one newly synthesized strand.*

is replicated in exactly the manner predicted by Watson and Crick.

The American biochemist Arthur Kornberg approached this problem in quite another manner by investigating the enzymatic mechanism of DNA replication. The Watson–Crick model suggested that DNA acts as a template, inducing an enzyme to synthesize new DNA. Kornberg reasoned that the starting materials for this enzyme-catalyzed DNA synthesis must be the four deoxyribonucleoside triphosphates. Accordingly he and his colleagues incubated dATP, dGTP, dTTP, and dCTP with crude extracts of *E. coli* and were able to demonstrate the synthesis of new DNA. The enzyme catalyzing this reaction is named DNA polymerase.

$$\begin{matrix} n_1\text{dATP} \\ n_2\text{dGTP} \\ n_3\text{dCTP} \\ n_4\text{dTTP} \end{matrix} \xrightarrow[\text{DNA polymerase}]{\text{DNA (template)}} \text{DNA}\left\{\begin{matrix} \text{dAMP} \\ \text{dGMP} \\ \text{dCMP} \\ \text{dTMP} \end{matrix}\right. + (n_1 + n_2 + n_3 + n_4)\text{pyrophosphate}$$

The striking and characteristic property of DNA polymerase is that it requires the presence of some preexisting DNA and that the newly synthesized DNA has a base composition nearly identical to that of the added DNA. After painstaking and laborious purification procedures, Kornberg was able to isolate a small amount of crystalline DNA polymerase. More recently he has shown that DNA polymerase is capable of inducing the replication of an infectious bacterial virus and that the synthetic product has the infecting capacity of the original material.

Thus, the experiments of Meselson and Stahl on the synthesis of DNA in living, intact *E. coli* cells, and those of Kornberg on the enzymatic mechanism involving DNA polymerase have provided convincing evidence that DNA is replicated in accord with the prediction of the Watson–Crick model.

14.8 RIBONUCLEIC ACID—RNA

Ribonucleic acids (RNAs) are similar to deoxyribonucleic acids in that they too consist of long, unbranched chains of nucleotides joined by phosphodiester bonds between the 3′-hydroxyl of one pentose and the 5′-hydroxyl of the next. Thus, their structure is much the same as that of DNA shown in Figure 14.3 except that (1) the pentose in RNA is ribose rather than deoxyribose, and (2) one of the four major bases normally is uracil rather than thymine. Like thymine, uracil can form a hydrogen-bonded base pair with adenine. However, it lacks the methyl group present in thymine.

Unlike DNA, RNA is distributed throughout the cell; it is present in the nucleus, the cytoplasm, and even in subcellular particles such as the mitochondria. Further, cells contain three types of RNA: ribosomal RNA, transfer RNA, and messenger RNA. These three types differ in molecular weight and, as their descriptive names imply, they perform different functions within the cell.

Ribosomal RNA (rRNA) molecules have molecular weights of 0.5–1.0 million and comprise up to 85–90% of the total cellular ribonucleic acid. In the cytoplasm of the cell are located a large number of subcellular particles called ribosomes, which contain about 60% rRNA and 40% protein. The intact ribosomal particles are an essential component of the protein-synthesizing machinery of the cell.

Transfer RNA (tRNA) has the smallest structure of all the nucleic acids. There are some 20 known tRNAs, each corresponding to one of the 20 amino acids found in proteins. Transfer RNAs have molecular weights of about 25,000 and contain 75–80 mononucleotide residues per molecule. The function of tRNA during protein synthesis is to carry a specific amino acid to a site on the ribosome. For this transportation, an amino acid is joined to a specific tRNA in an enzyme-catalyzed ester formation between the α-carbonyl group of the amino acid and the 3′-hydroxyl of the ribose from the terminal tRNA nucleotide. Amino acids so esterified are said to be "activated." Several tRNAs have been isolated and their base sequences decoded.

Messenger RNA (mRNA) is present in the cell in relatively small amounts and has a high turnover rate. This RNA fraction has an average molecular weight of several hundred thousand and a base composition like that of the DNA of the organism from which it is isolated. For this reason, its discoverers called it DNA-like RNA. Messenger RNA is synthesized in a cell using DNA as a template and the reaction is catalyzed by an enzyme named RNA polymerase. For the synthesis of mRNA the cell requires (1) the four major ribonucleoside triphosphates, (2) DNA as a template, and (3) RNA polymerase.

$$\begin{matrix} n_1\text{ATP} \\ n_2\text{GTP} \\ n_3\text{CTP} \\ n_4\text{UTP} \end{matrix} \xrightarrow[\text{RNA polymerase}]{\text{DNA (template)}} \text{RNA}\left\{\begin{matrix} \text{AMP} \\ \text{GMP} \\ \text{CMP} \\ \text{UMP} \end{matrix}\right. + (n_1 + n_2 + n_3 + n_4)\text{pyrophosphate}$$

The name "messenger" RNA was coined by the French scientists F. Jacob and J. Monod because this RNA, which is made in the nucleus of the cell on a DNA template, migrates into the cytoplasm and to the ribosome where it is involved in protein synthesis.

14.9 THE GENETIC CODE

Given our understanding of the structure of DNA and its function as an archive of genetic information for the entire organism, the next questions to be asked are: In what manner is this genetic information coded on the DNA molecule? What is it coded for? How is this code read and expressed? In answer to the first two questions, it is now clear that the sequence of bases in the DNA molecule constitutes the store of genetic information and that this sequence of bases serves to direct protein synthesis. However, the statement that the sequence of bases is the genetic information and that this sequence directs the synthesis of proteins presents a paradox. How can a molecule consisting of only four variable units (adenine, cytosine, guanine, thymine) direct the synthesis of a protein in which there are as many as 20 different units? How can a 4-letter alphabet specify the sequence of the 20-letter alphabet that occurs in proteins?

One obvious answer is that it is not one base but some combination of bases that codes for each amino acid. If the code consists of nucleoside pairs, there are 16 (4^2) combinations, a more extensive code, but still not extensive enough to code for 20 amino acids. If the nucleosides are considered in groups of three, there are 64 (4^3) possible combinations, more than enough to specify the primary sequence of a protein. This appears to be a very simple solution to a system that must have taken eons of evolutionary trial and error to develop. Yet there is convincing evidence that nature does indeed use this simple 3-letter code to store genetic information. One of the triumphs of molecular biology is that the code has been deciphered. But before we analyze the code itself, let us examine how the code directs protein synthesis.

14.10 PROTEIN SYNTHESIS

The process of information transfer in protein synthesis is formulated as follows. Each gene (a section of a DNA molecule) contains a sequence of bases that dictates the sequence of amino acids in a protein. For example, to code for the sequence of 146 amino acids in the β chain of normal adult hemoglobin (p. 304), there must be a run of 438 nucleotides (146 triplets) on a section of DNA. This section serves as a template for the synthesis of a complementary strand of mRNA. Thus the sequence —CGAATTA— in DNA directs the synthesis of the complementary sequence —GCUUAAU— in mRNA, a process called transcription.

The newly synthesized mRNA migrates to the cytoplasm and attaches to a ribosome, thus initiating the process in which the information encoded in mRNA is expressed as a precise sequence of amino acids in a polypeptide molecule. This process is called translation. Each of the 20 amino acids is brought to the site of protein synthesis as an "activated" ester of a specific tRNA. Each tRNA has in its structure a triplet complementary to three bases in mRNA. The tRNA triplet that is the complement of a mRNA codon is called an anticodon. These complementary triplets interact by hydrogen bonding, and in so doing position the activated amino acid in its proper sequence along the polypeptide chain. Finally, enzyme-catalyzed peptide bond formation between the carboxyl group of one amino acid and the amino group of the next initiates growth of the polypeptide chain. The strand of mRNA is "read off," triplet by triplet, until the polypeptide chain is complete.

14.11 THE GENETIC CODE DECIPHERED

The next question of course is: Which particular triplets code for each amino acid? At one time it seemed that the only hope of answering this question was to isolate a section of a gene coding for a particular protein and then compare the sequence of amino acids on the protein with the sequence of bases on the corresponding section of DNA. However, this was not experimentally possible even as late as 1960 because the base sequences of genes were unknown.

Fortunately, the young biochemist Marshall Nirenberg provided a simple and very direct experimental approach to the problem. It was based on the observation that synthetic polynucleotides will direct polypeptide synthesis in much the same manner as mRNA. Therefore Nirenberg incubated ribosomes, amino acids, tRNA, and the appropriate protein-synthesizing enzymes. With only these components, there was essentially no polypeptide synthesis. However, when Nirenberg added synthetic polyuridylic acid (poly U), a polypeptide of high molecular weight was formed. The exciting result of this experiment was that the polypeptide contained only phenylalanine. Poly U had served as a synthetic messenger RNA. With this discovery, the first element of the genetic code had been deciphered. The triplet UUU codes for the amino acid phenylalanine.

This same type of experiment was carried out with different polyribonucleotides. It was found that polyadenylic acid (poly A) led to the synthesis of polylysine and that polycytidylic acid (poly C) led to the synthesis of polyproline.

codon	amino acid
UUU	phenylalanine
AAA	lysine
CCC	proline

This strategy was extended and by 1966 all 64 codons had been deciphered. Table 14.3 lists these codons and the amino acid that each one codes.

Table 14.3 *The genetic code: the mRNA codons and the amino acid whose incorporation each codon directs.*

UUU	Phe	UCU	Ser	UAU	Tyr	UGU	Cys
UUC	Phe	UCC	Ser	UAC	Tyr	UGC	Cys
UUA	Leu	UCA	Ser	UAA	Stop	UGA	Stop
UUG	Leu	UCG	Ser	UAG	Stop	UGG	Trp
CUU	Leu	CCU	Pro	CAU	His	CGU	Arg
CUC	Leu	CCC	Pro	CAC	His	CGC	Arg
CUA	Leu	CCA	Pro	CAA	Gln	CGA	Arg
CUG	Leu	CCG	Pro	CAG	Gln	CGG	Arg
AUU	Ile	ACU	Thr	AAU	Asn	AGU	Ser
AUC	Ile	ACC	Thr	AAC	Asn	AGC	Ser
AUA	Ile	ACA	Thr	AAA	Lys	AGA	Arg
AUG	Met	ACG	Thr	AAG	Lys	AGG	Arg
GUU	Val	GCU	Ala	GAU	Asp	GGU	Gly
GUC	Val	GCC	Ala	GAC	Asp	GGC	Gly
GUA	Val	GCA	Ala	GAA	Glu	GGA	Gly
GUG	Val	GCG	Ala	GAG	Glu	GGG	Gly

A number of features of the genetic code are evident from Table 14.3.

1. Only 61 triplets code for amino acids. The remaining three (UAA, UAG, and UGA) are signals for chain termination, i.e., they are signals to the protein-synthesizing machinery of the cell that the primary structure of the protein is complete.
2. The code is degenerate. Since there are 20 amino acids and 61 triplets to code for them, obviously many amino acids must be coded for by more than one triplet. If you count the number of triplets coding for each amino acid, you will find that only methionine and tryptophan are coded by just one triplet. The other 18 are coded by two or more triplets.
3. In all cases where an amino acid is coded by two, three or four triplets, the degeneracy is only in the last base of the triplet. In other words, in the codons for these 15 amino acids, it is only the third letter of the code that varies. For example, glycine is coded by the triplets GGA, GGG, GGC, and GGU.
4. Finally, there is no ambiguity in the code. Each triplet codes for one and only one amino acid.

We must ask one last question about the genetic code, namely is the code universal—is it the same for all organisms? Every bit of experimental evidence available today from the study of viruses, bacteria, and higher animals including man indicates that the code is the same for all organisms and that it is universal. Further, the fact that it is the same in all these organisms means that it has been the same over billions of years of evolution.

Problems

14.1 Examine the structure of purine. Would you predict this molecule to be planar (or nearly so) or puckered; to exist as a number of interconvertible conformations (as in the case of cyclohexane) or to be rigid and inflexible? Explain the basis for your answers.

14.2 Draw structural formulas for 5-methylcytosine, N^6-methyladenine, and N^2-methylguanine.

14.3 An important drug in the chemotherapy of leukemia is 6-mercaptopurine. Draw the structural formula of this compound.

14.4 Compare and contrast the structural formulas of:

a ribose and deoxyribose
b nucleoside and nucleotide
c nucleotide and nucleic acid

14.5 Draw structural formulas for:

a uridine monophosphate
b guanosine triphosphate
c adenosine diphosphate
d deoxycytidine monophosphate
e deoxythymidylic acid

14.6 Show by structural formulas the hydrogen bonding between thymine and adenine; between uracil and adenine.

14.7 Cyclic-AMP (adenosine-3′,5′-cyclic monophosphate), first isolated in 1959, is involved in many diverse biological processes as a regulator of metabolic and physiological activity. In it, a single phosphate group is esterified with both the 3′- and 5′-hydroxyls of adenosine. Draw a structural formula for this substance.

14.8 Compare and contrast the α-helix found in proteins with the double helix of DNA in the following ways:

a The units that repeat in the backbone of the chain.
b The projection in space of the backbone substituents (R– groups in the case of amino acids, purine and pyrimidine bases in the case of DNA) relative to the axis of the helix.
c The importance of the backbone substituents in stabilizing the helix.
d The number of backbone units per complete turn of the helix.

14.9 List the postulates of the Watson–Crick model of DNA structure. This model is based on certain experimental observations of base composition and molecular dimensions. Describe these observations and show how the model accounts for each.

14.10 Describe the process of DNA replication.

14.11 Compare and contrast DNA and RNA in the following ways:

a primary structure **b** the major purine and pyrimidine bases present
c location in the cell **d** function in the cell

14.12 Compare and contrast ribosomal RNA, transfer RNA, and messenger RNA as follows:

a molecular weight
b function in protein synthesis

14.13 List all amino acids that are specified by

a only a single codon **b** two codons
c three codons **d** four codons
e five codons **f** six codons

14.14 List all amino acids that have either U or C (both pyrimidines) as the second base of a codon. Also list all amino acids that have either A or G (both purines) as the second base of a codon. What generalization can you draw from these lists?

14.15 What peptide sequences are coded for by the following mRNA sequences?

a —G-C-U-G-A-A-U-G-G— **b** —U-C-A-G-C-A-A-U-C—
c —G-U-C-G-A-G-G-U-G— **d** —G-C-U-U-C-U-U-A-A—

14.16 There are two principal types of mutations: (1) substitution of one mononucleotide for another, as for example A for C in mRNA, and (2) insertion or deletion of a mononucleotide. On p. 345 is shown a mRNA sequence which is read from left to right beginning with the codon UCC. Below it are three substitution mutations and one insertion mutation. For what polypeptide sequence does the normal mRNA sequence code and what is the effect of each mutation on the resulting polypeptide?

a normal —U-C-C-C-A-G-G-C-U-U-A-C-A-A-A-G-U-A—

b substitution of A for C —U-C-C-A-A-G-G-C-U-U-A-C-A-A-A-G-U-A—

c substitution of A for C —U-C-C-C-A-G-G-C-U-U-A-A-A-A-A-G-U-A—

d substitution of A for C —U-C-A-C-A-G-G-C-U-U-A-C-A-A-A-G-U-A—

e insertion of A —U-C-C-C-A-G-G-C-U-A-U-A-C-A-A-A-G-U-A—

14.17 In HbS, the abnormal human hemoglobin found in individuals with sickle-cell anemia (p. 304), glutamic acid at position 6 in the β-chain is replaced by valine.

a List the two codons for glutamic acid and the four codons for valine.
b Show that a glutamic acid codon can be converted into a valine codon by a single mononucleotide substitution.

14.18 Examine Table 1 of the mini-essay "Abnormal Human Hemoglobins" and show that each amino acid substitution is consistent with the change of a single nucleotide in a codon.

15

THE FLOW OF ENERGY IN THE BIOLOGICAL WORLD

15.1 INTRODUCTION

During the introduction to organic chemistry we concentrated on four major themes: (1) the structure of organic molecules, (2) the relationships between structure and physical properties, (3) typical reactions of key functional groups, and (4) the three-dimensional shapes of biologically important macromolecules.

Now, as we proceed to an introduction to biochemistry and to the study of organic molecules in the living cell, we shall develop two more major themes. At this point let us state them as questions which we will try to answer in this and the remaining chapters. First, how do cells extract energy from foodstuffs? Second, how do cells use this energy to synthesize other small molecules, including the simple building blocks of macromolecules; that is, how do cells use this energy of foodstuffs for growth and development?

15.2 OXIDATION OF FOODSTUFFS AND THE GENERATION OF ATP

The uniqueness of living systems rests in their ability to capture energy from the environment, to store it at least temporarily, and to use it to power the vast number of biological processes vital to life. This energy for biological processes comes ultimately from the sun, whose enormous energy is derived from the conversion of hydrogen atoms into helium atoms. A portion of this energy streams toward us as sunlight and is absorbed by the chlorophyll pigments in plant cells. There, through the process of photosynthesis, it is transformed into a chemical storage form of energy, namely glucose and polysaccharides.

$$6CO_2 + 6H_2O + \text{energy} \xrightarrow{\text{photosynthesis}} \underset{\text{glucose}}{C_6H_{12}O_6} + 6O_2$$

Glucose and plant polysaccharides are the immediate products of photosynthesis. In secondary steps, plants convert carbohydrates into lipids, proteins, and other foodstuffs.

Our first question in this chapter is: How do cells extract this energy from foodstuffs? The basic strategy used by all cells is to oxidize foodstuffs and to convert a portion of the energy released into another, more immediately useful form of chemical energy, namely into adenosine triphosphate (ATP). This oxidation of foodstuffs and the generation of ATP is accomplished in four main stages. In the first stage (Figure 15.1), the cell hydrolyzes the vast number of different proteins, fats, and polysaccharides into fewer than 30 smaller molecules. These include the 20 amino acids, glucose, a few fatty acids, and glycerol.

$$\text{polysaccharides} \xrightarrow{\text{hydrolysis}} \text{monosaccharides}$$
$$\text{proteins} \xrightarrow{\text{hydrolysis}} \text{amino acids}$$
$$\text{fats} \xrightarrow{\text{hydrolysis}} \text{fatty acids + glycerol}$$

Figure 15.1 *Stage I of the oxidation of foodstuffs and the generation of ATP.*

In the second stage (Figure 15.2), these few small molecules are further degraded to fewer, even smaller molecules. The carbon skeletons of glucose, fatty acids, glycerol, and several amino acids are degraded to acetate in the form of acetyl coenzyme A (acetyl CoA). The carbon skeletons of the other amino acids are degraded to different small molecules including succinate and α-ketoglutarate.

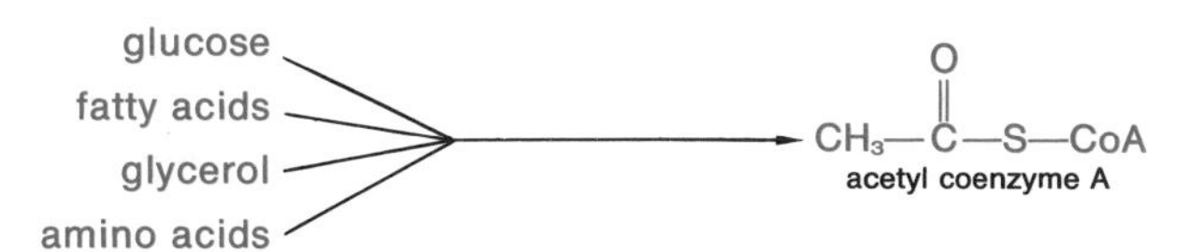

Figure 15.2 *Stage II of the oxidation of foodstuffs and the generation of ATP.*

Acetyl coenzyme A plays a key role in metabolism. In addition to being formed in the degradation of carbohydrates, fatty acids, and certain amino acids, it is used directly in stage III of the oxidation of foodstuffs. Furthermore, acetyl CoA is a starting material for the biosynthesis of many other biomolecules including steroids, new fatty acids, and terpenes. The structural formula of coenzyme A (CoA–SH) is shown in Figure 15.3.

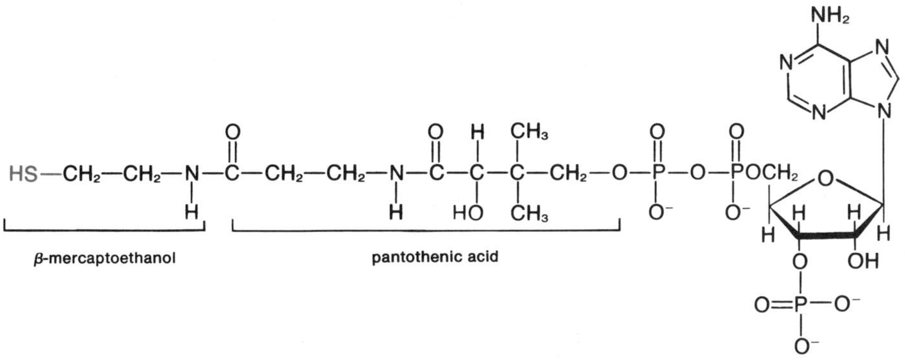

β-mercaptoethanol

pantothenic acid

Figure 15.3 *Coenzyme A. Pantothenic acid is one of the vitamins of the B group. The acetylated form of this coenzyme, designated acetyl coenzyme A, or acetyl CoA, is the thioester of acetic acid and the terminal sulfhydryl group.*

Stage III consists of a set of reactions known alternatively as the tricarboxylic acid cycle, the citric acid cycle, or the Krebs cycle (Figure 15.4). In stage III, acetyl coenzyme A and the other small molecules derived from stages I and II are completely oxidized to carbon dioxide.

$$CH_3-\overset{O}{\overset{\|}{C}}-S-CoA \xrightarrow[\text{acid cycle}]{\text{tricarboxylic}} 2CO_2 + CoA-SH$$

acetyl CoA — coenzyme A

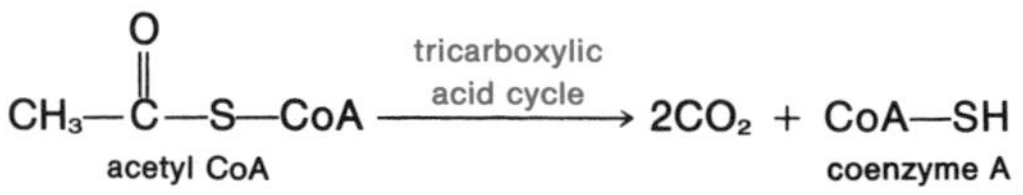

Figure 15.4 Stage III of the oxidation of foodstuffs and the generation of ATP. Small molecules derived from stages I and II are oxidized to carbon dioxide.

The biological oxidizing agents in stage III are nicotinamide adenine dinucleotide (NAD^+) and flavin adenine dinucleotide (FAD). The first of these, NAD^+ (Figure 15.5), is the major acceptor of electrons in the oxidation of fuel molecules.

nicotinamide unit

D-ribose unit

adenine unit

D-ribose unit

a phosphate ester in $NADP^+$

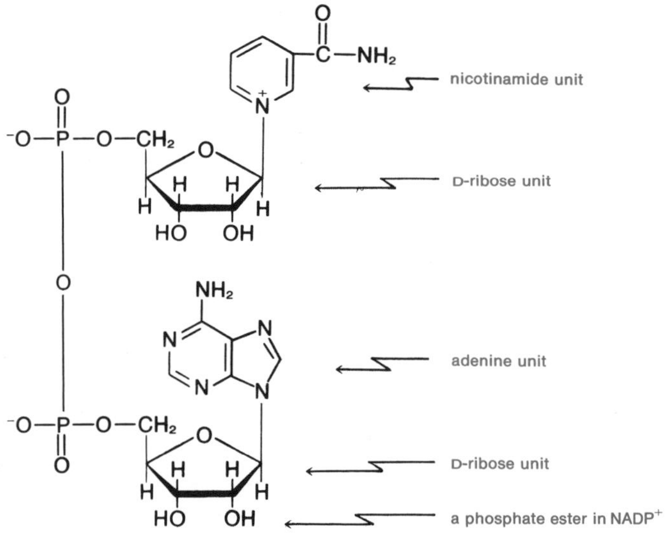

Figure 15.5 Nicotinamide adenine dinucleotide, NAD^+. Nicotinamide is one of the water-soluble vitamins. In nicotinamide adenine dinucleotide phosphate, $NADP^+$, the 2′-hydroxyl of a D-ribose unit is esterified with phosphoric acid.

As we saw in Section 5.12, the reactive group in this coenzyme is the pyridine ring, which can accept two electrons and one proton to form the reduced coenzyme, NADH.

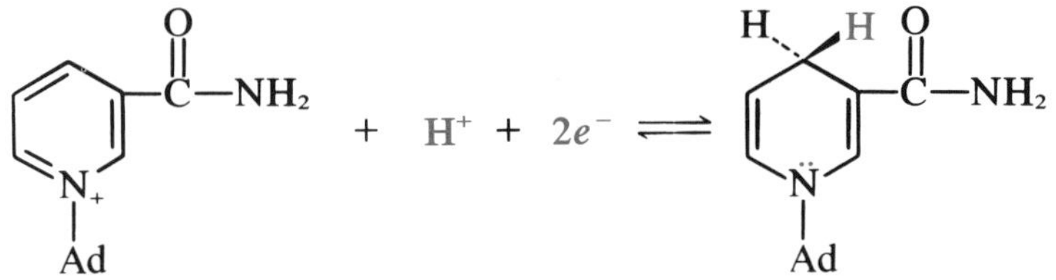

$$NAD^+ + H^+ + 2e^- \rightleftharpoons NADH$$

NAD⁺ — NADH

The other electron acceptor in the oxidation of fuel molecules is FAD, shown in Figure 15.6.

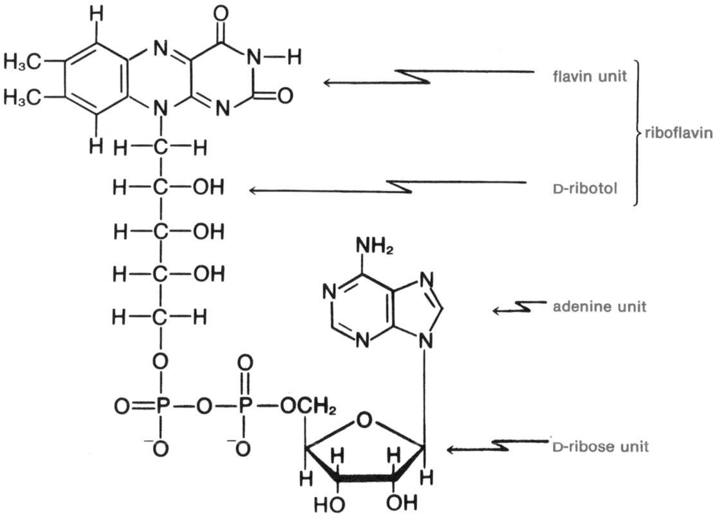

flavin unit
D-ribotol
riboflavin
adenine unit
D-ribose unit

Figure 15.6 *Flavin adenine dinucleotide, FAD. Riboflavin is one of the B group vitamins.*

The reactive group of this coenzyme is the flavin group, which accepts two electrons and two protons in forming the reduced coenzyme, $FADH_2$.

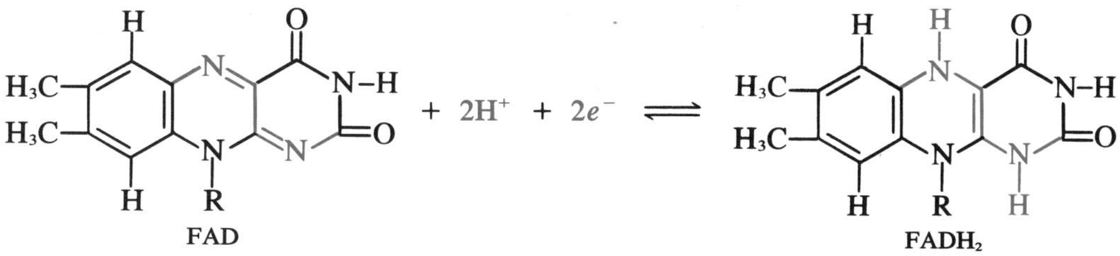

$$\text{FAD} + 2H^+ + 2e^- \rightleftharpoons \text{FADH}_2$$

Finally, in stage IV the reduced coenzymes (NADH and $FADH_2$) accumulated from stages II and III are reoxidized by molecular oxygen. This reoxidation is coupled with the phosphorylation of ADP to ATP, and for this reason stage IV is called oxidative phosphorylation. The net reactions of stage IV are shown in Figure 15.7.

$$2\text{NADH} + O_2 + 2H^+ \rightarrow 2\text{NAD}^+ + 2H_2O$$

$$2\text{FADH}_2 + O_2 \rightarrow 2\text{FAD} + 2H_2O$$

$$\text{ADP} + HPO_4^{2-} \rightarrow \text{ATP} + H_2O$$

Figure 15.7 *Stage IV of the oxidation of foodstuffs and the generation of ATP. The reoxidation of NADH and $FADH_2$ is coupled with the phosphorylation of ADP.*

Several very important things are accomplished by the cell in the reactions we have just outlined. First, the cell has recovered a portion of the chemical energy from glucose and other fuel molecules and has converted it into ATP, a more immediately useful form of chemical energy. The cell can then use ATP to drive other biochemical processes that require the input of energy, for example, muscle contraction and the synthesis of macromolecules. Second, the cell has generated a source of reducing power in the form of NADH and $FADH_2$. These in turn can be used as reducing agents for the synthesis of larger, more complex biomolecules. Third, the cell has created a pool of small molecules including acetyl coenzyme A, succinate, and α-ketoglutarate, which can serve as raw materials for building other biomolecules required for maintenance, growth, and development.

15.3 ATP AND THE TRANSFER OF BIOLOGICAL ENERGY

The central role of adenosine triphosphate in the transfer of biological energy was first realized by Herman Kalcker and Fritz Lipman about 1940. They observed that during the oxidation of glucose by incubated suspensions of liver or ground muscle, the concentration of free phosphate ions in solution decreased. Further investigation revealed that phosphate had become bound to an organic molecule later identified as ATP.

A second clue to the importance of ATP was the observation by Albert Szent-Györgyi that ATP is involved in muscle contraction. Muscle is made up of long fibers composed chiefly of the protein myosin. To free myosin fibers, Szent-Györgyi added some ATP and found that the fibers curled up on contact with ATP. This demonstration was a milestone in the history of science, for in Szent-Györgyi's words: "Motion is one of the most basic biological phenomena and has always been looked upon as the index of life. Now we could produce motion in a test tube with the constituents of the cell." With this demonstration, the production of ATP in the metabolic machinery of the cell and the use of ATP in the performance of cellular work were linked. From this and many other experiments it became clear that ATP is the central molecule in the performance of biological work.

An examination of thermodynamic data on the free energy of hydrolysis of ATP and other phosphate-containing compounds will help us to understand the unique role of ATP in the transfer of biological energy. Recall from Section 14.3 that ATP contains the heterocyclic base adenine bonded to D-ribose by a β-glycosidic linkage. This combination of adenine and ribose is called adenosine. Attachment of one phosphate group to the 5′-hydroxyl of ribose by a phosphate ester bond forms adenosine monophosphate (AMP); attachment of a second phosphate group by an anhydride bond forms adenosine diphosphate (ADP); and attachment of a third phosphate by an anhydride bond forms adenosine triphosphate (ATP), as shown in Figure 15.8.

Figure 15.8 *Structural formulas of AMP, ADP, and ATP. Adenosine consists of adenine linked to ribose by a β-glycosidic bond.*

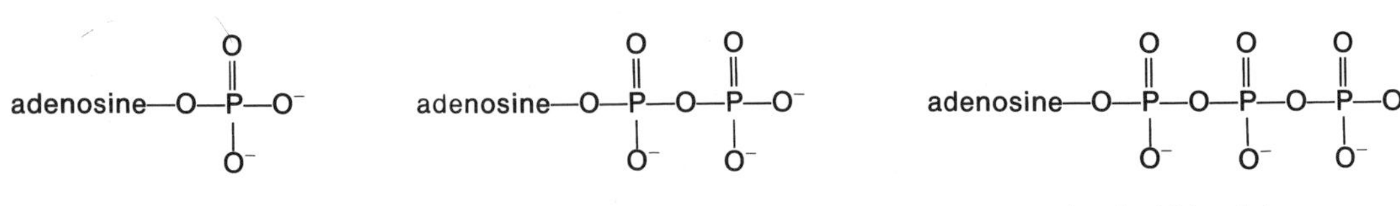

In discussing the central role of ATP in the transfer of biological energy, we can focus on just the triphosphate end of the molecule. The terminal phosphate anhydride of ATP undergoes hydrolysis (either uncatalyzed or catalyzed by an enzyme) to form ADP and inorganic phosphate. This hydrolysis is accompanied by a large decrease in free energy.

$$\text{ATP} + H_2O \longrightarrow \text{ADP} + HPO_4^{2-} \qquad \Delta G^{0\prime} = -7{,}300 \text{ cal/mole}$$

Table 15.1 shows the standard free energy of hydrolysis of ATP and several other phosphate-containing biomolecules. Notice from the data in this table that the free energy of hydrolysis of ATP is higher than that of simple phosphate esters like glucose 6-phosphate and glycerol 1-phosphate. It is for this reason that ATP is called a high-energy compound. High-energy compounds have $\Delta G^{0\prime}$ values of −5,000 to −15,000 cal/mole. Low-energy compounds have $\Delta G^{0\prime}$ values from −1,000 to −5,000 cal/mole. From Table 15.1, we see that there is no sharp line between the two classes of compounds, but rather a graded change in the $\Delta G^{0\prime}$ values. This classification into high-energy and low-energy compounds is somewhat arbitrary, but nonetheless useful.

Table 15.1 *Standard free energy of hydrolysis of some phosphate compounds present in biological systems.*

Compound		Product	$\Delta G^{0\prime}$ (cal/mole)
phosphoenolpyruvate + H_2O	⟶	pyruvate + phosphate	−14,800
1,3-diphosphoglycerate + H_2O	⟶	3-phosphoglycerate + phosphate	−11,800
creatine phosphate + H_2O	⟶	creatine + phosphate	−10,300
ATP + H_2O	⟶	ADP + phosphate	−7,300
glucose 1-phosphate + H_2O	⟶	glucose + phosphate	−5,000
fructose 6-phosphate + H_2O	⟶	fructose + phosphate	−3,800
glucose 6-phosphate + H_2O	⟶	glucose + phosphate	−3,300
glycerol 1-phosphate + H_2O	⟶	glycerol + phosphate	−2,200

Why is the $\Delta G^{0\prime}$ of hydrolysis of a phosphate anhydride (such as ATP) so much larger than that of a phosphate ester (such as glucose 6-phosphate)? There are two basic structural features which contribute to the relatively high free energy of hydrolysis of ATP. First, at pH 7 the three phosphate groups are fully ionized, resulting in four negative charges very close together. Hydrolysis of the terminal phosphate anhydride relieves some of this electrostatic strain. Second, the resonance stabilization of the products of hydrolysis is greater than that of the starting material. By contrast, the hydrolysis of the phosphate ester in glucose 6-phosphate or glycerol 1-phosphate to an alcohol and phosphate ion offers neither the large release of electrostatic strain nor the enhanced resonance stabilization of the product.

A bond that undergoes hydrolysis with a large decrease in f- energy is called a high-energy bond and is indicated by a squig

Accordingly, ATP may be represented as

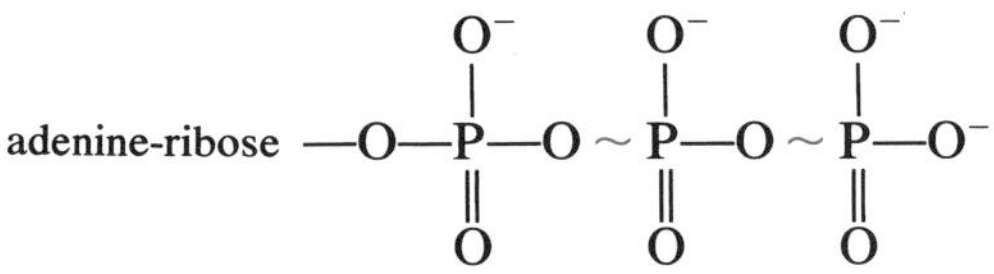

$$\text{adenine-ribose} \;-\mathrm{O}-\underset{\underset{\displaystyle \mathrm{O}}{\|}}{\overset{\overset{\displaystyle \mathrm{O}^-}{|}}{\mathrm{P}}}-\mathrm{O}\sim\underset{\underset{\displaystyle \mathrm{O}}{\|}}{\overset{\overset{\displaystyle \mathrm{O}^-}{|}}{\mathrm{P}}}-\mathrm{O}\sim\underset{\underset{\displaystyle \mathrm{O}}{\|}}{\overset{\overset{\displaystyle \mathrm{O}^-}{|}}{\mathrm{P}}}-\mathrm{O}^-$$

or

$$\text{adenine-ribose} \;-\mathrm{O}-\text{Ⓟ}-\mathrm{O}\sim\text{Ⓟ}-\mathrm{O}\sim\text{Ⓟ}-\mathrm{O}^-$$

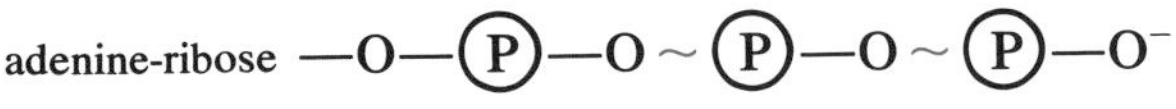

In the second, more abbreviated representation of ATP, the symbol Ⓟ indicates the —(PO_2^-)— group of atoms. Note that ATP contains two high-energy (pyrophosphate) bonds and one low-energy (phosphate ester) bond.

We must be careful not to confuse the term high-energy bond with the term bond energy. To the physical chemist, bond energy (or bond dissociation energy, BDE) is that energy required to disrupt a covalent bond into separate atoms (Table 1.5). This is quite a different process from the hydrolysis of an ester or an anhydride.

ATP is not the only high-energy compound found in the cell (but it is the most important one). Two of the four high-energy compounds listed in Table 15.1 are phosphate anhydrides: ATP and 1,3-diphosphoglycerate (see Problem 8.16 for the structural formula of 1,3-diphosphoglycerate.) The third compound, phosphoenolpyruvate, is a phosphate ester of the enol form of pyruvate. Hydrolysis of this phosphate ester gives the enol form of pyruvate.

$$\underset{\text{phosphoenolpyruvate}}{CH_2{=}\overset{\overset{\displaystyle {}^-O-\overset{\overset{\displaystyle O}{\|}}{P}-O^-}{|}}{\overset{\displaystyle O}{\underset{|}{}}}C-CO_2^-} + H_2O \longrightarrow \underset{\text{pyruvate (enol form)}}{CH_2{=}\overset{\overset{\displaystyle OH}{|}}{C}-CO_2^-} + HPO_4^{2-}$$

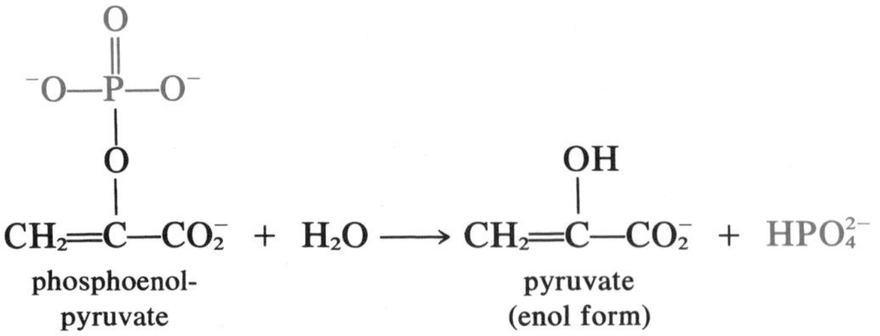

This enol form of pyruvate is in tautomeric equilibrium with the keto form.

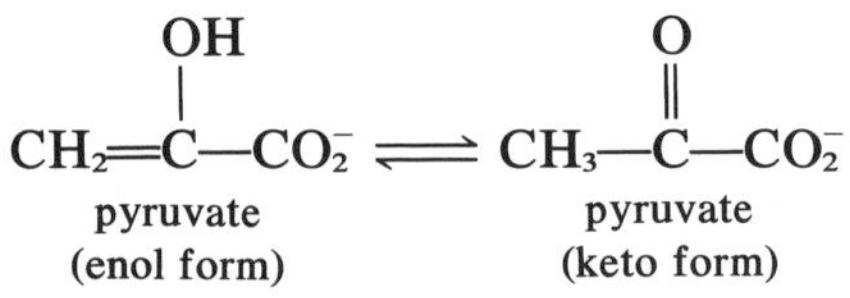

$$\underset{\text{pyruvate (enol form)}}{CH_2{=}\overset{\overset{\displaystyle OH}{|}}{C}-CO_2^-} \rightleftharpoons \underset{\text{pyruvate (keto form)}}{CH_3-\overset{\overset{\displaystyle O}{\|}}{C}-CO_2^-}$$

The equilibrium between this pair of keto–enol tautomers lies almost completely on the side of the keto form and accordingly the conversion to the keto form is spontaneous and accompanied by a large decrease in free energy.

The fourth high-energy compound, creatine phosphate, is an amide of phosphoric acid and creatine. The hydrolysis of this amide is also accompanied by a large decrease in free energy.

$$\begin{array}{c} \text{O} \\ \| \\ ^{-}\text{O—P—NH—C(=}^{+}\text{NH}_2\text{)—N(CH}_3\text{)—CH}_2\text{—CO}_2^- \\ | \\ \text{O}^- \end{array} + H_2O \longrightarrow H_2N\text{—C(=}^{+}NH_2\text{)—N(CH}_3\text{)—CH}_2\text{—CO}_2^- + HPO_4^{2-}$$

creatine phosphate — creatine

The fact that 1,3-diphosphoglycerate, phosphoenolpyruvate, and creatine phosphate all have higher free energies of hydrolysis than ATP means that each of these compounds can readily transfer a phosphate group to ADP to form ATP. Let us calculate the free energy change for the reaction

$$\text{phosphoenolpyruvate} + \text{ADP} \longrightarrow \text{pyruvate} + \text{ATP}$$

The free energy change for this reaction is not given in Table 15.1. However, we can find it by dividing this reaction into two separate steps, since one of the principles of thermodynamics is that the overall change in free energy for a reaction is equal to the sum of the free energy changes for the individual steps.

Reaction	$\Delta G^{0\prime}$ (cal/mole)
phosphoenolpyruvate + H_2O ⟶ pyruvate + HPO_4^{2-}	−14,800
ADP + HPO_4^{2-} ⟶ ATP + H_2O	+7,300
phosphoenolpyruvate + ADP ⟶ pyruvate + ATP	−7,500

This reaction is precisely one of the ways that ATP is generated in the metabolism of glucose.

While several high-energy compounds are found in biological systems, it is ATP which is central to the transfer of energy. The free energy liberated on hydrolysis of the terminal phosphate anhydride of ATP can be used to drive biochemical reactions that require the input of energy. These include the mechanical work of muscle contraction, the osmotic work of transporting inorganic and organic molecules and ions across membranes, and the chemical work of biosynthesis. The supply of ATP is in turn replenished by the oxidation of foodstuffs and the coupling of this oxidation with the phosphorylation of ADP. Obviously, the ATP/ADP + HPO_4^{2-} cycle is central in the exchange of energy and the performance of work in biological systems. These relationships are summarized in Figure 15.9.

15.4 COUPLED REACTIONS

Now let us take up the second question posed at the beginning of this chapter: How do cells use the chemical energy of ATP to drive other biochemical reactions? They do not do it by hydrolyzing ATP and using the heat liberated, for living cells cannot use heat in the performance of biological work. Heat can be used for work only if it flows from a warmer region to a cooler region (the principle of the heat engine). For all practical purposes, there is no temperature differential in the cell

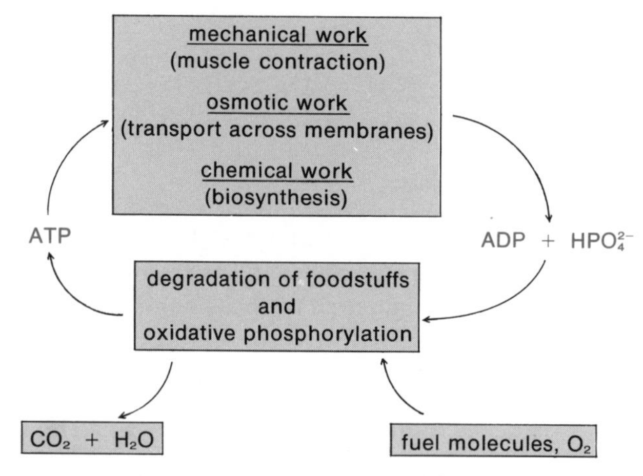

Figure 15.9 *The central role of ATP and ADP + HPO_4^{2-} in the transfer of energy in biological systems.*

Rather, the chemical energy of ATP is used in coupled reactions. As an example, let us look at how the cell uses the energy of ATP for the synthesis of sucrose from glucose and fructose.

Joining glucose and fructose to form sucrose involves formation of a glycoside bond and requires the input of 5,500 calories per mole of sucrose formed.

$$\text{glucose} + \text{fructose} \longrightarrow \text{sucrose} + H_2O \qquad \Delta G^{0\prime} = +5{,}500 \text{ cal/mole}$$

Because this reaction is endergonic, it is not spontaneous in the thermodynamic sense and will not go unless energy to drive it is supplied from some other reaction. A reaction that can supply enough energy to drive this reaction is the hydrolysis of ATP.

$$\text{ATP} + H_2O \longrightarrow \text{ADP} + HPO_4^{2-} \qquad \Delta G^{0\prime} = -7{,}300 \text{ cal/mole}$$

Adding these two reactions together gives a net reaction which has a decrease in free energy and therefore will be spontaneous in the thermodynamic sense.

	$\Delta G^{0\prime}$ (cal/mole)
glucose + fructose $\longrightarrow$ sucrose + H_2O	+5,500
ATP + H_2O $\longrightarrow$ ADP + HPO_4^{2-}	−7,300
glucose + fructose + ATP $\longrightarrow$ sucrose + ADP + HPO_4^{2-}	−1,800

Thus, if it were possible to capture the energy released on the hydrolysis of ATP and channel it into glycoside bond formation, the energetically favorable net reaction resulting in the synthesis of sucrose could be achieved. This is effectively done by two sequential enzyme-catalyzed reactions involving a common intermediate.

$$\begin{array}{rcl} \text{glucose} + \text{ATP} & \longrightarrow & \text{glucose 1-phosphate} + \text{ADP} \\ \text{glucose 1-phosphate} + \text{fructose} & \longrightarrow & \text{sucrose} + HPO_4^{2-} \\ \hline \text{glucose} + \text{fructose} + \text{ATP} & \longrightarrow & \text{sucrose} + \text{ADP} + HPO_4^{2-} \end{array}$$

Glucose 1-phosphate is the intermediate common to the two consecutive reactions, which together have a net free energy change of −1,800 cal/mole. Thus, instead of the energy of ATP being released into the environment as heat, a portion of it is captured in the common intermediate, which then participates in the formation of the glycoside bond. We will see many examples of coupled reactions in the next three chapters.

Problems

15.1 List the three major classes of foodstuffs from which organisms extract energy.

15.2 Outline the four stages in the basic strategy by which all cells extract energy from foodstuffs. Of these four stages, which are concerned primarily with the degradation of fuel molecules; which with the generation of ATP?

15.3 In stage I, fuel molecules are hydrolyzed to smaller molecules, the vast bulk of which can be accounted for by only 25 substances. List these 25 major hydrolysis products.

15.4 What is meant by the term free energy; by the term $\Delta G^{0\prime}$?

15.5 How is the sign of $\Delta G^{0\prime}$ related to the spontaneity of a chemical reaction; to the rate of a chemical reaction?

15.6 Draw structural formulas for ADP and ATP. In each, label all phosphate anhydride (pyrophosphate) bonds, all phosphate ester bonds, and all glycosidic bonds.

15.7 Define high-energy compounds and high-energy bonds as the terms are commonly used in biochemistry.

15.8 What is the difference between the terms high-energy bond and bond energy (bond dissociation energy)?

15.9 Draw structural formulas for each phosphate-containing biomolecule listed in Table 15.1. Which of these molecules are high-energy phosphates; which are low-energy phosphates?

15.10 Calculate the $\Delta G^{0\prime}$ for the following reactions:

a phosphoenolpyruvate + ADP → pyruvate + ATP
b creatine phosphate + ADP → creatine + ATP
c 1,3-diphosphoglycerate + ADP → 3-phosphoglycerate + ATP
d glucose + ATP → glucose 6-phosphate + ADP
e glucose 1-phosphate + ADP → glucose + ATP
f glucose 1-phosphate → glucose 6-phosphate

15.11 What is meant by the term "coupled reaction"? Give an example of the coupling of two biochemical reactions by a common intermediate.

15.12 Explain the significance of the following statements.

a ATP is a universal carrier of free energy.
b NADH is a universal carrier of hydrogens and electrons. As you think abo·· this second statement, review especially Sections 6.10 and 15.2.

16

THE METABOLISM OF CARBOHYDRATES

16.1 METABOLIC PATHWAYS

As noted in Chapter 15, man and other organisms ingest as foodstuffs a complex mixture of plant and animal carbohydrates, fats, and proteins. During digestion, these are broken down into a relatively small number of simpler substances including glucose, the 20 protein-derived amino acids, a few different fatty acids, and glycerol. These smaller molecules, or metabolites as they are more properly called, enter the various cells of the body via the blood stream and there they become the raw materials for maintenance, growth, and development of the organism.

The manner in which any cell or organism uses its foodstuffs is organized into an orderly, very carefully regulated series of reaction steps and sequences known as metabolic pathways. The full range of metabolic pathways in even a one-celled organism such as the bacterium *E. coli* may include as many as a thousand different reactions. To support growth, a culture medium for *E. coli* needs to contain nothing more than glucose as a source of carbon atoms and energy, inorganic salts as sources of nitrogen and phosphorus, and a few other simple substances. The fact that *E. coli* can grow on this medium means that each cell has the biochemical machinery necessary to synthesize all of the proteins, enzymes, lipids, coenzymes, nucleic acids, carbohydrates, and other biomolecules it needs for growth. Furthermore, the growing cells can make these new and more complex biomolecules from just a few simple precursors. All of this is truly remarkable when you consider the complexity of some of the biomolecules of the living cell.

If there are as many as a thousand different biochemical reactions in the metabolic program of such a simple organism as *E. coli*, what of the number and complexity of the biochemical reactions in the cells of higher organisms including man? Fortunately for those who study the biochemistry of the living cell, there are a great many similarities between the major metabolic pathways in man, in *E. coli*, and for that matter, in most other organisms. The number of individual reactions is large, but the number of different kinds of reactions is small. In other words, there are many common themes in the design of metabolic sequences. For example, the basic features of how each type of cell extracts energy from foodstuffs, then uses this energy to synthesize other biomolecules, and even the means of self-regulation are surprisingly similar. Because of these similarities or common themes in the design of biochemical processes, scientists can study the metabolism of simple organisms and then use these results and conclusions to help understand the corresponding cellular processes in man and other more complex organisms. In this and the following chapters, we shall concentrate on human biochemistry, but it should be realized that much of what we say about human metabolism can be applied equally well to the metabolism of most other organisms.

16.2 THE CENTRAL ROLE OF GLUCOSE AS AN ENERGY SOURCE

The major function of dietary carbohydrate is as a source of energy, and in the typical American diet, carbohydrates provide about 50–60% of the daily energy needs. The remainder is supplied by fats and proteins. Many tissues can use both glucose and fatty acids as energy sources. However, glucose is the only fuel that can be used by the central nervous system, including the brain.

The brain, muscle, and other tissues are supplied with glucose via the blood stream. Under normal circumstances, the concentration of glucose in the blood is about 70–100 mg per 100 ml. This level usually rises following a meal and then falls until it hits a fasting level, a point which usually is associated with the onset of hunger. If blood glucose falls below about 60 mg per 100 ml, the condition is known as hypoglycemia. In hypoglycemia, there is danger that the cells of the central nervous system and other tissues which depend on the blood stream for nourishment will not receive adequate supplies of glucose. When blood glucose levels rise to 160 mg per 100 ml or higher, the condition is known as hyperglycemia.

Because of the needs of cells, especially those of the brain, for glucose, the body has developed a set of interrelated metabolic pathways designed to use glucose efficiently for the production of energy and also to insure a continued and adequate supply of glucose in the blood stream. In this chapter we shall look at these pathways and describe the major functions of each. Several of them function to oxidize glucose to carbon dioxide and water and to generate energy, either in the form of ATP or reducing agents for biosynthesis. Other pathways serve to "buffer" the concentration of glucose in the blood, i.e., they maintain blood glucose levels within a rather narrow range. In addition to giving an overview of these pathways and their major functions, we shall also discuss two of these pathways in some detail: anaerobic glycolysis, and the Krebs or tricarboxylic acid cycle.

16.3 THE METABOLISM OF GLUCOSE—AN OVERVIEW

Glucose is derived largely from the hydrolysis of dietary disaccharides and polysaccharides, including plant starches and animal glycogen.

$$\text{polysaccharides} \xrightarrow{\text{hydrolysis}} \text{glucose}$$
$$\text{sucrose} \xrightarrow{\text{hydrolysis}} \text{glucose} + \text{fructose}$$
$$\text{lactose} \xrightarrow{\text{hydrolysis}} \text{glucose} + \text{galactose}$$

Almost immediately in their metabolism, fructose and galactose are converted into the same intermediates that are involved in the metabolism of glucose.

When glucose is completely oxidized in air to carbon dioxide and water, a total of 686,000 calories per mole is liberated.

$$C_6H_{12}O_6 + 6O_2 \longrightarrow 6CO_2 + 6H_2O \qquad \Delta G^{0\prime} = -686{,}000 \text{ cal/mole}$$

Recall from Section 15.4 that cells cannot use heat for the performance of biological work. Rather, they use ATP to drive all energy-requiring reactions. Therefore, the problem for the cell is how to degrade glucose in a controlled manner so that at least a portion of the 686,000 cal/mole derived from the oxidation of glucose can be channeled into the generation of ATP. The combination of glycolysis, the tricarboxylic acid cycle, and oxidative phosphorylation accomplish just this because each of these metabolic pathways includes one or more energy-releasing steps that is coupled directly with the production of ATP.

Glycolysis, the first of the major metabolic pathways of glucose, is a series of reactions which degrades glucose to two molecules of pyruvate.

$$\underset{\text{glucose}}{C_6H_{12}O_6} \xrightarrow{\text{glycolysis}} \underset{\text{pyruvate}}{2CH_3-\overset{\overset{O}{\|}}{C}-CO_2^-}$$

Glycolysis involves oxidation of glucose and uses NAD^+ as the oxidizing agent. Furthermore, as we shall see in Section 16.4, two molecules of ATP are produced for each molecule of glucose converted to pyruvate. The net reaction for the transformation of glucose into pyruvate is given in Figure 16.1.

$$\underset{\text{glucose}}{C_6H_{12}O_6} + 2NAD^+ + 2HPO_4^{2-} + 2ADP \xrightarrow{\text{glycolysis}} \underset{\text{pyruvate}}{2CH_3\overset{\overset{O}{\|}}{C}CO_2^-} + 2NADH + 2ATP$$

Figure 16.1 *The net reaction of glycolysis.*

Next, in an enzyme-catalyzed oxidation involving several coenzymes, pyruvate is converted to acetyl coenzyme A and carbon dioxide.

$$\underset{\text{pyruvate}}{CH_3-\overset{\overset{O}{\|}}{C}-CO_2^-} + NAD^+ + CoA-SH \longrightarrow \underset{\text{acetyl CoA}}{CH_3-\overset{\overset{O}{\|}}{C}-SCoA} + CO_2 + NADH$$

This oxidative decarboxylation of pyruvate prepares the degradation products of glycolysis for entry into the tricarboxylic acid (TCA) cycle, the second major metabolic pathway in the degradation of glucose. In the reactions of the TCA cycle, the two-carbon acetyl group of acetyl CoA is oxidized to two molecules of carbon dioxide. There are four oxidation steps in this pathway, three requiring NAD^+ and one requiring FAD. Further, there is production of one molecule of ATP for each molecule of acetyl CoA entering the cycle. The net reaction for the tricarboxylic acid cycle is shown in Figure 16.2.

$$CH_3\text{—}\overset{\overset{\displaystyle O}{\|}}{C}\text{—}SCoA + 3NAD^+ + FAD + HPO_4^{2-} + ADP \xrightarrow{\text{TCA cycle}} 2CO_2 + 3NADH + FADH_2 + ATP + CoA\text{—}SH$$

Figure 16.2 *The net reaction of the tricarboxylic acid cycle.*

The combination of glycolysis, the oxidation of pyruvate to acetyl CoA, and the tricarboxylic acid cycle brings about the complete oxidation of glucose to carbon dioxide and water, and generates 2 moles of $FADH_2$, 10 moles of NADH, and 4 moles of ATP for each mole of glucose oxidized.

Note that glycolysis, oxidation of pyruvate to acetyl CoA, and the tricarboxylic acid cycle are completely anaerobic for they do not involve molecular oxygen. Rather, there is a build-up of NADH and $FADH_2$. Oxidation of these reduced coenzymes by O_2 is coupled with the phosphorylation of ADP to ATP in a pathway called oxidative phosphorylation (Figure 16.3). It is oxidative phosphorylation that generates the major share of the ATP produced during degradation of glucose.

$$10NAD^+ + 2FADH_2 + 6O_2 + 32ADP + 32HPO_4^{2-} \xrightarrow{\text{oxidative phosphorylation}} 10NAD^+ + 2FAD + 32ATP$$

Figure 16.3 *The net reaction of oxidative phosphorylation following degradation of 1 mole of glucose to carbon dioxide and water.*

Another major metabolic pathway of glucose is called the hexose monophosphate shunt (Figure 16.4). This pathway can bring about the complete oxidation of glucose to carbon dioxide and water.

$$C_6H_{12}O_6 + 12NADP^+ \xrightarrow{\text{hexose monophosphate shunt}} 6CO_2 + 12NADPH$$

Figure 16.4 *The net reaction of the hexose monophosphate shunt.*

At first glance, the hexose monophosphate shunt appears to accomplish the same thing as the combination of glycolysis and the tricarboxylic acid cycle, namely oxidation of glucose to carbon dioxide and water. While it is true that both sets of pathways bring about the oxidation of glucose, there are two important differences between them. For one thing, the hexose monophosphate shunt degrades glucose by a different series of steps than those of glycolysis and the tricarboxylic acid cycle reactions, and in so doing generates a different set of small molecules. These small molecules can then be used as starting materials for the biosynthesis of larger, more complex biomolecules. Among the small molecules generated by the hexose monophosphate shunt is D-ribose, an essential building block of RNA. The second major difference between these glucose-degradation pathways is the biological oxidizing agents used in each case. Glycolysis and the tricarboxylic acid cycle require primarily NAD^+ as an oxidizing agent and generate NADH. The major use of NADH is for the generation of ATP through the process of oxidative phosphorylation. In contrast, the hexose monophosphate shunt requires $NADP^+$ (a phosphorylated form of NAD^+) as the oxidizing agent and generates NADPH. The major function of NADPH is as a reducing agent for the biosynthesis of other molecules. For example, tissues like muscle, which have high demands for energy, are rich in NAD^+. Those such as adipose tissue, which have high demands for reducing power to support biosynthesis, are rich in $NADP^+$.

Now let us turn to some of the metabolic pathways that are involved in the regulation of glucose concentration. When the dietary intake of glucose exceeds immediate needs, man and other animals can convert the excess to the polysaccharide glycogen (Section 10.9), which is stored in either the liver or muscle tissue. In normal adults the liver can store about 108 grams of glycogen and muscles about 245 grams. The pathway that converts glucose into glycogen is called glycogenesis (Figure 16.5) and is the fourth major pathway in the metabolism of glucose.

$$\text{glycogen} \underset{\text{glycogenesis}}{\overset{\text{glycogenolysis}}{\rightleftharpoons}} \text{glucose}$$

Figure 16.5 *Glycogenesis and glycogenolysis.*

Liver and muscle glycogen are storage forms of glucose, and when there is need for additional blood glucose, glycogen is hydrolyzed and released into the blood stream. Glycogenolysis (Figure 16.5), the hydrolysis of glycogen, is the fifth major metabolic pathway of glucose. This process is stimulated by the pancreatic hormone glucagon (Section 12.7). Glycogenesis and glycogenolysis are probably the most important processes contributing to the maintenance of relatively constant blood glucose levels. Note by way of comparison that the major functions of glycolysis and the tricarboxylic acid cycle are the generation of energy in the form of ATP.

The counterbalancing actions of glucagon and insulin in the regulation of normal, resting levels of blood glucose are shown schematically in Figure 16.6. Under normal circumstances, the rate of glucagon-stimulated hydrolysis of glycogen and release of glucose into the blood stream is balanced by the insulin-stimulated uptake and metabolism of glucose by liver, fat, muscle, and brain tissue.

Although the body can store glucose as glycogen, this storage capacity is limited to about 350 grams in a normal adult. When carbohydrate intake is greater than the body's immediate needs for energy and its capacity to store glycogen, the excess is converted into fat, a substance that can be stored in almost unlimited quantities. To be stored as fat, the carbon atoms of glucose are first metabolized to

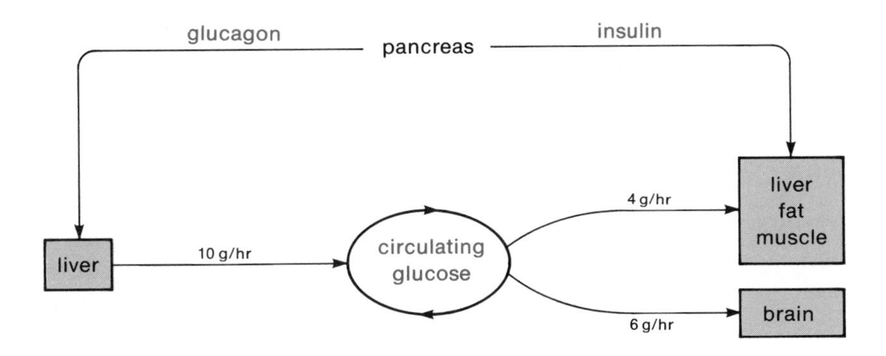

Figure 16.6 *The counterbalancing roles of glucagon and insulin in regulating normal, resting levels of blood glucose.*

acetyl coenzyme A and carbon dioxide, and it is the acetyl CoA that provides the carbon skeleton for the synthesis of fatty acids.

$$C_6H_{12}O_6 \longrightarrow CH_3{-}\overset{\overset{\large O}{\|}}{C}{-}SCoA + CO_2$$

$$\Updownarrow$$

$$\text{fatty acids}$$

These fatty acids then combine with glycerol to form neutral fats. The synthesis of fatty acids from acetyl CoA thus represents a link between the metabolism of glucose and that of fatty acids. We shall discuss the biochemistry of fatty acid synthesis and degradation in Chapter 17.

The total supply of glucose (in the form of liver and muscle glycogen and also blood glucose) can be depleted after about 12–18 hours of fasting. In fact, these stores of glucose often are not sufficient for the duration of an overnight fast between dinner and breakfast. Further, they also can be depleted in a short time during work or strenuous exercise (Figure 16.7).

Without any way to provide additional supplies, nerve tissue including the brain would soon be deprived of glucose. Fortunately,

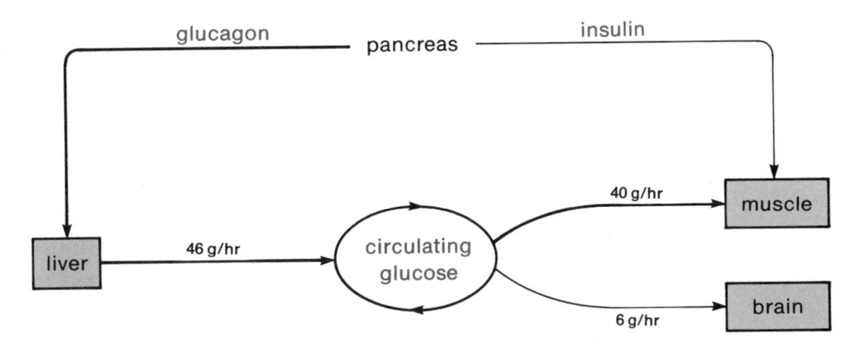

Figure 16.7 *The actions of insulin and glucagon in regulating glucose flow during extreme exercise.*

the body has developed a metabolic pathway to overcome this problem. Between meals, the liver begins to draw on the fats and protein of tissue for the building blocks from which to synthesize the required glucose. This sixth major pathway in the metabolism of glucose results in the synthesis of glucose from noncarbohydrate molecules and is called gluconeogenesis. In this process, body fats are first hydrolyzed to fatty acids and glycerol, and proteins are hydrolyzed to amino acids. The carbon skeletons of glycerol and certain of the 20 protein-derived amino acids are then channeled into the synthesis of glucose (Figure 16.8).

Figure 16.8 *Gluconeogenesis, the synthesis of glucose from noncarbohydrate precursors.*

Those amino acids which can be used by the body for the synthesis of glucose are called glycogenic amino acids. Thus, glycerol and the glycogenic amino acids provide an alternative source of glucose during periods of low carbohydrate intake and when carbohydrate stores are being rapidly depleted. (We shall discuss the glycogenic amino acids more fully in Section 18.6.) The glucocorticoid hormones (Section 11.9) play a major role in controlling these aspects of carbohydrate and protein metabolism. In fact, the primary metabolic effect of the glucocorticoid hormones is to provide a supply of carbohydrate, blood glucose, and glycogen at the expense of body protein, primarily skeletal muscle.

The six major metabolic pathways of glucose are summarized in Figure 16.9. Note that this figure does not show the production of ATP and, with the exception of NADH and NADPH, does not show the involvement of cofactors.

In the remainder of this chapter we shall look in detail at glycolysis and the tricarboxylic acid cycle, the two major pathways that degrade glucose to carbon dioxide and water. We shall pay particular attention to the conservation of chemical energy in the form of ATP.

16.4 GLYCOLYSIS

Although the net reaction of glycolysis is simple to write (Figure 16.1), it took several decades of very patient, intensive research by scores of scientists to discover the separate steps by which glucose is degraded to pyruvate and to understand how this degradation is coupled with the production of ATP. But by 1940, the steps in the glycolytic pathway had been worked out. Glycolysis is frequently called the Embden–Meyerhof pathway in honor of the two German biochemists, Gustav Embden and Otto Meyerhof, who contributed so greatly to our present knowledge of it.

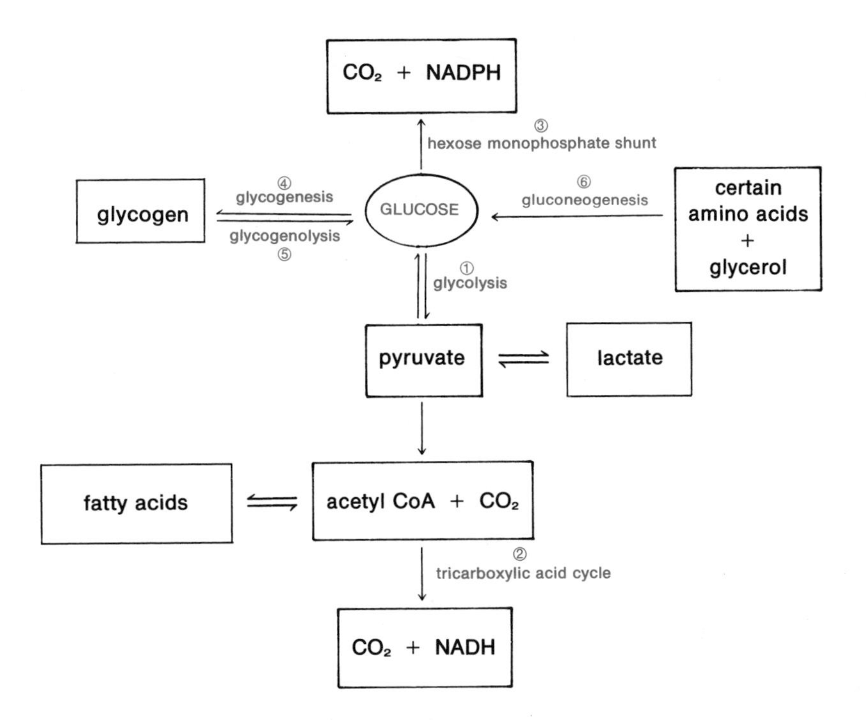

Figure 16.9 *Six major metabolic pathways of glucose: glycolysis, the tricarboxylic acid cycle, the hexose monophosphate shunt, glycogenesis, glycogenolysis, and gluconeogenesis.*

The first reaction of glycolysis is phosphorylation of glucose by ATP to yield the phosphate ester glucose 6-phosphate.

$$\text{glucose} + ATP \xrightarrow{\text{hexokinase}} \text{glucose 6-phosphate} + ADP$$

The transfer of a phosphate group from ATP to an organic molecule is one of the basic reactions of biochemistry and any enzyme that catalyzes this kind of reaction is called a kinase. The enzyme that catalyzes the phosphorylation of glucose by ATP is called hexokinase.

The next step of glycolysis is the isomerization of glucose 6-phosphate to fructose 6-phosphate.

$$\text{glucose 6-phosphate} \xrightleftharpoons[\text{isomerase}]{\text{phosphoglucose}} \text{fructose 6-phosphate}$$

This isomerization probably involves formation of an enol tautomer, or more precisely an enediol tautomer, because there are two hydroxyl groups rather than one on the double bond. (Recall that you used this same type of intermediate in Problem 6.21 to account for the conversion of glyceraldehyde to dihydroxyacetone.) The enediol intermediate from glucose 6-phosphate then ketonizes to generate fructose 6-phosphate.

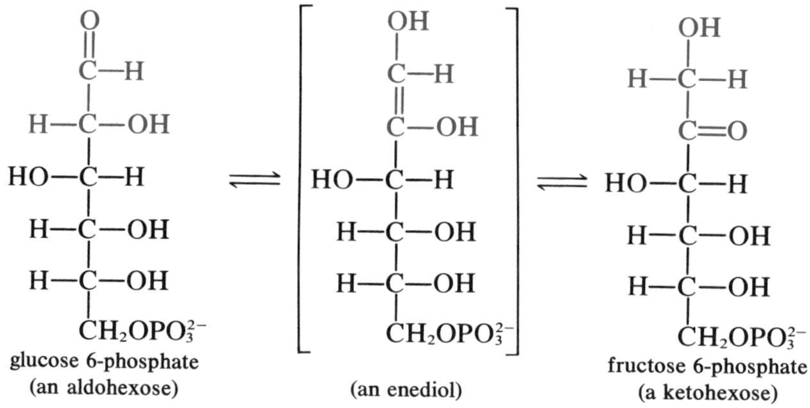

O=C—H
H—C—OH
HO—C—H
H—C—OH
H—C—OH
$CH_2OPO_3^{2-}$
glucose 6-phosphate
(an aldohexose)

⇌

[OH
C—H
‖
C—OH
HO—C—H
H—C—OH
H—C—OH
$CH_2OPO_3^{2-}$]
(an enediol)

⇌

OH
H—C—H
C=O
HO—C—H
H—C—OH
H—C—OH
$CH_2OPO_3^{2-}$
fructose 6-phosphate
(a ketohexose)

At this point, a second mole of ATP is used to convert fructose 6-phosphate into fructose 1,6-diphosphate.

$^{2-}O_3POCH_2$ O CH_2OH HO OH OH
fructose 6-phosphate

+ ATP $\xrightleftharpoons{\text{phosphofructokinase}}$

$^{2-}O_3POCH_2$ O $CH_2OPO_3^{2-}$ HO OH OH
fructose 1,6-diphosphate

+ ADP

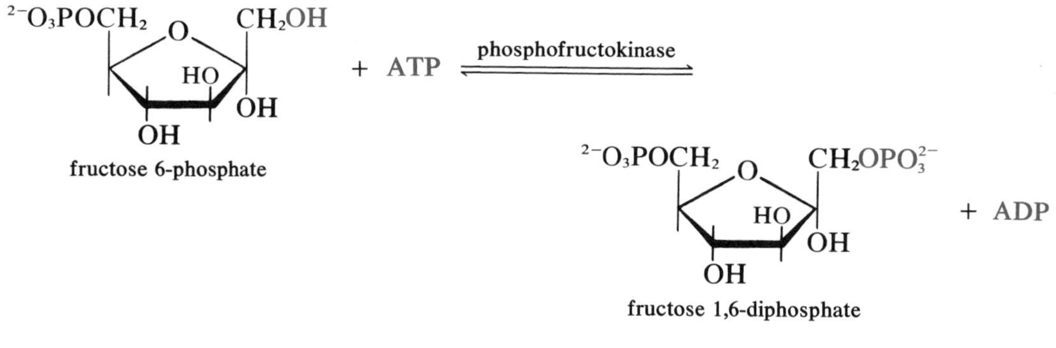

Note that the enzyme that catalyzes this reaction is also a kinase. Because it is specific for fructose 6-phosphate, it is called phosphofructokinase.

The net effect of these first three reactions of glycolysis is to convert glucose into a molecule which can be split into two three-carbon fragments. This cleavage is catalyzed by the enzyme aldolase and is most easily seen by writing fructose 1,6-diphosphate in the open-chain form. Aldolase derives its name from the fact that the reverse reaction, namely the condensation of dihydroxyacetone phosphate and glyceraldehyde 3-phosphate to form fructose 1,6-diphosphate, is an aldol condensation (Section 6.13).

Dihydroxyacetone phosphate and glyceraldehyde 3-phosphate are readily interconverted by the same type of enzyme-catalyzed isomerization we just saw for the isomerization of glucose 6-phosphate

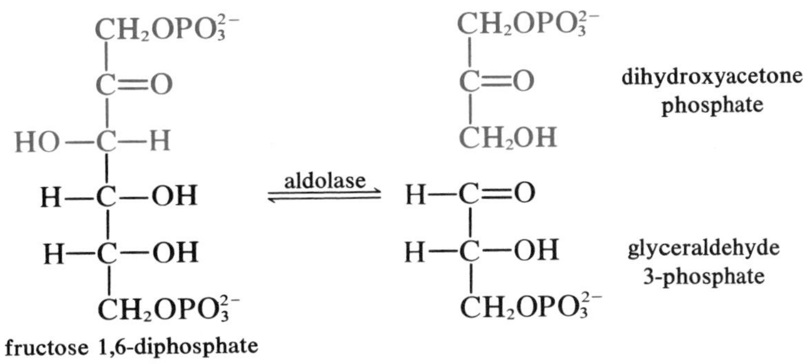

$$\begin{array}{c} CH_2OPO_3^{2-} \\ | \\ C{=}O \\ | \\ HO{-}C{-}H \\ | \\ H{-}C{-}OH \\ | \\ H{-}C{-}OH \\ | \\ CH_2OPO_3^{2-} \end{array} \underset{}{\overset{\text{aldolase}}{\rightleftharpoons}} \begin{array}{c} CH_2OPO_3^{2-} \\ | \\ C{=}O \\ | \\ CH_2OH \\ \\ H{-}C{=}O \\ | \\ H{-}C{-}OH \\ | \\ CH_2OPO_3^{2-} \end{array} \begin{array}{c} \text{dihydroxyacetone phosphate} \\ \\ \\ \text{glyceraldehyde 3-phosphate} \end{array}$$

fructose 1,6-diphosphate

and fructose 6-phosphate. This isomerization is catalyzed by triose phosphate isomerase.

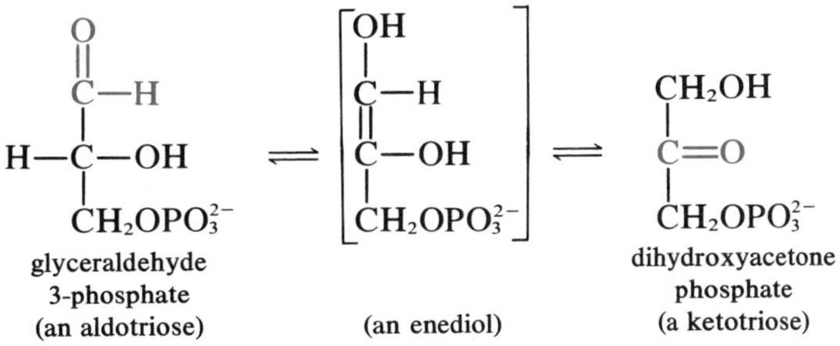

$$\begin{array}{c} O \\ \| \\ C{-}H \\ | \\ H{-}C{-}OH \\ | \\ CH_2OPO_3^{2-} \end{array} \rightleftharpoons \left[\begin{array}{c} OH \\ | \\ C{-}H \\ \| \\ C{-}OH \\ | \\ CH_2OPO_3^{2-} \end{array}\right] \rightleftharpoons \begin{array}{c} CH_2OH \\ | \\ C{=}O \\ | \\ CH_2OPO_3^{2-} \end{array}$$

glyceraldehyde 3-phosphate (an aldotriose) — (an enediol) — dihydroxyacetone phosphate (a ketotriose)

The next step is one of the most important in glycolysis because it couples the oxidation of the aldehyde group of glyceraldehyde 3-phosphate with the synthesis of ATP. In the first phase of this coupled oxidation and phosphorylation, glyceraldehyde 3-phosphate is oxidized by NAD^+. In this process catalyzed by glyceraldehyde 3-phosphate dehydrogenase, one inorganic phosphate ion is required and the immediate product of the aldehyde oxidation is the mixed anhydride 1,3-diphosphoglycerate.

$$\begin{array}{c} O \\ \| \\ C{-}H \\ | \\ H{-}C{-}OH \\ | \\ CH_2OPO_3^{2-} \end{array} + NAD^+ + HPO_4^{2-} \rightleftharpoons \begin{array}{c} O \qquad\quad O \\ \| \qquad\quad \| \\ C{-}O{-}P{-}O^- \\ | \qquad\quad | \\ H{-}C{-}OH \quad O^- \\ | \\ CH_2OPO_3^{2-} \end{array} + NADH$$

glyceraldehyde 3-phosphate — 1,3-diphosphoglycerate

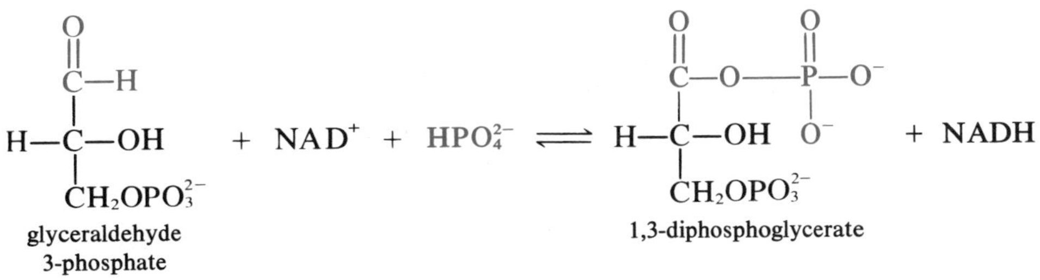

Transfer of a phosphate group from 1,3-diphosphoglycerate to ADP produces the first molecule of ATP generated in glycolysis. Note that up to this point, two molecules of ATP have been consumed in the conversion of glucose to fructose 1,6-diphosphate; now, in the oxidation of two molecules of glyceraldehyde 3-phosphate (remember that the original glucose molecule has been cleaved into two three-carbon fragments), two molecules of ATP have been generated.

$$\underset{\text{1,3-diphosphoglycerate}}{\begin{array}{c} O=C-O-PO_3^{2-} \\ | \\ H-C-OH \\ | \\ CH_2OPO_3^{2-} \end{array}} + ADP \longrightarrow \underset{\text{3-phosphoglycerate}}{\begin{array}{c} CO_2^- \\ | \\ H-C-OH \\ | \\ CH_2OPO_3^{2-} \end{array}} + ATP$$

In the final stage of glycolysis, 3-phosphoglycerate is isomerized into 2-phosphoglycerate, which is then dehydrated to form phosphoenolpyruvate.

$$\underset{\text{3-phosphoglycerate}}{\begin{array}{c} CO_2^- \\ | \\ H-C-OH \\ | \\ CH_2OPO_3^{2-} \end{array}} \rightleftharpoons \underset{\text{2-phosphoglycerate}}{\begin{array}{c} CO_2^- \\ | \\ H-C-OPO_3^{2-} \\ | \\ CH_2OH \end{array}} \xrightleftharpoons{-H_2O} \underset{\text{phosphoenolpyruvate}}{\begin{array}{c} CO_2^- \\ | \\ C-OPO_3^{2-} \\ \| \\ CH_2 \end{array}}$$

Phosphoenolpyruvate is a high-energy compound (Section 15.3) and can transfer a phosphate group to ADP to form ATP.

$$\underset{\text{phosphoenolpyruvate}}{\begin{array}{c} CO_2^- \\ | \\ C-OPO_3^{2-} \\ \| \\ CH_2 \end{array}} + ADP \longrightarrow \underset{\text{pyruvate}}{\begin{array}{c} CO_2^- \\ | \\ C=O \\ | \\ CH_3 \end{array}} + ATP$$

These ten steps in the conversion of glucose to pyruvate, including those that consume and generate ATP and NADH, are summarized in Figure 16.10.

16.5 THE FATES OF PYRUVATE

The sequence of reactions that converts glucose into pyruvate is very similar in all cells and all organisms. In contrast, the fate of pyruvate may be quite different depending on the organism and the particular environment of the cell. In the presence of an adequate supply of NAD^+, pyruvate undergoes oxidation and decarboxylation and is transformed into the acetyl group of acetyl coenzyme A.

$$\underset{\text{pyruvate}}{CH_3-\overset{O}{\overset{\|}{C}}-CO_2^-} + NAD^+ + CoA-SH \longrightarrow \underset{\text{acetyl CoA}}{CH_3-\overset{O}{\overset{\|}{C}}-SCoA} + NADH + CO_2$$

The two-carbon acetyl group of acetyl CoA is then further oxidized by the reactions of the tricarboxylic acid cycle to carbon dioxide. But before we consider the reactions of the TCA cycle, let us

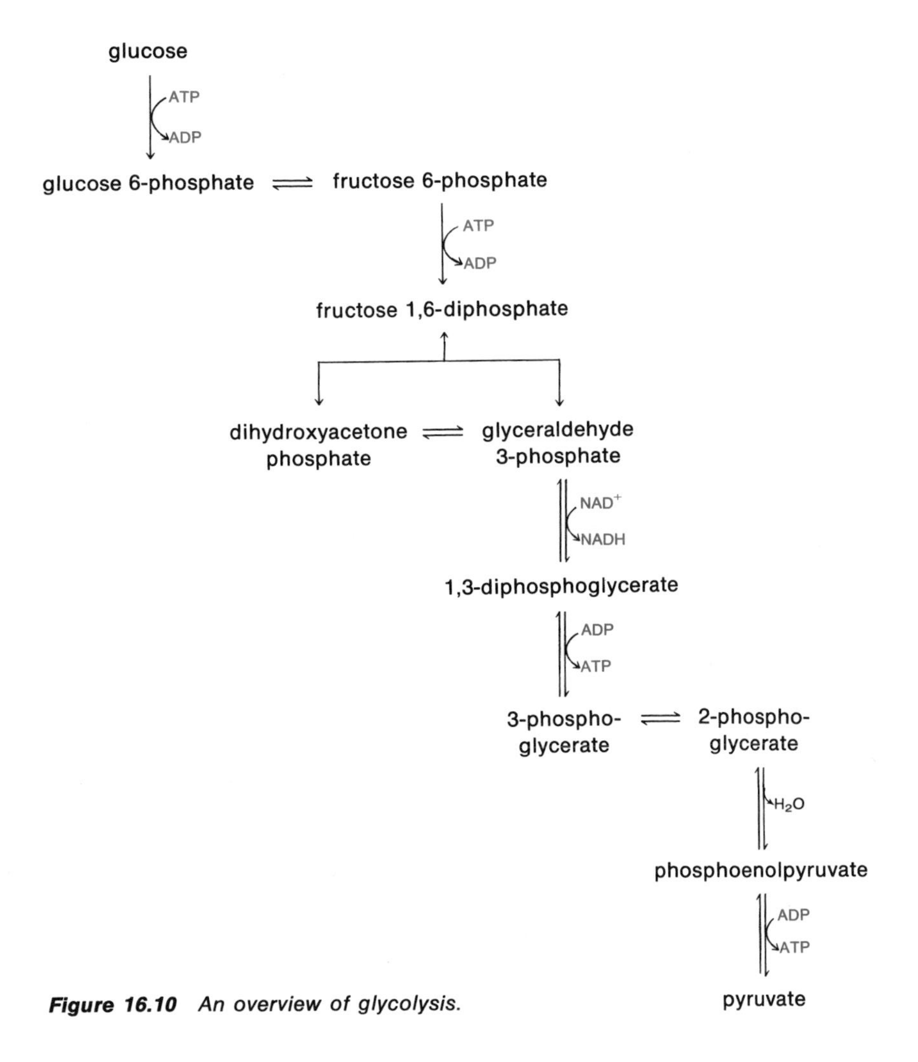

Figure 16.10 *An overview of glycolysis.*

look at the fate of pyruvate under anaerobic conditions. Glycolysis involves a partial oxidation of glucose to pyruvate, and in the process NAD^+ is reduced to NADH. Under conditions where there is an adequate supply of oxygen to the cell, NADH is reoxidized by O_2 and the supply of NAD^+ is replenished. However, if the oxygen supply is not adequate to reoxidize the accumulated NADH, glycolysis would soon consume the available NAD^+. At this point, glycolysis would cease if there were not some alternative way for the cell to replenish its supply of NAD^+. Fortunately there is such an alternative pathway. During strenuous muscle activity or under other conditions when the supply of oxygen is not adequate to meet the needs of the cell for reoxidation of NADH, the cell turns to the reduction of pyruvate as a means of regenerating NAD^+.

$$\underset{\text{pyruvate}}{CH_3-\overset{\overset{\displaystyle O}{\|}}{C}-CO_2^-} + NADH + H^+ \longrightarrow \underset{\text{lactate}}{CH_3-\overset{\overset{\displaystyle OH}{|}}{C}H-CO_2^-} + NAD^+$$

Adding this net reaction to the net reaction of glycolysis gives an overall reaction for the metabolic pathway called anaerobic glycolysis.

$$\underset{\text{glucose}}{C_6H_{12}O_6} + 2ADP + 2HPO_4^{2-} \xrightarrow[\text{glycolysis}]{\text{anaerobic}} \underset{\text{lactate}}{2CH_3\text{—}\overset{\overset{\displaystyle OH}{|}}{C}H\text{—}CO_2^-} + 2ATP$$

Note that the overall reaction for the conversion of glucose to lactate involves neither oxidation nor reduction. Yet there are two oxidations along the way (two molecules of glyceraldehyde 3-phosphate are oxidized to 3-phosphoglycerate) and there are also two compensating reductions (two molecules of pyruvate are reduced to lactate).

While this anaerobic process allows glycolysis to continue and generates some ATP to power muscle contraction, it also results in an increase in the concentration of lactate in muscle and in the blood. This build-up of lactate is associated with fatigue and at the point when blood lactate reaches a concentration of about 0.4 mg per 100 ml, muscle tissue is almost completely exhausted. During resting and recovery, the accumulated lactate must be reoxidized to pyruvate. Thus, although muscle tissue can continue glycolysis and the generation of ATP under anaerobic conditions, it pays a price for using glucose in this way. First, the build-up of lactate must be dealt with eventually. Second, anaerobic glycolysis extracts only a small fraction of the energy potentially available from the complete oxidation of glucose.

A major portion of the lactate formed in active skeletal muscle is transported by the blood stream to the liver, where it is converted to glucose by the reactions of gluconeogenesis. This newly synthesized glucose is then returned to contracting skeletal muscles for further anaerobic glycolysis and the generation of ATP. In this way, a part of the metabolic burden of active skeletal muscle is shifted, at least temporarily, to the liver. The transport of lactate from muscle to the liver, resynthesis of glucose by gluconeogenesis, and the return of glucose to muscle tissue is called the Cori cycle (Figure 16.11).

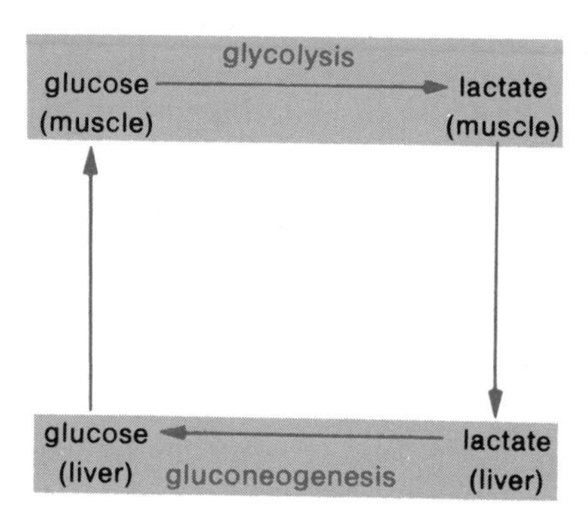

Figure 16.11 *The Cori cycle.*

There is one more fate of pyruvate we should mention. Yeast and several other microorganisms are able to decarboxylate pyruvate to form acetaldehyde and carbon dioxide.

$$\underset{\text{pyruvate}}{CH_3-\overset{\overset{O}{\|}}{C}-CO_2^-} + H^+ \xrightarrow[\text{decarboxylase}]{\text{pyruvate}} \underset{\text{acetaldehyde}}{CH_3-\overset{\overset{O}{\|}}{C}-H} + CO_2$$

Note that this reaction is balanced as it stands and hence involves neither oxidation nor reduction. Pyruvate decarboxylase, the enzyme that catalyzes this decarboxylation reaction, requires the coenzyme thiamine pyrophosphate for activity. Acetaldehyde is now reduced to ethanol by NADH.

$$\underset{\text{acetaldehyde}}{CH_3-\overset{\overset{O}{\|}}{C}-H} + NADH \xrightarrow[\text{dehydrogenase}]{\text{alcohol}} \underset{\text{ethanol}}{CH_3CH_2OH} + NAD^+$$

Adding the reactions for the decarboxylation of pyruvate and the reduction of acetaldehyde to the net reaction of glycolysis gives an overall reaction:

$$\underset{\text{glucose}}{C_6H_{12}O_6} + 2ADP + 2HPO_4^{2-} \xrightarrow[\text{fermentation}]{\text{alcoholic}} \underset{\text{ethanol}}{2CH_3CH_2OH} + 2CO_2 + 2ATP$$

The series of biochemical reactions that converts glucose to ethanol and carbon dioxide is called alcoholic fermentation. Note that both alcoholic fermentation in yeast and the production of lactate in skeletal muscle represent ways in which a cell can continue the reactions of glycolysis under anaerobic conditions, i.e., under conditions where NADH cannot be reoxidized by O_2.

16.6 THE TRICARBOXYLIC ACID CYCLE

Under aerobic conditions, the final pathway in the oxidation of glucose and the generation of energy is the tricarboxylic acid (TCA) or Krebs cycle. The latter name is in honor of Sir Hans Adolph Krebs, the biochemist who in 1937 first proposed the cyclic nature of the process. Through the reactions of this cycle, the carbon atoms of the acetyl group of acetyl CoA are oxidized to carbon dioxide. As you can see from looking at the balanced half-reaction, this is an eight-electron oxidation.

$$CH_3-\overset{\overset{O}{\|}}{C}-SCoA + 3H_2O \longrightarrow 2CO_2 + CoA-SH + 8H^+ + 8e^-$$

As we study the individual reactions of the cycle, we shall concentrate on the four reactions that involve oxidations and the two that produce carbon dioxide.

The remaining carbon atoms of glucose enter the tricarboxylic acid cycle by condensation of acetyl CoA with oxaloacetate. This Claisen-like condensation (Section 8.13) is catalyzed by the enzyme citrate synthetase.

$$\begin{array}{c} CH_3-\overset{O}{\overset{\|}{C}}-SCoA \\ O=\underset{\text{oxaloacetate}}{\overset{|}{C}-CO_2^- \atop \underset{|}{} CH_2-CO_2^-} \end{array} + H_2O \xrightarrow[\text{synthetase}]{\text{citrate}} \underset{\text{citrate}}{\begin{array}{c} CH_2-CO_2^- \\ | \\ HO-C-CO_2^- \\ | \\ CH_2-CO_2^- \end{array}} + CoA-SH$$

Notice that this carbonyl condensation reaction is coupled with the hydrolysis of the thioester to form coenzyme A and citrate, the tricarboxylic acid from which this cycle derives its name.

In the next stage of the cycle, citrate is converted into an isomer, isocitrate.

$$\underset{\text{citrate}}{\begin{array}{c} CH_2-CO_2^- \\ | \\ HO-C-CO_2^- \\ | \\ H-CH-CO_2^- \end{array}} \xrightarrow{-H_2O} \underset{\text{aconitate}}{\begin{array}{c} CH_2-CO_2^- \\ | \\ C-CO_2^- \\ \| \\ CH-CO_2^- \end{array}} \xrightarrow{+H_2O} \underset{\text{isocitrate}}{\begin{array}{c} CH_2-CO_2^- \\ | \\ H-C-CO_2^- \\ | \\ HO-CH-CO_2^- \end{array}}$$

This isomerization is accomplished in two steps. First, citrate is dehydrated to aconitate, and then this tricarboxylic acid is hydrated to isocitrate.

Oxidation of isocitrate by NAD^+ produces oxalosuccinate, a β-ketoacid which then undergoes decarboxylation (Section 8.14) to produce α-ketoglutarate.

$$\underset{\text{isocitrate}}{\begin{array}{c} CH_2-CO_2^- \\ | \\ H-C-CO_2^- \\ | \\ HO-C-CO_2^- \\ | \\ H \end{array}} \xrightarrow{NAD^+ \quad NADH} \underset{\text{oxalosuccinate}}{\begin{array}{c} CH_2-CO_2^- \\ | \\ H-C-CO_2^- \\ | \\ O=C-CO_2^- \end{array}} \longrightarrow \underset{\alpha\text{-ketoglutarate}}{\begin{array}{c} CH_2-CO_2^- \\ | \\ CH_2 \\ | \\ O=C-CO_2^- \end{array}} + CO_2$$

The conversion of isocitrate to α-ketoglutarate is the first oxidation reaction of the cycle and it also produces the first molecule of carbon dioxide.

The second molecule of carbon dioxide is produced from α-ketoglutarate by the same type of oxidative decarboxylation reaction as we saw in Section 16.5 for the conversion of pyruvate to acetyl CoA and carbon dioxide.

$$\underset{\alpha\text{-ketoglutarate}}{\begin{array}{l} CH_2{-}CO_2^- \\ | \\ CH_2 \\ | \\ O{=}C{-}CO_2^- \end{array}} + NAD^+ + CoA{-}SH \longrightarrow \underset{\text{succinyl coenzyme A}}{\begin{array}{l} CH_2{-}CO_2^- \\ | \\ CH_2 \\ | \\ O{=}C{-}SCoA \end{array}} + NADH + CO_2$$

In a series of coupled reactions, succinyl coenzyme A, HPO_4^{2-}, and guanosine diphosphate (GDP) react to form succinate and guanosine triphosphate (GTP).

$$\underset{\text{succinyl coenzyme A}}{\begin{array}{l} CH_2{-}CO_2^- \\ | \\ CH_2{-}C{-}SCoA \\ \quad\quad \| \\ \quad\quad O \end{array}} + GDP + HPO_4^{2-} \longrightarrow \underset{\text{succinate}}{\begin{array}{l} CH_2{-}CO_2^- \\ | \\ CH_2{-}CO_2^- \end{array}} + GTP + CoA$$

The terminal phosphate group of GTP can be transferred to ADP according to the reaction

$$GTP + ADP \rightleftharpoons GDP + ATP$$

Thus, one molecule of high-energy compound (either GTP or ATP) is produced for each molecule of acetyl CoA entering the tricarboxylic acid cycle. This is the only reaction of the cycle that conserves energy as ATP.

In the third oxidation of the cycle, succinate is converted to fumarate, with FAD as the oxidizing agent. Hydration of fumarate generates malate, and in the fourth and final oxidation, malate is converted to oxaloacetate. The oxidizing agent for the conversion of malate to oxaloacetate is NAD^+.

$$\underset{\text{succinate}}{\begin{array}{c} CO_2^- \\ | \\ CH_2 \\ | \\ CH_2 \\ | \\ CO_2^- \end{array}} \xrightarrow{FAD \quad FADH_2} \underset{\text{fumarate}}{\begin{array}{c} CO_2^- \\ | \\ CH \\ \| \\ CH \\ | \\ CO_2^- \end{array}} \xrightarrow{+H_2O} \underset{\text{malate}}{\begin{array}{c} CO_2^- \\ | \\ HO{-}C{-}H \\ | \\ H{-}C{-}H \\ | \\ CO_2^- \end{array}} \xrightarrow{NAD^+ \quad NADH} \underset{\text{oxaloacetate}}{\begin{array}{c} CO_2^- \\ | \\ O{=}C \\ | \\ CH_2 \\ | \\ CO_2^- \end{array}}$$

With the production of oxaloacetate, the reactions of the tricarboxylic acid cycle are complete and the cycle is ready to accept another molecule of acetyl coenzyme A. Figure 16.12 summarizes the steps in the tricarboxylic acid cycle, including those that consume and generate high-energy phosphates and reduced coenzymes.

Up to this point, except for the formation of one molecule of ATP coupled with the oxidative decarboxylation of α-ketoglutarate, we have said nothing about the flow of energy in the tricarboxylic acid cycle. Actually, the cycle itself describes the fate of the carbon atoms

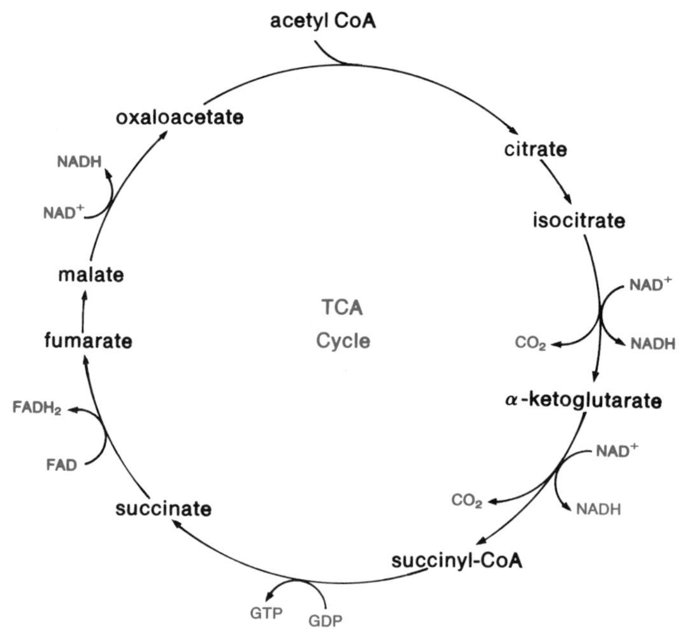

Figure 16.12 *The tricarboxylic acid cycle.*

of acetate and shows how they are converted to carbon dioxide. In addition to serving as a degradative and NADH-generating cycle, it has another function. It creates a supply of key intermediates which the cell can use for the synthesis of other biomolecules. For example, in the presence of a suitable source of nitrogen atoms, α-ketoglutarate can be transformed into the important amino acid glutamic acid.

$$\underset{\alpha\text{-ketoglutarate}}{^{-}O_2C{-}CH_2{-}CH_2{-}\overset{\overset{\displaystyle O}{\|}}{C}{-}CO_2^-} \xrightleftharpoons[\text{source}]{\text{nitrogen}} \underset{\text{glutamate}}{^{-}O_2C{-}CH_2{-}CH_2{-}\overset{\overset{\displaystyle NH_3^+}{|}}{CH}{-}CO_2^-}$$

Thus, the tricarboxylic acid cycle accomplishes two very important things for the cell:

1. It oxidizes acetate units to carbon dioxide and in the process generates NADH and $FADH_2$. The reoxidation of these reduced coenzymes by oxygen is coupled with the generation of ATP.
2. It continually accepts acetate as a means of generating a pool of intermediates which can be used by the cell for the synthesis of other biomolecules.

16.7 ELECTRON TRANSPORT AND OXIDATIVE PHOSPHORYLATION

In the preceding sections we have described glycolysis, the oxidation of pyruvate to acetyl CoA, and the tricarboxylic acid cycle. Although these pathways are grouped together under the heading of aerobic carbohydrate metabolism, the term is not altogether accurate because the oxidizing agents are NAD^+ and FAD rather than molecular oxygen.

Molecular oxygen serves as the ultimate oxidizing agent for $FADH_2$ and NADH.

$$2NADH + O_2 + 2H^+ \longrightarrow 2NAD^+ + 2H_2O$$

Although this balanced redox equation correctly indicates the stoichiometry for the oxidation of NADH by oxygen, by no means does it suggest a mechanism for the reaction. In fact, the oxidations of NADH and $FADH_2$ involve a series of redox reactions, the major features of which are well understood at the present time. This series of redox reactions constitutes the respiratory chain, the final common pathway by which electrons from the oxidation of all cellular fuels ultimately reduce molecular oxygen.

The carriers of electrons in the respiratory chain are at least four structurally related proteins known as cytochromes, and one flavoprotein. Cytochrome c, the most thoroughly investigated of these electron carriers, is a globular protein of molecular weight 12,400 and consists of a single polypeptide chain of 104 amino acids folded around a single heme group (Figure 16.13). Central to the structure of heme are four substituted pyrrole rings joined by one-carbon bridges into a larger ring called a porphyrin.

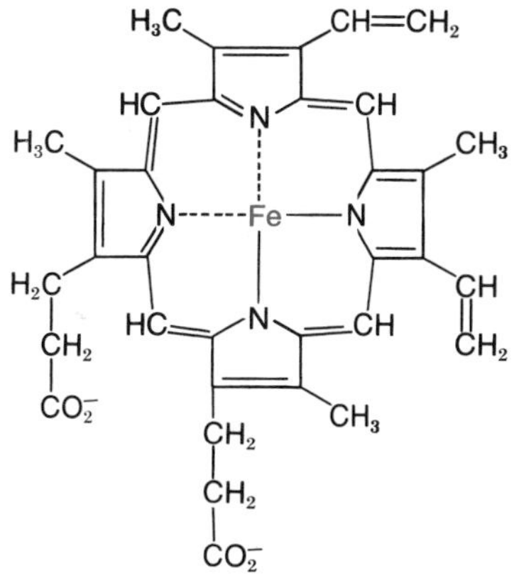

Figure 16.13 *Heme, a complex of a substituted porphyrin and* Fe^{2+}. *The heme shown here is found in hemoglobin, myoglobin, and most of the cytochromes.*

The iron atom can exist in either the Fe^{2+} (ferrous) or Fe^{3+} (ferric) oxidation state. Thus each cytochrome in the oxidized form can accept an electron and become reduced to Fe^{2+}, which in turn gives up an electron to reduce the next cytochrome in the chain. In the last step in the chain, the electrons are transferred to a molecule of oxygen and water is formed. Cyanide and carbon monoxide are both powerful inhibitors of respiratory-chain phosphorylation because they combine with the iron to form stable complexes and thus prevent the normal functioning of the cytochromes.

An examination of the energetics of electron transport in the respiratory chain shows that the transport of one pair of electrons from NADH to oxygen is accompanied by a sufficiently large decline in free energy to make possible the synthesis of several molecules of ATP from ADP, providing that a coupling mechanism is available.

$$NADH + H^+ + \tfrac{1}{2}O_2 \rightarrow NAD^+ + H_2O \qquad \Delta G^{0\prime} = -52{,}000 \text{ cal/mole}$$

Such a mechanism is available in the cell and results in the production of three molecules of ATP during the oxidation of each molecule of NADH.

$$3ADP + 3HPO_4^{2-} \rightarrow 3ATP + 3H_2O \qquad \Delta G^{0\prime} = +21{,}900 \text{ cal/mole}$$

The overall equation for respiratory-chain phosphorylation can be written as the sum of the exergonic oxidation and the endergonic phosphorylation. The coupling of these two reactions is called oxidative phosphorylation, or more accurately, respiratory-chain phosphorylation.

$$NADH + H^+ + \tfrac{1}{2}O_2 + 3ADP + 3HPO_4^{2-} \rightarrow NAD^+ + 3ATP + 4H_2O$$
$$\Delta G^{0\prime} = -30{,}100 \text{ cal/mole}$$

Coupling the oxidation and phosphorylation of three molecules of ADP thus conserves about 22/52 or a little more than 42% of the total free energy decrease in the reoxidation of NADH. The reoxidation of one molecule of $FADH_2$ by oxygen through the cytochrome system is coupled with the formation of two molecules of ATP from ADP.

16.8 ENERGY BALANCE FOR GLUCOSE METABOLISM

Now that we have examined each of the steps in the metabolism of glucose, we can sum up the net results of the transformation of chemical energy. The balanced half-reaction for the conversion of glucose to carbon dioxide and water shows a change of 24 electrons:

$$C_6H_{12}O_6 + 6H_2O \rightarrow 6CO_2 + 24H^+ + 24e^-$$

These electrons are removed in twelve separate two-electron oxidations, ten of which involve NAD^+ and two involve FAD. The reoxidation of these reduced coenzymes is coupled with the production of 32 moles of ATP.

Note that glycolysis takes place in the cytosol and therefore the NADH produced during the oxidation of 3-phosphoglyceraldehyde must be transported from the cytosol to the mitochondria before it can undergo oxidative phosphorylation. Because of the mechanism of this transport system, each mole of NADH generated in glycolysis yields only 2 moles of ATP. Each mole of NADH generated in the oxidation of pyruvate and the reactions of the tricarboxylic acid cycle yields 3 moles of ATP. Further, there is a net profit of 2 moles of ATP from glycolysis, and an additional 2 moles are produced during the hydrolysis of 2 moles of succinyl CoA. These reactions are tabulated in Table 16.1.

The total, then, for the aerobic oxidation of one mole of glucose to carbon dioxide and water is 36 moles of ATP. This represents a net conservation of chemical energy of $36 \times 7{,}300$ or 263,000 cal/mole. The efficiency of the energy conservation during glucose metabolism is

$$\frac{263{,}000}{686{,}000} \quad \text{or} \quad \text{approximately } 38\%$$

Table 16.1 *Yield of ATP from the complete oxidation of glucose.*

Reaction	Process	Net Yield of ATP (moles)
Glycolysis		
glucose → glucose 6-phosphate	phosphorylation	−1
glucose 6-phosphate → fructose 1,6-diphosphate	phosphorylation	−1
glyceraldehyde 3-phosphate → 3-phosphoglycerate	oxidation by NAD^+	+4
phosphoenolpyruvate → pyruvate	phosphorylation	+2
Oxidation of Pyruvate		
pyruvate → acetyl CoA + CO_2	oxidation by NAD^+	+6
Tricarboxylic Acid Cycle		
isocitrate → α-ketoglutarate + CO_2	oxidation by NAD^+	+6
α-ketoglutarate → succinyl CoA + CO_2	oxidation by NAD^+	+6
succinyl CoA → succinate	phosphorylation	+2
succinate → fumarate	oxidation by FAD	+6
malate → oxaloacetate	oxidation by NAD^+	+6
Net reaction: glucose ⟶ $6CO_2$		+36

It is an impressive feat for the living cell to trap this amount of energy as ATP. In fact, the efficiency may be even larger than this. If we were to adjust the $\Delta G^{0\prime}$ values to reflect the actual concentrations of ADP, ATP, and HPO_4^{2-}, the efficiency of the process in the intact cell might be as high as 60%.

Problems

16.1 Discuss the factors that operate in the body to maintain blood glucose concentration at a relatively constant level.

16.2 Write balanced half-reactions for the following conversions.

a glucose → pyruvate **b** glucose → lactate
c lactate → pyruvate **d** pyruvate → ethanol + carbon dioxide

16.3 In what ways are glycolysis and alcoholic fermentation similar; in what ways do these processes differ?

16.4 In the Embden–Meyerhof pathway:

a What is the oxidation step?
b What coenzyme functions as the oxidizing agent?
c How is the reduced form of this coenzyme normally oxidized?
d How is it reoxidized under anaerobic conditions in muscle cells?
e How is it reoxidized in yeast cells?

16.5 The total amount of energy that can be obtained from complete oxidation of glucose is 686,000 calories per mole. What fraction of this energy is conserved as ATP in glycolysis? (Note that although the fraction of energy conserved as ATP is small, it is sufficient for the survival of anaerobic cells.)

16.6 Number the carbon atoms of glucose 1 through 6. Show the fate of each atom in (a) alcoholic fermentation and (b) glycolysis.

16.7 Write equations for the two steps in glycolysis which are coupled with the formation of ATP.

16.8 Show by structural formulas the sequence of reactions in the tricarboxylic acid cycle.

16.9 What are the major functions of the Krebs cycle in carbohydrate metabolism?

16.10 The conversion of acetate to carbon dioxide in the tricarboxylic acid cycle involves four separate oxidation steps. Write a balanced half-reaction for each of these steps.

16.11 A high-energy phosphate is produced directly in only one step of the tricarboxylic acid cycle. Write an equation for this step.

16.12 If aconitate undergoes hydration according to Markovnikov's rule, would you predict the product to be citrate or isocitrate? Explain.

16.13 What is the function of the electron transport system?

16.14 A maximum of 36 moles of ATP can be formed as the result of complete metabolism of one mole of glucose to carbon dioxide and water. How many of these are formed in:

a glycolysis?
b the tricarboxylic acid cycle?
c the electron transport system?

16.15a Name three coenzymes involved in the metabolism of carbohydrates.
b What vitamin is associated with each of these coenzymes?

16.16 The degradation of carbohydrates provides the cell with three things: energy, NADPH as reducing power for biosynthesis, and a pool of small-molecular intermediates. Which of these three are produced by the following pathways?

a the conversion of glucose to lactate
b the tricarboxylic acid cycle
c the hexose monophosphate shunt

16.17 Based on your knowledge of the steps in glycolysis, the tricarboxylic acid cycle, and alcoholic fermentation, propose a series of steps for the following biochemical reactions:

a glycerol → lactate
b phosphoglycerol → ethanol + carbon dioxide
c 3-phosphoglyceraldehyde → glucose 6-phosphate
d glycerol → acetyl CoA
e ethanol → carbon dioxide

16.18 The first reactions of the hexose monophosphate shunt convert glucose 6-phosphate to ribulose 5-phosphate and carbon dioxide. This process requires

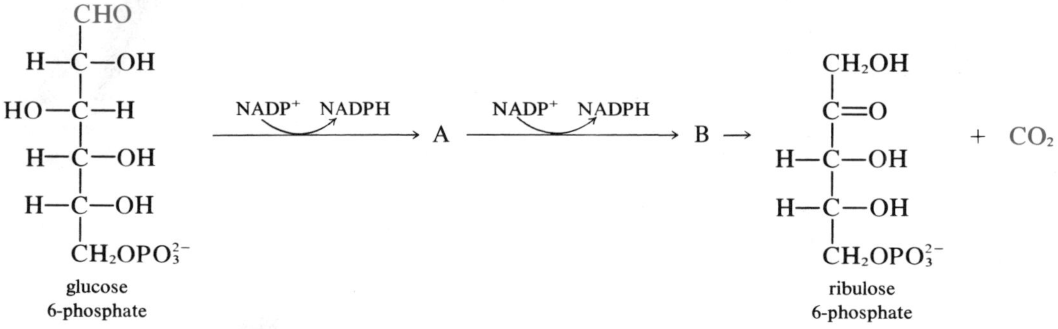

CHO
|
H—C—OH
|
HO—C—H
|
H—C—OH
|
H—C—OH
|
$CH_2OPO_3^{2-}$
glucose
6-phosphate

$\xrightarrow{NADP^+ \to NADPH}$ A $\xrightarrow{NADP^+ \to NADPH}$ B →

CH_2OH
|
C=O
|
H—C—OH
|
H—C—OH
|
$CH_2OPO_3^{2-}$
ribulose
6-phosphate

+ CO_2

two molecules of $NADP^+$ and generates two molecules of NADPH. This sequence involves formation of two intermediate compounds, shown on p. 376 as A and B. Propose structural formulas for both A and B.

16.19 What is the function of the Cori cycle?

16.20 Although glucose is the principal source of carbohydrate for glycolysis, etc., fructose and galactose also are metabolized for energy.

a What is the major dietary source of fructose; of galactose?
b Fructose is converted in one step into one of the intermediates of glycolysis. Draw the structural formula for this intermediate and write a balanced equation for its formation from fructose.
c Galactose is first converted to α-galactose 1-phosphate and then isomerized to α-glucose 1-phosphate. Draw structural formulas for these two hexose phosphates.

17

THE METABOLISM OF FATS

17.1 INTRODUCTION

As we saw in Chapters 15 and 16, carbohydrates, fats, and proteins are the three major nutrients for man. Of these, fats constitute about 30–40% of the calories in the average American diet. Further, fats in the form of triglycerides are the major storage form of energy since they are the only energy supply that can be stored in large quantities. Adipose tissue contains specialized cells called adipocytes whose sole function is to store fat. While carbohydrates in the form of glycogen can be stored in the liver and in skeletal muscle, there are no specialized cells for glycogen storage. For this reason the body's capacity to store carbohydrates is limited to about 350 grams. There is no such limitation on the body's ability to store fat.

In terms of available energy, fats, or more properly fatty acids, have a higher caloric value than carbohydrates; in fact, fatty acids have the highest caloric value of any food. This is obvious from a comparison of the energies of oxidation of a fatty acid and a carbohydrate (Table 17.1).

Table 17.1 *Comparison of glucose and palmitic acid as energy sources.*

Energy Source	Formula	Mol Wt (g/mole)	$\Delta G^{0\prime}$ (cal/mole)	$\Delta G^{0\prime}$ (cal/gram)
glucose	$C_6H_{12}O_6$	180	−686,000	−3,800
palmitic acid	$CH_3(CH_2)_{14}CO_2H$	256	−2,340,000	−9,300

Gram for gram, palmitic acid has more than twice the caloric value of glucose. This larger yield of energy per gram stems from the fact that the hydrocarbon chain of the fatty acid is a more highly reduced biological material than the oxygenated carbon chain of a carbohydrate. This difference in degree of oxygenation of the two fuel substances also can be seen by looking at the balanced equations for the complete oxidation of each to carbon dioxide and water.

$$C_6H_{12}O_6 + 6O_2 \longrightarrow 6CO_2 + 6H_2O$$

$$CH_3(CH_2)_{14}CO_2H + 23O_2 \longrightarrow 16CO_2 + 16H_2O$$

In this process, one molecule of oxygen is consumed per carbon atom in glucose. In comparison, 23/16 or 1.44 molecules of oxygen are consumed per carbon atom of palmitic acid.

In this chapter we shall discuss the metabolic pathway for the degradation of fatty acids and how this degradation is coupled with the generation of ATP. In addition, we shall discuss the pathway for the biosynthesis of fatty acids and then compare and contrast the steps by which nature degrades and synthesizes these vital substances. Finally, we will show some of the interrelationships between the metabolism of fatty acids and carbohydrates.

17.2 MOBILIZATION OF FATTY ACIDS

Fatty acids are stored largely in adipose tissue as triglycerides. The first step in the mobilization of fatty acids is hydrolysis of triglycerides catalyzed by a group of enzymes called lipases.

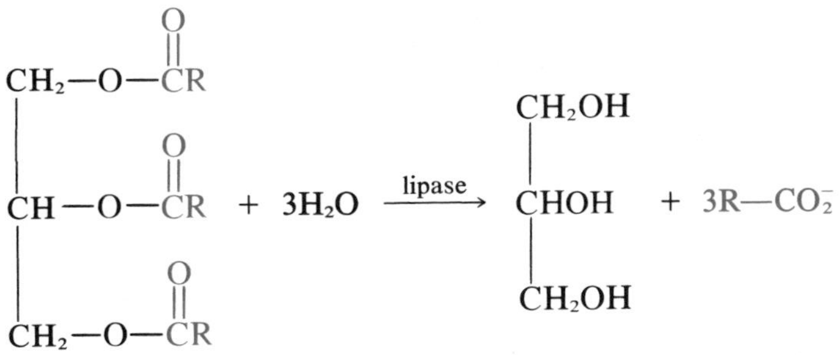

$$\begin{array}{l} CH_2{-}O{-}\overset{O}{\overset{\|}{C}}R \\ | \\ CH{-}O{-}\overset{O}{\overset{\|}{C}}R \\ | \\ CH_2{-}O{-}\overset{O}{\overset{\|}{C}}R \end{array} + 3H_2O \xrightarrow{\text{lipase}} \begin{array}{l} CH_2OH \\ | \\ CHOH \\ | \\ CH_2OH \end{array} + 3R{-}CO_2^-$$

a triglyceride — fatty acids

The release of fatty acids from adipose tissue into the blood stream is stimulated by several hormones including epinephrine, adrenocorticotropic hormone (Figure 12.9), growth hormone (Section 12.7), and thyroxine (Table 12.2). Conversely, fatty acid accumulation in adipose tissue and storage as triglycerides is stimulated by high levels of glucose and insulin in the blood stream.

17.3 DEGRADATION OF FATTY ACIDS

It was long suspected that both the degradation and synthesis of fatty acids occurred by the addition or subtraction of two-carbon fragments. This hypothesis was based on several observations. For one thing, most fatty acids in animal and plant fats and oils have an even number of carbon atoms in an unbranched hydrocarbon chain. For another, in fasting mammals, or under circumstances where normal carbohydrate metabolism is impaired (as in the disease diabetes mellitus), there is an accumulation in the urine and blood of two four-carbon acids, acetoacetate and β-hydroxybutyrate. In addition, there is an accumulation of acetone, the decarboxylation product of acetoacetate.

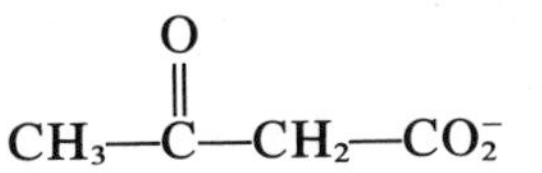

$$CH_3{-}\overset{O}{\overset{\|}{C}}{-}CH_2{-}CO_2^- \qquad CH_3{-}\overset{OH}{\overset{|}{C}H}{-}CH_2{-}CO_2^- \qquad CH_3{-}\overset{O}{\overset{\|}{C}}{-}CH_3$$

acetoacetate — β-hydroxybutyrate — acetone

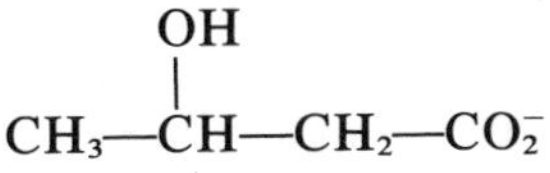

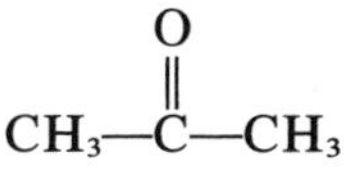

Clinically, these three compounds are referred to as ketone bodies. (Note, however, that β-hydroxybutyrate is not a ketone at all.) The accumulation of these ketone bodies in diabetes mellitus suggested that fatty acid as well as carbohydrate metabolism is impaired in this disease. It was further suggested that acetoacetate represents a terminal four-carbon fragment of normal fatty acid metabolism which, for some reason, is not oxidized beyond that point.

$$\underset{\text{palmitate}}{CH_3{-}CH_2{-}CH_2{-}CH_2{-}(CH_2)_{11}CO_2^-} \longrightarrow \underset{\text{acetoacetate}}{CH_3{-}\overset{\overset{\Large O}{\|}}{C}{-}CH_2{-}CO_2^-} + 12CO_2$$

This observation and reasoning played an important role in early theorizing about fatty acid metabolism. Later experiments, however, have shown that while ketone bodies certainly do accumulate during impaired carbohydrate and fatty acid metabolism, they are formed in quite another way than that suggested above. We shall return to the formation of ketone bodies in Section 17.6.

In 1904, Franz Knoop conceived the idea of a chemical "tag" or "label" as a way to determine the metabolic fate of at least a part of the fatty acid chain. He knew that a phenyl group is not readily metabolized by the body, and therefore he planned a series of carefully designed experiments using this group as a label in order to follow the degradation of a fatty acid chain. He first determined that benzoate and phenylacetate are not metabolized in the body. Instead they are excreted in the urine. He then discovered that when dogs were fed 3-phenylpropanoate, benzoate was excreted, and when the dogs were fed 4-phenylbutanoate, phenylacetate was excreted.

$$\underset{\text{3-phenylpropanoate}}{C_6H_5{-}CH_2{-}CH_2{-}CO_2^-} \longrightarrow \underset{\text{benzoate}}{C_6H_5{-}CO_2^-} + C_2 \text{ fragment}$$

$$\underset{\text{4-phenylbutanoate}}{C_6H_5{-}CH_2{-}CH_2{-}CH_2{-}CO_2^-} \longrightarrow \underset{\text{phenylacetate}}{C_6H_5{-}CH_2{-}CO_2^-} + C_2 \text{ fragment}$$

From these observations, Knoop concluded that fatty acid degradation involves oxidation of the carbon atom beta to the carboxylate group and cleavage of a two-carbon fragment. To account for this oxidation and cleavage, Knoop proposed the β-oxidation pathway (Figure 17.1), which outlined the essential features of metabolic degradation of fatty acids. Yet it took another fifty years of research before his brilliant conclusions were fully confirmed.

By 1950 it had become clear that both ATP and coenzyme A are required for the degradation of fatty acids. Further, through the use of cell-free extracts, biochemists were able to identify and study each of the various enzymes involved in this process. As we now know,

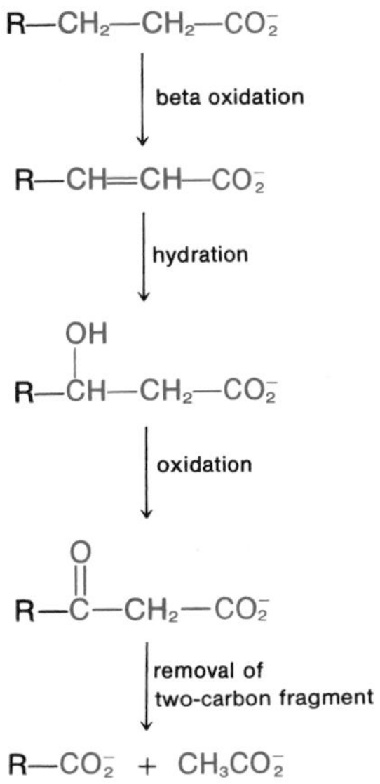

$$R{-}CH_2{-}CH_2{-}CO_2^-$$

↓ beta oxidation

$$R{-}CH{=}CH{-}CO_2^-$$

↓ hydration

$$R{-}CH(OH){-}CH_2{-}CO_2^-$$

↓ oxidation

$$R{-}C(=O){-}CH_2{-}CO_2^-$$

↓ removal of two-carbon fragment

$$R{-}CO_2^- + CH_3CO_2^-$$

Figure 17.1 *Fatty acid oxidation according to the β-oxidation theory of Knoop.*

β-oxidation is the normal pathway for the oxidative degradation of fatty acids. However, before they can be oxidized, fatty acids must first be "activated." This activation involves formation of a thioester with coenzyme A.

$$\underset{\text{fatty acid}}{R{-}CH_2{-}CH_2{-}\overset{O}{\overset{\|}{C}}{-}O^-} + HS{-}CoA + ATP \longrightarrow$$

$$\underset{\text{acyl CoA}}{R{-}CH_2{-}CH_2{-}\overset{O}{\overset{\|}{C}}{-}SCoA} + AMP + 2HPO_4^{2-}$$

The formation of the thioester in acyl CoA is coupled with the hydrolysis of ATP to AMP and two molecules of HPO_4^{2-}. Note that this activation of a fatty acid prior to degradation is analogous to the activation of glucose prior to its degradation. The phosphorylation of glucose to fructose 1,6-diphosphate requires two molecules of ATP, and in this sense is coupled with the hydrolysis of two phosphate anhydride bonds. The activation of a fatty acid molecule also is coupled with the hydrolysis of two phosphate anhydride bonds, except in this instance they are both part of the same molecule of ATP. Yet the cost in energy is the same for the activation of glucose and a fatty acid.

The activated fatty acid molecule is next metabolized by a series of four steps (oxidation, hydration, oxidation, and cleavage) until the entire hydrocarbon chain is degraded to acetyl coenzyme A. These four steps constitute what is called the fatty acid oxidation spiral. Figure 17.2 illustrates this process with the degradation of palmitate to acetyl CoA by the β-oxidation pathway.

In the first of these four steps, the acyl thioester of palmitate is

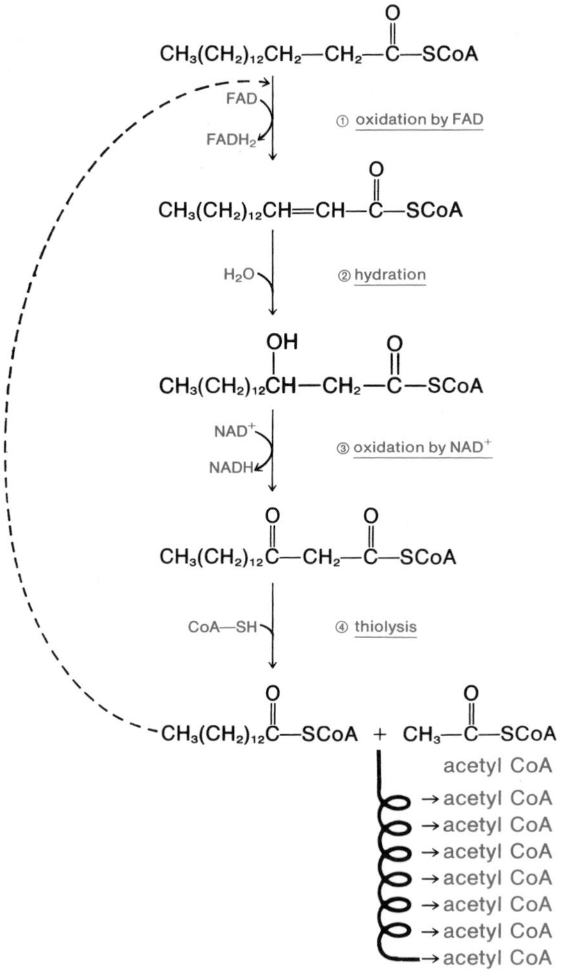

$$CH_3(CH_2)_{12}CH_2{-}CH_2{-}\overset{O}{\overset{\|}{C}}{-}SCoA$$

FAD → $FADH_2$ ① oxidation by FAD

$$CH_3(CH_2)_{12}CH{=}CH{-}\overset{O}{\overset{\|}{C}}{-}SCoA$$

H_2O ② hydration

$$CH_3(CH_2)_{12}\overset{OH}{\overset{|}{C}}H{-}CH_2{-}\overset{O}{\overset{\|}{C}}{-}SCoA$$

NAD^+ → NADH ③ oxidation by NAD^+

$$CH_3(CH_2)_{12}\overset{O}{\overset{\|}{C}}{-}CH_2{-}\overset{O}{\overset{\|}{C}}{-}SCoA$$

CoA—SH ④ thiolysis

$$CH_3(CH_2)_{12}\overset{O}{\overset{\|}{C}}{-}SCoA + CH_3{-}\overset{O}{\overset{\|}{C}}{-}SCoA$$

acetyl CoA
→acetyl CoA
→acetyl CoA
→acetyl CoA
→acetyl CoA
→acetyl CoA
→acetyl CoA
→acetyl CoA

Net reaction: $CH_3(CH_2)_{14}\overset{O}{\overset{\|}{C}}{-}SCoA + 7FAD + 7NAD^+ + 8CoA{-}SH \longrightarrow$

$$8CH_3\overset{O}{\overset{\|}{C}}{-}SCoA + 7FADH_2 + 7NADH + 7H^+$$

Figure 17.2 *The fatty acid oxidation spiral. Degradation of palmitate to acetyl coenzyme A by the β-oxidation pathway.*

oxidized by FAD to form an α,β-unsaturated thioester. The enzyme that catalyzes this oxidation is stereospecific and produces only the *trans* geometric isomer. In step 2 of the oxidation spiral, the unsaturated thioester is hydrated to form a β-hydroxythioester. Step 3, the second oxidation, converts the secondary alcohol to a β-ketothioester, and here NAD^+ is the oxidizing agent. Finally, the β-ketothioester is cleaved by coenzyme A to form a molecule of acetyl CoA and a new thioester of coenzyme A.

The result of one spiral (four reactions) in fatty acid oxidation is the production of one molecule of acetyl CoA and a fatty acid thioester containing two carbons fewer than the starting acid. This new thioester is then the substrate for another round of reactions beginning with oxidation, hydration, oxidation, and terminating with the removal of a second two-carbon unit as acetyl CoA. This sequence of reactions continues until the entire carbon chain of palmitate (or any other fatty acid with an even number of carbon atoms) is degraded completely to acetyl CoA. We should note several important points about this stepwise metabolism of fatty acids.

1. Regardless of the length of the hydrocarbon chain, only one molecule of ATP is required to activate a molecule of fatty acid for its complete degradation to acetyl CoA.
2. Only derivatives of coenzyme A serve as substrates for fatty acid oxidation.
3. All enzymes for β-oxidation are located in the mitochondria, where the enzymes for the tricarboxylic acid cycle, electron transport, and oxidative phosphorylation also are located. This localization within the cell is of fundamental importance, for the acetyl CoA produced from the breakdown of either fatty acids or carbohydrates can be channeled directly into the Krebs cycle for oxidation to CO_2 and H_2O and the generation of ATP.
4. Finally, note the similarity between the first three steps of this cycle (oxidation, hydration, and oxidation) and the last three steps of the tricarboxylic acid cycle, which are compared in Figure 17.3.

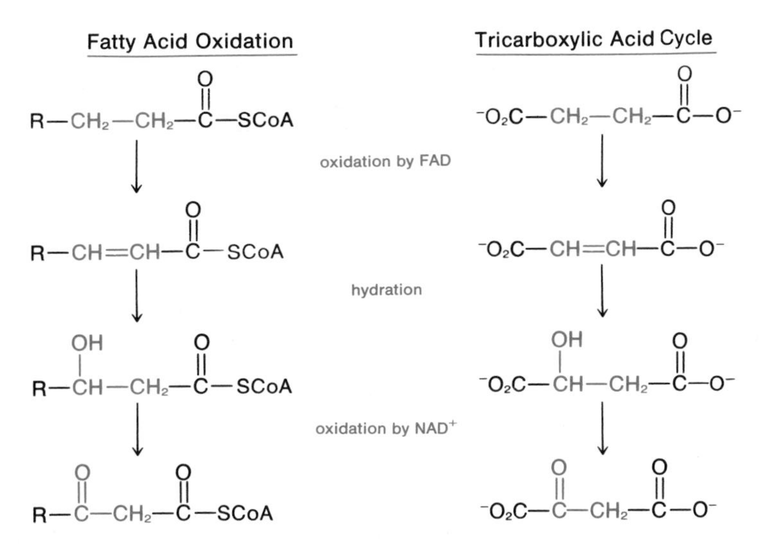

Figure 17.3 *Comparison of oxidation–hydration–oxidation steps in the degradation of fatty acids and carbohydrates.*

17.4 ENERGETICS OF FATTY ACID OXIDATION

Now that we have examined the steps in the fatty acid oxidation spiral, the next question to consider is: How much of the energy is conserved as ATP and thus made available to the cell for biochemical work? For convenience in tabulating the energetics, let us continue with the example of the oxidation of palmitate to carbon dioxide and water. Seven turns of the oxidation spiral converts 1 molecule of palmityl

CoA to 8 molecules of acetyl CoA and generates 7 molecules each of NADH and $FADH_2$. As we have already seen (Section 16.7), the reoxidation of each molecule of NADH is coupled with the formation of 3 molecules of ATP; the reoxidation of each molecule of $FADH_2$ is coupled with the formation of 2 molecules of ATP. Further, the oxidation of 1 molecule of acetyl CoA is coupled with the formation of 12 molecules of ATP.

Summing these processes for 1 mole of palmitate leads to the generation of 131 moles of ATP.

Reaction	Coupled ATP production
$7\,FADH_2 + 3\frac{1}{2}O_2 \rightarrow 7FAD + 7H_2O$	14 ATP
$7\,NADH + 3\frac{1}{2}O_2 + 7H^+ \rightarrow 7NAD^+ + 7H_2O$	21 ATP
$8\,\text{acetyl CoA} + 16O_2 \rightarrow 16CO_2 + 8H_2O + 8CoA$	96 ATP
	131 ATP

Since two phosphate anhydride bonds (equivalent to 2 moles of ATP) were required to activate palmitate, the net yield of ATP is 129 moles per mole of palmitate. This overall oxidation and the coupled formation of ATP can be expressed as the sum of an exergonic and an endergonic process.

Reaction	ΔG^0 (kcal/mole)
$\text{palmitate} + 23O_2 \rightarrow 16CO_2 + 16H_2O$	−2,340
$129ADP + 129HPO_4^{2-} \rightarrow 129ATP + 129H_2O$	+940
$\text{palmitate} + 23O_2 + 129ADP + 129HPO_4^{2-} \longrightarrow 16CO_2 + 145H_2O + 129ATP$	−1,400

Thus we see that some 940/2,340 or 40% of the standard free energy of oxidation of palmitate is conserved in the form of ATP and can be utilized by the cell for the performance of work. This fraction of energy conserved as ATP is comparable to that conserved in the complete oxidation of glucose to carbon dioxide and water (Section 16.8).

17.5 BIOSYNTHESIS OF FATTY ACIDS

From the time Knoop proposed the β-oxidation pathway in 1904 to account for the degradation of fatty acids, biochemists had speculated that these same reactions, but in reverse, would be used by the cell to synthesize fatty acids. Yet, soon after systematic studies of fatty acid biosynthesis were undertaken, it became clear that biosynthesis is not simply the reverse of degradation. Several observations led to this conclusion. First, the four enzymes catalyzing the β-oxidation spiral were isolated and it was demonstrated that these enzymes were not capable of synthesizing either palmitate or stearate from acetyl coenzyme A. Second, carbon dioxide (or bicarbonate) is an absolute requirement for fatty acid synthesis. This requirement was quite unexpected in view of the fact that fatty acid degradation to acetyl

CoA does not produce carbon dioxide. The requirement was all the more puzzling when it was shown using an isotope label that the carbon atom of carbon dioxide, although necessary for the reaction, is not incorporated into the fatty acid molecule. Third, mitochondria (the site of fatty acid degradation) show little tendency to synthesize fatty acids from acetate. However, the synthesis of fatty acids from acetate occurs at a rapid rate in the cytoplasm. Thus, acetate is used as a subunit for the synthesis of fatty acids. But how? This problem was not solved until 1959.

The key to the puzzle was the discovery that CO_2 (or bicarbonate), the absolute requirement for synthesis, is incorporated into one of the carboxyl groups of malonyl CoA (the thioester of malonic acid and coenzyme A).

$$CO_2 + \underset{\text{acetyl CoA}}{CH_3{-}\overset{\displaystyle O}{\overset{\|}{C}}{-}SCoA} \xrightarrow{\text{enzyme}} \underset{\text{malonyl CoA}}{{}^{-}O{-}\overset{\displaystyle O}{\overset{\|}{C}}{-}CH_2{-}\overset{\displaystyle O}{\overset{\|}{C}}{-}SCoA}$$

Once malonyl CoA was identified as the essential precursor, the overall pathway of fatty acid synthesis was quickly discovered. The series of reactions is catalyzed by a complex of enzymes called fatty acid synthetase, found in the soluble fraction of the cytoplasm. The various intermediates produced in the biosynthetic process are bound to the enzyme complex and the sequence of reactions is governed by the position of each enzyme on the complex. A key protein in this complex is the so-called acyl carrier protein (ACP). This low-molecular-weight protein first binds the acetyl and malonyl starting materials as thioesters.

$$CH_3{-}\overset{\displaystyle O}{\overset{\|}{C}}{-}S{-}CoA + ACP{-}SH \rightarrow CH_3{-}\overset{\displaystyle O}{\overset{\|}{C}}{-}S{-}ACP + CoA{-}SH$$

$$\underset{\text{malonyl CoA}}{{}^{-}O{-}\overset{\displaystyle O}{\overset{\|}{C}}{-}CH_2{-}\overset{\displaystyle O}{\overset{\|}{C}}{-}S{-}CoA} + ACP{-}SH \rightarrow \underset{\text{malonyl-ACP}}{{}^{-}O{-}\overset{\displaystyle O}{\overset{\|}{C}}{-}CH_2{-}\overset{\displaystyle O}{\overset{\|}{C}}{-}S{-}ACP} + CoA{-}SH$$

At this point, a pair of two-carbon fragments is activated, one as acetyl-ACP and the other as malonyl-ACP, and is prepared to enter into the series of reactions that leads to the elongation of the hydrocarbon chain two carbon atoms at a time. In the first of these steps, acetyl-ACP condenses with malonyl-ACP to form acetoacetyl-ACP.

$$\underset{\text{acetyl-ACP}}{CH_3{-}\overset{\displaystyle O}{\overset{\|}{C}}{-}S{-}ACP} + \underset{\text{malonyl-ACP}}{{}^{-}O{-}\overset{\displaystyle O}{\overset{\|}{C}}{-}CH_2{-}\overset{\displaystyle O}{\overset{\|}{C}}{-}S{-}ACP}$$

$$\longrightarrow \underset{\text{acetoacetyl-ACP}}{CH_3\overset{\displaystyle O}{\overset{\|}{C}}{-}CH_2{-}\overset{\displaystyle O}{\overset{\|}{C}}{-}S{-}ACP} + CO_2 + ACP{-}SH$$

Note that this enzyme-catalyzed condensation is analogous to the Claisen condensation (Section 8.13). Note also that this condensation reaction is coupled with the loss of carbon dioxide by decarboxylation. It is now clear why the carbon atom of CO_2 (or bicarbonate) never appears in the fatty acid finally formed. In effect, the carbon dioxide plays a catalytic role and is regenerated as the carbon chain is lengthened.

Following the carbonyl condensation reaction, reduction of acetoacetyl-ACP by NADPH produces β-hydroxybutyryl-ACP. Dehydration of this β-hydroxythioester, followed by reduction with a second molecule of NADPH, forms butyryl-ACP.

$$CH_3-\overset{\overset{O}{\|}}{C}-CH_2-\overset{\overset{O}{\|}}{C}-S-ACP \quad \text{acetoacetyl-ACP}$$

$$\downarrow \text{NADPH} \rightarrow \text{NADP}^+$$

$$CH_3-\overset{\overset{OH}{|}}{C}H-CH_2-\overset{\overset{O}{\|}}{C}-S-ACP \quad \beta\text{-hydroxybutyryl-ACP}$$

$$\downarrow \rightarrow H_2O$$

$$CH_3-CH=CH-\overset{\overset{O}{\|}}{C}-S-ACP \quad \textit{trans}\text{-crotonyl-ACP}$$

$$\downarrow \text{NADPH} \rightarrow \text{NADP}^+$$

$$CH_3-CH_2-CH_2-\overset{\overset{O}{\|}}{C}-S-ACP \quad \text{butyryl-ACP}$$

The formation of butyryl-ACP completes the first cycle and this compound is now the substrate for a second cycle beginning with condensation of a second molecule of malonyl-ACP and followed by reduction, dehydration, and reduction. A total of seven such cycles leads to the formation of 1 molecule of palmitoyl-ACP from 1 molecule of acetyl-ACP and 7 of malonyl-ACP. Since the synthesis of palmitate begins with acetyl-ACP and proceeds by the addition of carbon atoms two at a time in each of the condensation steps, we can easily account for the fact that most fatty acids in nature contain an even number of carbon atoms in an unbranched hydrocarbon chain.

At this point, let us stop to compare the enzyme-catalyzed reactions for the synthesis and degradation of fatty acids. First, synthesis takes place in the cytoplasm and degradation takes place in the mitochondria. In other words, the enzymes catalyzing these two pathways are separated into different compartments within the cell. Second, biosynthesis uses NADPH as a reducing agent and generates $NADP^+$. In contrast, degradation used NAD^+ as an oxidizing agent and generates NADH. This illustrates a general principle that biosynthetic sequences involve NADPH/$NADP^+$ whereas energy-producing sequences involve NADH/NAD^+. Third, the two-carbon units are added in the form of malonyl-ACP but are removed in the form of acetyl CoA.

17.6 FORMATION OF KETONE BODIES

As we saw in Section 17.3, acetoacetate, β-hydroxybutyrate, and acetone are classed together as ketone bodies. When the production of these substances exceeds the capacity of the body to use them, the condition is known as ketosis. Of the three ketone bodies, acetoacetic acid and β-hydroxybutyric acid are the more important physiologically and clinically for both are moderately strong acids and must be buffered in blood and other tissue fluids to prevent their accumulation from disrupting normal acid–base balance. The acidosis that results from the production of excessive amounts of ketone bodies is called ketoacidosis and may be fatal in uncontrolled diabetes mellitus.

In man and most other animals, the liver is the only organ that produces any significant amounts of ketone bodies. Most other tissues, with the notable exception of the brain, have the capacity to use them as energy sources. The direct precursor of ketone bodies is acetyl coenzyme A which, as we have seen already, is formed during the metabolism of both carbohydrates and fats. Under normal circumstances (Figure 17.4), where carbohydrate and fat degradation are appropriately balanced and there is an adequate supply of oxaloacetate, acetyl CoA in the liver either is channeled into the tricarboxylic acid cycle (a degradative pathway) for oxidation to carbon dioxide and water, or it is converted into fatty acids (a synthetic pathway) and in turn into fats. Production of ketone bodies is minimal.

Several conditions lead to a drastic reduction in carbohydrate metabolism. These include starvation, an unbalanced diet (i.e., one low in carbohydrate and high in fat), and of course diabetes mellitus. Under these circumstances, fats become the primary source of energy for all tissues, with the exception of the brain, and the formation of ketone

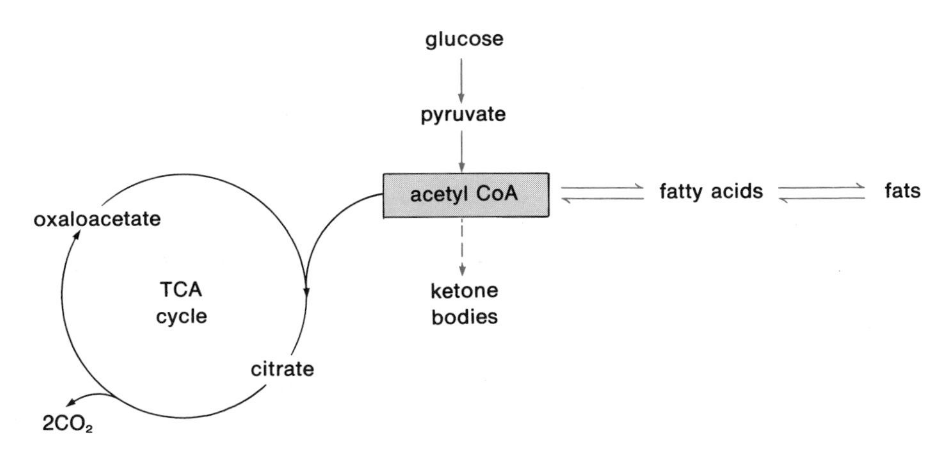

Figure 17.4 The fates of acetyl CoA in the liver under normal circumstances. Formation of ketone bodies is minimal.

bodies is increased. While each of these conditions will produce ketosis and ketoacidosis, they are most severe in diabetes mellitus.

In the case of diabetes mellitus, carbohydrate metabolism is reduced because of the lack of insulin, the pancreatic hormone that promotes cellular uptake and oxidation of glucose. In addition to regulating glucose metabolism, insulin also plays a role in the regulation of fat metabolism since this hormone promotes the synthesis of triglycerides in the liver and inhibits the hydrolysis of triglycerides in adipose tissue. These actions of insulin are opposed and balanced by the action of glucagon, another hormone of the pancreas, which stimulates hydrolysis of triglycerides in adipose tissue and the oxidation of fatty acids in the liver. The insulin–glucagon ratio is a key means by which the body regulates and balances carbohydrate and fat metabolism.

In diabetes mellitus, insulin production is drastically reduced and the insulin–glucagon ratio is thrown seriously out of balance. As a result of this imbalance, the hydrolysis of triglycerides in adipose tissue is stimulated and increased quantities of free fatty acids are released into the blood stream and transported to the liver. The liver responds by absorbing these fatty acids and generating considerably increased quantities of acetyl CoA. This acetyl CoA in the liver can be channeled into three metabolic pathways: the tricarboxylic acid cycle for oxidation to carbon dioxide and the generation of ATP; the synthesis of fatty acids and fats for storage in the liver; and finally the synthesis of ketone bodies. While both the tricarboxylic acid cycle and fatty acid synthesis can bear some of the burden, the major fraction of the increased supply of acetyl CoA is channeled into the synthesis of ketone bodies (Figure 17.5).

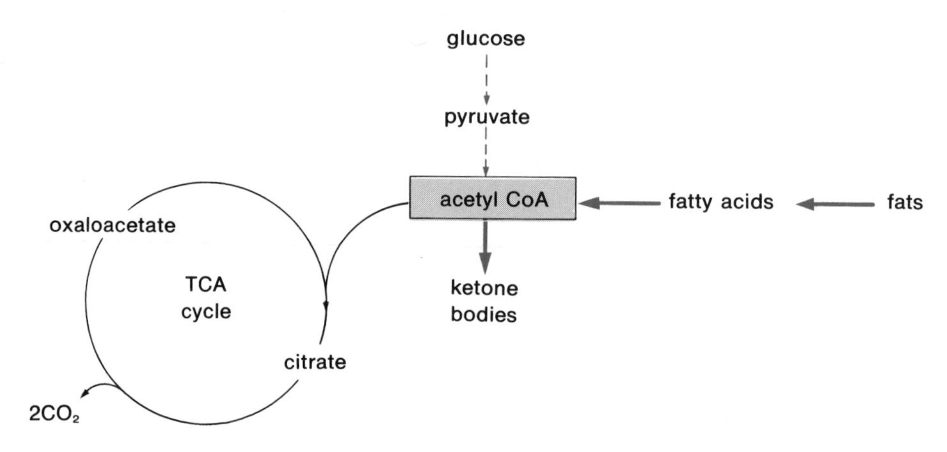

Figure 17.5 *The generation of ketone bodies as a result of impaired carbohydrate metabolism.*

In the synthesis of ketone bodies, acetyl CoA is converted into acetoacetate in a series of three enzyme-catalyzed reactions:

$$2CH_3—\overset{\overset{\displaystyle O}{\|}}{C}—SCoA + H_2O \rightarrow \rightarrow \rightarrow \underset{\text{acetoacetate}}{CH_3—\overset{\overset{\displaystyle O}{\|}}{C}—CH_2—CO_2^-} + 2CoA—SH$$

Acetoacetate is in turn reduced by NADH to β-hydroxybutyrate.

$$\underset{\text{acetoacetate}}{CH_3\text{—}\overset{\overset{\displaystyle O}{\|}}{C}\text{—}CH_2\text{—}CO_2^-} + NADH + H^+ \longrightarrow \underset{\beta\text{-hydroxybutyrate}}{CH_3\text{—}\overset{\overset{\displaystyle OH}{|}}{C}H\text{—}CH_2\text{—}CO_2^-} + NAD^+$$

These ketone bodies are then transported by the blood from the liver to other tissues where they are used for the generation of energy. However, when the supply exceeds the body's capacity to use them, they accumulate in the blood and cause ketoacidosis. Under these circumstances, the effectiveness of hemoglobin to transport oxygen is reduced and a deficiency in the supply of oxygen, in the extreme, can result in a fatal coma. The presence of ketone bodies in the urine indicates an advanced state of ketoacidosis and is a clear signal that medical attention is essential.

17.7 THE INTERRELATIONSHIP OF CARBOHYDRATES AND FATTY ACIDS

As we have pointed out several times during our discussions of carbohydrate and fat metabolism, glucose can be used as a source of carbon atoms for the synthesis of fatty acids. Let us now look at the biochemical pathways for the synthesis of a fatty acid from glucose and see exactly which atoms of the fatty acid are derived from which atoms of glucose. Figure 17.6 traces this flow of carbon atoms from glucose to palmitate.

Cleavage of glucose during glycolysis produces one molecule each of glyceraldehyde 3-phosphate and dihydroxyacetone phosphate. Each triose phosphate is next converted to pyruvate and finally to acetyl CoA and carbon dioxide. Note that the carbons of one molecule of pyruvate are derived from atoms 1, 2, and 3 of glucose; the carbons of the other molecule of pyruvate are derived from atoms 4, 5, and 6 of glucose. In the oxidative decarboxylation of pyruvate to acetyl CoA, carbon atoms 3 and 4 of the original glucose molecule appear as carbon

Figure 17.6 *The synthesis of palmitate from glucose. The carbon atoms of glucose are numbered 1 through 6.*

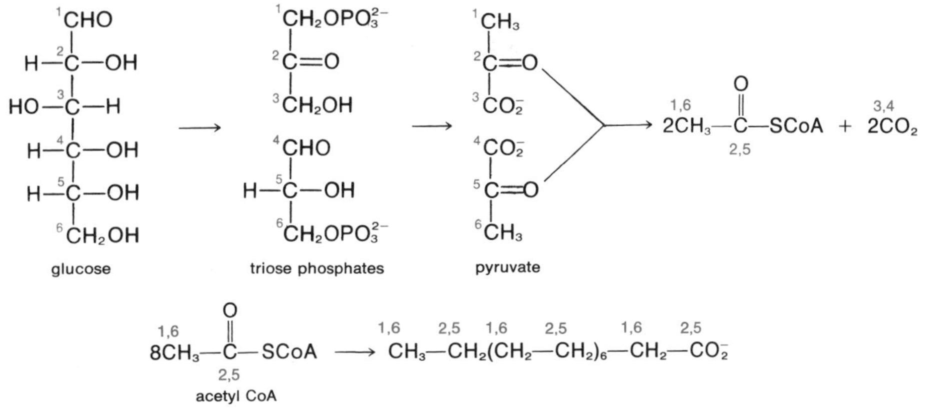

$$\underset{\text{glucose}}{\begin{array}{c} {}^{1}CHO \\ H\text{—}{}^{2}C\text{—}OH \\ HO\text{—}{}^{3}C\text{—}H \\ H\text{—}{}^{4}C\text{—}OH \\ H\text{—}{}^{5}C\text{—}OH \\ {}^{6}CH_2OH \end{array}} \longrightarrow \underset{\text{triose phosphates}}{\begin{array}{c} {}^{1}CH_2OPO_3^{2-} \\ {}^{2}C{=}O \\ {}^{3}CH_2OH \\ {}^{4}CHO \\ H\text{—}{}^{5}C\text{—}OH \\ {}^{6}CH_2OPO_3^{2-} \end{array}} \longrightarrow \underset{\text{pyruvate}}{\begin{array}{c} {}^{1}CH_3 \\ {}^{2}C{=}O \\ {}^{3}CO_2^- \\ {}^{4}CO_2^- \\ {}^{5}C{=}O \\ {}^{6}CH_3 \end{array}} \longrightarrow 2\overset{1,6}{CH_3}\text{—}\underset{2,5}{\overset{\overset{\displaystyle O}{\|}}{C}}\text{—}SCoA + 2\overset{3,4}{CO_2}$$

$$\underset{\text{acetyl CoA}}{8\overset{1,6}{CH_3}\text{—}\underset{2,5}{\overset{\overset{\displaystyle O}{\|}}{C}}\text{—}SCoA} \longrightarrow \overset{1,6}{CH_3}\text{—}\overset{2,5}{CH_2}(\overset{1,6}{CH_2}\text{—}\overset{2,5}{CH_2})_6\text{—}\overset{1,6}{CH_2}\text{—}\overset{2,5}{CO_2^-}$$

18

AMINO ACID METABOLISM

18.1 INTRODUCTION

In the broadest sense, amino acids serve three vital functions in the human body. They are (1) building blocks for the synthesis of proteins, (2) sources of carbon and nitrogen atoms for the synthesis of other biomolecules, and (3) sources of energy.

The most important of these functions, at least in terms of total amino acid utilization, is as building blocks for the synthesis of polypeptides and proteins. It is estimated that about 75% of the amino acid metabolism in normal, healthy adults is devoted to this purpose, which stems from the fact that there is constant "wear-and-tear" destruction, repair, and rebuilding of body protein. The use of radioisotopes has given us some idea of the extent of this metabolic turnover. For example, the half-life of liver proteins is about 10 days, plasma proteins about 10 days, hemoglobin about 120 days, and muscle protein about 180 days. The half-life of collagen is much longer. However, other proteins, particularly enzymes and polypeptide hormones, have considerably shorter half-lives. That of insulin, once it is released from the pancreas, is estimated to be only 7–10 minutes. Clearly, the apparent stability of body proteins in fact represents a dynamic balance between degradation and synthesis. It is estimated that up to 400 grams of body protein turns over per day. At least 75% of this amount is salvaged and reused for protein synthesis. The remainder must be replaced by dietary protein.

Amino acids also provide a source of carbon and nitrogen atoms used in the synthesis of nonprotein biomolecules. These include the porphyrin rings of myoglobin, hemoglobin, and the cytochromes; the choline and ethanolamine building blocks of phospholipids; glucosamine and other amino sugars required for the synthesis of connective tissue; dopamine, norepinephrine, serotonin, and other neurotransmitters (see the mini-essay "Biogenic Amines and Emotion); the purine and pyrimidine bases required for the synthesis of nucleic acids; and certain hormones including thyroxine, etc.

Unlike carbohydrates and fatty acids, amino acids in excess of those needed for biosynthesis cannot be stored for later use. Rather, their carbon skeletons are degraded to pyruvate, acetyl coenzyme A, or one of the intermediates of the tricarboxylic acid cycle. These carbon skeletons can then be used as metabolic fuels in one of two ways. They may be oxidized directly to carbon dioxide by the reactions of the Krebs cycle, or alternatively, they may be converted first to

7 — qualitative analysis
1* — acid-base chemistry
2* —
6 —
8 — nickel complexes
3* — aspirin

prelab is O.K.

U

Experiment I - Inorganic Qual.

1st day - Do all of unknown A (using procedures 1-5 in IV B discarding the supernatant above the chloride ppte in proc. I)

- Using unknown C, do procedures 1-5 in IV B again discarding the supernatant above the chloride ppte in proc. I - even tho there are hydroxides present.

2nd day - Do all of unknown B (using procedures 1-11 in V B.

3rd day - Using unknown C procedure 1 in IV B, this time discarding the chloride ppte and saving the supernatant. Use the supernatant (10 drops) and Do procedures 1-11 in V B.

glucose by the reactions of gluconeogenesis. Of course, the carbon skeletons first converted to glucose are ultimately oxidized to carbon dioxide via the tricarboxylic acid cycle. Oxidation of amino acids, either directly or via glucose, supplies 15–20% of the total energy requirement of the average adult.

While the carbon skeletons of amino acids may be stored as either carbohydrate or fat, there is no storage site for the nitrogen atoms derived from amino acid degradation. For the most part, nitrogen atoms in excess of immediate needs are converted to urea and excreted in the urine. These various pathways of amino acid metabolism are summarized in Figure 18.1.

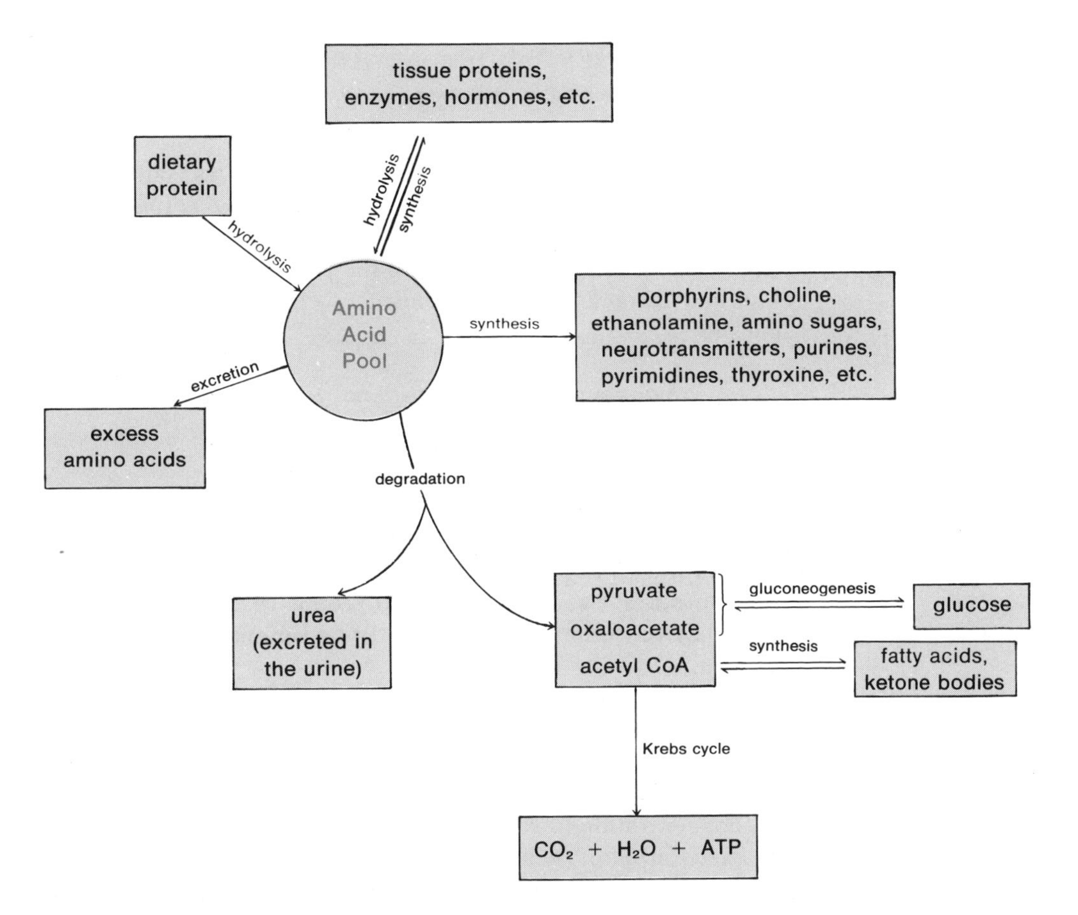

Figure 18.1 *An overview of amino acid metabolism. Average daily turnover of amino acids is estimated to be about 400 g.*

Let us begin our discussion of amino acid metabolism by following the degradation of amino groups and the eventual excretion of amino acid-derived nitrogen as urea and ammonium ions. This involves three metabolic processes: transamination, oxidative deamination, and the urea cycle.

18.2 TRANSAMINATION

Of all the reactions involved in the metabolism of amino groups, transamination is one of the most important for it is generally the first reaction in the degradation of amino acids and the last reaction in their biosynthesis. Transamination is illustrated in Figure 18.2 by the reaction of glutamate and pyruvate. Note that in this process, the donor amino acid is transformed into an α-ketoacid (i.e., glutamate is transformed into α-ketoglutarate), and the acceptor α-ketoacid is transformed into an α-amino acid (pyruvate is transformed into alanine).

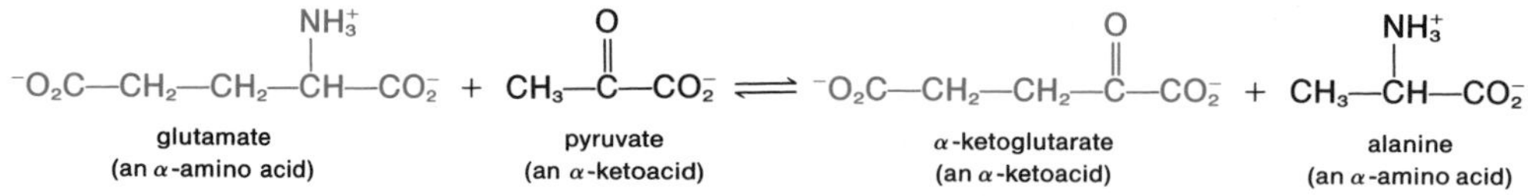

$$^{-}O_2C{-}CH_2{-}CH_2{-}\underset{}{\overset{NH_3^+}{\overset{|}{C}H}}{-}CO_2^- + CH_3{-}\overset{O}{\overset{\|}{C}}{-}CO_2^- \rightleftharpoons {}^{-}O_2C{-}CH_2{-}CH_2{-}\overset{O}{\overset{\|}{C}}{-}CO_2^- + CH_3{-}\overset{NH_3^+}{\overset{|}{C}H}{-}CO_2^-$$

glutamate (an α-amino acid) + pyruvate (an α-ketoacid) ⇌ α-ketoglutarate (an α-ketoacid) + alanine (an α-amino acid)

Figure 18.2 *Transamination between glutamate and pyruvate.*

Three different α-ketoacids participate in transamination reactions of amino acids. These are pyruvate, α-ketoglutarate, and oxaloacetate. Thus, during the transamination phase of amino acid degradation, all amino groups (with the possible exception of those of lysine and threonine) are channeled to either alanine, glutamate, or aspartate.

$$\text{amino acid} + \text{pyruvate} \rightleftharpoons \alpha\text{-ketoacid} + \text{alanine}$$
$$\text{amino acid} + \alpha\text{-ketoglutarate} \rightleftharpoons \alpha\text{-ketoacid} + \text{glutamate}$$
$$\text{amino acid} + \text{oxaloacetate} \rightleftharpoons \alpha\text{-ketoacid} + \text{aspartate}$$

While pyruvate and oxaloacetate do serve as acceptors of amino groups as shown above, it is α-ketoglutarate which serves as the final collection point, even for the amino groups of alanine and aspartate, before they are ultimately converted to ammonium ions and urea for excretion.

Transaminations are catalyzed by a specific group of enzymes called aminotransferases, or more commonly, transaminases. The transamination in Figure 18.2 is catalyzed by the enzyme glutamate pyruvate transaminase (GPT). Transaminases are found in all cells. However, their concentrations are particularly high in heart and liver tissue. Damage to either of these organs leads to release of transaminases into the blood, and determination of serum levels of these enzymes can provide the clinician with valuable information about the extent of heart or liver damage. The two transaminases most commonly assayed for this purpose are serum glutamate oxaloacetate transaminase (SGOT) and serum glutamate pyruvate transaminase (SGPT).

For catalytic activity, all transaminases require pyridoxal phosphate (Figure 18.3), a coenzyme derived from pyridoxine (vitamin B_6). In its role as a catalyst, this coenzyme undergoes reversible transformations between an aldehyde (pyridoxal phosphate) and a primary amine (pyridoxamine phosphate) as shown in Figure

pyridoxine (vitamin B_6) — pyridoxal phosphate (PLP) — pyridoxamine phosphate (PMP)

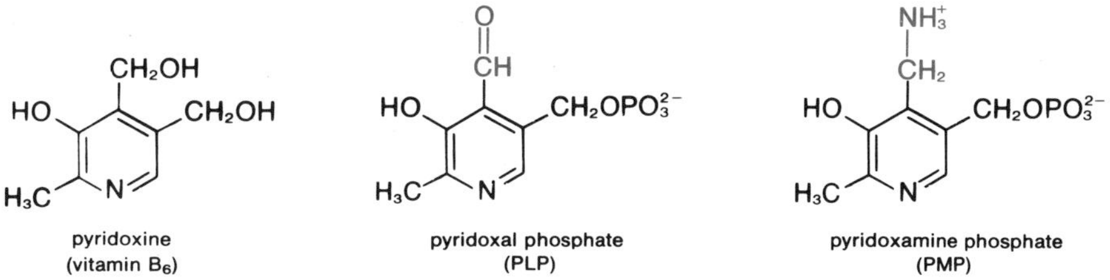

Figure 18.3 Pyridoxine or vitamin B_6. Pyridoxal phosphate (PLP) and pyridoxamine phosphate (PMP) are coenzymes derived from pyridoxine.

18.4. Note that in these reactions, the enzyme–pyridoxal phosphate complex is indicated by the symbol Ⓟ—CHO and the enzyme–pyridoxamine phosphate complex by Ⓟ—CH_2NH_2.

During transamination, the aldehyde group of pyridoxal phosphate reacts with the incoming amino acid to form a tetrahedral carbonyl addition intermediate (not shown), which in turn loses a molecule of water to form a Schiff base. Shift of a proton and rearrangement of the carbon–nitrogen double bond leads to the formation of an isomeric Schiff base. Reaction of this isomeric Schiff base with a molecule of water forms a new tetrahedral carbonyl addition intermediate (also not shown), which breaks apart to generate pyridoxamine phosphate and a new α-ketoacid. Pyridoxamine phosphate can in turn donate its amino group to another α-ketoacid by reversal of the steps just described. Thus, by alternating between the aldehyde and amino forms, this coenzyme acts as a carrier of amino groups from amino acids to α-ketoacids.

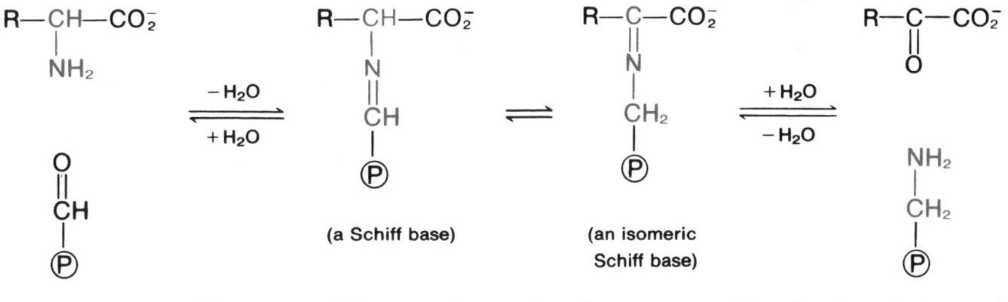

(a Schiff base) — (an isomeric Schiff base)

Figure 18.4 The reversible transformation between pyridoxal phosphate and pyridoxamine phosphate during transamination.

Transamination reactions serve two general functions. First, they provide a means of readjusting the relative proportions of a number of amino acids to meet the particular needs of the cell and the organism for, in most diets, the amino acid blend does not correspond precisely to what is needed. Second, transaminations channel all amino groups to alanine, aspartate, and ultimately to glutamate. When the supply of glutamate exceeds the organism's immediate needs for it, the amino groups are released by oxidative deamination for eventual excretion as either ammonium ion or urea.

18.3 OXIDATIVE DEAMINATION

The second major pathway by which amino groups are removed from amino acids is oxidative deamination of glutamate in the liver to form ammonium ion and α-ketoglutarate. The oxidation requires either NAD^+ or $NADP^+$ and is catalyzed by the enzyme glutamate dehydrogenase.

$$\underset{\text{glutamate}}{{}^-O_2C{-}CH_2{-}CH_2{-}\overset{\overset{\displaystyle NH_3^+}{|}}{C}H{-}CO_2^-} + NAD^+ + H_2O \rightleftharpoons \underset{\alpha\text{-ketoglutarate}}{{}^-O_2C{-}CH_2{-}CH_2{-}\overset{\overset{\displaystyle O}{\|}}{C}{-}CO_2^-} + NADH + NH_4^+ + H^+$$

In this way, the amino groups collected from other amino acids are converted to ammonium ions.

Normal concentrations of ammonium ion in plasma are 0.025 to 0.04 mg/liter. Since ammonium ion is extremely toxic, it must be detoxified and eliminated. There are three principal metabolic pathways for detoxification and disposal: reaction with glutamate to form glutamine, reaction with α-ketoglutarate to form glutamate, and conversion to urea and excretion in the urine.

The liver has the capability to catalyze the conversion of α-ketoglutarate and ammonium ions to glutamate by reversing the glutamate dehydrogenase reaction just described. In this way, nitrogen atoms derived from the degradation of amino acids can be salvaged and reused for the synthesis of amino acids and other nitrogen-containing biomolecules.

Many tissues, particularly the kidneys and brain, dispose of ammonium ions through the formation of glutamine.

$$\underset{\text{glutamate}}{{}^-O{-}\overset{\overset{\displaystyle O}{\|}}{C}{-}CH_2{-}CH_2{-}\overset{\overset{\displaystyle NH_3^+}{|}}{C}H{-}CO_2^-} + NH_4^+ \rightleftharpoons \underset{\text{glutamine}}{H_2N{-}\overset{\overset{\displaystyle O}{\|}}{C}{-}CH_2{-}CH_2{-}\overset{\overset{\displaystyle NH_3^+}{|}}{C}H{-}CO_2^-} + H_2O$$

However, glutamine formed in this manner is only a temporary, nontoxic transport form of amino acid-derived nitrogen atoms.

Formation of urea, a neutral nontoxic compound, in the liver is the third and most important route by which man detoxifies ammonium ions and it is also the major pathway by which amino acid-derived nitrogen atoms are excreted.

18.4 UREA SYNTHESIS

Urea synthesis in mammals occurs exclusively in the liver, and the urea is then transported by the blood to the kidneys for excretion in the urine. The metabolic pathway which catalyzes the formation of urea is called the urea cycle or the Krebs–Henseleit cycle after Hans Krebs and Kurt Henseleit who proposed it in 1932. This cycle accepts one carbon atom in the form of bicarbonate (or carbon dioxide) and two nitrogen atoms, one from ammonium ion directly and the other from aspartate, and in a cyclic process requiring five steps generates urea and fumarate. The net reaction of the urea cycle is shown in Figure 18.5.

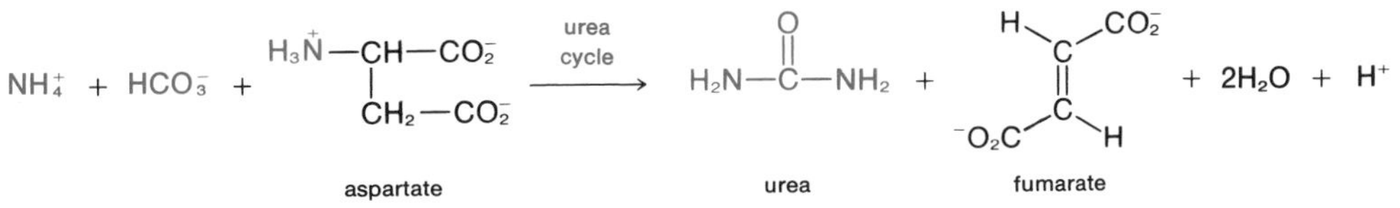

$$NH_4^+ + HCO_3^- + \underset{\text{aspartate}}{H_3\overset{+}{N}{-}CH(CH_2{-}CO_2^-){-}CO_2^-} \xrightarrow{\text{urea cycle}} \underset{\text{urea}}{H_2N{-}C(=O){-}NH_2} + \underset{\text{fumarate}}{{}^-O_2C{-}CH{=}CH{-}CO_2^-} + 2H_2O + H^+$$

Figure 18.5 *The net reaction of the urea cycle.*

The step in the cycle in which urea is formed (step 5) involves the hydrolysis of arginine catalyzed by the enzyme arginase.

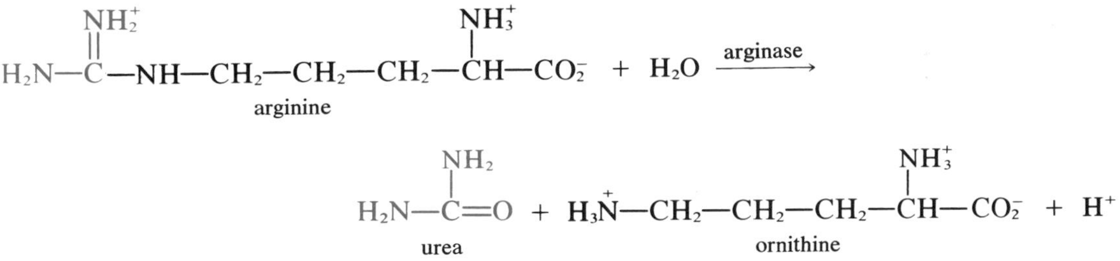

$$\underset{\text{arginine}}{H_2N{-}C(=NH_2^+){-}NH{-}CH_2{-}CH_2{-}CH_2{-}CH(NH_3^+){-}CO_2^-} + H_2O \xrightarrow{\text{arginase}}$$

$$\underset{\text{urea}}{H_2N{-}C(NH_2){=}O} + \underset{\text{ornithine}}{H_3\overset{+}{N}{-}CH_2{-}CH_2{-}CH_2{-}CH(NH_3^+){-}CO_2^-} + H^+$$

The other four reactions of the urea cycle catalyze the resynthesis of arginine from ornithine, using ammonium ion and aspartate as sources of nitrogen and bicarbonate as a source of carbon. The five separate reactions of the urea cycle are shown in Figure 18.6.

The first step in the formation of urea is the synthesis of carbamoyl phosphate from bicarbonate, ammonium ion, and inorganic phosphate. This reaction is catalyzed by the enzyme carbamoyl phosphate synthetase and is coupled with the hydrolysis of two molecules of ATP to ADP. In step 2, carbamoyl phosphate reacts directly with ornithine to form citrulline. The third step, condensation of citrulline with aspartate, is coupled with the hydrolysis of a third molecule of ATP and incorporates the second nitrogen atom into the cycle. Cleavage of argininosuccinate produces arginine and fumarate (step 4). Finally, hydrolysis of arginine yields one molecule of urea and regenerates ornithine. Note that the fumarate produced in the urea cycle can enter the tricarboxylic acid cycle, where it is transformed first to malate, then to oxaloacetate. Transamination of oxaloacetate regenerates another molecule of aspartate. Thus, the reactions of the urea-forming pathway are truly cyclical. Of the two molecules necessary for the complete operation of the cycle, one (ornithine) is regenerated within the cycle itself and the other (aspartate) is regenerated through the reactions of the tricarboxylic acid cycle and transamination.

The amount of urea excreted in the urine of a normal adult is about 25–30 grams per day and represents about 90% of the total nitrogen-

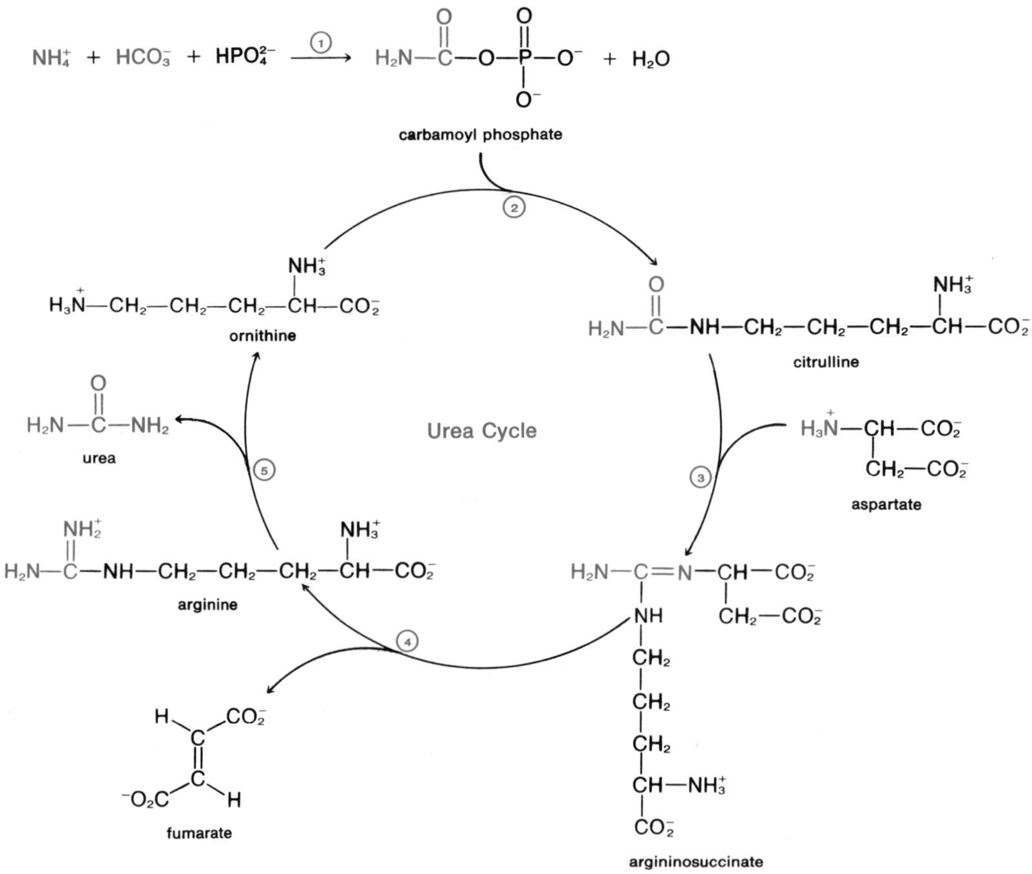

$$NH_4^+ + HCO_3^- + HPO_4^{2-} \xrightarrow{1} H_2N{-}C(=O){-}O{-}P(=O)(O^-){-}O^- + H_2O$$

carbamoyl phosphate

Urea Cycle

ornithine · citrulline · aspartate · argininosuccinate · fumarate · arginine · urea

Figure 18.6 The urea cycle. The enzyme-catalyzed steps are numbered 1–5.

containing substances excreted. Of course, this amount will increase or decrease in direct proportion to the protein content of the diet. The direct excretion of ammonium ion accounts for only about 2.5–4.0% of the total urinary nitrogen. However, excretion of nitrogen in this form does serve as one means of controlling acid–base balance. Ammonia can be formed in the kidneys and, in combination with hydrogen ions, is excreted as ammonium ions. Thus, the NH_4^+ concentration in urine increases during acidosis and decreases during alkalosis.

Virtually all tissues produce ammonia, and as we have already indicated, this substance is highly toxic, especially to the nervous system. The brain detoxifies ammonium ions by converting them to glutamine for transport to the liver. In man, however, the synthesis of urea in the liver is the only major route for the detoxification and elimination of this poison. Failure of the urea-synthesizing pathway for any reason, including liver malfunction or inherited defects in any of the five enzymes of the urea cycle, results in serious illness. For example, impaired activity or a deficiency of the enzyme that catalyzes the formation of carbamoyl phosphate (step 1 of the urea cycle), or of the enzyme that catalyzes the condensation of carbamoyl phosphate with ornithine (step 2), results in an increase in blood, liver, and urinary ammonium ion—a condition that produces ammonia intoxication.

Symptoms of ammonia intoxication are protein-induced vomiting, blurred vision, tremors, slurred speech, and ultimately coma and death. In treating ammonia intoxication, it is essential to decrease the intake of dietary protein in order to decrease ammonium ion formation. Genetic defects in each of the five enzymes of the urea cycle do exist, but fortunately these inborn errors of metabolism are rare.

18.5 THE FATES OF CARBON SKELETONS

As mentioned in the introduction to this chapter, the final pathway for the disposal of the carbon skeletons derived from amino acid metabolism is the tricarboxylic acid cycle. There are five points of entry for these skeletons into the cycle: acetyl CoA, α-ketoglutarate, succinyl CoA, fumarate, and oxaloacetate. The specific entry points for carbon fragments from each of the 20 amino acids are summarized in Figure 18.7.

As you study Figure 18.7, note the following about the points at which the carbon skeletons of amino acids enter the tricarboxylic acid cycle.

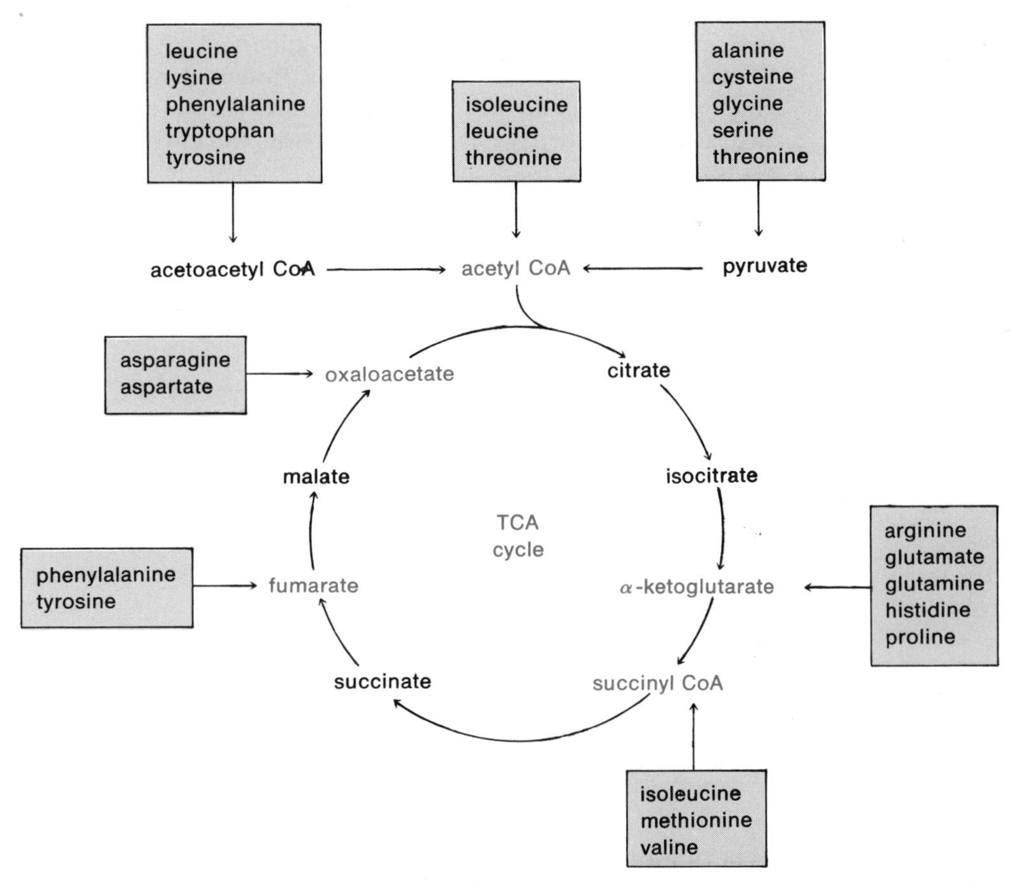

Figure 18.7 Pathways by which carbon skeletons from amino acid degradation enter the tricarboxylic acid cycle.

1. Eleven (alanine, cysteine, glycine, isoleucine, leucine, lysine, phenylalanine, serine, threonine, tryptophan, and tyrosine) enter via acetyl CoA. Thus acetyl CoA is the major point of entry.
2. Five (arginine, glutamate, glutamine, histidine, and proline) enter via α-ketoglutarate.
3. Three (isoleucine, methionine, and valine) enter via succinyl CoA.
4. Two (phenylalanine and tyrosine) enter via fumarate.
5. Two (asparagine and aspartate) enter via oxaloacetate.
6. The carbon skeletons of three amino acids are each degraded to two fragments, both of which enter the cycle at different points. Phenylalanine and tyrosine are degraded in part to acetoacetyl CoA, which then enters the cycle via acetyl CoA, and in part to fumarate, which enters directly. Isoleucine is degraded to acetyl CoA and succinyl CoA, both of which enter the cycle directly.

The pathways for the degradation of the individual amino acids are often long and complex and therefore we will not study them in any detail. (For a brief discussion of the metabolism of phenylalanine and tyrosine together with that of three hereditary diseases associated with the degradation of these amino acids, see the mini-essay "Inborn Errors of Metabolism.") However, we will look at the separate steps in the degradation of two amino acids, valine and glycine, to illustrate several important reaction types already presented in this and previous chapters. In addition, we shall discuss the structure and function of two additional coenzymes, cobalamin (or vitamin B_{12}) and tetrahydrofolate.

Let us look first at valine. The initial step in the degradation of valine (and of most other amino acids as well) is transamination with α-ketoglutarate serving as the acceptor of the amino group.

$$CH_3—CH(CH_3)—CH(NH_3^+)—CO_2^- + \alpha\text{-ketoglutarate} \rightarrow CH_3—CH(CH_3)—C(=O)—CO_2^- + \text{glutamate}$$

valine α-ketoisovalerate

Oxidative decarboxylation of α-ketoisovalerate by NAD^+, coupled with thioester formation, yields isobutyryl CoA.

$$CH_3—CH(CH_3)—C(=O)—CO_2^- + NAD^+ + CoA—SH \rightarrow CH_3—CH(CH_3)—C(=O)—SCoA + NADH + CO_2$$

α-ketoisovalerate isobutyryl CoA

This step is analogous to the conversion of pyruvate to acetyl CoA (Section 16.5). Isobutyryl CoA is converted in two steps to β-hydroxyisobutyryl CoA.

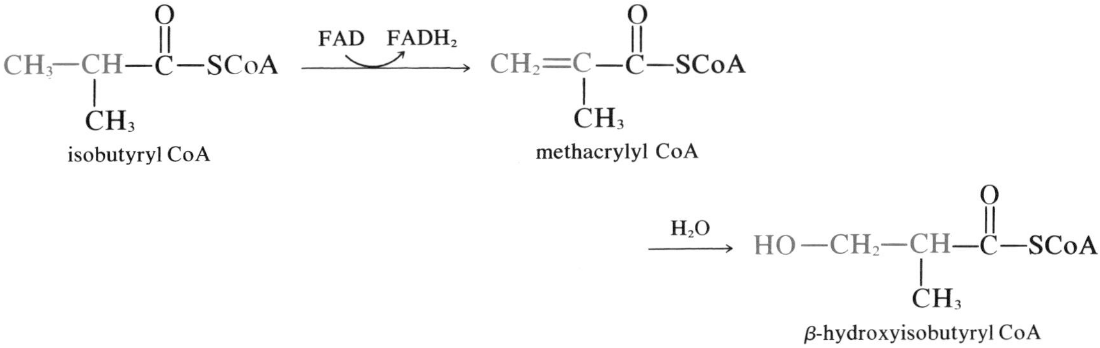

$$CH_3-\underset{\underset{CH_3}{|}}{CH}-\overset{\overset{O}{\|}}{C}-SCoA \xrightarrow{FAD \quad FADH_2} CH_2=\underset{\underset{CH_3}{|}}{C}-\overset{\overset{O}{\|}}{C}-SCoA$$

isobutyryl CoA — methacrylyl CoA

$$\xrightarrow{H_2O} HO-CH_2-\underset{\underset{CH_3}{|}}{CH}-\overset{\overset{O}{\|}}{C}-SCoA$$

β-hydroxyisobutyryl CoA

Next, a four-electron oxidation involving two molecules of NAD^+ converts β-hydroxyisobutyryl CoA to methylmalonyl CoA.

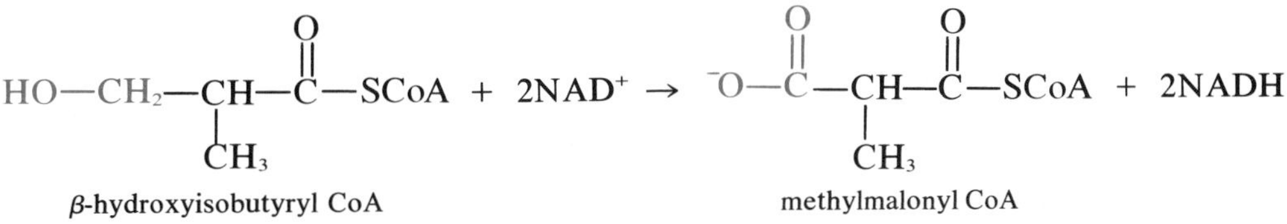

$$HO-CH_2-\underset{\underset{CH_3}{|}}{CH}-\overset{\overset{O}{\|}}{C}-SCoA + 2NAD^+ \rightarrow {}^-O-\overset{\overset{O}{\|}}{C}-\underset{\underset{CH_3}{|}}{CH}-\overset{\overset{O}{\|}}{C}-SCoA + 2NADH$$

β-hydroxyisobutyryl CoA — methylmalonyl CoA

In addition to being formed from valine by the pathway just described, methylmalonyl CoA is also formed in the degradation of methionine, isoleucine, and threonine. Further, although most fatty acids are composed of an even number of carbon atoms and are degraded directly to acetyl CoA, small amounts of fatty acids containing an odd number of carbon atoms also are present in fats and these also are degraded to methylmalonyl CoA.

Methylmalonyl CoA is isomerized to succinyl CoA in an unusual reaction catalyzed by the enzyme methylmalonyl CoA mutase.

$${}^-O-\overset{\overset{O}{\|}}{C}-\underset{\underset{CH_3}{|}}{CH}-\overset{\overset{O}{\|}}{C}-SCoA \xrightarrow{\text{methylmalonyl CoA mutase}} {}^-O-\overset{\overset{O}{\|}}{C}-CH_2-CH_2-\overset{\overset{O}{\|}}{C}-SCoA$$

methylmalonyl CoA — succinyl CoA

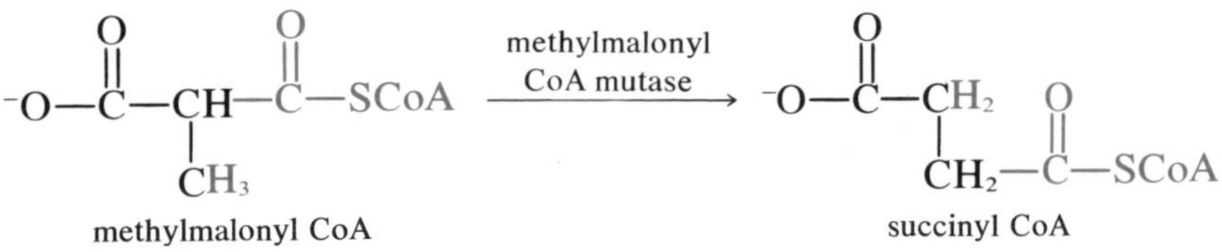

Studies using radioisotopes have revealed that this isomerization involves migration of the entire thioester group to the methyl carbon and exchange of a hydrogen atom (both shown here in color). Methylmalonyl CoA mutase requires vitamin B_{12} as a cofactor.

The importance of vitamin B_{12} was first realized when it was discovered that liver or liver extracts could be used to treat patients with pernicious anemia. The active principal was isolated, purified, and crystallized in 1948 and its complete three-dimensional structure was worked out by Dorothy Hodkin in 1956 (Figure 18.8). Central to the structure of vitamin B_{12} is an atom of cobalt embedded in a corrin ring which, like the porphyrin rings of hemoglobin and myoglobin, has four pyrrole rings joined together. For her work in determining the structure of this complex substance, Dorothy Hodkin received the Nobel Prize in 1964.

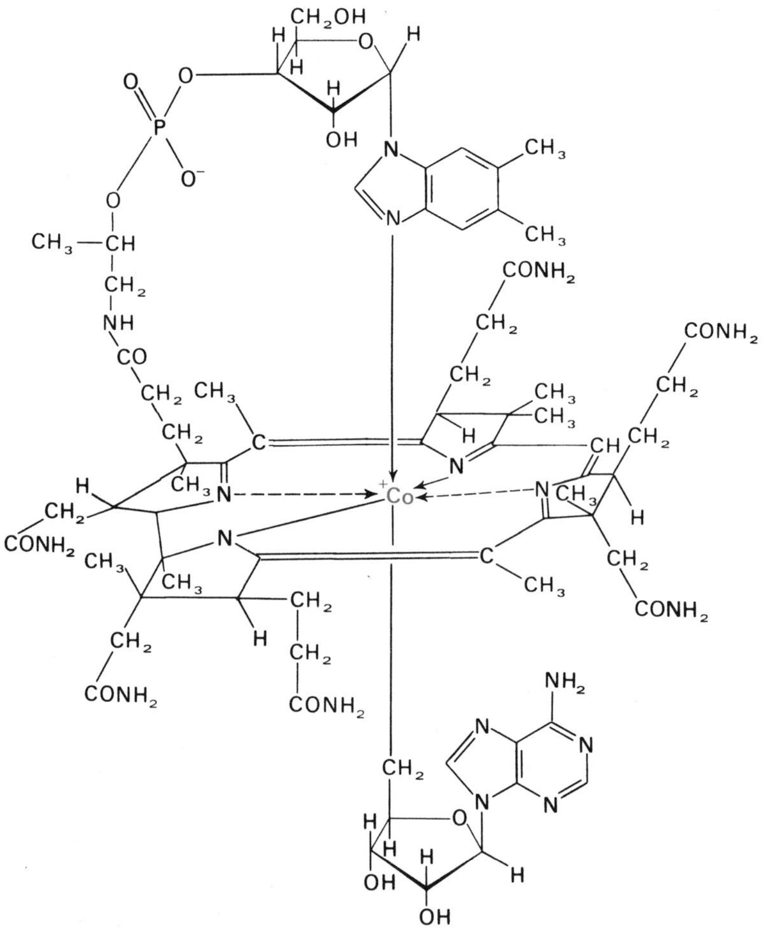

Figure 18.8 Structure of the coenzyme vitamin B_{12}. From David S. Page, Principles of Biological Chemistry (Willard Grant Press, Boston, 1976).

Pernicious anemia is a disease produced by a deficiency of vitamin B_{12}. However, most cases of pernicious anemia are not due to a lack of B_{12} in the diet but rather to a deficiency of a substance called intrinsic factor, which normally is present in gastric juice. Intrinsic factor is necessary for absorption of B_{12} through the walls of the gastrointestinal tract and into the blood stream. Since most B_{12}-deficiency diseases are due to this lack of intrinsic factor and reduced absorption of the vitamin, B_{12} taken orally often has little effect. Therefore the most common means of administration is by direct intramuscular injection. While the minimum daily requirement for humans is not known, in 1974 the Food and Nutrition Board of the National Research Council set the recommended daily allowance at 3 micrograms (0.000003 g). Vitamin B_{12} is synthesized only by microorganisms, especially anaerobic bacteria, including those of the flora normally present in the intestinal tract.

Now let us look at the pathway for the conversion of the carbon skeleton of glycine to carbon dioxide. According to Figure 18.7, glycine is degraded to pyruvate. When you first saw this, you may have thought it unusual and may even have asked: How can glycine with only two carbon atoms be degraded to pyruvate, which has three carbon atoms?

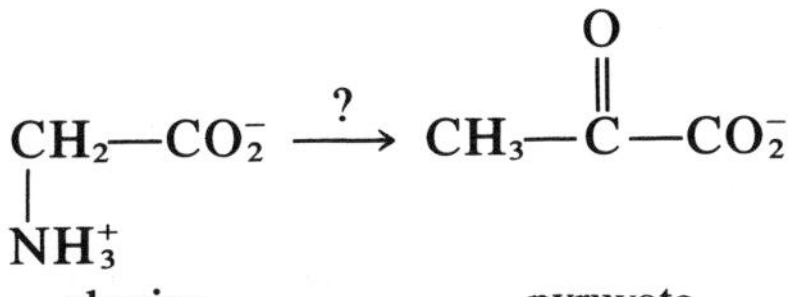

$$\underset{\text{glycine}}{CH_2(NH_3^+)—CO_2^-} \xrightarrow{?} \underset{\text{pyruvate}}{CH_3—C(=O)—CO_2^-}$$

The answer is that glycine is first converted into serine and then to pyruvate. This transformation requires the coenzyme tetrahydrofolate (THF), a derivative of the vitamin folic acid (Figure 18.9). Folic acid consists of three groups: glutamate, *p*-aminobenzoic acid, and a substituted pteridine ring. Reduction of folic acid by two molecules of NADPH generates tetrahydrofolate, the active coenzyme.

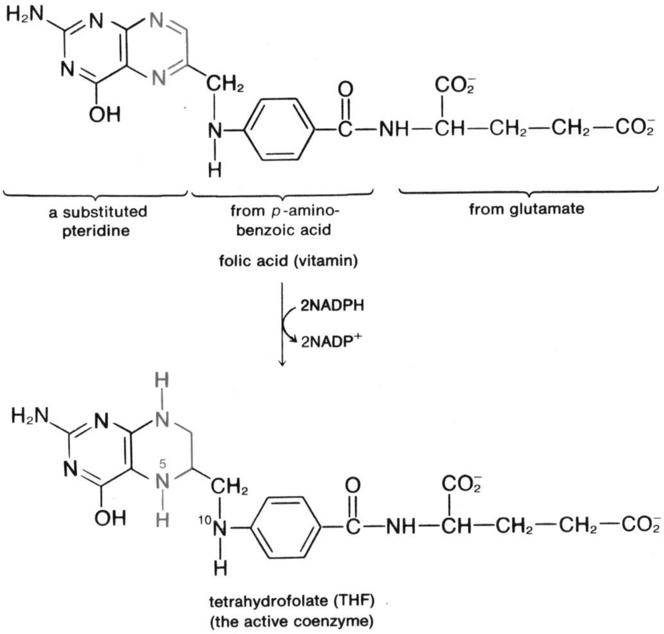

a substituted pteridine | from *p*-aminobenzoic acid | from glutamate

folic acid (vitamin)

2NADPH → 2NADP⁺

tetrahydrofolate (THF) (the active coenzyme)

Figure 18.9 *Folic acid and its reduction to tetrahydrofolate.*

Tetrahydrofolate is a carrier of one-carbon units. These may be in a form equivalent to CH_3OH, CH_2O, or HCO_2H. Because these substances all are toxic and cannot be stored or transported as such, they are carried as derivatives of tetrahydrofolate. The form of the coenzyme which donates the hydroxymethyl group ($HO—CH_2—$) to glycine is N^5,N^{10}-methylenetetrahydrofolate, whose partial structural formula is shown on the next page.

N^5,N^{10}-methylenetetrahydrofolate

The potential hydroxymethyl group is bound to nitrogen-5 and nitrogen-10 of tetrahydrofolate. Transfer of this one-carbon unit to glycine produces serine and regenerates tetrahydrofolate.

$$CH_2(NH_3^+)\text{—}CO_2^- + N^5,N^{10}\text{-methylenetetrahydrofolate} \underset{}{\overset{+H_2O}{\rightleftharpoons}} HOCH_2\text{—}CH(NH_3^+)\text{—}CO_2^- + \text{tetrahydrofolate}$$

glycine — N^5,N^{10}-methylene-tetrahydrofolate — serine — tetrahydrofolate

In fact, this reaction is readily reversible, and for this reason glycine and serine are freely interconvertible.

In addition to serving as both a donor and an acceptor of hydroxymethyl groups, tetrahydrofolate can also transfer one-carbon units as methyl groups (CH_3—) and formyl groups (O=CH—). The various folic acid coenzymes are involved in a variety of metabolic pathways including the degradation of glycine, serine, histidine, and tryptophan along with the biosynthesis of the purine and pyrimidine bases necessary for the production of nucleic acids.

The generic name for the folic acid vitamins is folacin. Folacin is widely distributed in nature, particularly in the foliage of plants, hence its name. Yeast, cauliflower, beef, liver, and wheat are all good sources of the vitamin. In 1974, the National Research Council established the recommended allowance of folacin as 400 mcg per day.

18.6 CONVERSION TO GLUCOSE AND KETONE BODIES

As we have just seen, one fate of amino acid-derived carbon skeletons is entry into the tricarboxylic acid cycle for immediate degradation to carbon dioxide and water. Alternatively, certain amino acids (i.e., those that can be degraded to pyruvate, α-ketoglutarate, succinyl CoA, and oxaloacetate) can be converted to phosphoenolpyruvate, and then into glucose. Those amino acids that can be used as carbon sources for the net synthesis of glucose are said to be glycogenic, and the process of glucose synthesis from them is called gluconeogenesis (p. 362). Recall from Section 16.3 that these glycogenic amino acids along with glycerol provide alternative sources of glucose during periods of low carbohydrate intake or when stores are being rapidly depleted.

Those amino acids degraded to acetyl CoA or acetoacetyl CoA cannot be converted into glucose because man and other animals have no biochemical pathways for the net synthesis of glucose from either of these intermediates. It is for this reason that fatty acids cannot serve as sources of carbon atoms for the net synthesis of glucose. However, the reverse process, the synthesis of fatty acids from glucose, is well established. Acetyl CoA and acetoacetyl CoA derived from the degradation of amino acids can be transformed into ketone bodies, as we saw in Section 17.6. Accordingly, those amino acids which are degraded to either acetyl CoA or acetoacetyl CoA and then transformed into ketone bodies are said to be ketogenic.

The first experimental attempts to classify amino acids as glycogenic or ketogenic were carried out on test animals at a time when the metabolic pathways for the degradation of the individual amino acids were only poorly understood. In the earliest studies, laboratory dogs were made diabetic either by selective chemical destruction of the insulin-producing capability of the pancreas, or by removal of the pancreas itself. Blood glucose levels were then controlled by injections of insulin. These diabetic dogs excreted glucose in the urine even when glycogen and fat stores had been depleted and when they were fed a diet containing protein as the sole source of metabolic fuel. They also excreted urea, the means by which amino acid-derived nitrogen atoms are detoxified and eliminated. Since both glucose and urea are excreted in the urine in the absence of insulin, the molar ratios of these two substances give an indication of the extent to which the amino acids or proteins can be transformed into glucose. These studies revealed that in diabetic dogs, a maximum of 58 grams of glucose can be derived from 100 grams of protein. In other words, 58% of protein is glycogenic.

To determine which of the 20 amino acids are glycogenic, diabetic test animals were fed pure amino acids, one at a time. If glucose was excreted in the urine following such a feeding, the amino acid was classified as glycogenic. Similarly, if acetoacetate, β-hydroxybutyrate, or acetone was excreted, the amino acid was classified as ketogenic. However, these methods were time-consuming and rather limited in sensitivity.

More recently, through the use of radioisotopes and labelled amino acids, scientists have been able to determine quite accurately the fates of the amino acid carbon skeletons, and thereby classify glycogenic and ketogenic amino acids more accurately. Furthermore, we now have a good understanding of the metabolic pathways for the degradation of each of the 20 amino acids. With all of these techniques and results, we can now classify amino acids as glycogenic, ketogenic, or both with a high degree of accuracy. This classification is given in Table 18.1.

Of the 20 protein-derived amino acids, only leucine is purely ketogenic, for on degradation it yields acetyl CoA and acetoacetyl CoA (Figure 18.7), both of which are ketogenic. Six amino acids are both glycogenic and ketogenic. For example, both phenylalanine and

Table 18.1 *Glycogenic and ketogenic amino acids.*

glycogenic		glycogenic and ketogenic	ketogenic
alanine	glycine	isoleucine	leucine
arginine	histidine	lysine	
asparagine	methionine	phenylalanine	
aspartic acid	proline	threonine	
cysteine	serine	tryptophan	
glutamic acid	valine	tyrosine	
glutamine			

tyrosine are degraded in part to fumarate and in part to acetoacetyl CoA. Fumarate is glycogenic for it can be converted to malate and oxaloacetate by the reactions of the tricarboxylic acid cycle and then to phosphoenolpyruvate for conversion to glucose. The remaining thirteen amino acids are purely glycogenic.

In Chapters 16–18 we have provided but a glimpse at the metabolic capabilities of the cell, but even from this limited perspective one must surely marvel at the design that permits such economy and efficiency in the use of cellular resources and at the remarkable coordination of the many metabolic pathways at the disposal of the cell.

Problems

18.1 List the three vital functions served by amino acids in the body.

18.2 Define the term half-life as it is applied in this chapter to tissue and plasma proteins.

18.3 Compare the degree to which carbohydrates, fats, and proteins can be stored in the body for later use.

18.4 What percentage of the total energy requirement of the average adult is supplied by carbohydrates; by fats; by proteins?

18.5 Draw structural formulas for the reactants and products of the following transamination reactions.

a tyrosine + α-ketoglutarate →
b glutamate + oxaloacetate →
c phenylalanine + pyruvate →

18.6 Draw structural formulas for pyridoxal phosphate and pyridoxamine phosphate. From what vitamin is each coenzyme derived?

18.7 What is the function of pyridoxal phosphate in transamination reactions?

18.8a Write an equation for the reaction catalyzed by the enzyme serum glutamate oxaloacetate transaminase (SGOT); for the reaction catalyzed by serum glutamate pyruvate transaminase (SGPT).
b Describe how an assay for the presence of these enzymes in blood can provide information about possible heart and liver damage.

18.9 In nutritional studies on rats, it has been found that certain α-ketoacids may substitute for essential amino acids. Shown here are structural formulas for three such α-ketoacids.

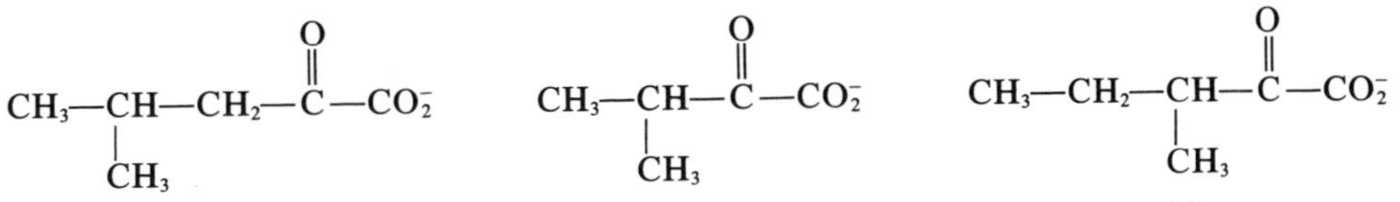

$$CH_3{-}\underset{\underset{CH_3}{|}}{CH}{-}CH_2{-}\overset{\overset{O}{\|}}{C}{-}CO_2^-$$
(a)

$$CH_3{-}\underset{\underset{CH_3}{|}}{CH}{-}\overset{\overset{O}{\|}}{C}{-}CO_2^-$$
(b)

$$CH_3{-}CH_2{-}\underset{\underset{CH_3}{|}}{CH}{-}\overset{\overset{O}{\|}}{C}{-}CO_2^-$$
(c)

a Account for the fact that, in certain instances, these α-ketoacids may substitute for essential amino acids.
b For which essential amino acid might each substitute?

18.10 The following reaction is the first step in the degradation of ornithine. Propose a metabolic pathway to account for this transformation.

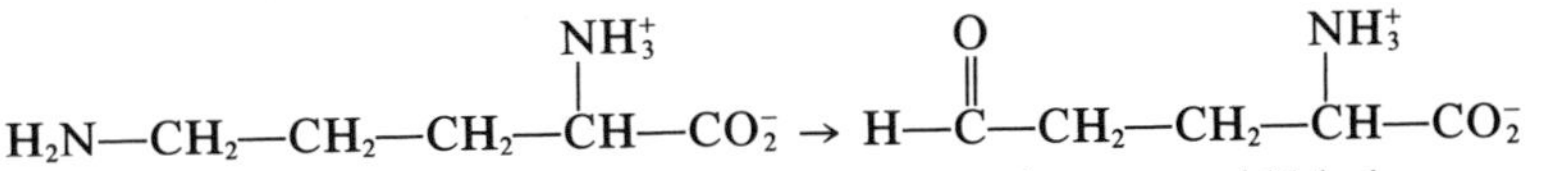

$$\underset{\text{ornithine}}{H_2N{-}CH_2{-}CH_2{-}CH_2{-}\overset{\overset{NH_3^+}{|}}{CH}{-}CO_2^-} \rightarrow \underset{\text{glutamate semialdehyde}}{H{-}\overset{\overset{O}{\|}}{C}{-}CH_2{-}CH_2{-}\overset{\overset{NH_3^+}{|}}{CH}{-}CO_2^-}$$

18.11 Write balanced equations to illustrate two major pathways for the removal of amino groups from amino acids. Name the coenzymes essential for each reaction and the vitamin from which each coenzyme is derived.

18.12 Describe three principal metabolic pathways for the detoxification of ammonium ions. Which of these is most important for the ultimate disposal of amino acid-derived nitrogen atoms?

18.13 What is the function of the urea cycle?

18.14 Write a balanced equation for the reaction of the urea cycle.

18.15 Which reactions of the urea cycle involve formation of carbon-nitrogen bonds; which involve cleavage of carbon-nitrogen bonds?

18.16 Explain how the excretion of nitrogen atoms as ammonium ions provides the body with one means of regulating acid–base balance.

18.17 What is meant by the term ammonia intoxication?

18.18 List the five points at which carbon skeletons derived from amino acids enter the tricarboxylic acid cycle.

18.19 The conversion of an α-ketoacid to a thioester of coenzyme A is a common theme in degradative pathways. Review Chapters 16 and 18 and give three specific examples of this type of reaction. Also name two coenzymes essential for this type of reaction and the vitamin precursor for each coenzyme.

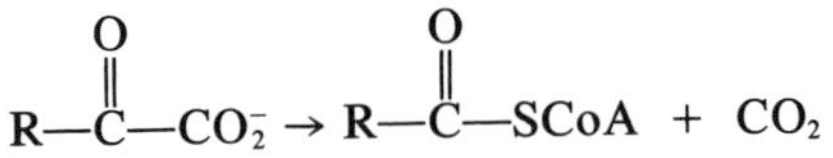

$$\underset{\alpha\text{-ketoacid}}{R{-}\overset{\overset{O}{\|}}{C}{-}CO_2^-} \rightarrow \underset{\text{thioester of coenzyme A}}{R{-}\overset{\overset{O}{\|}}{C}{-}SCoA} + CO_2$$

18.20a Draw structural formulas for the reactant and product in the following isomerization.

methylmalonyl CoA $\rightarrow$ succinyl CoA

b What cofactor is required for this isomerization?
c Large quantities, up to 50 to 90 mg per day, of a dicarboxylic acid appear in the urine of patients with pernicious anemia. What do you think this dicarboxylic acid might be and how would you account for its formation?

18.21 What is the function of tetrahydrofolate? From what vitamin is this cofactor derived?

18.22a What does it mean to say that an amino acid is glycogenic?
b What does it mean to say that an amino acid is ketogenic?
c Is it possible for an amino acid to be both glycogenic and ketogenic? Explain.

18.23 Propose biochemical pathways to explain how the cell might carry out the following transformations. Name each type of reaction (e.g., hydration, dehydration, oxidation, hydrolysis, etc.).

a phenylalanine → phenylacetate
b 3-phosphoglycerate → serine
c citrate → glutamate
d ornithine → glutamate
e methylmalonyl CoA → oxaloacetate

18.24 Propionyl CoA is produced from the degradation of several amino acids including leucine, and also from the degradation of fatty acids containing an odd number of carbon atoms. Propionyl CoA is in turn converted to methylmalonyl CoA. To what reaction in fatty acid metabolism is this analogous?

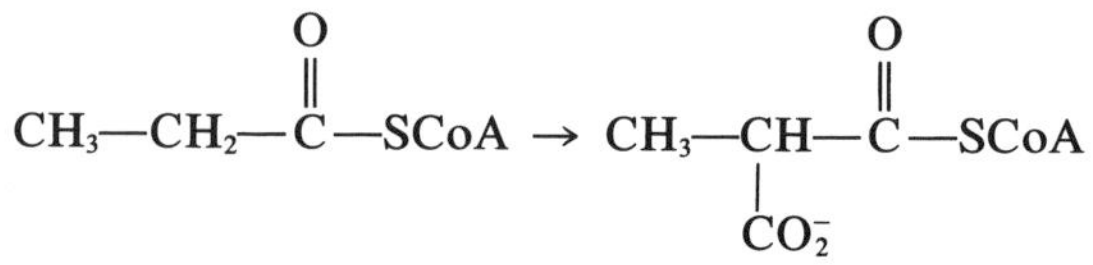

$$\underset{\text{propionyl CoA}}{CH_3{-}CH_2{-}\overset{\overset{\displaystyle O}{\|}}{C}{-}SCoA} \rightarrow \underset{\text{methylmalonyl CoA}}{CH_3{-}\underset{\underset{\displaystyle CO_2^-}{|}}{CH}{-}\overset{\overset{\displaystyle O}{\|}}{C}{-}SCoA}$$

18.25 Shown below is the metabolic pathway for the conversion of isoleucine to acetyl CoA and propionyl CoA. You have already studied each type of reaction shown here, though not necessarily in this chapter. For each step in this sequence, name the type of reaction and specify any coenzymes involved.

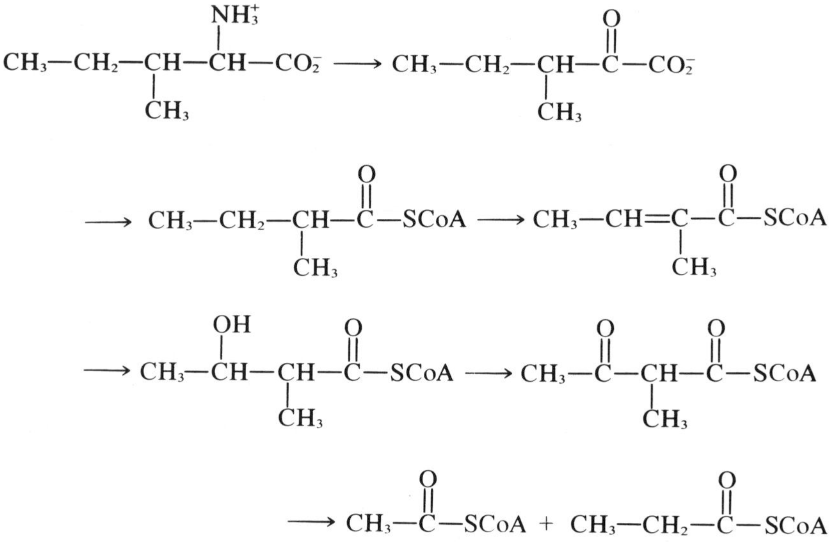

$$CH_3{-}CH_2{-}\underset{\underset{\displaystyle CH_3}{|}}{CH}{-}\overset{\overset{\displaystyle NH_3^+}{|}}{CH}{-}CO_2^- \longrightarrow CH_3{-}CH_2{-}\underset{\underset{\displaystyle CH_3}{|}}{CH}{-}\overset{\overset{\displaystyle O}{\|}}{C}{-}CO_2^-$$

$$\longrightarrow CH_3{-}CH_2{-}\underset{\underset{\displaystyle CH_3}{|}}{CH}{-}\overset{\overset{\displaystyle O}{\|}}{C}{-}SCoA \longrightarrow CH_3{-}CH{=}\underset{\underset{\displaystyle CH_3}{|}}{C}{-}\overset{\overset{\displaystyle O}{\|}}{C}{-}SCoA$$

$$\longrightarrow CH_3{-}\overset{\overset{\displaystyle OH}{|}}{CH}{-}\underset{\underset{\displaystyle CH_3}{|}}{CH}{-}\overset{\overset{\displaystyle O}{\|}}{C}{-}SCoA \longrightarrow CH_3{-}\overset{\overset{\displaystyle O}{\|}}{C}{-}\underset{\underset{\displaystyle CH_3}{|}}{CH}{-}\overset{\overset{\displaystyle O}{\|}}{C}{-}SCoA$$

$$\longrightarrow \underset{\text{acetyl CoA}}{CH_3{-}\overset{\overset{\displaystyle O}{\|}}{C}{-}SCoA} + \underset{\text{propionyl CoA}}{CH_3{-}CH_2{-}\overset{\overset{\displaystyle O}{\|}}{C}{-}SCoA}$$

18.26 Shown on p. 409 is the metabolic pathway for the conversion of proline to glutamate. Name the type of reaction involved in each step of this transformation and specify any coenzymes you think might be involved.

proline → [pyrroline ring with N and CO_2^-] ⟶ $H-\overset{O}{\overset{\|}{C}}-CH_2-CH_2-\overset{NH_3^+}{\overset{|}{C}H}-CO_2^-$ ⟶ glutamate

18.27 The vast majority of substances in the biological world are built from just thirty or so smaller molecules. Review the biochemistry we have discussed in this text and make up your own list of these thirty or so fundamental building blocks of nature.

MINI-
ESSAY

INBORN ERRORS OF METABOLISM

> ... the anomalies of which I propose to treat ... may be classed together as inborn errors of metabolism. Some of them are certainly, and all of them are probably, present from birth .... They are characterized by wide departures from the normal of the species far more conspicuous than any ordinary individual variations, and one is tempted to regard them as metabolic sports, the chemical analogues of structural malfunctions .... It may well be that the intermediate products formed at the various stages (in metabolism) have only momentary existence as such, being subjected to further change almost as soon as they are formed; and that the course of metabolism along any particular path should be pictured as in continuous movement rather than as a series of distinct steps. If any one step in the process fails, the intermediate product in being at the point of arrest will escape further change, just as when the film of a biograph is brought to a standstill the moving figures are left foot in the air. (Sir Archibald Garrod, *Inborn Errors of Metabolism*, 1909.)

In this remarkably prescient passage, Sir Archibald Garrod anticipated many of the developments in the rapidly unfolding study of the metabolic basis of inherited diseases. Garrod studied four inborn errors of metabolism—albinism, alcaptonuria, cystinuria, and pentosuria—and postulated that each results from a congenital failure of the body to carry out a particular reaction in a normal metabolic sequence. His predictions have been fully corroborated and the inborn errors of metabolism of which he spoke have been traced to a partial or total lack of activity of a particular enzyme. In this essay we shall consider two of the diseases studied by Garrod, albinism and alcaptonuria, and a third recognized more recently, phenylketonuria. These three involve closely related metabolic pathways and illustrate the important principles of metabolic disorders.

Figure 1 summarizes several important pathways in the metabolism of phenylalanine and tyrosine. Under normal physiological conditions, the bulk of phenylalanine is oxidized to tyrosine, a reaction catalyzed by phenylalanine hydroxylase. In a second and minor oxidative pathway, phenylalanine is converted to phenylpyruvate, which accumulates in the blood and is excreted in the urine. This conversion of an α-amino acid to an α-ketoacid is called transamination and requires pyridoxal phosphate and pyridoxamine phosphate, forms of vitamin B_6, as coenzymes (Figure 18.3).

Two metabolic pathways for tyrosine are shown. In the first, tyrosine undergoes transamination to *p*-hydroxyphenylpyruvate followed by oxidation to homogentisate. Oxidation of homogentisate, initiated by the enzyme homogentisate oxidase, leads ultimately to fumarate and acetoacetate. A second major pathway of tyrosine metabolism is oxidation to 3,4-dihydroxyphenylalanine (DOPA). Next,

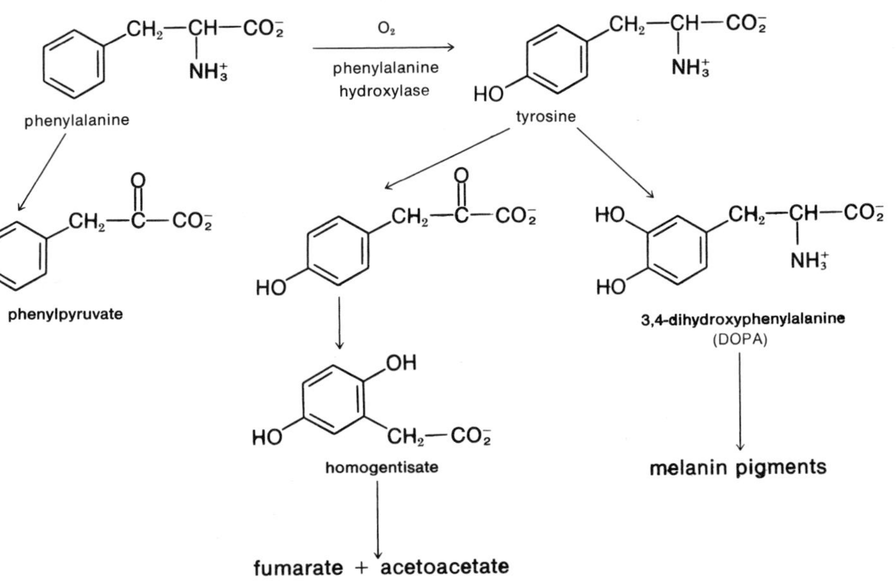

Figure 1 *Important pathways in the metabolism of phenylalanine and tyrosine.*

in a complex series of reactions initiated by tyrosinase, DOPA is converted into melanin, the polymeric pigment of hair, skin, and retina.

The characteristic feature of alcaptonuria is excretion of abnormally large amounts of homogentisate in the urine. This condition occurs about once in 200,000 births and is often discovered early in life from observation that infant's diapers turn black on exposure to air. The darkening results from oxidation of homogentisate by atmospheric oxygen to give the pigment alcapton. Alcaptonuric patients experience no other symptoms early in life, but later there is a darkening of tendons and cartilage due to deposition of pigments which are often associated with the development of arthritis.

At the time Garrod took up the study of this disease, it was generally assumed that the large quantity of homogentisate excreted in the urine was the product of some type of infection in the intestine. Garrod, however, felt this hypothesis was inadequate to explain the symptoms. In 1901 he made a critical discovery. Healthy and apparently normal parents of his alcaptonuric patients often were blood relations. He realized immediately that this parental consanguinity was more than chance and wrote, "The facts here brought forward lend support to the view that alcaptonuria is what may be described as a 'freak' of metabolism. They can hardly be reconciled with the theory that it (alcaptonuria) results from an infection of the alimentary canal."

The time was right for this discovery. The work of Gregor Mendel had been rediscovered only a few years before and the challenge of contemporary biologists was to learn how Mendel's laws applied in the animal kingdom and to humans in particular. William Bateson, one of the pioneers of genetics, was quick to point out that the characteristic familial distribution of alcaptonuria discovered by Garrod was precisely what would be expected if the abnormality were determined by a rare recessive Mendelian factor. Garrod was the first to suggest that the basic defect in the disease is the inability to metabolize homogentisate. This was fully confirmed in 1958 with the demonstration that the liver and kidneys of alcaptonurics are deficient in the enzyme homogentisate oxidase.

Albinism was another of the inborn errors of metabolism considered by Garrod, who correctly concluded that the condition is due to lack of the enzyme necessary for the formation of melanin. Complete (universal) albinism is characterized by a total lack of melanin pigment in the hair, skin, and retina. In partial (generalized) albinism, the lack of melanin is restricted to particular areas of the hair, skin, or retina. The key enzyme involved in the formation of melanin from tyrosine is tyrosinase. It is the lack of this enzyme that results in albinism.

Of the three diseases discussed here, phenylketonuria (PKU) is by far the most crippling. It is characterized by mental retardation and by excretion in the urine of large amounts of phenylpyruvate. It is estimated that this disease occurs once in every 10,000–20,000 births. About 40 years ago, before PKU was identified, Dr. Asbjörn Fölling of the University of Oslo School of Medicine discovered that the musty-smelling urine of two mentally retarded siblings turned olive-green, rather than the expected red-brown, when tested with 10% ferric chloride solution. Fölling speculated that the substance responsible for this unusual color reaction was somehow related to the children's mental retardation. Soon after he discovered that the substance was phenylpyruvate, formed in the kidneys from phenylalanine. From the time of this discovery until the mid-1950s, this simple test—the color reaction of ferric chloride solution on a wet diaper—was the standard method of identifying PKU.

However, this test sometimes is unreliable during the first few months after birth when many phenylketonurics are not yet secreting phenylpyruvate. A more sensitive test, known as the Guthrie test after its discoverer, Dr. Robert Guthrie of the State University of New York at Buffalo, requires only a few drops of blood and measures high serum levels of phenylalanine. This test is a successful screening technique in the first few days of life. In most countries, the Guthrie test has replaced the analysis of urine for PKU detection and in some states of the United States the test is required by law.

The disease results from a block in the conversion of phenylalanine to tyrosine. The enzyme phenylalanine hydroxylase appears to be present but its concentration is drastically below normal levels. This reduction in enzyme activity causes the accumulation of phenylalanine, which is then metabolized by transamination to phenylpyruvate. Almost all of the chemical abnormalities of the disease, e.g., excretion of phenylpyruvate in the urine, can be improved and in many cases even eliminated by restricting dietary intake of phenylalanine to the minimum level required for normal growth and development. In addition, the diet must be supplemented in tyrosine, which is an essential amino acid for phenylketonurics. If PKU is diagnosed and dietary restrictions are begun at birth, it is possible to prevent the mental retardation associated with the disease.

These three inborn errors of metabolism illustrate the typical consequences of a metabolic block. First, the normal products of a particular metabolic pathway are not formed, or are formed at a reduced rate, and it is the failure to form a product which is the significant feature of each disease. For example, failure to form melanin pigment is the characteristic feature of albinism. The alternative consequence of the enzyme deficiency is the accumulation of substrate of an enzyme, and it is this accumulation of substrate that gives rise to the pathological symptoms. In PKU, phenylalanine is not degraded in the normal manner and instead accumulates and is converted to phenylpyruvate. In many of these diseases the primary metabolic block may also lead to secondary biochemical changes, as in the mental retardation seen in PKU. The nature of these secondary changes is often obscure. For example, it is not at all clear how the absence of phenylalanine hydroxylase leads to the secondary biochemical changes that eventuate in mental retardation.

Beginning with the seminal observations of Garrod nearly 70 years ago, advances in the understanding of inborn errors of metabolism have been truly impressive. In the decades ahead, we can expect many more diseases to be understood in genetic and biochemical terms, perhaps even certain aspects of mental diseases.

References

Garrod, A. E., *Inborn Errors of Metabolism*, 1909. Reprinted with a supplement by H. Harris (Oxford University Press, New York, 1963).

Stanbury, J. B., Wyngaarden, A., and Fredrickson, D. S., Editors, *The Metabolic Basis of Inherited Diseases*, 2nd ed. (McGraw-Hill Book Company, New York, 1966).

INDEX